Advanced Control Theory

Third Edition

for BE, BTech, ME, MTech Courses

Advanced Control Theory

Third Edition

for BE, BTech, ME, MTech Courses

A Nagoor Kani

CBS

CBS Publishers & Distributors Pvt Ltd

New Delhi • Bengaluru • Chennai • Kochi • Kolkata • Lucknow • Mumbai
Hyderabad • Jharkhand • Nagpur • Patna • Pune • Uttarakhand

Advanced
Control Theory
Third Edition
for BE, BTech, ME, MTech Courses

ISBN: 978-93-89396-29-4

Third Edition: 2020
Reprint: 2022
First Edition: 1998
Second Edition: 1999

Published by Satish Kumar Jain and produced by Varun Jain for
CBS Publishers & Distributors Pvt Ltd
4819/XI Prahlad Street, 24 Ansari Road, Daryaganj, New Delhi–110 002, India.
Ph: 23289259, 23266861, 23266867   Website: www.cbspd.com
Fax: 011-23243014   e-mail: delhi@cbspd.com; cbspubs@airtelmail.in
Corporate Office: 204 FIE, Industrial Area, Patparganj, Delhi–110 092
Ph: 4934 4934   Fax: 4934 4935   e-mail: publishing@cbspd.com; publicity@cbspd.com

Branches

- **Bengaluru:** Seema House 2975, 17th Cross, K.R. Road, Banasankari 2nd Stage, Bengaluru 560 070, Karnataka, India
  Ph: +91-80-26771678/79   Fax: +91-80-26771680   e-mail: bangalore@cbspd.com
- **Chennai:** 7, Subbaraya Street, Shenoy Nagar, Chennai 600 030, Tamil Nadu, India
  Ph: +91-44-26680620, 26681266   Fax: +91-44-42032115   e-mail: chennai@cbspd.com
- **Kochi:** 42/1325, 1326, Power House Road, Opposite KSEB, Power House, Ernakulum-682018, Kochi, Kerala, India
  Ph: +91-484-4059061–67   Fax: +91-484-4059065   e-mail: kochi@cbspd.com
- **Kolkata:** 147, Hind Ceramics Compound, 1st Floor, Nilgunj Road, Belghoria, Kolkata 700056, West Bengal, India
  Ph: +91-9096713055/7798394118, 9836841399   e-mail: kolkata@cbspd.com
- **Lucknow:** Basement, Khushuma Complex, 7 Meerabai Marg (Behind Jawahar Bhawan), Lucknow 226001, UP, India
  Ph: +91-522-40000032   e-mail: tiwari.lucknow@cbspd.com
- **Mumbai:** PWD Shed, Gala No. 25/26, Ramchandra Bhatt Marg, Next JJ Hospital Gate No. 2, Opp. Union Bank of India, Noorbaug, Mumbai-400009, Maharashtra, India
  Ph: +91-22-66661880/89   e-mail: mumbai@cbspd.com

Representatives

• **Hyderabad**	0-9885175004	• **Jharkhand**	0-9811541605	• **Nagpur**	0-9421945513
• **Patna**	0-9334159340	• **Pune**	0-9623451994	• **Uttarakhand**	0-9716462459

Printed at Mudrak, Noida, UP, India.

to

my beloved sister
Ms A Yuhanitha Beevi

PREFACE

Advanced control theory primarily deals with modern concepts of control systems analysis and design. This text is designed for an advance study of control systems as an elective course for undergraduate and postgraduate engineering students and also for research scholars.

This book is organized with 5 chapters. The design of linear, nonlinear and discrete time control systems by conventional methods and the modern state space methods are presented in a very easiest and elaborative manner. Throughout the book, carefully chosen examples are presented so that the reader will have a clear understanding of the concepts discussed.

Chapter 1 presents the detailed conventional methods of linear control systems design using controllers and compensators in both frequency and time domain in order to meet the desired performance specifications. The design using lag, lead and lag-lead compensators, PI, PD and PID controllers are presented with appropriate examples.

Chapter 2 focuses on analysis and design of nonlinear control systems. The development of describing functions of various nonlinearities like dead-zone, saturation, hysteresis and backlash are presented with clear steps.

Chapter 3 elaborates on sampled data or discrete time control systems. The fundamental concepts of modeling and analysis of sampled data control systems are presented in this chapter. An introduction to Z-transform is included in this chapter, which is an important prerequirement for modeling and analysis of sampled data control systems. The stability of sampled data control systems is also presented with clear examples.

Chapter 4 deals with state space analysis of continuous time and discrete time control systems. State space representation and solution of state equations are given with explanation and examples.

Chapter 5 is concerned with controllability and observability of state models of control systems. The design of control systems using state feedback is also presented with illustrated design examples.

Several concepts and procedures are illustrated through simple examples rather than generalized formulation. Each chapter provides the foundation and practical implications of their own topic with more number of solved problems for better understanding.

Special care has been taken in presenting the concept, choosing examples and solved problems. I hope that the teaching and student community will welcome the book. The readers can feel free to convey their suggestions or criticism for further improvement of the book.

A Nagoor Kani

ACKNOWLEDGEMENTS

I express my heartfelt thanks to my wife Mrs C Gnanaparanjothi Nagoor Kani, and my sons N Bharath Raj alias Chandrakani Allaudeen and N Vikram Raj for their support, encouragement and cooperation they have extended to me throughout my career. I thank Ms T A Benazir, Manager, RBA Group, and all my office staff for their cooperation in carrying out my day-to-day activities.

I sincerely acknowledge the contributions of our technical editors, Ms C Mohana Priya and Ms S Saranya and Ms P Kanimozhi, for editing, proof-reading and type-setting of the manuscript and preparing the layout of the book.

My sincere thanks to all reviewers for their valuable suggestions and comments which helped me to explore the subject to a greater depth.

I am also grateful to Mr Satish K Jain, CMD, CBS Publishers & Distributors, for his keen interest in publishing this work in CBS banner. My sincere thanks to all team members of CBS Publishers & Distributors, for their concern and care in publishing this work.

Finally, a special note of appreciation is due to my sisters, brothers, relatives, friends, students and the entire teaching community, for their overwhelming support and encouragement to my writing.

A Nagoor Kani

CONTENTS

Chapter 1

Linear System Design

1.1 INTRODUCTION TO DESIGN USING COMPENSATORS

The control systems are designed to perform specific tasks. The requirements of a control system are usually specified as performance specifications. The specifications are generally related to accuracy, relative stability and speed of response.

In time domain, the transient state performance specifications are given in terms of rise time, maximum overshoot, settling time and/or damping ratio. In frequency domain, the transient state performance specifications are given in terms of phase margin, gain margin, resonant peak and/or bandwidth. The steady state requirement are given in terms of error constants.

When performance specifications are given for single input, single output linear time invariant systems, then the system can be designed by using root locus or frequency response plots. To be more precise, when time domain specifications are given, root locus technique is employed in designing the system. If frequency domain specifications are given, frequency response plots like Bode plots are used in designing the system.

The first step in design is the adjustment of gain to meet the desired specifications. In practical systems, adjustment of gain alone will not be sufficient to meet the given specifications. In many cases, increasing the gain may result in poor stability or instability. In such cases, it is necessary to introduce additional devices or components in the system to alter the behaviour and to meet the desired specifications.

Such a redesign or addition of a suitable device is called compensation. A device inserted into the system for the purpose of satisfying the specifications is called compensator. The compensators basically introduce pole and/or zero in open loop transfer function to modify the performance of the system.

The design problem may be stated as follows : *When a set of specifications are given for a system, then a suitable compensator should be designed so that the overall system will meet the given specification.*

The compensation schemes used for feedback control system is either series compensation or parallel (feedback) compensation.

In series compensation, the compensator with transfer function, $G_c(s)$ is placed in series with plant. In feedback compensation, the signal from some element is feedback to the input and a compensator with transfer function $G_c(s)$ is placed in the resulting inner feedback path. The series and parallel compensation schemes are shown in figs 1.1 and 1.2 respectively.

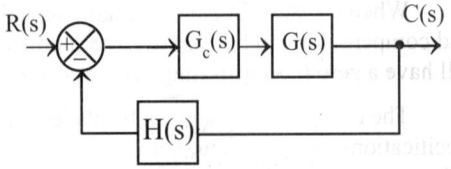

Fig 1.1 : Series compensation.

The choice between series compensation and parallel compensation depends on

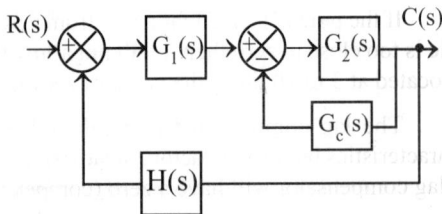

Fig 1.2 : Feedback/parallel compensation.

1. *Nature of signals in the system.*
2. *Power levels at various points.*
3. *Components available.*
4. *Designer's experience..*
5. *Economic considerations and so on.*

The compensator may be electrical, mechanical, hydraulic, pneumatic or other type of device or network. Usually, an electric network or electronic device serves as compensator in many control systems. The different types of electrical or electronic compensators used are ***Lag compensator, Lead compensator*** *and **Lag-lead compensator.***

In control systems, compensation is required in the following situations,

1.   *When the system is absolutely unstable, then compensation is required to stabilize the system and also to meet the desired performance.*

2.   *When the system is stable, compensation is provided to obtain the desired performance.*

The systems with type number 2 and above are usually absolutely unstable systems. Hence for systems with type number 2 and above, lead compensation is required, because the lead compensator increases the margin of stability.

In systems with type number 1 or 0, stability is achieved by adjusting the gain. In such cases any of the three compensators-lag, lead and lag-lead may be used to obtain the desired performance. The particular choice of compensator is based on the factors that are discussed in the following sections.

ROOT LOCUS APPROACH TO CONTROL SYSTEM DESIGN

In design using root locus, the desired behaviour is specified in terms of transient response specifications and steady - state error requirement. The steady - state error is usually specified in terms of error constants for standard inputs, while the transient response requirement is specified in terms of peak overshoot, settling time, rise time, etc., for a step input. The transient response specifications can be translated into desired locations for a pair of dominant closed loop poles.

In order to meet the desired specifications, the root loci are reshaped so that they pass through the points where the dominant closed loop poles are located. The root loci are reshaped by introducing a compensator. The compensator will add a pole and/or a zero in the open loop transfer function of the system.

The addition of a pole to the open-loop transfer function has the effect of pulling the root locus to the right, which reduce the relative stability of the system and increase the settling time. The addition of a zero to the open-loop transfer function has the effect of pulling the root locus to the left which make the system more stable and reduce the settling time.

When a system is either unstable or stable but has undesirable transient response characteristics a lead compensator can be employed to modify the root locus. The transfer function of lead compensator will have a zero (compensating zero) and a pole (compensating pole).

The compensator zero can be placed on the real axis by trial-and-error to satisfy transient response specifications. The introduction of zero will amplify high frequency noise which is eliminated by the compensating pole. The compensating pole is located on real axis such that it makes negligible effect on the root locus in the region where the two dominant closed loop poles are located.

If the pole is located far away from zero then it will not be effective in suppressing the noise. If the pole is too close to zero then it will not allow the zero to do its job. In order to avoid this conflict, the pole is located at 3 to 10 times the value of zero location.

The lag compensator is employed when a stable system has satisfactory transient response characteristics but unsatisfactory steady state characteristics, i.e., error requirement. The transfer function of lag compensator will have a zero (compensating zero) and a pole (compensating pole).

In order to preserve the transient response characteristics the compensating pole and zero should have negligible effect on shape of root locus. This is achieved by placing the compensating pole and zero very close to each other. If the pole and zero are located close to the origin then the error constant will increase which will reduce the steady state error.

The lag-lead compensator is employed when both the transient and steady-state characteristics are not satisfactory. The lead compensation will improve the transient response and lag compensation will reduce the steady state error.

The advantage in design using root locus technique is that the information about closed loop transient response and frequency response are directly obtained from the pole-zero configuration of the system in the s-plane.

FREQUENCY RESPONSE APPROACH TO CONTROL SYSTEM DESIGN

The objective of frequency domain design is to reshape the frequency-response characteristics so that the desired specifications are satisfied. The frequency domain design can be carried out using Nyquist plot, Bode plot or Nichols chart. But Bode plot are popularly used for design because they are easier to draw and modify.

In design using bode plots the desired performance specifications are given in terms of frequency domain specifications and steady-state error requirement.

The stability requirement is specified in terms of phase margin and resonant peak. The transient response requirements are specified in terms of gain crossover frequency, bandwidth and resonant frequency. The error requirement is specified in terms of static error constants (K_p, K_v or K_a).

Note : In case the transient response specifications are given in time domain, we can translate them into frequency domain specifications using the following formulae.

Phase Margin, $\gamma = \tan^{-1} \dfrac{2\zeta}{[\sqrt{4\zeta^4 + 1} - 2\zeta^2]^{\frac{1}{2}}} \approx 100\zeta$ Resonant peak, $M_r = 1/[2\zeta\sqrt{1 - \zeta^2}]$

Gain crossover frequency, $\omega_{gc} = \omega_n[\sqrt{4\zeta^4 + 1} - 2\zeta^2]^{\frac{1}{2}}$ Resonant frewquency, $\omega_r = \omega_n\sqrt{1 - 2\zeta^2}$

Bandwidth, $\omega_b = \omega_n[(1 - 2\zeta^2) + \sqrt{4\zeta^4 - 4\zeta^2 + 2}]^{\frac{1}{2}}$

The low frequency region of bode plot provides information regarding the steady state performance and high frequency region provides information regarding the transient-state performance. The medium frequency (or mid-frequency) range provides information regarding relative stability. Therefore, the low frequency region of the bode plot is reshaped by lag compensation to improve steady state performance. The high frequency region of the bode plot is reshaped by lead compensation to improve transient-state performance.

When the system requires improvement in both steady-state and transient state, lag-lead compensation can be employed to alter both the low and high frequency regions of bode plot.

The primary function of lead compensator is to reshape the frequency response curve to provide sufficient phase lead angle to offset the excessive phase lag associated with the components of the plant. The primary function of lag compensator is to provide attenuation in the high frequency region to achieve sufficient phase margin.

The advantage in frequency-domain design is that the effects of disturbances, sensor noise and plant uncertainties are relatively easy to visualize and assess in frequency domain. Another advantage of using frequency response is the ease with which experimental information can be used for design purposes.

A disadvantage of frequency-response design is that it gives us information on closed-loop system's transient response indirectly, while the root locus design gives this information directly.

1.2   LAG COMPENSATOR

A compensator having the characteristics of a lag network is called a lag compensator. If a sinusoidal signal is applied to a lag network, then in steady state the output will have a phase lag with respect to input.

Lag compensation results in a large improvement in steady state performance but results in slower response due to reduced bandwidth. The attenuation due to the lag compensator will shift the gain crossover frequency to a lower frequency point where the phase margin is acceptable. Thus, the lag compensator will reduce the bandwidth of the system and will result in slower transient response.

Lag compensator is essentially a low pass filter and so high frequency noise signals are attenuated. If the pole introduced by the compensator is not cancelled by a zero in the system , then lag compensator increases the order of the system by one.

S-PLANE REPRESENTATION OF LAG COMPENSATOR

The lag compensator has a pole at $s = -1/\beta T$ and a zero at $s = -1/T$. The pole-zero plot of lag compensator is shown in fig 1.3. Here, $\beta > 1$, so the zero is located to the left of the pole on the negative real axis. The general form of lag compensator transfer function is given by equation (1.1).

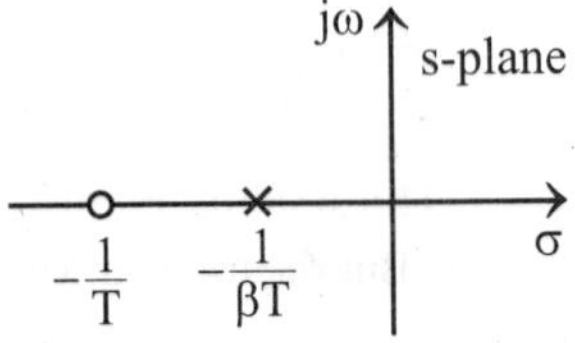

Fig 1.3 : *Pole-zero plot of lag compensator.*

Transfer function of lag compensator, $G_c(s) = \dfrac{s + z_c}{s + p_c} = \dfrac{s + \dfrac{1}{T}}{s + \dfrac{1}{\beta T}}$ .....(1.1)

where, $T > 0$ and $\beta > 1$

The zero of lag compensator, $z_c = \dfrac{1}{T}$ .....(1.2)

The pole of lag compensator, $p_c = \dfrac{1}{\beta T} = \dfrac{1}{\beta} z_c$ .....(1.3)

From equation(1.2) we get, $T = \dfrac{1}{Z_c}$ .....(1.4)

From equation(1.3) we get, $\beta = \dfrac{Z_c}{p_c}$ .....(1.5)

REALISATION OF LAG COMPENSATOR USING ELECTRICAL NETWORK

The lag compensator can be realised by the R-C network shown in fig 1.4.

Let, $E_i(s)$ = Input voltage

$E_o(s)$ = Output voltage

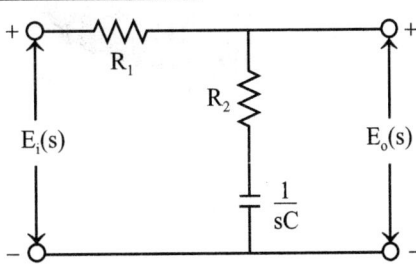

In the network shown in fig 1.4, the input voltage is applied to the series combination of R_1, R_2 and C. The output voltage is obtained across series combination of R_2 and C.

Fig 1.4 : Electrical lag compensator.

By voltage division rule,

$$E_o(s) = E_i(s) \frac{\left(R_2 + \dfrac{1}{sC}\right)}{\left(R_1 + R_2 + \dfrac{1}{sC}\right)} = E_i(s) \frac{(sCR_2 + 1)/sC}{[sC(R_1 + R_2) + 1]/sC}$$

$$= E_i(s) \frac{(sCR_2 + 1)}{[sC(R_1 + R_2) + 1]}$$

The transfer function of the electrical network is the ratio of output voltage to input voltage,

$$\left.\begin{array}{c}\text{Transfer function}\\ \text{of electrical network}\end{array}\right\} \frac{E_o(s)}{E_i(s)} = \frac{CR_2\left(s + \dfrac{1}{CR_2}\right)}{C(R_1 + R_2)\left[s + \dfrac{1}{C(R_1 + R_2)}\right]}$$

$$= \frac{\left(s + \dfrac{1}{R_2 C}\right)}{\left(\dfrac{R_1 + R_2}{R_2}\right)\left[s + \dfrac{1}{((R_1 + R_2)/R_2)R_2 C}\right]}$$ .....(1.6)

But the transfer function of lag compensator is given by,

$$G_c(s) = \frac{\left(s + \dfrac{1}{T}\right)}{\left(s + \dfrac{1}{\beta T}\right)}$$ .....(1.7)

On comparing equations (1.6) and (1.7) we get,

$$\frac{E_o(s)}{E_i(s)} = \frac{1}{\beta} \frac{\left(s + \dfrac{1}{T}\right)}{\left(s + \dfrac{1}{\beta T}\right)}$$ .....(1.8)

where, $T = R_2 C$ and $\beta = (R_1 + R_2)/R_2$

The transfer function of RC network as given by equation(1.8) is similar to the general form with an attenuation of $1/\beta$ (since $\beta > 1$, $(1/\beta) < 1$). If the attenuation is not required then an amplifier with gain β can be connected in cascade with RC network to nullify the attenuation.

FREQUENCY RESPONSE OF LAG COMPENSATOR

Consider the general form of lag compensator,

$$G_c(s) = \frac{\left(s + \frac{1}{T}\right)}{\left(s + \frac{1}{\beta T}\right)} = \frac{(sT + 1)/T}{(s\beta T + 1)/\beta T} = \beta \frac{(1 + sT)}{(1 + s\beta T)} \qquad .....(1.9)$$

Sinusoidal transfer function of lag compensator is obtained by letting $s = j\omega$ in equation (1.9).

$$\therefore\ G_c(j\omega) = \frac{\beta(1 + j\omega T)}{(1 + j\omega\beta T)} \qquad .....(1.10)$$

When $\omega = 0$, $G_c(j\omega) = \beta$ $\qquad .....(1.11)$

From equation(1.11) we can say that the lag compensator provides a dc gain of β (here $\beta > 1$). If the dc gain of the compensator is not desirable then it can be eliminated by a suitable attenuation.

Let us assume that the gain β is eliminated by a suitable attenuation network. Now, $G_c(j\omega)$ is given by,

$$G_c(j\omega) = \frac{1 + j\omega T}{1 + j\omega\beta T} = \frac{\sqrt{1 + (\omega T)^2}\ \angle \tan^{-1}\omega T}{\sqrt{1 + (\omega\beta T)^2}\ \angle \tan^{-1}\omega\beta T} \qquad .....(1.12)$$

The sinusoidal transfer function shown in equation(1.12) has two corner frequencies and they are denoted as ω_{c1} and ω_{c2}.

Here, $\omega_{c1} = \dfrac{1}{\beta T}$ and $\omega_{c2} = \dfrac{1}{T}$

Since, $\beta T > T$, $\omega_{c1} < \omega_{c2}$

Let, $A = |G_c(j\omega)|$ in db $= 20 \log \dfrac{\sqrt{1 + (\omega T)^2}}{1 + (\omega\beta T)^2}$ $\qquad .....(1.13)$

At very low frequencies i.e., upto ω_{c1}, $\omega T \ll 1$ and $\omega\beta T \ll 1$.

$\qquad \therefore\ A \approx 20 \log 1 = 0$

In the frequency range from ω_{c1} to ω_{c2}, $\omega T \ll 1$ and $\omega\beta T \gg 1$.

$\qquad \therefore\ A \approx 20 \log \dfrac{1}{\sqrt{(\omega\beta T)^2}} = 20 \log \dfrac{1}{\omega\beta T}$

At very high frequencies i.e., after ω_{c2}, $\omega T \gg 1$ and $\omega\beta T \gg 1$.

$\qquad \therefore\ A \approx 20 \log \dfrac{\sqrt{(\omega T)^2}}{\sqrt{(\omega\beta T)^2}} = 20 \log \dfrac{1}{\beta}$

The approximate magnitude plot of lag compensator is shown in fig 1.5. The magnitude plot of Bode plot of $G_c(j\omega)$ is a straight line through 0 db upto ω_{c1}, then it has a slope of -20 db/dec upto ω_{c2} and after ω_{c2} it is a straight line with a constant gain of $20 \log (1/\beta)$.

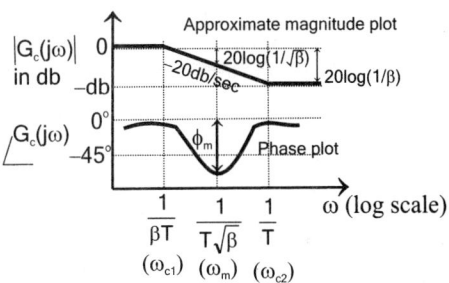

Fig 1.5 : *Bode plot of lag compensator.*

Let, $\phi = \angle G_c(j\omega)$

$\therefore \phi = \tan^{-1} \omega T - \tan^{-1} \omega \beta T$

As $\omega \to 0$, $\phi \to 0$

As $\omega \to \infty$, $\phi \to 0$

As ω is varied from 0 to ∞, the phase angle decreases from 0 to a negative maximum value of ϕ_m at $\omega = \omega_m$, then increases from this maximum value to 0. The phase plot of lag compensator is shown in fig 1.5. It can be shown that the frequency at which maximum phase lag occurs is the geometric mean of the two corner frequencies.

Frequency of maximum phase lag, $\omega_m = \sqrt{\omega_{c1}\omega_{c2}} = \sqrt{\dfrac{1}{\beta T}\cdot\dfrac{1}{T}} = \dfrac{1}{T\sqrt{\beta}}$

From the bode plot of lag compensator , we observe that lag compensator has a dc gain of unity while it offers a high frequency gain of $(1/\beta)$ [In decibels, it is 20 log $(1/\beta)$]. It means that the high frequency noise is attenuated in passing through the network and so the signal to noise ratio is improved. A typical choice of $\beta = 10$.

DETERMINATION OF ω_m AND ϕ_m

The frequency ω_m can be determined by differentiating ϕ with respect to ω and equating $d\phi/d\omega$ to zero as shown below.

From equation (1.12) we get,

Phase of $G_c(j\omega)$, $\phi = \angle G_c(j\omega) = \tan^{-1} \omega T - \tan^{-1}\omega\beta T$ ..... (1.14)

On differentiating the equation (1.14) we get,

$$\frac{d\phi}{d\omega} = \frac{1}{1+(\omega T)^2}\, T - \frac{1}{1+(\omega\beta T)^2}\,\beta T \qquad\qquad .....(1.15)$$

When $\omega = \omega_m$, $d\phi/d\omega = 0$

$$\boxed{\dfrac{d}{d\theta}(\tan^{-1}\theta) = \dfrac{1}{1+\theta^2}}$$

Hence, replace ω by ω_m in equation (1.15) and equate to zero.

$$\therefore \frac{1}{1+(\omega_m T)^2}\, T - \frac{1}{1+(\omega_m\beta T)^2}\,\beta T = 0$$

$$\frac{T}{1+(\omega_m T)^2} = \frac{\beta T}{1+(\omega_m \beta T)^2}$$

On cross multiplication we get,

$$1 + (\omega_m \beta T)^2 = \beta[1 + (\omega_m T)^2]$$

$$(\omega_m \beta T)^2 - \beta(\omega_m T)^2 = \beta - 1$$

$$\beta(\omega_m T)^2 (\beta - 1) = (\beta - 1)$$

$$\omega_m^2 = \frac{1}{T^2 \beta} \qquad\qquad \therefore\ \omega_m = \frac{1}{T\sqrt{\beta}}$$

Frequency corresponding to maximum phase lag, $\therefore\ \omega_m = \dfrac{1}{T\sqrt{\beta}}$ \qquad\qquad .....(1.16)

The maximum phase angle ϕ_m can be calculated from the knowledge of β and viceversa. The relations between ϕ_m and β are derived below.

From equation (1.14) we get, $G_c(j\omega) = \phi = \tan^{-1} \omega T - \tan^{-1} \omega\beta T$

$$\boxed{\tan(A - B) = \frac{\tan A - \tan B}{1 + \tan A \tan B}}$$

On taking tan on either side we get,

$$\tan \phi = \tan [\tan^{-1} \omega T - \tan^{-1} \omega\beta T]$$

$$\boxed{\omega_m = \frac{1}{T\sqrt{\beta}}}$$

$$= \frac{\tan(\tan^{-1}\omega T) - \tan(\tan^{-1}\omega\beta T)}{1 + \tan(\tan^{-1}\omega T).\tan(\tan^{-1}\omega\beta T)} = \frac{\omega T - \omega\beta T}{1 + \omega^2 T^2.\beta} = \frac{\omega T(1 - \beta)}{1 + \beta(\omega T)^2}$$

At $\omega = \omega_m,\ \phi \to \phi_m,\quad \therefore \tan\phi_m = \dfrac{\omega_m T(1 - \beta)}{1 + \beta(\omega_m T)^2} = \dfrac{(1 - \beta)/\sqrt{\beta}}{1 + 1} = \dfrac{(1 - \beta)}{2\sqrt{\beta}}$

$$\therefore\ \text{Maximum lag angle, } \phi_m = \tan^{-1}\left[\frac{1 - \beta}{2\sqrt{\beta}}\right] \qquad\qquad .....(1.17)$$

To find the value of β from ϕ_m

From the equation (1.17), it is evident that $(1-\beta)$ and $2\sqrt{\beta}$ are the two sides of right angled triangle. Hence construct a right angle triangle as shown in fig 1.6.

With reference to Fig 1.6, $\sin\phi_m = \dfrac{1 - \beta}{1 + \beta}$

$\sin\phi_m(1 + \beta) = (1 - \beta)$

$\sin\phi_m + \beta \sin\phi_m = 1 - \beta$

$\beta \sin\phi_m + \beta = 1 - \sin\phi_m$

$\beta(\sin\phi_m + 1) = 1 - \sin\phi_m$

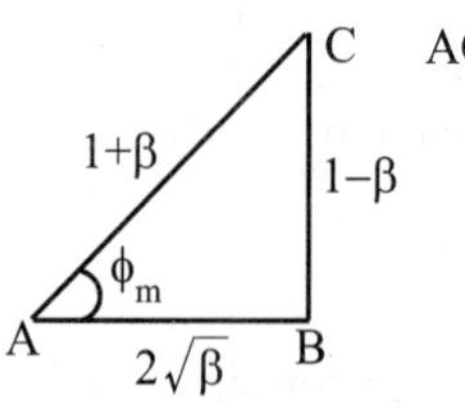

$AC = \sqrt{BC^2 + AB^2}$

$= \sqrt{(1-\beta)^2 + (2\sqrt{\beta})^2}$

$= \sqrt{1 + \beta^2 - 2\beta + 4\beta}$

$= \sqrt{1 + \beta^2 + 2\beta}$

$= \sqrt{(1+\beta)^2} = 1 + \beta$

Fig 1.6

$$\therefore\ \beta = \frac{1 - \sin\phi_m}{1 + \sin\phi_m} \qquad\qquad .....(1.18)$$

PROCEDURE FOR THE DESIGN OF LAG COMPENSATOR USING BODE PLOT

The following steps may be followed to design a lag compensator using bode plot and to be connected in series with transfer function of uncompensated system, G(s).

Step-1 : Choose the value of K in uncompensated system to meet the steady state error requirement.

Step-2 : Sketch the bode plot of uncompensated system. *[Refer Appendix-I for the procedure to sketch bode plot].*

Step-3 : Determine the phase margin of the uncompensated system from the bode plot. If the phase margin does not satisfy the requirement then lag compensation is required.

Step-4 : Choose a suitable value for the phase margin of the compensated system.

Let γ_d = Desired phase margin as given in specifications.

and γ_n = Phase margin of compensated system.

Now, $\gamma_n = \gamma_d + \epsilon$

where ϵ = Additional phase lag to compensate for shift in gain crossover frequency.

Choose an initial value of $\epsilon = 5°$.

Step-5 : Determine the new gain crossover frequency, ω_{gcn}. The new ω_{gcn} is the frequency corresponding to a phase margin of γ_n on the bode plot of uncompensated system.

Let, ϕ_{gcn} = Phase of G(jω) at new gain crossover frequency, ω_{gcn}

Now, $\gamma_n = 180° + \phi_{gcn}$ (or) $\phi_{gcn} = \gamma_n - 180°$

The new gain crossover frequency, ω_{gcn} is given by the frequency at which the phase of G(jω) is ϕ_{gcn}.

Step-6 : Determine the parameter, β of the compensator. The value of β is given by the magnitude of G(jω) at new gain crossover frequency, ω_{gcn}. Find the db gain (A_{gcn}) at new gain crossover frequency, ω_{gcn}.

Now, $A_{gcn} = 20\log \beta$ (or) $\dfrac{A_{gcn}}{20} = \log \beta$, $\therefore \beta = 10^{A_{gcn}/20}$

Step-7 : Determine the transfer function of lag compensator.

Place the zero of the compensator arbitrarily at $1/10^{th}$ of the new gain crossover frequency, ω_{gcn}.

$\therefore$ Zero of the lag compensator, $z_c = \dfrac{1}{T} = \dfrac{\omega_{gcn}}{10}$

Now, $T = \dfrac{10}{\omega_{gcn}}$

Pole of the lag compensator, $p_c = 1/\beta T$

$\left.\begin{array}{l} \text{Transfer function} \\ \text{of lag compensator} \end{array}\right\} G_c(s) = \dfrac{s + \dfrac{1}{T}}{s + \dfrac{1}{\beta T}} = \beta\left(\dfrac{1 + sT}{1 + s\beta T}\right)$

Step-8 : Determine the open loop transfer function of compensated system. The lag compensator is connected in series with plant as shown in fig 1.7.

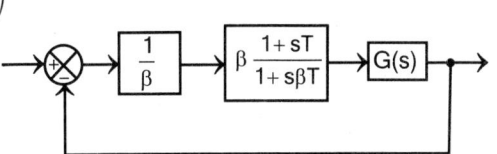

Fig 1.7 : *Block diagram of lag compensated system.*

When the lag compensator is inserted in series with plant, the open loop gain of the system is amplified by the factor β ($\because \beta > 1$). If the gain produced is not required then attenuator with gain $1/\beta$ can be introduced in series with the lag compensator to nullify the gain produced by lag compensator.

The open loop transfer function of the compensated system,

$$G_o(s) = \frac{1}{\beta}.G_c(s).G(s) = \frac{1}{\beta}.\beta\frac{(1+sT)}{(1+s\beta T)}.G(s) = \frac{(1+sT)}{(1+s\beta T)}.G(s)$$

Step-9 : Determine the actual phase margin of compensated system. Calculate the actual phase angle of the compensated system using the compensated transfer function at new gain crossover frequency, ω_{gcn}.

Let, ϕ_{gco} = Phase of $G_o(j\omega)$ at $\omega = \omega_{gcn}$

Actual phase margin of the compensated system, $\gamma_o = 180° + \phi_{gco}$

If the actual phase margin satisfies the given specification then the design is accepted. Otherwise repeat the procedure from step 4 to 9 by taking $\in$ as 5° more than previous design.

PROCEDURE FOR DESIGN OF LAG COMPENSATOR USING ROOT LOCUS

The following steps may be followed to design a lag compensator using root locus and to be connected in series with the transfer function of uncompensated system.

Step-1 : Draw the root locus of uncompensated system. *[Refer Appendix-II for the procedure to construct root locus].*

Step-2 : Determine the dominant pole, s_d. Draw a straight line through the origin with an angle $\cos^{-1}\zeta$ with respect to negative real axis. The intersection point of the straight line with root locus gives the dominant pole, s_d.

Step-3 : Determine the open loop gain of the uncompensated system at $s = s_d$. Let this gain be K. The open loop gain K at $s = s_d$ on root locus is given by,

$$K = \frac{\text{Product of vector lengths from } s_d \text{ to open loop poles}}{\text{Product of vector lengths from } s_d \text{ to open loop poles}} \quad \text{(vector length measured to scale)}$$

Step-4 : Calculate the parameter, β of the compensator.

Let, K_{vu} = Velocity error constant of uncompensated system.

and K_{vd} = Desired velocity error constant.

$$K_{vu} = \underset{s \to 0}{Lt}\ sG(s)$$

Let A be the factor by which the velocity error constant of the system has to be increased,

where, $A = K_{vd}/K_{vu}$

Choose β such that it is 10 to 20% greater than A.

$\therefore \beta = (1.1 \text{ to } 1.2) \times A$.

Step-5 : Determine the transfer function of lag compensator. The zero of the lag compensator (1/T) is chosen to be 10% of the second pole of uncompensated system.

$\therefore$ Zero of the compensator, $z_c = (-1/T) = 0.1 \times ($ second pole of G(s))

Now, $T = 1/[-0.1 \times ($ second pole of G(s))$]$

Pole of the lag compensator, $p_c = -1/\beta T$.

Transfer function of lag compensator $\Big\}$ $G_c(s) = \dfrac{s + \dfrac{1}{T}}{s + \dfrac{1}{\beta T}} = \beta \dfrac{(1+sT)}{(1+s\beta T)}$

Step-6 : Determine the open loop transfer function of the compensated system. The lag compensator is connected in series with the plant as shown in fig 1.8

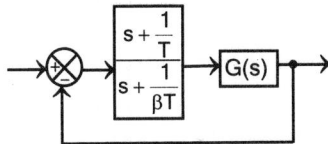

Fig 1.8 : *Block diagram of lag compensated system.*

Open loop transfer function of compensated system $\Big\}$ $G_o(s) = G_c(s).G(s) = \dfrac{\left(s + \dfrac{1}{T}\right)}{\left(s + \dfrac{1}{\beta T}\right)}.G(s)$

Step-7 : Check whether the compensated system satisfies the steady state error requirement. If it is satisfied, then the design is accepted otherwise repeat the design by modifying the locations of poles and zeros of the compensator.

EXAMPLE 1.1

A unity feedback system has an open loop transfer function, $G(s) = K/s(1+2s)$. Design a suitable lag compensator so that phase margin is 40° and the steady state error for ramp input is less than or equal to 0.2.

SOLUTION

Step-1: Calculation of gain, K.

Given that, $e_{ss} \le 0.2$ for ramp input. Let $e_{ss} = 0.2$

We know that, $e_{ss} = 1/K_v$ for ramp input.

$\therefore$ Velocity error constant, $K_v = \dfrac{1}{e_{ss}} = \dfrac{1}{0.2} = 5.$

By definition of velocity error constant, $K_v = \underset{s \to 0}{Lt}\ s\,G(s)\,H(s).$

Since the system is unity feedback system, $H(s) = 1$.

$\therefore K_v = \underset{s \to 0}{Lt}\ s\,G(s)\,H(s) = \underset{s \to 0}{Lt}\ s\dfrac{K}{s(1+2s)} = K \qquad \therefore K = 5$

Step-2: Bode plot of uncompensated system.

Given that, $G(s) = 5/s(1+2s)$

Let $s = j\omega, \quad \therefore G(j\omega) = 5/j\omega(1+j2\omega).$

MAGNITUDE PLOT

The corner frequency is, $\omega_c = 1/2 = 0.5$ rad/sec

The various terms of $G(j\omega)$ are listed in table-1. Also the table shows the slope contributed by each term and the change in slope at the corner frequency.

TABLE-1

Term	Corner frequency rad/sec	Slope db/dec	Change in slope db/dec
$\dfrac{5}{j\omega}$	—	−20	—
$\dfrac{1}{1+j2\omega}$	$\omega_c = \dfrac{1}{2} = 0.5$	−20	$-20-20 = -40$

Choose a low frequency ω_l such that $\omega_l < \omega_c$ and choose a high frequency ω_h such that $\omega_h > \omega_c$.

Let $\omega_l = 0.1$ rad/sec and $\omega_h = 10$ rad/sec

Let $A = |G(j\omega)|$ in db

At $\omega = \omega_l$, $A = 20 \log \left|\dfrac{5}{j\omega}\right| = 20 \log \dfrac{5}{0.1} = 34$ db

At $\omega = \omega_c$, $A = 20 \log \left|\dfrac{5}{j\omega}\right| = 20 \log \dfrac{5}{0.5} = 20$ db

At $\omega = \omega_h$, $A = \left[\text{slope from } \omega_c \text{ to } \omega_h \times \log \dfrac{\omega_h}{\omega_c}\right] + A_{(\text{at } \omega = \omega_c)}$

$$= -40 \times \log \dfrac{10}{0.5} + 20 = -32 \text{ db}$$

Let the points a, b and c be the points corresponding to frequencies ω_l, ω_c and ω_h respectively on the magnitude plot. In a semilog graph sheet choose appropriate scales and fix the points a, b and c. Join the points by straight lines and mark the slope on respective region. Magnitude plot is shown in fig 1.1.1.

PHASE PLOT

The phase angle of $G(j\omega)$ as a function of ω is given by,

$$\phi = \angle G(j\omega) = -90° - \tan^{-1} 2\omega$$

The phase angle of $G(j\omega)$ are calculated for various values of ω and listed in table-2.

TABLE-2

ω rad/sec	0.1	0.5	6.0	5	10
ϕ deg	−101	−135	−153	−174	−177

On the same semilog sheet take another y-axis, choose appropriate scale and draw phase plot as shown in fig 1.1.1.

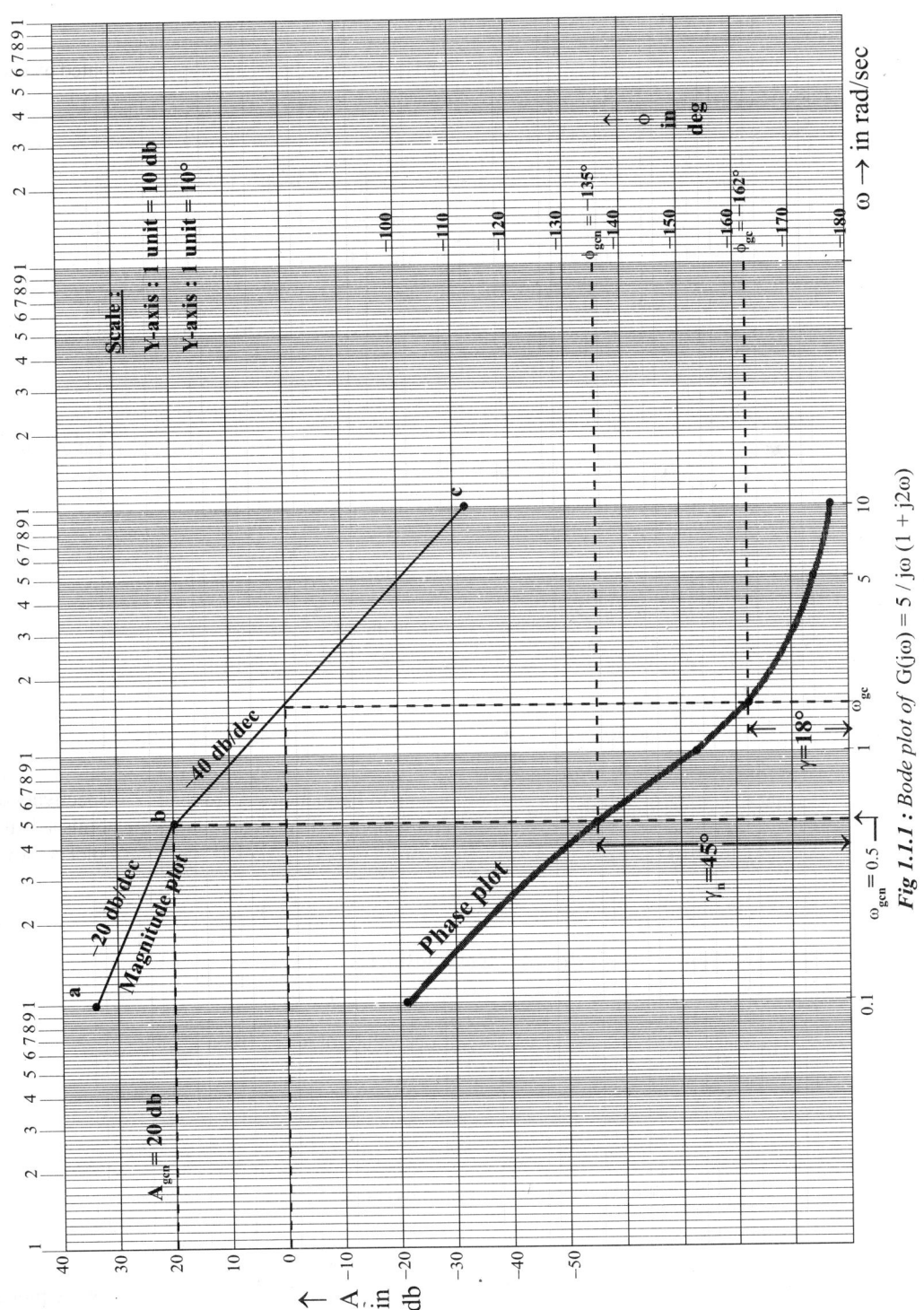

Fig 1.1.1 : Bode plot of $G(j\omega) = 5 / j\omega (1 + j2\omega)$

Step-3 : Determination of phase margin of uncompensated system.

Let, ϕ_{gc} = Phase of $G(j\omega)$ at gain crossover frequency (ω_{gc}).

and γ = Phase margin of uncompensated system.

From the bode plot of uncompensated system we get, ϕ_{gc} = –162°.

Now, $\gamma = 180° + \phi_{gc} = 180° – 162° = 18°$

The system requires a phase margin of 40°, but the available phase margin is 18° and so lag compensation should be employed to improve the phase margin.

Step-4 : Choose a suitable value for the phase margin of compensated system.

The desired phase margin, $\gamma_d = 40°$.

∴ Phase margin of compensated system, $\gamma_n = \gamma_d + \epsilon$

Let initial choice of $\epsilon = 5°$

∴ $\gamma_n = \gamma_d + \epsilon = 40° + 5° = 45°$

Step 5 : Determine new gain crossover frequency.

Let ω_{gcn} = New gain crossover frequency and ϕ_{gcn} = Phase of $G(j\omega)$ at ω_{gcn}

Now, $\gamma_n = 180° + \phi_{gcn}$

∴ $\phi_{gcn} = \gamma_n – 180° = 45° – 180° = –135°$

From the bode plot we found that, the frequency corresponding to a phase of –135° is 0.5 rad/sec.

∴ New gain crossover frequency, ω_{gcn} = 0.5 rad/sec.

Step-6 : Determine the parameter, β

From the bode plot we found that, the db magnitude at ω_{gcn} is 20 db.

∴ $|G(j\omega)|$ in db at ($\omega = \omega_{gcn}$) = A_{gcn} = 20 db

Also, $A_{gcn} = 20 \log \beta$; ∴ $\beta = 10^{A_{gcn}/20} = 10^{20/20} = 10$.

Step-7 : Determine the transfer function of lag compensator.

The zero of the compensator is placed at a frequency one-tenth of ω_{gcn}.

∴ Zero of the lag compensator, $z_c = \dfrac{1}{T} = \dfrac{\omega_{gcm}}{10}$

Now, $T = \dfrac{10}{\omega_{gcm}} = \dfrac{10}{0.5} = 20$

Pole of the lag compensator, $p_c = \dfrac{1}{\beta T} = \dfrac{1}{10 \times 20} = \dfrac{1}{200} = 0.005$

Transfer function of lag compensator, $G_c(s) = \dfrac{s + \dfrac{1}{T}}{s + \dfrac{1}{\beta T}} = \beta \dfrac{1 + sT}{1 + s\beta T} = 10 \dfrac{(1 + 20s)}{(1 + 200s)}$

Step-8 : Determine the open loop transfer function of compensated system.

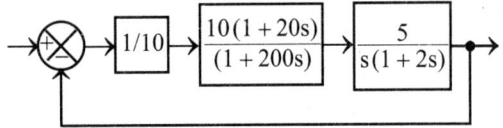

The block diagram of the compensated system is shown in fig 1.1.2. The gain of compensator is nullified by introducing an attenuator in series with compensator, as shown in fig 1.1.2.

Fig 1.1.2 : Block diagram of lag compensated system.

$$\text{Open loop transfer function of compensated system}\bigg\} G_o(s) = \frac{1}{10} \times \frac{10(1+20s)}{(1+200s)} \times \frac{5}{s(1+2s)}$$

$$= \frac{5(1+20s)}{s(1+200s)(1+2s)}$$

Step-9 : Determine the actual phase margin of compensated system.

On substituting $s = j\omega$ in $G_o(s)$ we get, $G_o(j\omega) = \dfrac{5(1+j20\omega)}{j\omega(1+j200\omega)(1+j2\omega)}$

Let, ϕ_o = Phase of $G_o(j\omega)$

and ϕ_{gco} = Phase of $G_o(j\omega)$ at $\omega = \omega_{gcn}$

$\phi_o = \tan^{-1} 20\omega - 90° - \tan^{-1} 200\omega - \tan^{-1} 2\omega$

At $\omega = \omega_{gcn}$, $\phi_o = \phi_{gco} = \tan^{-1} 20\,\omega_{gcn} - 90° - \tan^{-1} 200\omega_{gcn} - \tan^{-1} 2\,\omega_{gcn}$

$\therefore \phi_{gco} = \tan^{-1}(20 \times 0.5) - 90° - \tan^{-1}(200 \times 0.5) - \tan^{-1}(2 \times 0.5) = -140°$.

Actual phase margin of compensated system, $\gamma_o = 180° + \phi_{gco} = 180° - 140° = 40°$

CONCLUSION

The actual phase margin of the compensated system satisfies the requirement. Hence the design is acceptable.

RESULT

Transfer function of lag compensator, $G_c(s) = \dfrac{10(1+20s)}{(1+200s)} = \dfrac{(s+0.05)}{(s+0.005)}$

Open loop transfer function of compensated system, $G_o(s) = \dfrac{5(1+20s)}{s(1+200s)(1+2s)}$

EXAMPLE 1.2

The open loop transfer function of certain unity feedback control system is given by $G(s) = K/s(s+4)\,(s+80)$. It is desired to have the phase margin to be atleast 33° and the velocity error constant $K_v = 30 \text{ sec}^{-1}$. Design a phase lag series compensator.

SOLUTION

Step-1 : Calculation of gain, K

Given that, $K_v = 30 \text{ sec}^{-1}$

By definition of velocity error constant, $K_v = \underset{s\to 0}{Lt}\ sG(s)\,H(s)$

Since the system is unity feedback system, $H(s) = 1$.

$$\therefore\ K_v = \underset{s\to 0}{Lt}\ sG(s) = \underset{s\to 0}{Lt}\ s\frac{K}{s(s+4)(s+80)} = \frac{K}{4\times 80}$$

i.e., $\dfrac{K}{4\times 80} = 30$, $\therefore\ K = 30\times 80\times 4 = 9600.$

Step-2 : Bode plot of uncompensated system

$$G(s) = \frac{9600}{s(s+4)(s+80)} = \frac{9600/4\times 80}{s\left(1+\frac{s}{4}\right)\left(1+\frac{s}{4}\right)} = \frac{30}{s(1+0.25s)(1+0.0125s)}$$

Let $s = j\omega$, $\therefore\ G(j\omega) = \dfrac{30}{j\omega(1+j0.25\omega)(1+j0.0125\omega)}$

MAGNITUDE PLOT

The corner frequencies are, $\omega_{c1} = 1/0.25 = 4$ rad/sec and $\omega_{c2} = 1/0.0125 = 80$ rad/sec. The various terms of $G(j\omega)$ are listed in table-1. Also the table shows the slope contributed by each term and the change in slope at the corner frequency.

TABLE-1

Term	Corner frequency rad/sec	Slope db/dec	Change in slope db/dec
$\dfrac{30}{j\omega}$	—	−20	—
$\dfrac{1}{1+j0.25\omega}$	$\omega_{c1}=\dfrac{1}{0.25}=4$	−20	$-20-20=-40$
$\dfrac{1}{1+j0.0125\omega}$	$\omega_{c2}=\dfrac{1}{0.0125}=80$	−20	$-40-20=-60$

Choose a low frequency ω_l such that $\omega_l < \omega_{c2}$ and choose a high frequency ω_h such that $\omega_h > \omega_{c2}$.

Let $\omega_l = 1$ rad/sec and $\omega_h = 100$ rad/sec.

Let $A = |G(j\omega)|$ in db

At $\omega = \omega_l$, $A = 20\log\left|\dfrac{30}{j\omega}\right| = 20\log\dfrac{30}{1} = 29.5$ db ≈ 30 db

At $\omega = \omega_{c1}$, $A = 20\log\left|\dfrac{30}{j\omega}\right| = 20\log\dfrac{30}{4} = 17.5$ db ≈ 18 db

At $\omega = \omega_{c2}$, $A = \left[\text{slope from }\omega_{c1}\text{ to }\omega_{c2}\times\log\dfrac{\omega_{c2}}{\omega_{c1}}\right] + A_{(at\ \omega = \omega_{c1})}$

$$= -40\log\frac{80}{4} + 18 = -34\ db$$

At $\omega = \omega_h$, $\quad A = \left[\text{slope from } \omega_{c2} \text{ to } \omega_h \times \log \dfrac{\omega_h}{\omega_{c2}}\right] + A_{(at\ \omega\ =\ \omega_{c2})}$

$$= -60 \log \frac{100}{80} + (-34) = -40 \text{ db}$$

Let the points a, b, c and d be the points corresponding to frequencies ω_l, ω_{c1}, ω_{c2} and ω_h respectively on the magnitude plot. In a semilog graph sheet choose appropriate scales and fix the points a, b, c and d. Join the points by straight lines and mark the slope on the respective region. The magnitude plot is shown in fig 1.2.1.

PHASE PLOT

The phase angle of $G(j\omega)$ as a function of ω is given by,

$$\phi = \angle G(j\omega) = -90° - \tan^{-1}0.25\omega - \tan^{-1}0.0125\omega$$

The phase angle of $G(j\omega)$ are calculated for various values of ω and listed in table-2.

TABLE-2

ω rad/sec	1	4	10	50	80	100
ϕ deg	-104	-138	$-165 \approx -164$	$-207 \approx -208$	-222	$-229 \approx -230$

On the same semilog sheet take another y-axis, choose appropriate scale and draw phase plot as shown in fig 1.2.1.

Step-3: Determination of phase margin of uncompensated system.

Let, ϕ_{gc} = Phase of $G(j\omega)$ at gain crossover frequency (ω_{gc}).

and γ = Phase margin of uncompensated system.

From the bode plot of uncompensated system we found that, $\phi_{gc} = -168°$.

Now, $\gamma = 180° + \phi_{gc} = 180° - 168° = 12°$

The system requires a phase margin of atleast 33°, but the available phase margin is 12° and so lag compensation should be employed to improve the phase margin.

Step-4 : Choose a suitable value for the phase margin of compensated system.

The desired phase margin, $\gamma_d = 33°$.

$\therefore$ Phase margin of compensated system, $\gamma_n = \gamma_d + \epsilon$

Let initial choice of $\epsilon = 5°$; $\quad \therefore \gamma_n = \gamma_d + \epsilon = 33° + 5° = 38°$

Step 5: Determine new gain crossover frequency.

Let ω_{gcn} = New gain crossover frequency and ϕ_{gcn} = Phase of $G(j\omega)$ at ω_{gcn}

Now, $\gamma_n = 180° + \phi_{gcn}$

$\therefore \phi_{gcn} = \gamma_n - 180° = 38° - 180° = -142°$

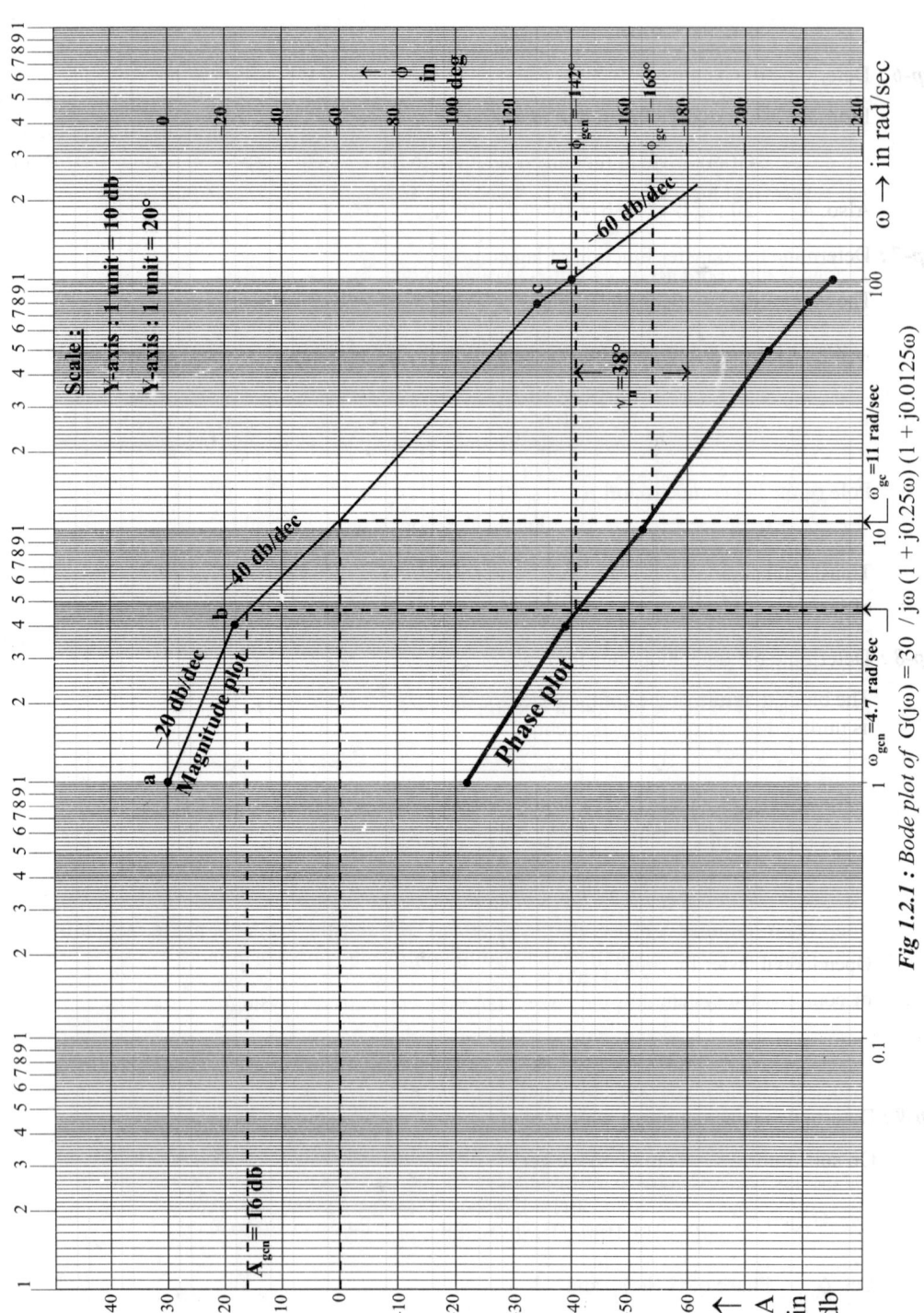

Fig 1.2.1 : Bode plot of $G(j\omega) = 30 / j\omega \, (1 + j0.25\omega) \, (1 + j0.0125\omega)$

From the bode plot we found that, the frequency corresponding to a phase of $-142°$ is 4.7 rad/sec.

$\therefore$ New gain crossover frequency, $\omega_{gcn} = 4.7$ rad/sec.

Step-6 : Determine the parameter, β

From the bode plot we found that, the db magnitude at ω_{gcn} as 16 db.

$\therefore |G(j\omega)|$ in db at $(\omega = \omega_{gcn}) = A_{gcn} = 16$ db

Also, $A_{gcn} = 20 \log \beta$; $\therefore \beta = 10^{A_{gcn}/20} = 10^{16/20} = 6.3$.

Step-7 : Determine the transfer function of lag compensator.

The zero of the compensator is placed at a frequency one-tenth of ω_{gcn}.

$\therefore$ Zero of the lag compensator, $z_c = \dfrac{1}{T} = \dfrac{\omega_{gcn}}{10}$

$$\text{Now, } T = \dfrac{10}{\omega_{gcn}} = \dfrac{10}{4.7} = 2.13$$

Pole of the lag compensator, $p_c = \dfrac{1}{\beta T} = \dfrac{1}{6.3 \times 2.13} = \dfrac{1}{13.419}$

$\left.\begin{array}{l}\text{Transfer function}\\\text{of lag compensator}\end{array}\right\}G_c(s) = \dfrac{s+\dfrac{1}{T}}{s+\dfrac{1}{\beta T}} = \beta\dfrac{1+sT}{1+s\beta T} = 6.3\dfrac{(1+2.13s)}{(1+13.419s)}$

Step-8 : Determine the open loop transfer function of the compensated system.

The block diagram of the compensated system is shown in fig 1.2.2. The gain of the compensator is nullified by introducing an attenuator in series with the compensator, as shown in fig 1.2.2.

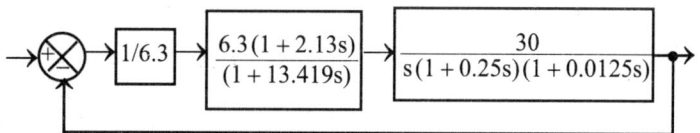

Fig 1.2.2 : *Block diagram of lag compensated system.*

$\left.\begin{array}{l}\text{Open loop transfer function}\\\text{of compensated system}\end{array}\right\}G_o(s) = \dfrac{1}{6.3} \times \dfrac{6.3(1+2.13s)}{(1+13.419s)} \times \dfrac{30}{s(1+0.25s)(1+0.0125s)}$

$$= \dfrac{30(1+2.13s)}{s(1+13.419s)(1+0.25s)(1+0.0125s)}$$

Step-9 : Determine the actual phase margin of compensated system.

On substituting $s = j\omega$ in $G_o(s)$ we get,

$$G_o(j\omega) = \dfrac{30(1+j2.13\omega)}{j\omega(1+j13.419\omega)(1+j0.25\omega)(1+j0.0125\omega)}$$

Let ϕ_o = Phase of $G_o(j\omega)$ and ϕ_{gco} = Phase of $G_o(j\omega)$ at $\omega = \omega_{gcn}$

$\phi_o = \tan^{-1} 2.13\omega - 90° - \tan^{-1} 13.419\omega - \tan^{-1} 0.25\omega - \tan^{-1} 0.0125\omega$

At $\omega = \omega_{gcn}$, $\phi_o = \phi_{gco} = \tan^{-1} 2.13\ \omega_{gcn} - 90° - \tan^{-1} 13.419\omega_{gcn}$

$\qquad\qquad - \tan^{-1} 0.25\omega_{gcn} - \tan^{-1} 0.0125\ \omega_{gcn}$

$\therefore\ \phi_{gco} = \tan^{-1} (2.13 \times 4.7) - 90° - \tan^{-1} (13.419 \times 4.7)$

$\qquad\qquad - \tan^{-1} (0.25 \times 4.7) - \tan^{-1} (0.0125\times4.7) = -147°.$

Actual phase margin of cmpensated system, $\gamma_o = 180° + \phi_{gco} = 180° - 147° = 33°.$

CONCLUSION

The actual phase margin of the compensated system satisfies the requirement. Hence the design is acceptable.

RESULT

Transfer function of lag compensator, $G_c(s) = \dfrac{6.3(1+2.13s)}{(1+13.419s)} = \dfrac{(s+0.469)}{(s+0.074)}$

$\left.\begin{array}{l}\text{Transfer function of}\\ \text{lag compensated system}\end{array}\right\}G_o(s) = \dfrac{30(1+2.13s)}{s(1+13.419s)(1+0.25\omega)(1+0.0125\omega)}$

EXAMPLE 1.3

The forward path transfer function of a certain unity feedback control system is given by $G(s) = K/s\ (s+2)\ (s+8)$. Design a suitable lag compensator so that the system meets the following specifications. (i) Percentage overshoot $\leq 16\%$ for unit step input, (ii) Steady state error ≤ 0.125 for unit ramp input.

SOLUTION

Step-1 : Sketch the root locus of uncompensated system.

To find poles of open loop system

Given that, $G(s) = K/s(s+2)\ (s+8)$.

The poles of open loop transfer function are the roots of the equation, $s(s+2)\ (s+8) = 0$.

$\therefore$ The poles are lying at $s = 0, -2, -8$.

Let us denote poles by, p_1, p_2 and p_3. Here, $p_1 = 0$, $p_2 = -2$ and $p_3 = -8$.

To find root locus on real axis

The segments of real axis between $s = 0$ and $s = -2$ and the segment of real axis between $s = -8$ and $s = -\infty$ will be part of root locus. Because if we choose a test point in this segment then to the right of this point we have odd number of real poles and zeros.

To find angles of asymptotes and centroid

Since there are three poles, the number of root locus branches are three. There is no finite zero and so all the three root locus branches will meet the zeros at infinity. Hence the number of asymptotes required is three.

Angles of asymptotes $= \dfrac{\pm 180°(2q + 1)}{n - m}$; $\quad q = 0, 1, 2, \ldots\ldots\ldots n - m$.

Here $n = 3$ and $m = 0$. $\quad\therefore q = 0, 1, 2, 3$.

$\therefore$ If q = 0, Angles $= \dfrac{\pm 180°}{3} = \pm 60°$

If q = 1, Angles $= \dfrac{\pm 180°(2+1)}{3} = \pm 180°$

$\therefore$ Angles of asymptotes are, $+60°, -60°$ and $\pm 180°$

Centroid $= \dfrac{\text{Sum of poles} - \text{Sum of zeros}}{n - m} = \dfrac{0 - 2 - 8}{3} = -3.33$

To find breakaway point

Closed loop transfer function, $\dfrac{C(s)}{R(s)} = \dfrac{G(s)}{1+G(s)} = \dfrac{\dfrac{K}{s(s+2)(s+8)}}{1 + \dfrac{K}{s(s+2)(s+8)}} = \dfrac{K}{s(s+2)(s+8)+K}$

The characteristic equation is, $s(s+2)(s+8) + K = 0$.

$\therefore K = -s(s+2)(s+8) = -(s^3 + 10s^2 + 16s)$

On differentiating K with respect to s we get,

$dK/ds = -(3s^2 + 20s + 16)$

Put $dK/ds = 0$, $\therefore 3s^2 + 20s + 16 = 0$

$$s = \dfrac{-20 \pm \sqrt{20^2 - 4 \times 3 \times 16}}{2 \times 3} = -0.9 \text{ or } -5.7$$

When s = -0.9, $K = -(s^3+10s^2+16s) = -((-0.9)^3 +10 (-0.9)^2 +16(-0.9)) = 7$

When s = -5.7, $K = -(s^3+10s^2+16s) = -((-5.7)^3 +10 (-5.7)^2 +16(-5.7)) = -48$

For s = -0.9, the value of K is positive and real and so it is actual breakaway point.

To find crossing point on imaginary axis

The characteristic equation is,

$s(s+2)(s+8) + K = 0$.

$s^3 + 10s^2 + 16s + K = 0$

Put s = jω.

$(j\omega)^3 + 10(j\omega)^2 + 16(j\omega) + K = 0$

$-j\omega^3 - 10\omega^2 + j16\omega + K = 0$

On equating imaginary part to zero we get,

$-j\omega^3 + j16\omega = 0$

$-j\omega^3 = -j16\omega$

$\omega^2 = 16$

$\omega = \pm\sqrt{16} = \pm 4$

Hence the root locus crosses the imaginary axis at $+j4$ and $-j4$. The complete root locus sketch is shown in fig1.3.1.

Step-2 : Determine the dominant pole s_d.

Given that, $\%M_p = 16\%$

We know that, $\% \, M_p \,=\, e^{-\zeta\pi\sqrt{1-\zeta^2}}$

$$\therefore \, e^{-\zeta\pi\sqrt{1-\zeta^2}} = 0.16$$

On taking natural log we get, $\dfrac{-\zeta\pi}{\sqrt{1-\zeta^2}} \,=\, ln \, 0.16 \,=\, -1.83$

On squaring we get, $\dfrac{\zeta^2\pi^2}{1-\zeta^2} \,=\, 3.3489$

On cross multiplication we get,

$\zeta^2\pi^2 \,=\, 3.3489(1-\zeta^2)$

$\zeta^2\pi^2 + 3.3489\zeta^2 \,=\, 3.3489$

$\zeta^2(\pi^2 + 3.3489) \,=\, 3.3489 \qquad \therefore \, \zeta \,=\, \sqrt{\dfrac{3.3489}{\pi^2 + 3.3489}} \,=\, 0.5$

$\cos^{-1}\zeta \,=\, \cos^{-1}0.5 \,=\, 60°$

Draw a straight line at an angle of 60° with respect to real axis as shown in fig 1.3.1. The meeting point of this line with root locus is the dominant pole, s_d.

From fig 1.3.1. we get, $s_d = -0.75 \pm j6.35$ (Dominant poles occur as conjugate poles).

Step 3 : To find gain K, at $s = s_d$

$$K \,=\, \frac{\text{Product of vector lengths from open loop poles to } s_d}{\text{Product of vector lengths from open loop zeros to } s_d}$$

Product of vector lengths from poles $= l_1 \times l_2 \times l_3$

From root locus plot of fig 1.3.1. we get, $l_1 = 1.5$; $l_2 = 1.8$ and $l_3 = 7.35$

Note : Vector lengths are measured to scale.

Since there is no finite zero, the product of vector lengths from zeros is unity.

$$\therefore K = l_1 \times l_2 \times l_3 = 1.5 \times 1.8 \times 7.35 = 19.845 \approx 20$$

Step 4 : To find parameter, β

Given that, $G(s) = K/s(s+2)(s+8) = 20/s(s+2)(s+8)$

$$\left.\begin{array}{l}\text{Velocity error constant of}\\ \text{uncompensated system}\end{array}\right\} K_{vu} \,=\, \underset{s \to 0}{Lt} \, sG(s) \,=\, \underset{s \to 0}{Lt} \, s\,\frac{20}{s(s+2)(s+8)} \,=\, \frac{20}{2 \times 8} \,=\, 1.25$$

It is given that steady state error, $e_{ss} \leq 0.125$ for ramp input.

$\therefore$ The desired velocity error constant, $K_{vd} \,=\, \dfrac{1}{e_{ss}} \,=\, \dfrac{1}{0.125} \,=\, 8$

Let A be the factor by which K_v is increased.

$$\therefore \, A \,=\, \frac{K_{vd}}{K_{vu}} \,=\, \frac{8}{1.25} \,=\, 6.4$$

Let, $\beta = 1.2 \times A = 1.2 \times 6.4 = 7.68$

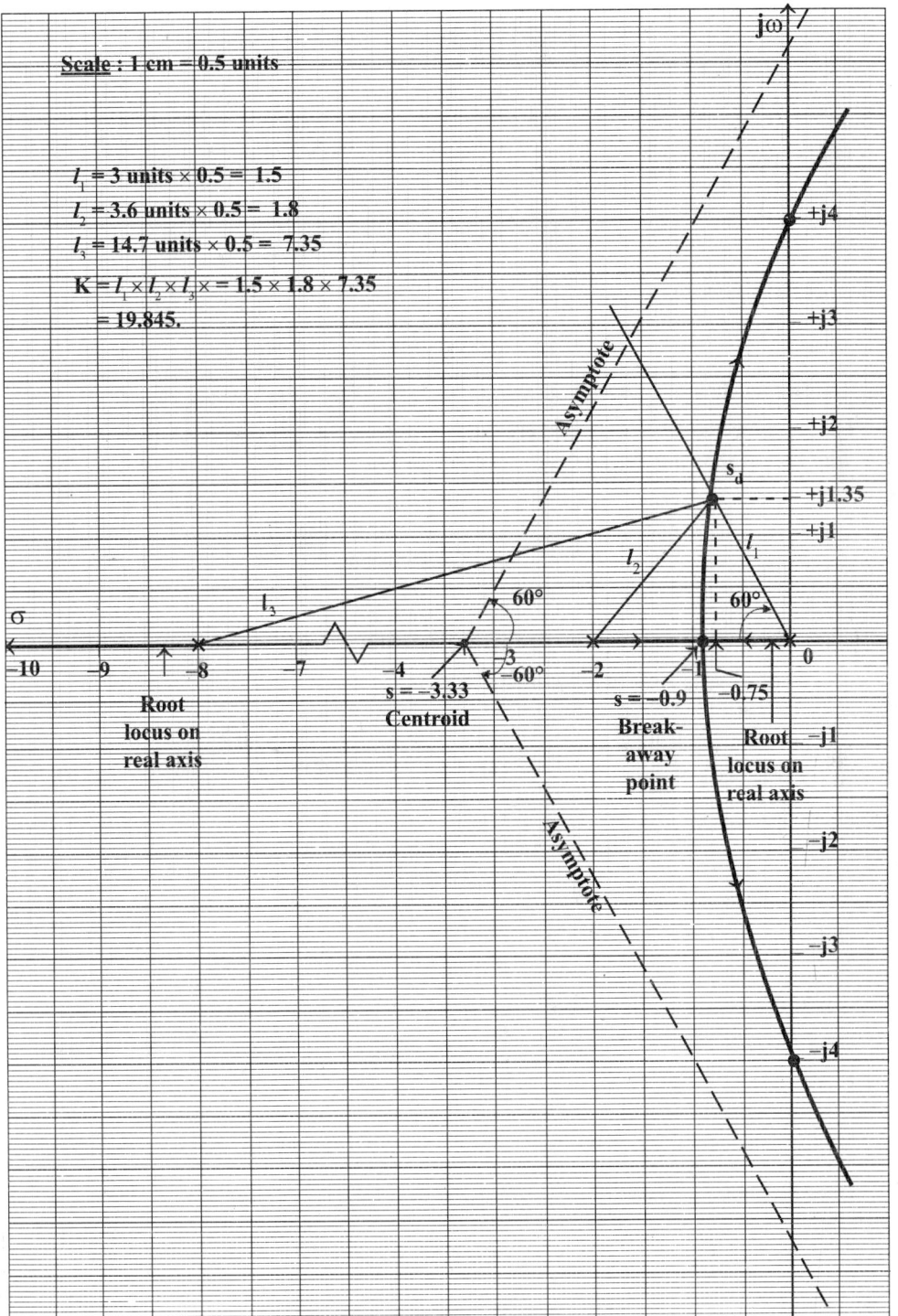

Scale : 1 cm = 0.5 units

$l_1 = 3 \text{ units} \times 0.5 = 1.5$

$l_2 = 3.6 \text{ units} \times 0.5 = 1.8$

$l_3 = 14.7 \text{ units} \times 0.5 = 7.35$

$K = l_1 \times l_2 \times l_3 \times = 1.5 \times 1.8 \times 7.35$

$= 19.845.$

Fig 1.3.1. : Root locus sketch of $1 + G(s) = 1 + K/s(s+2)(s+8)$.

Step-5 : Determine the transfer function of lag compensator.

Zero of the compensator, $z_c = -1/T = 0.1 \times$ second pole of G(s)

$$= 0.1 \times (-2) = -0.2$$

Now, $T = 1/0.2 = 5$

Pole of the compensator, $P_c = \dfrac{-1}{\beta T} = \dfrac{-1}{7.68 \times 5} = -\dfrac{1}{38.4} = -0.026$

$$\therefore \dfrac{1}{\beta T} = 0.026 \quad \text{and} \quad \beta T = 38.4$$

Transfer function of $\Big\}$ $G_c(s) = \dfrac{s + 1/T}{s + 1/\beta T} = \dfrac{(s + 0.2)}{(s + 0.026)}$
lag compensator

Step-6 : Transfer function of compensated system.

The lag compensator is connected in series with G(s) as shown in fig 1.3.2.

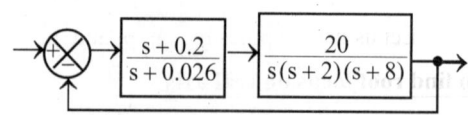

Fig 1.3.2 : Block diagram of lag compensated system.

Open loop transfer function $\Big\}$ $G_o(s) = \dfrac{(s + 0.2)}{(s + 0.026)} \times \dfrac{20}{s(s + 2)(s + 8)}$
of compensated system

$$= \dfrac{20 (s + 0.2)}{s(s + 2)(s + 8)(s + 0.026)}$$

Step-7 : Check steady state error of compensated system

Velocity error constant $\Big\}$ $K_{vc} = \underset{s \to 0}{Lt}\ sG_o(s) = \underset{s \to 0}{Lt}\ s\,\dfrac{20(s + 0.2)}{s(s + 2)(s + 8)(s + 0.026)}$
of compensated system

$$= \dfrac{20 \times 0.2}{2 \times 8 \times 0.026} = 9.615$$

Steady state error of compensated $\Big\}$ $e_{ss} = \dfrac{1}{K_{vc}} = \dfrac{1}{9.615} = 0.104$
system for unit ramp input

CONCLUSION

Since the steady state error of the compensated system is less than 0.125, the design is acceptable.

RESULT

Transfer function of lag compensator, $G_c(s) = (s+0.2)/(s+0.026)$.

Open loop transfer function $\Big\}$ $G_o(s) = \dfrac{20(s + 0.2)}{s(s + 2)(s + 8)(s + 0.026)}$
of lag compensated system

EXAMPLE 1.4

The controlled plant of a unity feedback system is $G(s) = K/s(s+10)^2$. It is specified that velocity error constant of the system be equal to 20, while the damping ratio of the dominant roots be 0.707. Design a suitable cascade compensation scheme to meet the specifications.

SOLUTION

Step-1 : Sketch the root locus of uncompensated system

To find poles of open loop system

Given that, $G(s) = K/s(s + 10)^2$

The poles of open loop transfer function are the roots of the equation $s(s +10)^2 = 0$.

$\therefore$ The poles are lying at $s = 0, -10, -10$.

Let us denote poles by, p_1, p_2 and p_3. Here, $p_1 = 0, p_2 = -10$ and $p_3 = -10$.

To find root locus on real axis

The entire negative real axis will be a part of root locus. Because if we choose a test point on negative real axis then to the right of this point we have odd number of real poles and zeros.

To find angles of asymptotes and centroid

Since there are three poles, the number of root locus branches are three. There is no finite zero and so all the three root locus branches will meet the zeros at infinity. Hence the number of asymptotes required is three.

Angles of asymptotes $= \dfrac{\pm 180°(2q+1)}{n-m}$; $q = 0, 1, 2, \dots n - m$.

Here $n = 3$ and $m = 0$. $\quad \therefore q = 0, 1, 2, 3$.

When, $q = 0$, Angles $= \dfrac{\pm 180°}{3} = \pm 60°$

When, $q = 1$, Angles $= \dfrac{\pm 180°(2+1)}{3} = \pm 180°$

$\therefore$ Angles of asymptotes are, $+60°, -60°$ and $\pm 180°$

Centroid $= \dfrac{\text{Sum of poles} - \text{Sum of zeros}}{n-m} = \dfrac{0-10-10}{3} = \dfrac{-20}{3} = -6.6$

To find breakaway point

Closed loop transfer function, $\dfrac{C(s)}{R(s)} = \dfrac{G(s)}{1+G(s)} = \dfrac{\dfrac{K}{s(s+10)^2}}{1 + \dfrac{K}{s(s+10)^2}} = \dfrac{K}{s(s+10)^2 + K}$

The characteristic equation is, $s(s + 10)^2 + K = 0$

$\therefore K = -s(s +10)^2 = -(s^3 + 20s^2 + 100s)$

On differentiating K with respect to s we get, $dK/ds = -(3s^2 + 40s +100)$.

Put $dK/ds = 0$, $\quad \therefore 3s^2 + 40s + 100 = 0$

$$s = \frac{-40 \pm \sqrt{40^2 - 4 \times 3 \times 100}}{2 \times 3} = \frac{-40 \pm 20}{6} = -3.33, -10.$$

When s = −3.33,   K = −[(−3.33)³ + 20 (−3.33)² + 100(−3.33)] = 148.14

When s = −10,   K = −[(−10)³ + 20 (−10)² + 100(−10)] = 0

For s = −3.33, the value of K is positive and real and so it is actual breakaway point.

To find crossing point of imaginary axis

The characteristic equation is,

$$s(s + 10)^2 + K = 0$$

$$s^3 + 20s^2 + 100s + K = 0$$

Put s = jω.

$$(j\omega)^3 + 20(j\omega)^2 + 100(j\omega) + K = 0$$

$$-j\omega^3 - 20\omega^2 + j100\omega + K = 0$$

On equating imaginary part to zero we get,

$$-j\omega^3 + j100\omega = 0$$

$$-j\omega^3 = -j100\omega$$

$$\omega^2 = 100$$

$$\omega = \pm\sqrt{100} = \pm 10$$

Hence the root locus crosses the imaginary axis at +j10 and −j10. The complete root locus sketch is shown in fig 1.4.1.

Step-2 : Determine the dominant pole s_d.

Given that, ζ = 0.707

$$\therefore \cos^{-1}\zeta = \cos^{-1}0.707 = 45.008 \approx 45°$$

Draw a straight line at an angle of 45° with respect to real axis as shown in fig 1.4.1. The meeting point of this line with root locus is the dominant pole, s_d.

From fig 1.4.1. we get, $s_d = -2.9 \pm j2.9$ (Dominant poles occur as conjugate poles).

Step-3 : To find gain K, at s = s_d

$$\dot{K} = \frac{\text{Product of vector length from open loop poles to } s_d}{\text{Product of vector length from open loop zeros to } s_d}$$

Product of vector lengths from poles = $l_1 \times l_2 \times l_3$

From root locus plot of fig 1.4.1. we get,

$l_1 = 4.1$; $l_2 = l_3 = 7.7$ | *Note : Lengths should be measured to scale.*

Since there is no finite zero, the product of vector lengths from zeros is unity.

$$\therefore K = l_1 \times l_2 \times l_3 = 4.1 \times 7.7 \times 7.7 = 243.089 \approx 240$$

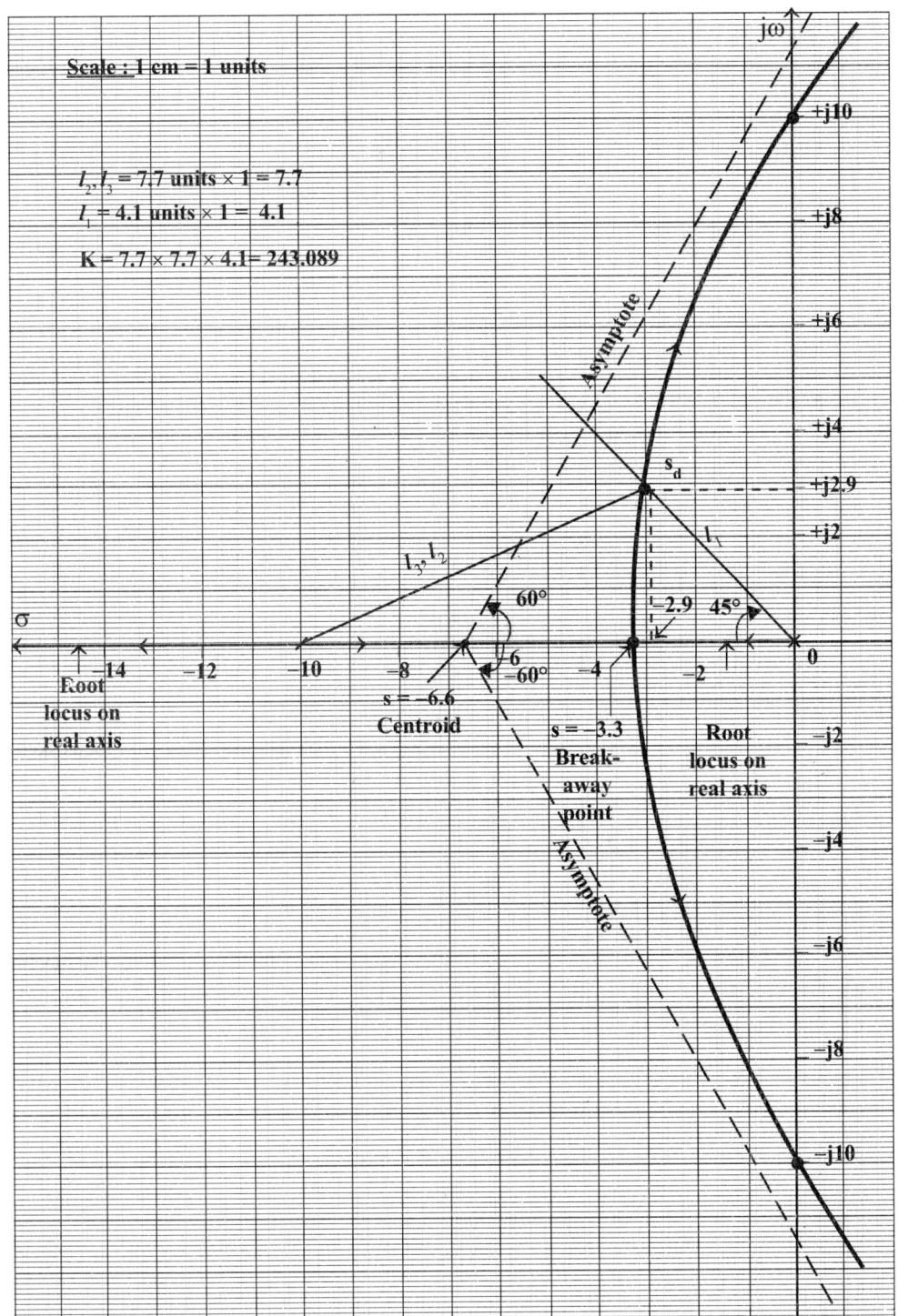

Fig 1.4.1. : Root locus sketch of $1 + G(s) = 1 + K/s(s+10)^2$.

Step-4 : To find parameter, β

Given that, $G(s) = K/s(s + 10)^2 = 240/s(s + 10)^2$

$$
\left.\begin{array}{l}\text{Velocity error constant of}\\ \text{uncompensated system}\end{array}\right\} K_{vu} = \underset{s \to 0}{\text{Lt}}\ sG(s) = \underset{s \to 0}{\text{Lt}}\ s\ \frac{240}{s(s+10)^2} = \frac{240}{10^2} = 2.4
$$

It is given that desired velocity error constant, K_{vd} should be 20. Let A be the factor by which K_v is increased.

$$
\therefore\ A = \frac{K_{vd}}{K_{vu}} = \frac{20}{2.4} = 8.33
$$

Let, $\beta = 1.2 \times A = 1.2 \times 8.33 = 9.996 \approx 10$

Step-5 : Determine the transfer function of lag compensator

Zero of the compensator, $z_c = -1/T = 0.1 \times$ second pole of $G(s) = 0.1 \times (-10) = -1.0$

$$
\therefore\ 1/T = 1.0\ \text{and}\ T = 1.0
$$

Pole of the compensator, $p_c = \dfrac{-1}{\beta T} = \dfrac{-1}{10 \times 1} = -0.1$

$$
\therefore\ \frac{1}{\beta T} = 0.1\ \text{and}\ \beta T = 10
$$

$$
\left.\begin{array}{l}\text{Transfer function of}\\ \text{lag compensator}\end{array}\right\} G_c(s) = \frac{s + \dfrac{1}{T}}{s + \dfrac{1}{\beta T}} = \frac{(s+1)}{(s+0.1)}
$$

Step-6 : Determine the open loop transfer function of compensated system.

The block diagram of lag compensated system is shown in fig 1.4.2 and in this, the lag compensator is connected in series with G(s).

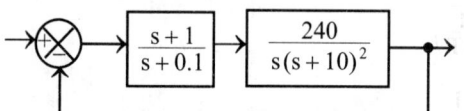

Fig 1.4.2 : *Block diagram of lag compensated system.*

$$
\left.\begin{array}{l}\text{Open loop transfer function}\\ \text{of compensated system}\end{array}\right\} G_o(s) = \frac{(s+1)}{(s+0.1)} \times \frac{240}{s(s+10)^2} = \frac{240(s+1)}{s(s+10)^2(s+0.1)}
$$

Step-7 : Check K_v of compensated system

$$
\left.\begin{array}{l}\text{Velocity error constant}\\ \text{of compensated system}\end{array}\right\} K_{vc} = \underset{s \to 0}{\text{Lt}}\ sG_o(s)
$$

$$
= \underset{s \to 0}{\text{Lt}}\ s\ \frac{240(s+1)}{s(s+10)^2(s+0.1)} = \frac{240}{10^2 \times 0.1} = 24.
$$

CONCLUSION

Since the velocity error constant of the compensated system is greater than the desired value, the design is accepted.

RESULT

$$\left.\begin{array}{l}\text{Transfer function of}\\ \text{lag compensator}\end{array}\right\} G_c(s) = \frac{(s+1)}{(s+0.1)}$$

$$\left.\begin{array}{l}\text{Open loop transfer function}\\ \text{of lag compensated system}\end{array}\right\} G_c(s) = \frac{240(s+1)}{s(s+10)^2(s+0.1)}$$

1.3   LEAD COMPENSATOR

A compensator having the characteristics of a lead network is called a lead compensator. If a sinusoidal signal is applied to the lead network, then in steady state the output will have a phase lead with respect to the input.

The lead compensation increases the bandwidth, which improves the speed of response and also reduces the amount of overshoot. Lead compensation appreciably improves the transient response, whereas there is a small change in steady state accuracy. Generally, lead compensation is provided to make an unstable system as a stable system.

A lead compensator is basically a high pass filter and so it amplifies high frequency noise signals. If the pole introduced by the compensator is not cancelled by a zero in the system, then lead compensation increases the order of the system by one.

S-PLANE REPRESENTATION OF LEAD COMPENSATOR

The lead compensator has a zero at $s = -1/T$ and a pole at $s = -1/\alpha T$. The pole-zero plot of lead compensator is shown in fig 1.9. Here, $\alpha < 1$, so the zero is closer to the origin than the pole. The general form of lead compensator transfer function is given by equation (1.19),

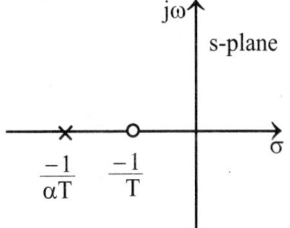

Fig 1.9 : Pole-zero plot of lead compensator.

$$G_c(s) = \frac{s+z_c}{s+p_c} = \frac{\left(s+\dfrac{1}{T}\right)}{\left(s+\dfrac{1}{\alpha T}\right)}$$

.....(1.19)

where, $T > 0$ and $\alpha < 1$

The zero of lead compensator, $z_c = \dfrac{1}{T}$ .....(1.20)

The pole of lead compensator, $p_c = \dfrac{1}{\alpha T}$ .....(1.21)

From equation (1.20) we get, $T = \dfrac{1}{z_c}$ ......(1.22)

From equation (1.21) we get, $\alpha = \dfrac{z_c}{p_c}$ .....(1.23)

REALISATION OF LEAD COMPENSATOR USING ELECTRICAL NETWORK

The lead compensator can be realised by the RC network shown in fig 1.10.

Let, $E_i(s)$ = Input voltage, and $E_0(s)$ = Output voltage

In the network shown in fig 1.10, the input voltage is applied to the series combination of $(R_1 \| C)$ and R_2. The output voltage is obtained across R_2.

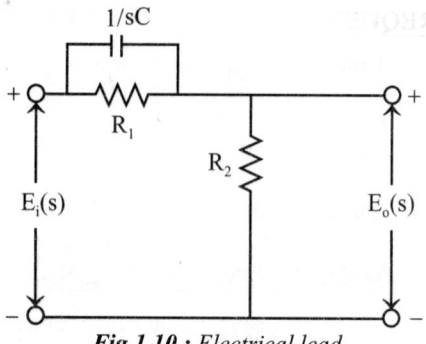

Fig 1.10 : *Electrical lead compensator.*

By voltage division rule,

Output voltage, $E_o(s) = E_i(s) \dfrac{R_2}{R_2 + \dfrac{\left(R_1 \times \dfrac{1}{sC}\right)}{\left(R_1 + \dfrac{1}{sC}\right)}}$

$E_o(s) = E_i(s) . \dfrac{R_2}{R_2 + \dfrac{R_1}{(R_1Cs + 1)}} = E_i(s) \dfrac{R_2}{\dfrac{R_2(R_1Cs + 1) + R_1}{R_1Cs + 1}}$

The transfer function of the electrical network is the ratio of output voltage to input voltage.

Transfer function of electrical network $\left.\begin{array}{l} \dfrac{E_o(s)}{E_i(s)} \end{array}\right\} = \dfrac{R_2(R_1Cs + 1)}{[R_1R_2Cs + R_2 + R_1]} = \dfrac{R_1CR_2\left[s + \dfrac{1}{R_1C}\right]}{R_1CR_2\left[s + \dfrac{(R_1 + R_2)}{R_1CR_2}\right]}$

$$= \dfrac{\left[s + \dfrac{1}{R_1C}\right]}{\left[s + \left(\dfrac{1}{R_2/(R_1 + R_2)}\right)\dfrac{1}{R_1C}\right]} \qquad\qquad .....(1.24)$$

The general form of lead compensator transfer function is,

$$G_c(s) = \dfrac{\left(s + \dfrac{1}{T}\right)}{\left(s + \dfrac{1}{\alpha T}\right)} \qquad\qquad .....(1.25)$$

On comparing equations (1.24) and (1.25) we get,

$$\dfrac{E_o(s)}{E_i(s)} = \dfrac{s + \dfrac{1}{T}}{\left(s + \dfrac{1}{\alpha T}\right)} \qquad\qquad .....(1.26)$$

where, $T = R_1C$ and $\alpha = \dfrac{R_2}{R_1 + R_2}$

The transfer function of the RC network is similar to the general form of transfer function of lead compensator.

FREQUENCY RESPONSE OF LEAD COMPENSATOR

Consider the general form of lead compensator,

$$G_c(s) = \frac{s + \dfrac{1}{T}}{\left(s + \dfrac{1}{\alpha T}\right)} = \frac{(1 + sT)/T}{(s\alpha T + 1)/\alpha T} = \alpha \frac{(1 + sT)}{(1 + \alpha sT)} \qquad .....(1.27)$$

The sinusoidal transfer function of lead compensator is obtained by letting $s = j\omega$ in equation (1.27).

$$\therefore \; G_c(j\omega) = \alpha \frac{(1 + j\omega T)}{(1 + j\omega \alpha T)} \qquad .....(1.28)$$

$$\text{When } \omega = 0, \; G_c(j\omega) = \alpha \qquad .....(1.29)$$

From equation (1.29) we can say that the lead compensator provides an attenuation of α (Here $\alpha < 1$). If the attenuation of the compensator is not desirable then it can be eliminated by a suitable amplifier.

Let us assume that the attenuation α is eliminated by a suitable amplifier network. Now, $G_c(j\omega)$ is given by,

$$G_c(j\omega) = \frac{(1 + j\omega T)}{(1 + j\omega \alpha T)} = \frac{\sqrt{1 + (\omega T)^2} \, \angle \tan^{-1}\omega T}{\sqrt{1 + (\omega \alpha T)^2} \, \angle \tan^{-1}\omega \alpha T} \qquad .....(1.30)$$

Sinusoidal transfer function shown in equation (1.30) has two corner frequencies ω_{c1} and ω_{c2}.

Here, $\omega_{c1} = \dfrac{1}{T}$ and $\omega_{c2} = \dfrac{1}{\alpha T}$. Since, $T > \alpha T$, $\omega_{c1} < \omega_{c2}$

Let $A = \left| G_c(j\omega) \right|$ in db $= 20 \log \dfrac{\sqrt{1 + (\omega T)^2}}{\sqrt{1 + (\omega T)^2}} \qquad .....(1.31)$

At very low frequencies i.e., upto ω_{c1}, $\omega T \ll 1$ and $\omega \alpha T \ll 1$

$$\therefore \; A \approx 20 \log 1 = 0$$

In the frequency range from ω_{c1} to ω_{c2}, $\omega T \gg 1$ and $\omega \alpha T \ll 1$

$$\therefore \; A \approx 20 \log \sqrt{(\omega T)^2} = 20 \log(\omega T)$$

At very high frequencies i.e., after ω_{c2}, $\omega T \gg 1$ and $\omega \alpha T \gg 1$

$$\therefore \; A \approx 20 \log \frac{\sqrt{(\omega T)^2}}{\sqrt{(\omega \alpha T)^2}} = 20 \log \frac{1}{\alpha}$$

The approximate magnitude plot of lead compensator is shown in fig 1.11. The magnitude plot of Bode plot of $G_c(j\omega)$ is a straight line through 0 db upto ω_{c1}, then it has a slope of +20 db/decade upto ω_{c2} and after ω_{c2} it is a straight line with a constant gain of 20 log (1/α).

Let, $\phi = \angle G_c(j\omega)$.

$$\therefore \; \phi = \tan^{-1} \omega T - \tan^{-1} \omega \alpha T$$

As $\omega \to 0$, $\phi \to 0$

As $\omega \to \infty$, $\phi \to 0$

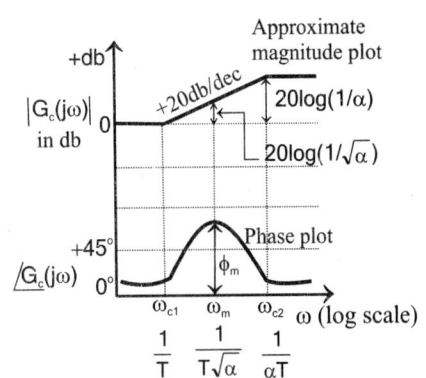

Fig 1.11 : Bode plot of lead compensator.

As ω is varied from 0 to ∞, the phase angle increases from 0 to a maximum value of ϕ_m at $\omega = \omega_m$, then decreases from this maximum value to 0.

It can be shown that the frequency at which maximum phase lead occurs is the geometric mean of the two corner frequencies,

Frequency of maximum phase lead, $\omega_m = \sqrt{\omega_{c1} \cdot \omega_{c2}} = \sqrt{\dfrac{1}{T} \cdot \dfrac{1}{\alpha T}} = \dfrac{1}{T\sqrt{\alpha}}$

The choice of α is governed by the inherent noise in control systems. From the Bode plot of the lead network, we observe that the high frequency noise signals are amplified by a factor $1/\alpha$, while the low frequency control signals undergo unit amplification. Thus the signal/noise ratio at the output of the lead compensator is poorer than at its input. To prevent the signal/noise ratio at the output from deteriorating excessively, it is recommended that the value of α should not be less than 0.07. A typical choice of $\alpha = 0.6$. Also it is advisable to provide two cascaded lead networks when ϕ_m required (i.e., phase lead required) is more than 60°.

Determination of ω_m, ϕ_m and α

The frequency ω_m can be determined by differentiating ϕ with respect to ω and equating $d\phi/d\omega$ to zero.

From equation (1.32) we get,

Phase of $G_c(j\omega)$, $\phi = \angle G_c(j\omega) = \tan^{-1}\omega T - \tan^{-1}\alpha\omega T$ .....(1.32)

On differentiating the equation (1.32) with respect to ω and equating $d\phi/d\omega$ to zero, we get the frequency corresponding to maximum phase lead as, $\omega_m = 1/T\sqrt{\alpha}$.

$\therefore$ Frequency corresponding to maximum phase lead, $\omega_m = \dfrac{1}{T\sqrt{\alpha}}$ .....(1.33)

Also we can express ϕ_m in terms of α and α in terms of ϕ_m as shown below.

$$\phi_m = \tan^{-1}\left(\dfrac{1-\alpha}{2\sqrt{\alpha}}\right)$$ .....(1.34)

$$\alpha = \dfrac{1-\sin\phi_m}{1+\sin\phi_m}$$ .....(1.35)

> *Note : The equations (1.33), (1.34) and (1.35) can be derived by a similar analysis shown in section 1.2 for lag compensator after replacing β by α.*

PROCEDURE FOR DESIGN OF LEAD COMPENSATOR USING BODE PLOT

The following steps may be followed to design a lead compensator using bode plot and to be connected in series with transfer function of uncompensated system, G(s).

Step-1 : The open loop gain K of the given system is determined to satisfy the requirement of the error constant.

Step-2 : The bode plot is drawn for the uncompensated system using the value of K, determined from the previous step. *[Refer Appendix-1 for the procedure to sketch bode plot].*

Step-3 : The phase margin of the uncompensated system is determined from the bode plot.

Step-4 : Determine the amount of phase angle to be contributed by the lead network by using the formula given below,

$$\phi_m = \gamma_d - \gamma + \in$$

where,

ϕ_m = Maximum phase lead angle of the lead compensator

γ_d = Desired phase margin

γ = Phase margin of the uncompensated system

$\in$ = Additional phase lead to compensate for shift in gain crossover frequency

Choose an initial choice of $\in$ as $5°$

> *Note : If ϕ_m is more than 60° then realize the compensator as cascade of two lead compensator with each compensator contributing half of the required angle).*

Step-5 : Determine the transfer function of lead compensator

Calculate α using the equation, $\alpha = \dfrac{1 - \sin\phi_m}{1 + \sin\phi_m}$

From the bode plot, determine the frequency at which the magnitude of $G(j\omega)$ is $-20 \log 1/\sqrt{\alpha}$ db. This frequency is ω_m.

Calculate T from the relation, $\omega_m = \dfrac{1}{T\sqrt{\alpha}}$ $\therefore$ T $= \dfrac{1}{\omega_m \sqrt{\alpha}}$

$\left.\begin{array}{l}\text{Transfer function of} \\ \text{lead compensator}\end{array}\right\rbrace G_c(s) = \dfrac{s + \dfrac{1}{T}}{s + \dfrac{1}{\alpha T}} = \dfrac{\alpha(1 + sT)}{(1 + \alpha sT)}$

Step-6 : Determine the open loop transfer function of compensated system.

The lag compensator is connected in series with G(s) as shown in fig 1.12. When the lead network is inserted in series with the plant, the open loop gain of the system is attenuated by the factor α ($\because \alpha < 1$), so an amplifier with the gain of $1/\alpha$ has to be introduced in series with the compensator to nullify the attenuation caused by the lead compensator.

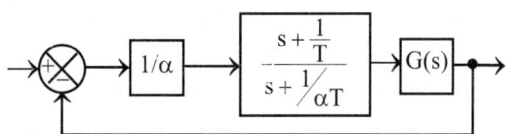

Fig 1.12 : Block diagram of lead compensated system.

$\left.\begin{array}{l}\text{Open loop transfer function} \\ \text{of the overall system}\end{array}\right\rbrace G_o(s) = \dfrac{1}{\alpha} \times \dfrac{s + \dfrac{1}{T}}{s + \dfrac{1}{\alpha T}} \times G(s)$

$= \dfrac{1}{\alpha} \times \dfrac{\alpha(1 + sT)}{(1 + s\alpha T)} \times G(s) = \dfrac{(1 + sT)}{(1 + s\alpha T)} \times G(s)$

Step-7 : Verify the design.

Finally the Bode plot of the compensated system is drawn and verify whether it satisfies the given specifications. If the phase margin of the compensated system is less than the required phase margin then repeat step 4 to 10 by taking $\in$ as 5° more than the previous design.

PROCEDURE FOR DESIGN OF LEAD COMPENSATOR USING ROOT LOCUS

The following steps may be followed to design a lead compensator using root locus and to be connected in series with transfer function of uncompensated system, G(s).

Step-1 : Determine the dominant pole, s_d from the given specifications.

Dominant pole, $s_d = -\zeta\omega_n \pm j\omega_n \sqrt{1-\zeta^2}$

> *Note : If ζ alone is specified and ω_n is not available, then draw the root locus and from the root locus find the dominant pole. Refer example 1.8.*

Step-2 : Mark the poles and zeros of open loop transfer function and the dominant pole on the s-plane. Let the dominant pole be point P.

Step-3 : Determine the angle to be contributed by lead network.

Let, ϕ = Angle to be contributed by lead network to make the point P as a point on root locus.

Draw vectors from all open loop poles and zeros to point P. Measure the angle contributed by the vectors. *[For procedure to find angle contribution by vectors refer root locus in Appendix-II].*

$$\text{Now, } \phi = \left(\begin{array}{c}\text{Sum of angles} \\ \text{contributed by poles} \\ \text{of uncompensated system}\end{array}\right) - \left(\begin{array}{c}\text{Sum of angles} \\ \text{contributed by zeros} \\ \text{of uncompensated system}\end{array}\right) \pm n\,180°$$

where n is an odd integer so that $n180°$ is nearest to the difference between angles contributed by poles and zeros.

> *Note : If the angle to be contributed is more than 60° then realise the compensator as cascade of two lead compensators with each compensator contributing half of the required angle.*

Step-4 : Determine the pole and zero of the lead compensator.

Let point O be the origin of s-plane and point P be the dominant pole. Draw straight lines OP and AP such that AP is parallel to x-axis as shown in fig 1.13.

Draw a line PC so as to bisect the angle APO [∠APO] where the point C is on the real axis. With line PC as reference, draw angles BPC and CPD such that each equal to $\phi/2$. Here the points B and D are located on the real axis.

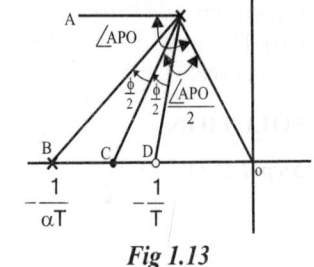

Now the point B is the location of the pole of the compensator $(-1/\alpha T)$ and the point D is the location of the zero of the compensator $(-1/T)$. From the values of point D and B compute T and α.

Fig 1.13

Step-5 : Determine the transfer function of lead compensator

$$\left.\begin{array}{l}\text{Transfer function of} \\ \text{lead compensator}\end{array}\right\} G_c(s) = \frac{\left(s+\dfrac{1}{T}\right)}{\left(s+\dfrac{1}{\alpha T}\right)}$$

$$= \alpha\,\frac{(1+sT)}{(1+s\alpha T)}$$

Step-6 : Determine open loop transfer function of lead compensated system.

The lead compensator is connected in series with the plant as shown in fig 1.14.

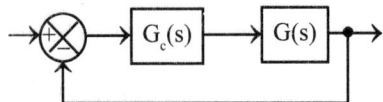

Fig 1.14 : Block diagram of
lead compensated system.

Open loop transfer function of compensated system, $G_0(s) = G_c(s) G(s)$.

The open loop gain K is given by the value of gain of $s = s_d$. The value of gain, K is determined from pole-zero plot of lead compensated system and by using the magnitude condition given below.

$$K = \frac{\text{Product of vector length from all poles to } s = s_d}{\text{Product of vector length from all zeros to } s = s_d}$$

Note : *The length of vectors should be measured to scale. For details of magnitude condition, refer root locus in chapter-5.*

Step-7 : Check whether the compensated system satisfies the error requirement. If the error requirement is satisfied then the design is accepted. Otherwise repeat the design by altering the location of poles and zeros by trial and error, without changing the value of ϕ.

Note : *If the open loop gain K is specified in the problem, then take the gain at $s = s_d$ as K_6. Find a parameter, A where $A = K_1/K$. Now introduce an amplifier with gain, A in cascade with compensator to account for reduction in gain due to attenuation by parameter, α. Now, $G_0(s) = A G_c(s) G(s)$.*

EXAMPLE 1.5

Design a phase lead compensator for the system shown in fig 1.5.1 to satisfy the following specifications. (i) The phase margin of the system $\geq 45°$. (ii) Steady state error for a unit ramp input $\leq 1/15$. (iii) The gain crossover frequency of the system must be less than 7.5 rad/sec.

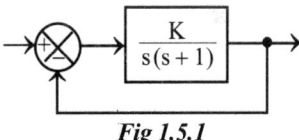

Fig 1.5.1

SOLUTION

Step-1 : Determine K.

Given that, steady state error, $e_{ss} \leq 1/15$ for unit ramp input

When the input is unit ramp, $e_{ss} = 1/K_v = 1/15$. $\therefore K_v = 15$

By definition of velocity error constant we get,

$$\therefore K_v = \underset{s \to 0}{\text{Lt}} \ s \ \frac{K}{s(s+1)} = K$$

Here, $G(s) = \dfrac{K}{s(s+1)}$ and $H(s) = 1$, $\therefore K_v = \underset{s \to 0}{\text{Lt}} \ s. \dfrac{K}{s(s+1)} = K$

Step-2 : Draw bode plot .

Given that, $G(s) = \dfrac{K}{s(s+1)} = \dfrac{15}{s(s+1)}$

Let $s = j\omega$, $\therefore G(j\omega) = \dfrac{15}{j\omega(1+j\omega)}$

MAGNITUDE PLOT

The corner frequency is, $\omega_{c1} = 1$ rad/sec.

The various terms of $G(j\omega)$ are listed in table-6. Also the table shows the slope contributed by each term and the change in slope at the corner frequency.

TABLE-1

Term	Corner frequency rad/sec	Slope db/dec	Change in slope db/dec
$\dfrac{15}{j\omega}$	—	−20	—
$\dfrac{1}{1+j\omega}$	$\omega_{c1} = 1$	−20	$-20 + (-20) = -40$

Choose a low frequency ω_l such that $\omega_l < \omega_{c1}$ and choose a high frequency ω_h such that $\omega_h > \omega_{c6}$.

Let, $\omega_l = 0.1$ rad/sec and $\omega_h = 10$ rad/sec.

Let, $A = |G(j\omega)|$ in db

Let us calculate A at ω_l, ω_{c1} and ω_h.

At $\omega = \omega_l = 0.1$ rad/sec, $A = 20 \log \left|\dfrac{15}{j\omega}\right| = 20 \log \dfrac{15}{0.1} = 43.5$ db $\approx$ 44 db

At $\omega = \omega_{c1} = 1$ rad/sec, $A = 20 \log \left|\dfrac{15}{j\omega}\right| = 20 \log \dfrac{15}{1} = 23.5$ db $\approx$ 24 db

At $\omega = \omega_h = 10$ rad/sec, $A = \left[\text{slope from } \omega_{c1} \text{ to } \omega_h \times \dfrac{\omega_h}{\omega_{c1}}\right] + A_{(\text{at } \omega\, =\, \omega_{c1})}$

$= -40 \times \log \dfrac{10}{1} + 24 = -16$ db

Let the points a, b and c be the points corresponding to frequencies ω_l, ω_{c1}, and ω_h respectively on the magnitude plot. In a semilog graph sheet choose appropriate scales and fix the points a, b and c. Join the points by straight lines and mark the slope on respective region. Magnitude plot is shown in fig 1.5.2.

PHASE PLOT

The phase angle of $G(j\omega)$ as a function of ω is given by,

$\phi = \angle G(j\omega) = -90° - \tan^{-1}\omega$

The phase angle of $G(j\omega)$ are calculated for various values of ω and listed in table-2.

TABLE-2

ω rad/sec	0.1	0.5	1	2	5	10
ϕ deg	−96	−117	−135	−153	−169	−174

On the same semilog sheet take another y-axis, choose appropriate scale and draw phase plot as shown in fig 1.5.2.

Step-3 : Determine the phase margin of uncompensated system.

Let, ϕ_{gc} = Phase of $G(j\omega)$ at gain crossover frequency (ω_{gc}).

and γ = Phase margin of uncompensated system.

From the bode plot of uncompensated system we get, $\phi_{gc} = -167°$.

Now, $\gamma = 180° + \phi_{gc} = 180° - 167° = 13°$

The system requires a phase margin of 45°, but the available phase margin is 13° and so lead compensation should be employed to improve the phase margin.

Step-4 : Find ϕ_m

The desired phase margin, $\gamma_d \geq 45°$

Let additional phase lead required, $\epsilon = 5°$

Maximum lead angle, $\phi_m = \gamma_d - \gamma + \epsilon = 45° - 13° + 5° = 37°$

Step-5 : Determine the transfer function of lead compensator.

$$\alpha = \frac{1 - \sin\phi_m}{1 + \sin\phi_m} = \frac{1 - \sin 37°}{1 + \sin 37°} = 0.2486 \approx 0.25$$

The db magnitude corresponding to $\left\} \omega_n = -20\log\frac{1}{\sqrt{\alpha}} = -20\log\frac{1}{\sqrt{0.25}} = -6\ db \right.$

From the bode plot of uncompensated system the frequency ω_m corresponding to a db gain of −6 db is found to be 5.6 rad/sec.

$\therefore \omega_m = 5.6$ rad/sec.

Now, $T = \dfrac{1}{\omega_n\sqrt{\alpha}} = \dfrac{1}{5.6\sqrt{0.25}} = 0.357 \approx 0.36$

Transfer function of the lead compensator $\left\} G_c(s) = \dfrac{s + \dfrac{1}{T}}{s + \dfrac{1}{\alpha T}} = \alpha\dfrac{(s + sT)}{(1 + s\alpha T)} \right.$

$$= 0.25\frac{(1 + 36s)}{(1 + 0.09s)}$$

Step-6 : Open loop transfer function of compensated system.

The block diagram of the lead compensated system is shown in fig 1.5.3.

The compensator will provide an attenuation of α. To compensate for that, an amplifier of gain $1/\alpha$ is introduced in series with compensator.

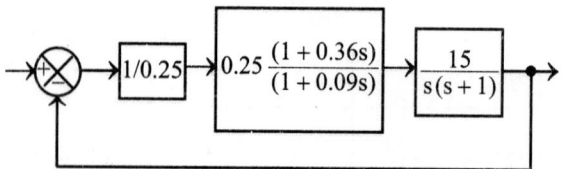

Fig 1.5.3 : Block diagram of lead compensated system.

$$\left.\begin{array}{l}\text{Open loop transfer function}\\\text{of compensated system}\end{array}\right\}G_o(s) = \frac{1}{0.25}\times\frac{0.25(1+0.36s)}{(1+0.09s)}\times\frac{15}{s(s+1)}$$

$$= \frac{15(1+0.36s)}{s(1+0.09s)(1+s)}$$

Step-7 : Draw the bode plot of compensated system to verify the design.

Put $s = j\omega$ in $G_o(s)$, $\therefore$ $G_o(j\omega) = \dfrac{15(1+j0.36\omega)}{j\omega(1+0.09\omega)(1+j\omega)}$

MAGNITUDE PLOT

The corner frequencies are ω_{c_1}, ω_{c_2} and ω_{c_3}.

$$\omega_{c1} = \frac{1}{1} = 1\,\text{rad/sec} \; ; \quad \omega_{c2} = \frac{1}{0.36} = 2.8\,\text{rad/sec} \; ; \quad \omega_{c3} = \frac{1}{0.09} = 11.1\,\text{rad/sec}$$

The various terms of $G_o(j\omega)$ are listed in table-3. Also the table shows the slope contributed by each term and the change in slope at the corner frequency.

Choose a low frequency ω_l such that $\omega_l < \omega_{c1}$ and choose a high frequency ω_h such that $\omega_h > \omega_{c3}$.

Let $\omega_l = 0.1$ rad/sec and $\omega_h = 50$ rad/sec

Let $A_o = |G_o(j\omega)|$ in db

At $\omega = \omega_l = 0.1$ rad/sec, $\quad A = 20\log\left|\dfrac{15}{j\omega}\right| = 20\log\dfrac{15}{0.1} = 43.5\,\text{db} \approx 44\,\text{db}$

At $\omega = \omega_{c1} = 1$ rad/sec, $\quad A = 20\log\left|\dfrac{15}{j\omega}\right| = 20\log\dfrac{15}{1} = 23.5\,\text{db} \approx 24\,\text{db}$

At $\omega = \omega_{c2} = 2.8$ rad/sec, $\quad A_o = \left[\text{slope from }\omega_{c1}\text{ to }\omega_{c2}\times\log\dfrac{\omega_{c2}}{\omega_{c1}}\right]+\left(\begin{array}{c}\text{gain at}\\\omega=\omega_{c1}\end{array}\right)$

$$= -40\times\log\frac{2.8}{1}+24 = 6\,\text{db}$$

At $\omega = \omega_{c3} = 11.1$ rad/sec, $\quad A_o = \left[\text{slope from }\omega_{c2}\text{ to }\omega_{c3}\times\log\dfrac{\omega_{c3}}{\omega_{c2}}\right]+\left(\begin{array}{c}\text{gain at}\\\omega=\omega_{c2}\end{array}\right)$

$$= -20\times\log\frac{11.1}{2.8}+6 = -6\,\text{db}$$

At $\omega = \omega_h = 50$ rad/sec, $\quad A_o = \left[\text{slope from }\omega_{c3}\text{ to }\omega_h\times\log\dfrac{\omega_h}{\omega_{c3}}\right]+\left(\begin{array}{c}\text{gain at}\\\omega=\omega_{c3}\end{array}\right)$

$$= -40\times\log\frac{50}{11.1}+(-6) = -32\,\text{db}$$

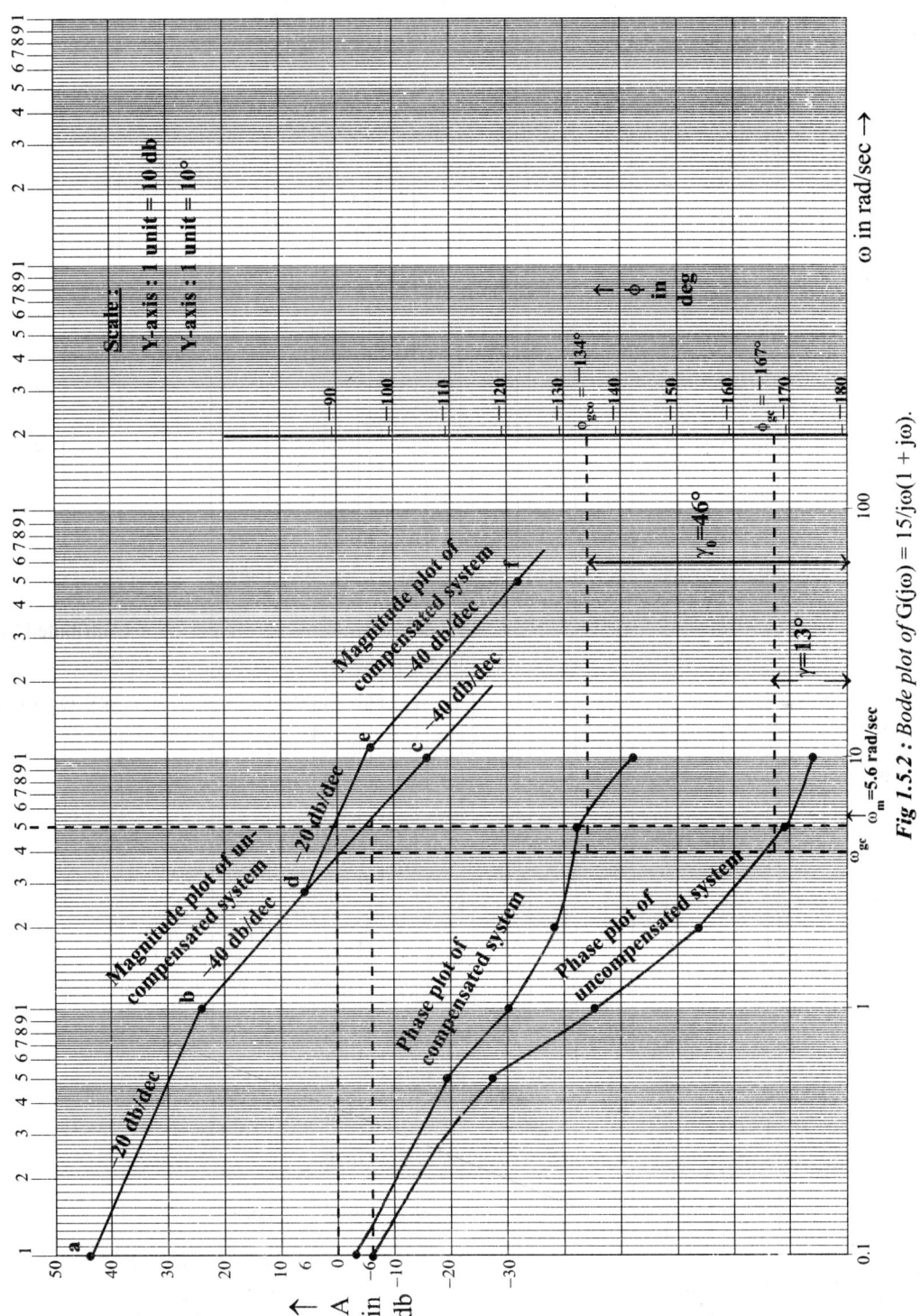

Scale :
Y-axis : 1 unit = 10 db
Y-axis : 1 unit = 10°

Magnitude plot of compensated system

−40 db/dec

−20 db/dec

Magnitude plot of un-compensated system

−40 db/dec

−20 db/dec

−20 db/dec

Phase plot of compensated system

Phase plot of uncompensated system

$\gamma_0 = 46°$

$\gamma = 13°$

$\omega_m = 5.6$ rad/sec

$\theta_{gco} = -134°$

$\phi_{gc} = -167°$

ϕ in deg

ω in rad/sec →

← A in db

Fig 1.5.2 : Bode plot of G(jω) = 15/jω(1 + jω).

TABLE-3

Term	Corner frequency rad/sec	Slope db/dec	Change in slope db/dec
$\dfrac{15}{j\omega}$	-	-20	-
$\dfrac{1}{1+j\omega}$	$\omega_{c1} = 1$	-20	$-20-20 = -40$
$1+j0.36\omega$	$\omega_{c2} = \dfrac{1}{0.36} = 2.8$	+20	$-40+20 = -20$
$\dfrac{1}{1+j0.09\omega}$	$\omega_{c3} = \dfrac{1}{0.09} = 11.1$	-20	$-20-20 = -40$

Let the points a, b, d, e and f be the points corresponding to frequencies ω_1, ω_{c1}, ω_{c2}, ω_{c3} and ω_h respectively on the magnitude plot of compensated system. The magnitude plot of compensated system is drawn on the same semilog graph sheet by using the same scales as shown in fig 1.5.2.

PHASE PLOT

The phase angle of $G_0(j\omega)$ as a function of ω is given by,

$$\phi_0 = \angle G_0(j\omega) = \tan^{-1}0.36\omega -90° - \tan^{-1}0.09\omega - \tan^{-1}\omega.$$

The phase angle of $G_0(j\omega)$ are calculated for various values of ω and listed in table- 4.

TABLE-4

ω rad/sec	0.1	0.5	1	2	5	10
ϕ_0 deg	-94	-109	-120	-128	-132	-142

In the same semilog sheet and by using the same scales, the phase plot of compensated system is sketched as shown in fig 1.5.2.

Let, ϕ_{gc0} = Phase of $G_0(j\omega)$ at new gain crossover frequency.

and γ_0 = Phase margin of compensated system.

From the bode plot of compensated system we get, $\phi_{gc0} = -134°$.

Now, $\gamma_0 = 180° + \phi_{gc0} = 180° - 134° = 46°$

CONCLUSION

The phase margin of the compensated system is satisfactory. Hence the design is acceptable.

RESULT

The transfer function of lead compensator, $G_c(s) = \dfrac{0.25(1 + 0.36s)}{(1 + 0.09s)} = \dfrac{(s + 2.78)}{(s + 11.11)}$

Open loop transfer function of lead compensated system $\Big\}$ $G_0(s) = \dfrac{15(1 + 0.36s)}{s(1 + 0.09s)(1 + s)}$

EXAMPLE 1.6

Design a lead compensator for a unity feedback system with open loop transfer function, $G(s) = K/s\,(s+1)(s+5)$ to satisfy the following specifications (i) Velocity error constant, $K_v \geq 50$ (ii) Phase margin is $\geq 20°$.

SOLUTION

Step-1 : Determine K

Given that, $K_v \geq 50$, Let $K_v = 50$

By definition of velocity error constant, K_v we get,

$$K_v = \underset{s \to 0}{Lt}\, sG(s) = \underset{s \to 0}{Lt}\, s\,\frac{K}{s(s+1)(s+5)} = \frac{K}{5}$$

$$\therefore K = 5 \times K_v = 5 \times 50 = 250.$$

Step-2 : Draw bode plot

Given that, $G(s) = \dfrac{K}{s(s+1)(s+5)} = \dfrac{250}{s(s+1)(s+5)}$

$$= \frac{250}{s(1+s) \times 5 \times (1 + s/5)} = \frac{50}{s(1+s)(1+0.2s)}$$

Let $s = j\omega$, $\therefore G(j\omega) = \dfrac{50}{j\omega(1 + j\omega)(1 + j0.2\omega)}$

MAGNITUDE PLOT

The corner frequency are, $\omega_{c1} = 1$ rad/sec and $\omega_{c2} = 1/0.2 = 5$ rad/sec.

The various terms of $G(j\omega)$ are listed in table-6. Also the table shows the slope contributed by each term and the change in slope at the corner frequencies.

TABLE-1

Term	Corner frequency rad/sec	Slope db/dec	Change in slope db/dec
$\dfrac{50}{j\omega}$	–	–20	–
$\dfrac{1}{1+j\omega}$	$\omega_{c1}=1$	–20	$-20+(-20)=-40$
$\dfrac{1}{1+j0.2\omega}$	$\omega_{c2}=\dfrac{1}{0.2}=5$	–20	$-40-20=-60$

Choose a low frequency ω_l such that $\omega_l<\omega_{c1}$ and choose a high frequency ω_h such that $\omega_h>\omega_{c2}$.

Let $\omega_l=0.5$ rad/sec and $\omega_h=10$ rad/sec

Let $A = |G(j\omega)|$ in db

At $\omega=\omega_l$, $A = 20\log\left|\dfrac{50}{j\omega}\right| = 20\log\dfrac{50}{0.5} = 40$ db

At $\omega=\omega_{c1}$, $A = 20\log\left|\dfrac{50}{j\omega}\right| = 20\log\dfrac{50}{1} = 34$ db

At $\omega=\omega_{c2}$, $A = \left[\text{slope from }\omega_{c1}\text{ to }\omega_{c2}\times\log\dfrac{\omega_{c2}}{\omega_{c1}}\right]+A_{(\text{at }\omega=\omega_{c1})}$

$\qquad\qquad = -40\times\log\dfrac{5}{1}+34 = 6$ db

At $\omega=\omega_h$, $A = \left[\text{slope from }\omega_{c2}\text{ to }\omega_h\times\log\dfrac{\omega_h}{\omega_{c2}}\right]+A_{(\text{at }\omega=\omega_{c2})}$

$\qquad\qquad = -60\times\log\dfrac{10}{5}+6 = -12$ db

Let the points a, b, c and d be the points corresponding to frequencies $\omega_l,\omega_{c1},\omega_{c2}$ and ω_h respectively on the magnitude plot. In a semilog graph sheet choose appropriate scales and fix the points a, b, c and d. Join the points by straight lines and mark the slope on the respective region. The magnitude plot is shown in fig 1.6.2.

PHASE PLOT

The phase angle of $G(j\omega)$ as a function of ω is given by,

$\phi = \angle G(j\omega) = -90° - \tan^{-1}\omega - \tan^{-1}0.2\omega$

The phase angle of $G(j\omega)$ are calculated for various values of ω and listed in table-2.

TABLE-2

ω rad/sec	0.1	0.5	1.0	5	10
ϕ deg	$-96°$	-122	-146	-214	-238

On the same semilog sheet take another y-axis, choose appropriate scale and draw phase plot as shown in fig 1.6.2.

Step-3 : Determine the phase margin

Let, ϕ_{gc} = Phase of $G(j\omega)$ at gain crossover frequency.

and γ = Phase margin of uncompensated system.

From the bode plot of uncompensated system we get, $\phi_{gc} = -224°$.

Now, $\gamma = 180° + \phi_{gc} = 180° - 224° = -44°$

The phase margin of the system is negative and so the system is unstable. Hence lead compensation is required to make the system stable and to have a phase margin of 20°.

Step-4 : Find ϕ_m

The desired phase margin, $\gamma_d \geq 20°$

Let additional phase lead required, $\epsilon = 5°$

Maximum lead angle, $\phi_m = \gamma_d - \gamma + \epsilon = 20° - (-44°) + 5° = 69°$

Since the lead angle required is greater than 60°, we have to realise the lead compensator as cascade of two lead compensators with each compensator providing half of the required phase lead angle.

$$\therefore \phi_m = \frac{69°}{2} = 34.5°$$

Step-5 : Determine the transfer function of lead compensator

$$\alpha = \frac{1 - \sin\phi_m}{1 + \sin\phi_m} = \frac{1 - \sin 34.5°}{1 + \sin 34.5°} = 0.28$$

The db magnitude corresponding to $\left. \right\}$ $\omega_m = -20\log\frac{1}{\sqrt{a}} = -20\log\frac{1}{\sqrt{0.28}} = -5.5$ db

From the bode plot of uncompensated system the frequency, ω_m corresponding to a db gain of -5.5 db is found to be 7.8 rad/sec.

$$\therefore \omega_m = 7.8 \text{ rad/sec.}$$

Now, $T = \dfrac{1}{\omega_m \sqrt{\alpha}} = \dfrac{1}{7.8\sqrt{0.28}} = 0.24$

Trasfer function of $\left.\begin{matrix}\end{matrix}\right\}$ $G_c(s) = \dfrac{\left(s + \dfrac{1}{T}\right)^2}{\left(s + \dfrac{1}{\alpha T}\right)^2} = \alpha^2 \dfrac{(1 + sT)^2}{(1 + s\alpha T)^2}$
the lead compensator

$$= (0.28)^2 \frac{(1 + 0.24s)^2}{(1 + 0.28 \times 0.24s)^2} = 0.0784 \frac{(1 + 0.24s)^2}{(1 + 0.067s)^2}$$

Step-6 : Open loop transfer function of compensated system.

The block diagram of the compensated system is shown in fig 1.6.1.

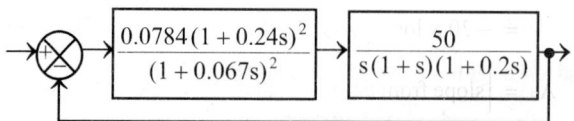

Fig 1.6.1 : Block diagram of lead compensated system.

The attenuation provided by the compensator can be retained to reduce the large value of open loop gain, so that the unstable system can be easily brought to stable region.

Let $G_o(s)$ be open loop transfer function of compensated system.

$$G_o(s) = \frac{0.0784(1 + 0.24s)^2}{(1 + 0.067s)^2} \times \frac{50}{s(1 + s)(1 + 0.2s)} = \frac{4(1 + 0.24s)^2}{s(1 + s)(1 + 0.2s)(1 + 0.067s)^2}$$

Step-7 : Draw the bode plot of compensated system to verify the design.

Put, $s = j\omega$ in $G_o(s)$, $\therefore$ $G_o(j\omega) = \dfrac{4(1 + j0.24\omega)^2}{j\omega(1 + j\omega)(1 + j0.2\omega)(1 + j0.067\omega)^2}$

MAGNITUDE PLOT

The corner frequencies are ω_{c1}, ω_{c2}, ω_{c3} and ω_{c4}.

$$\omega_{c1} = 1 \; ; \; \omega_{c2} = \frac{1}{0.24} = 4.2 \; ; \; \omega_{c3} = \frac{1}{0.2} = 5 \; ; \; \omega_{c4} = \frac{1}{0.067} = 15$$

The various terms of $G_o(j\omega)$ are listed in table-3. Also the table shows the slope contributed by each term and the change in slope at the corner frequency.

Choose a low frequency ω_l such that $\omega_l < \omega_{c1}$ and choose a high frequency ω_h such that $\omega_h > \omega_{c4}$.

Let $\omega_l = 0.5$ rad/sec and $\omega_h = 30$ rad/sec

Let $A_0 = |G_o(j\omega)|$ in db

At $\omega = \omega_l$, $\quad A_0 = 20\log\left|\dfrac{4}{j\omega}\right| = 20\log\dfrac{4}{0.5} = 18$ db

At $\omega = \omega_{c1}$, $\quad A_0 = 20\log\left|\dfrac{4}{j\omega}\right| = 20\log\dfrac{4}{1} = 12$ db

At $\omega = \omega_{c2}$, $A_0 = \left[\text{slope from } \omega_{c1} \text{ to } \omega_{c2} \times \log \frac{\omega_{c2}}{\omega_{c1}}\right] + A_0 \text{ at } (\omega = \omega_{c1})$

$$= -40 \times \log \frac{4.2}{1} + 12 = -13 \text{ db}$$

At $\omega = \omega_{c3}$, $A_0 = \left[\text{slope from } \omega_{c2} \text{ to } \omega_{c3} \times \log \frac{\omega_{c3}}{\omega_{c2}}\right] + A_0 \text{ at } (\omega = \omega_{c2})$

$$= 0 \times \log \frac{5}{4.2} + (-13) = -13 \text{ db}$$

At $\omega = \omega_{c4}$, $A_0 = \left[\text{slope from } \omega_{c3} \text{ to } \omega_{c4} \times \log \frac{\omega_{c4}}{\omega_{c3}}\right] + A_0 \text{ at } (\omega = \omega_{c3})$

$$= -20 \times \log \frac{15}{5} + (-13) = -22.5 \text{ db} \approx -23 \text{ db}$$

At $\omega = \omega_h$, $A_0 = \left[\text{slope from } \omega_{c4} \text{ to } \omega_h \times \log \frac{\omega_h}{\omega_{c4}}\right] + A_0 \text{ at } (\omega = \omega_{c4})$

$$= -60 \times \log \frac{30}{15} + (-23) = -41 \text{ db}$$

TABLE-3

Term	Corner frequency rad/sec	Slope db/dec	Change in slope db/dec
$4/j\omega$	—	-20	—
$\dfrac{1}{1+j\omega}$	$\omega_{c1} = 1$	-20	$-20 - 20 = -40$
$(1+j0.24\omega)^2$	$\omega_{c2} = \dfrac{1}{0.24} = 4.2$	$+40$	$-40 + 40 = 0$
$\dfrac{1}{1+j0.2\omega}$	$\omega_{c3} = \dfrac{1}{0.2} = 5$	-20	$0 - 20 = -20$
$\dfrac{1}{(1+j0.067\omega)^2}$	$\omega_{c4} = \dfrac{1}{0.067} = 15$	-40	$-20 - 40 = -60$

Let the points e, f, g, h, i and j be the points corresponding to frequencies ω_l, ω_{c1}, ω_{c2}, ω_{c3}, ω_{c4} and ω_h respectively on the magnitude plot of compensated system. The magnitude plot of compensated system is drawn on the same semilog graph sheet by using the same scales as shown in fig 1.6.2.

PHASE PLOT

The phase angle of $G_0(j\omega)$ as a function of ω is given by,

$$\phi_0 = \angle G_0(j\omega) = 2\tan^{-1}0.24\omega - 90° - \tan^{-1}\omega - \tan^{-1}0.2\omega - 2\tan^{-1}0.067\omega.$$

The phase angle of $G_0(j\omega)$ are calculated for various values of ω and listed in table - 4.

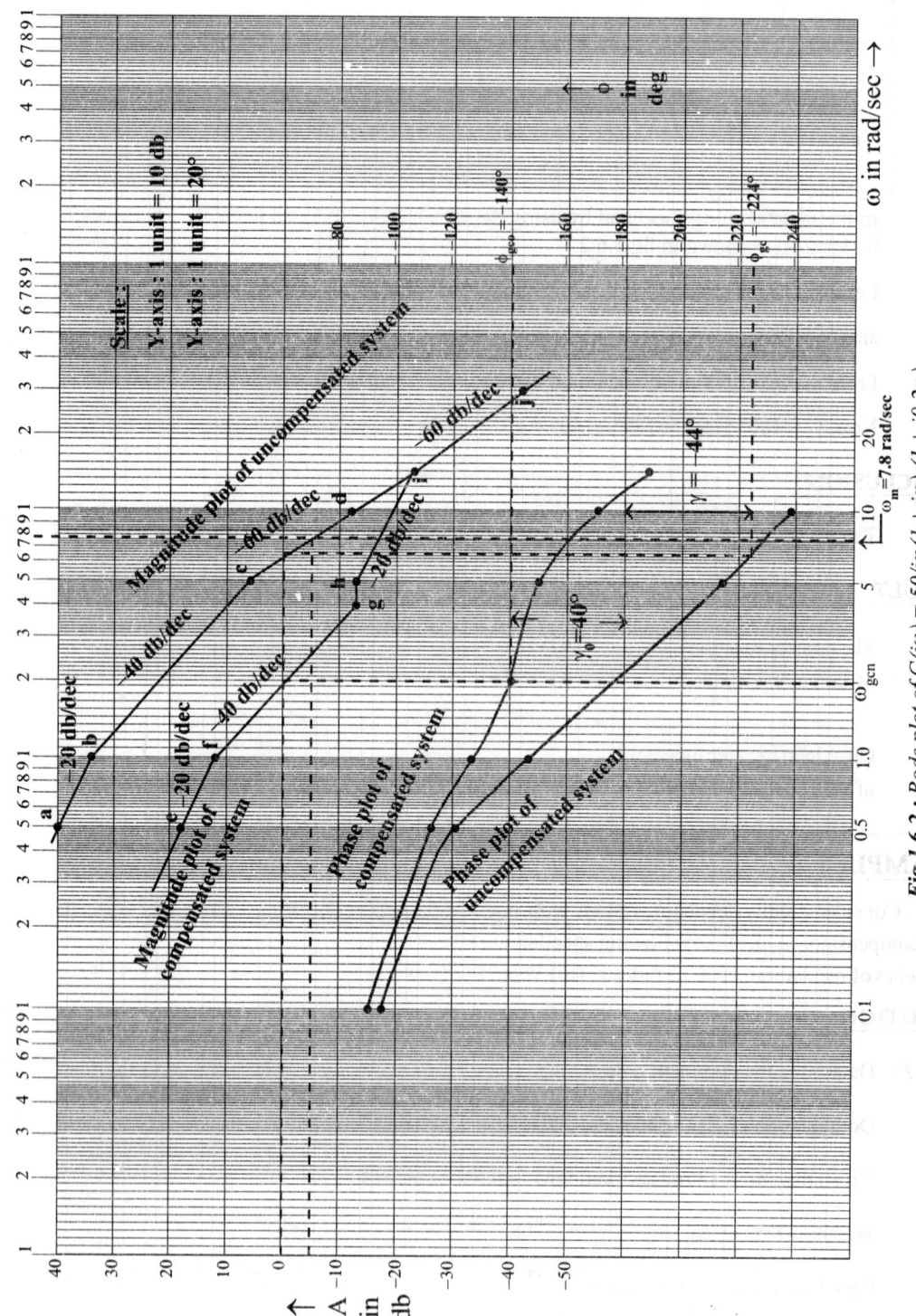

Fig 1.6.2 : Bode plot of $G(j\omega) = 50/j\omega(1 + j\omega)(1 + j0.2\omega)$.

TABLE-4

ω rad/sec	0.1	0.5	6.0	2.0	5	10	15
$\angle G_0(j\omega)$ deg	−94	−112	−127°	−139 ≈ −140	−150	−171°	−189 ≈ −188

In the same semilog sheet and by using the same scales, the phase plot of compensated system is sketched as shown in fig 1.6.2.

Let, ϕ_{gco} = Phase of $G_0(j\omega)$ at new gain crossover frequency (ω_{gcn}).

and γ_0 = Phase margin of compensated system.

From the bode plot of compensated system we get, $\phi_{gco} = -140°$.

Now, $\gamma_0 = 180° + \phi_{gco} = 180° - 140° = 40°$

CONCLUSION

The phase margin of the compensated system is satisfactory. Hence the design is acceptable.

RESULT

The transfer function of lead compensator } $G_c(s) = \dfrac{0.0784(1+0.24s)^2}{(1+0.67s)^2} = \dfrac{(s+4.17)^2}{(s+14.92)^2}$

Open loop transfer function of lead compensated system } $G_0(s) = \dfrac{4(1+0.24s)^2}{s(1+s)(1+0.2s)(1+0.067s)^2}$

EXAMPLE 1.7

Consider a unity feedback system with open loop transfer function, $G(s) = K/s(s+8)$. Design a lead compensator to meet the following specifications. (i) Percentage peak overshoot = 9.5%. (ii) Natural frequency of oscillation, $\omega_n = 12$ rad/sec. (iii) Velocity error constant, $K_v \geq 10$.

SOLUTION

Step-1 : Determine the dominant pole, s_d.

Dominant pole, $s_d = -\zeta\omega_n \pm j\omega_n\sqrt{1-\zeta^2}$.

Given that, $\omega_n = 12$ rad/sec and $\%M_p = 9.5\%$.

We know that, $\%M_p = e^{-\zeta\pi/\sqrt{1-\zeta^2}} \times 100$ $\therefore e^{-\zeta\pi/\sqrt{1-\zeta^2}} = 0.095$.

On taking natural log we get, $-\zeta\pi/\sqrt{1-\zeta^2} = ln(0.095)$

On squaring we get, $\dfrac{\zeta^2\pi^2}{1-\zeta^2} = (ln\,0.095)^2 = 5.54$.

$$\therefore \zeta^2\pi^2 = 5.54 - 5.54\zeta^2$$

$$\zeta^2\pi^2 + 5.54\zeta^2 = 5.54$$

$$\zeta^2(\pi^2 + 5.54) = 5.54 \qquad \therefore \zeta = \sqrt{\frac{5.54}{\pi^2 + 5.54}} = 0.6$$

$$\therefore s_d = -(0.6 \times 12) \pm j12 \times \sqrt{1 - 0.6^2} = -7.2 \pm j9.6$$

Step-2 : Draw the pole-zero plot

The pole-zero plot of open loop transfer function is shown fig 1.7.1. Poles are represented by the symbol "x". The pole at point P is the dominant pole, s_d.

Step-3 : To find the angle to be contributed by lead network.

Let ϕ = Angle to be contributed by lead network to make point, P as a point on root locus.

$$\text{Now, } \phi = \left(\begin{array}{c}\text{Sum of angles}\\\text{contributed by poles}\\\text{of uncompensated system}\end{array}\right) - \left(\begin{array}{c}\text{Sum of angles}\\\text{contributed by zeros}\\\text{of uncompensated system}\end{array}\right) \pm n180^\circ$$

From fig 1.7.1, we get,

$$\left.\begin{array}{l}\text{Sum of angles contributed}\\\text{by poles of uncompensated system}\end{array}\right\} = \theta_1 + \theta_2 = 127^\circ + 85^\circ = 212^\circ$$

Since there is no finite zero in uncompensated system, there is no angle contribution by zeros.

$$\therefore \phi = 212^\circ \pm n180^\circ$$

Let $n = 1$, $\qquad \therefore \phi = 212^\circ - 180^\circ = 32^\circ.$

Step-4 : To find the pole and zero of the compensator

Draw a line AP parallel to x-axis as shown in fig 1.7.1. The bisector PC is drawn to bisect the angle APO. The angles CPD and BPC are constructed as shown in fig 1.7.1. Here $\angle CPD = \angle BPC = \phi/2 = 32^\circ/2 = 16^\circ.$

From fig 1.7.1.

Pole of the compensator, $p_c = -16.25$

Zero of the compensator, $z_c = -9.1$

We know that, $z_c = -1/T$ $\qquad \therefore T = 1/9.1 = 0.11$

We know that, $p_c = -1/\alpha T$ $\qquad \therefore \alpha T = 1/16.25$ (or) $\alpha = 1/(T \times 16.25) = 0.56.$

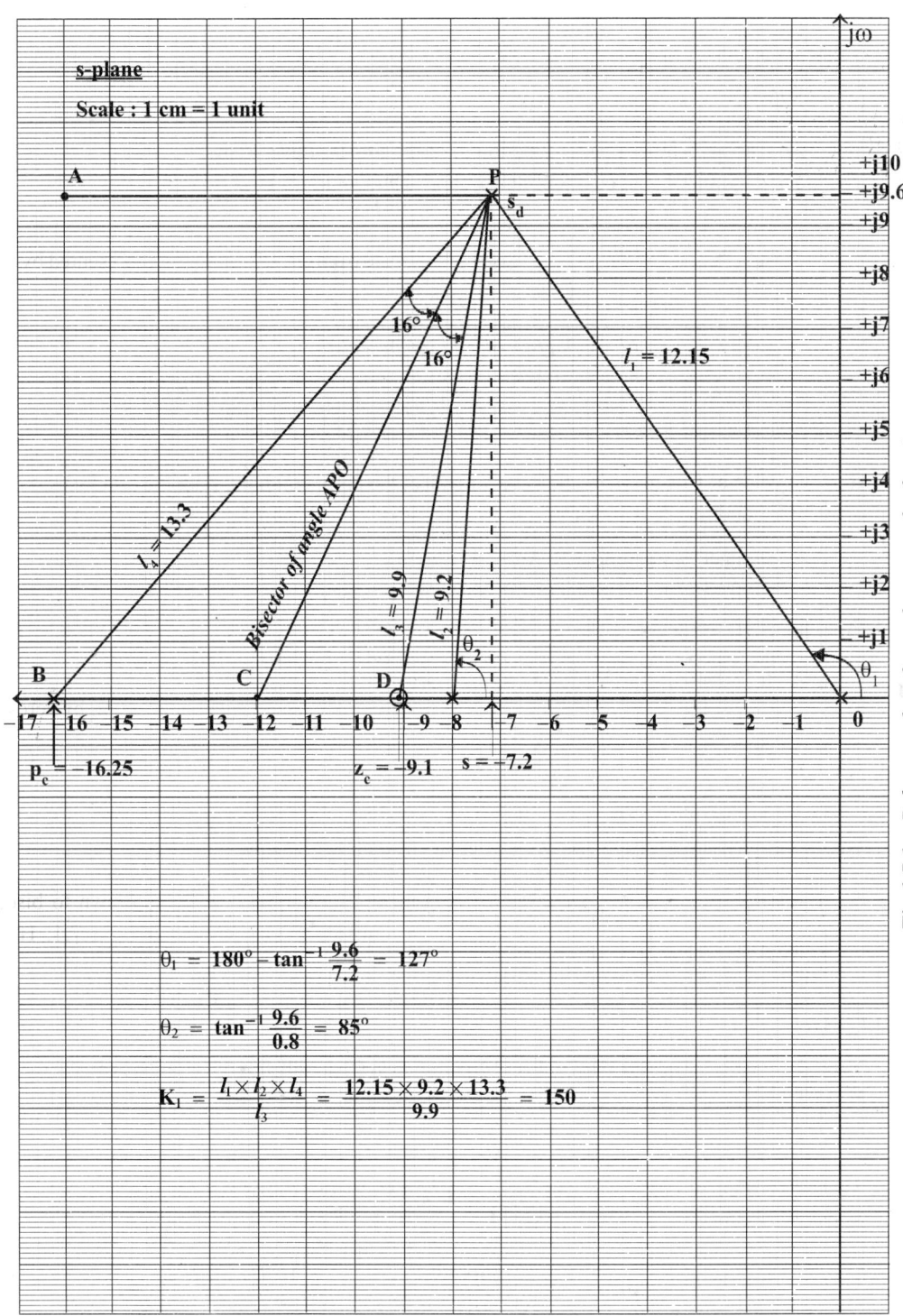

Fig 1.7.1. : Pole zero plot of open loop transfer function.

$$\theta_1 = 180° - \tan^{-1}\frac{9.6}{7.2} = 127°$$

$$\theta_2 = \tan^{-1}\frac{9.6}{0.8} = 85°$$

$$K_1 = \frac{l_1 \times l_2 \times l_4}{l_3} = \frac{12.15 \times 9.2 \times 13.3}{9.9} = 150$$

Step-5 : Determine the transfer function of lead compensator

$$\left.\begin{array}{l}\text{Transfer function of}\\ \text{lead compensator}\end{array}\right\} G_c(s) \ = \ \frac{\left(s+\frac{1}{T}\right)}{\left(s+\frac{1}{\alpha T}\right)} \ = \ \frac{(s+9.1)}{(s+16.25)}$$

Step-6 : Determine the open loop transfer function of lead compensated system.
The block diagram of lead compensated system is shown in fig 1.7.2.

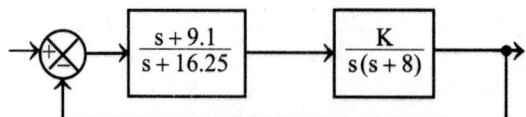

Fig 1.7.2 : *Block diagram of lead compensated system.*

$$\left.\begin{array}{l}\text{Open loop transfer function}\\ \text{of lead compensated system}\end{array}\right\} G_0(s) \ = \ \frac{(s+9.1)}{(s+16.25)} \times \frac{K}{s(s+8)} \ = \ \frac{K(s+9.1)}{s(s+8)(s+16.25)}$$

Here the value of K is given by the value of gain at the dominant pole, s_d on the root locus. From magnitude condition K is given by,

$$K \ = \ \frac{\text{Product of vector lengths from all poles to } s = s_d}{\text{Product of vector lengths from all zeros to } s = s_d}$$

From fig 1.7.1, we get,

$$K \ = \ \frac{l_1 \times l_2 \times l_4}{l_3} \ = \ \frac{12.15 \times 9.2 \times 13.3}{9.9} \ = \ 150$$

$$\therefore \ G_0(s) \ = \ \frac{150(s+9.1)}{s(s+8)(s+16.25)}$$

Step-7 : Check for error requirement

For the compensated system, the velocity error constant is given by,

$$K_v \ = \ \underset{s\to 0}{\text{Lt}} \ sG_0(s) \ = \ \underset{s\to 0}{\text{Lt}} \ s \ \frac{150(s+9.1)}{s(s+8)(s+16.25)}$$

$$= \ \frac{150 \times 9.1}{8 \times 16.25} \ = \ 10.5$$

CONCLUSION

Since the velocity error constant of the compensated system, satisfies the requirement, the design is acceptable.

RESULT

$$\left.\begin{array}{l}\text{Transfer function of}\\ \text{lead compensator}\end{array}\right\} G_c(s) \ = \ \frac{(s+9.1)}{(s+16.25)} \ = \ 0.56\frac{(1+0.11s)}{(1+0.06s)}$$

$$\left.\begin{array}{l}\text{Transfer function of}\\ \text{lead compensated system}\end{array}\right\} G_0(s) \ = \ \frac{150(s+9.1)}{s(s+8)(s+16.25)} \ = \ \frac{10.5(1+0.11s)}{s(1+0.125s)(1+0.06s)}$$

EXAMPLE 1.8

Design a lead compensator for a unity feedback system with open loop transfer function $G(s) = K/s \ (s + 4) \ (s + 7)$ to meet the following specifications. (i) % Peak overshoot = 12.63%. (ii) Natural frequency of oscillation, $\omega_n = 8$ rad/sec. (iii) Velocity error constant, $K_v \geq 2.5$.

SOLUTION

Step-1:   Determine the dominant pole, s_d.

Dominant pole, $s_d = -\zeta\omega_n \pm j\omega_n\sqrt{1-\zeta^2}$.

Given that, $\omega_n = 8$ rad/sec and $\%M_p = 12.63\%$.

We know that, $\%M_p = e^{-\zeta\pi\sqrt{1-\zeta^2}} \times 100$

$$\therefore e^{-\zeta\pi\sqrt{1-\zeta^2}} = 0.1263$$

On taking natural log we get, $-\zeta\pi/\sqrt{1-\zeta^2} = In(0.1263)$

On squaring we get, $\dfrac{\zeta^2\pi^2}{1-\zeta^2} = (In\,0.1263)^2 = 4.28.$

$$\therefore \zeta^2\pi^2 = 4.28 - 4.28\zeta^2$$

$$\zeta^2\pi^2 + 4.28\zeta^2 = 4.28$$

$$\zeta^2(\pi^2 + 4.28) = 4.28 \qquad \therefore \zeta = \sqrt{\dfrac{4.28}{\pi^2 + 4.28}} = 0.55$$

$$\therefore s_d = -(0.55 \times 8) \pm j8 \times \sqrt{1 - 0.55^2}$$

$$= -4.4 \pm j6.68$$

$$= -4.4 \pm j6.7$$

Step-2 :   Draw the pole-zero plot

The pole-zero plot of open loop transfer function is shown fig 1.8.1. Poles are represented by the symbol "x". The pole at point P is the dominant pole, s_d.

Step-3 :   To find the angle to be contributed by lead network.

Let $\phi =$ Angle to be contributed by lead network to make point, P as a point on root locus.

Now, $\phi = \left(\begin{array}{c}\text{Sum of angles}\\\text{contributed by poles}\\\text{of uncompensated system}\end{array}\right) - \left(\begin{array}{c}\text{Sum of angles}\\\text{contributed by zeros}\\\text{of uncompensated system}\end{array}\right) \pm n180°$

From fig 1.8.1, we get,

$\left.\begin{array}{c}\text{Sum of angles contributed}\\\text{by poles of uncompensated system}\end{array}\right\} = \theta_1 + \theta_2 + \theta_3 - 123° + 93° + 69° = 285°$

Since there is no finite zero in uncompensated system, there is no angle contribution by zeros.

$$\therefore \phi = 285° \pm n180°$$

Let $n = 1$, $\therefore \phi = 285° - 180° = 105°$.

Since the angle contribution is more than 60°, the lead compensator is realised as cascade of two compensators with each compensator, contributing half of the required angle.

$\therefore \phi = 105/2 = 52.5° \approx 52°$

Step-4 : To find the poles and zeros of the compensator

Draw a line AP parallel to x-axis as shown in fig 1.8.6. The bisector PC is drawn to bisect the angle APO. The angles CPD and BPC are constructed as shown in fig 1.8.6. Here $\angle CPD = \angle BPC$ $= \phi/2 = 52°/2 = 26°$.

From fig 1.8.1.

Pole of the compensator, $p_c = -13.55$

Zero of the compensator, $z_c = -4.65$

We know that, $z_c = -1/T$; $\therefore T = 1/4.65 = 0.215$

We know that, $p_c = -1/\alpha T$; $\therefore \alpha T = 1/13.55$ (or) $\alpha = 1/T \times 13.55 = 0.343$.

Step-5 : Determine the transfer function of lead compensator

$$\left.\begin{array}{l}\text{Transfer function of} \\ \text{lead compensator}\end{array}\right\} G_c(s) \;=\; \frac{\left(s + \dfrac{1}{T}\right)^2}{\left(s + \dfrac{1}{\alpha T}\right)^2} \;=\; \frac{(s + 4.65)^2}{(s + 13.55)^2}$$

Step-6 : Determine the open loop transfer function of lead compensated system.

Block diagram of lead compensated system is shown in fig 1.8.2.

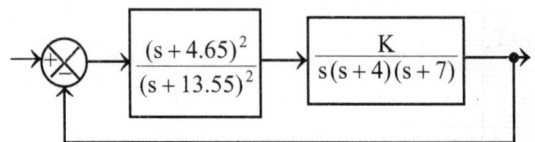

Fig 1.8.2 : *Block diagram of lead compensated system.*

$$\left.\begin{array}{l}\text{Open loop transfer function} \\ \text{of lead compensated system}\end{array}\right\} G_0(s) \;=\; \frac{(s + 4.65)^2}{(s + 13.55)^2} \times \frac{K}{s(s + 4)(s + 7)}$$

Here the value of K is given by the value of gain at the dominant pole, s_d on the root locus. From magnitude condition K is given by,

$$K = \frac{\text{Product of vector lengths from all poles to } s = s_d}{\text{Product of vector lengths from all zeros to } s = s_d}$$

From fig 1.8.1, we get,

$$K = \frac{l_1 \times l_2 \times l_4 \times 2l_5}{2l_3} = \frac{8.1 \times 6.8 \times 7.3 \times 2 \times 11.5}{2 \times 6.75} = 685$$

$$\therefore G_0(s) = \frac{685(s + 4.65)^2}{s(s + 4)(s + 7)(s + 13.55)^2}$$

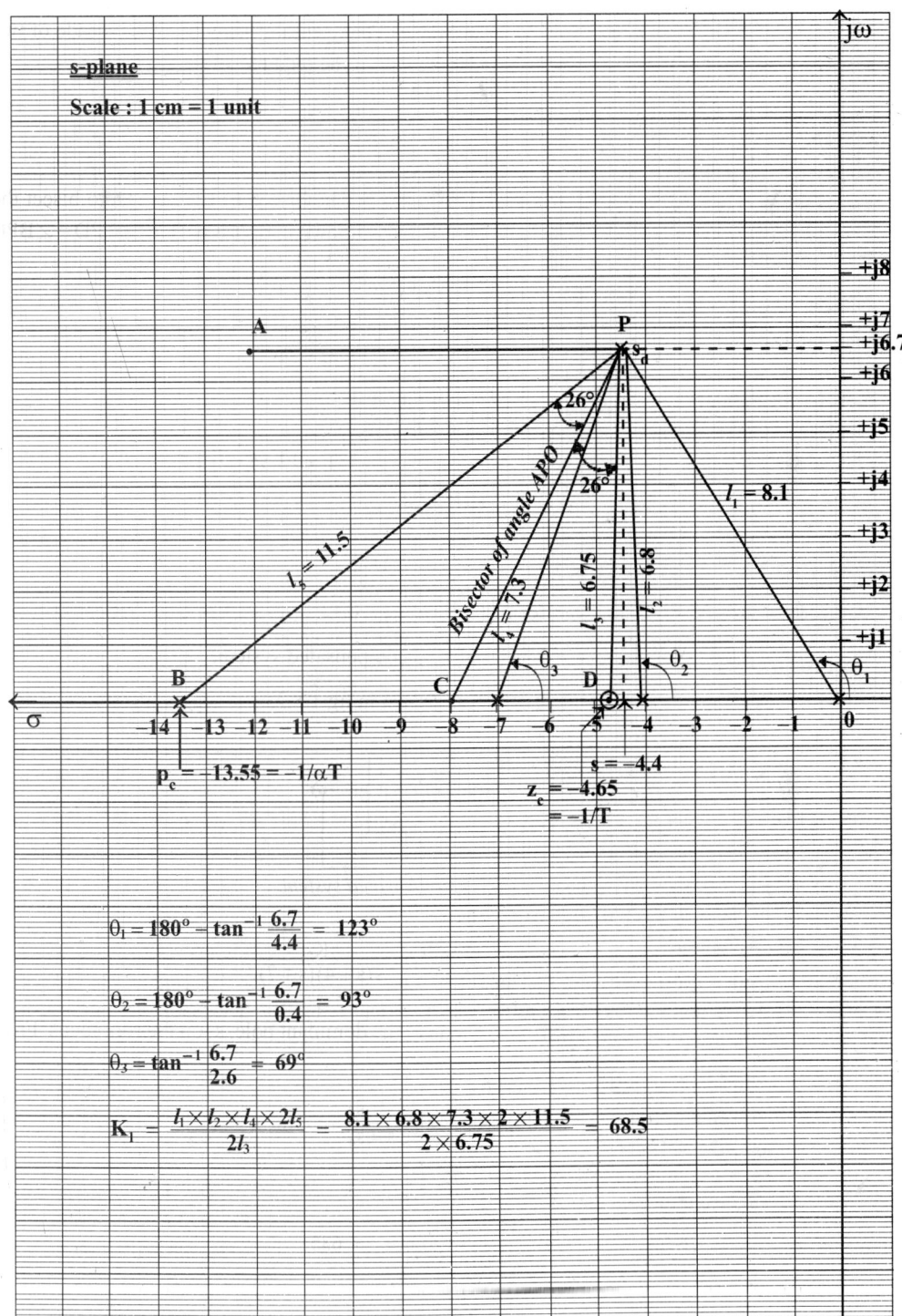

Fig 1.8.1. : Pole zero plot of lead compesated systerm.

$$\theta_1 = 180^\circ - \tan^{-1} \frac{6.7}{4.4} = 123^\circ$$

$$\theta_2 = 180^\circ - \tan^{-1} \frac{6.7}{0.4} = 93^\circ$$

$$\theta_3 = \tan^{-1} \frac{6.7}{2.6} = 69^\circ$$

$$K_1 = \frac{l_1 \times l_2 \times l_4 \times 2l_5}{2l_3} = \frac{8.1 \times 6.8 \times 7.3 \times 2 \times 11.5}{2 \times 6.75} = 68.5$$

Step-7 : Check for error requirement

For the compensated system, the velocity error constant is given by,

$$K_v = \underset{s \to 0}{Lt} \; sG_0(s) = \underset{s \to 0}{Lt} \; s \frac{685(s+4.65)^2}{s(s+4)(s+7)(s+13.55)^2} = \frac{685 \times 4.65^2}{4 \times 7 \times 13.55^2} = 2.88$$

CONCLUSION

Since the velocity error constant of the compensated system, satisfies the requirement, the design is acceptable.

RESULT

Transfer function of lead compensator $\left.\right\}$ $G_c(s) = \dfrac{(s+4.65)^2}{(s+13.55)^2} = 0.1178 \dfrac{(1+0.215s)^2}{(1+0.0738s)^2}$

Transfer function of lead compensated system $\left.\right\}$ $G_0(s) = \dfrac{685(s+4.65)^2}{s(s+4)(s+7)(s+13.55)^2}$

$$= \frac{2.88(1+0.215s)^2}{s(1+0.25s)(1+0.143s)(1+0.0738s)^2}$$

1.4 LAG-LEAD COMPENSATOR

A compensator having the characteristics of lag-lead network is called lag-lead compensator. In a lag-lead network when sinusoidal signal is applied, both phase lag and phase lead occurs in the output, but in different frequency regions. Phase lag occurs in the low frequency region and phase lead occurs in the high frequency region (i.e) the phase angle varies from lag to lead as the frequency is increased from zero to infinity.

A lead compensator basically increases bandwidth and speeds up the response and decreases the maximum overshoot in the step response. Lag compensation increases the low frequency gain and thus improves the steady state accuracy of the system, but reduces the speed of responses due to reduced bandwidth.

If improvements in both transient and steady state response are desired, then both a lead compensator and lag compensator may be used simultaneously, rather than introducing both a lead and lag compensator as separate elements. However it is economical to use a single lag-lead compensator.

A lag-lead compensation combines the advantages of lag and lead compensations. Lag-lead compensator possess two poles and two zeros and so such a compensation increases the order of the system by two, unless cancellation of poles and zeros occurs in the compensated system.

S-PLANE REPRESENTATION OF LAG-LEAD COMPENSATOR

The s-plane representation of lag-lead compensator is shown in fig 1.15. The lag section has one real pole and one real zero with pole to the right of zero. The lead section also has one real pole and one real zero but the zero is to the right of the pole.

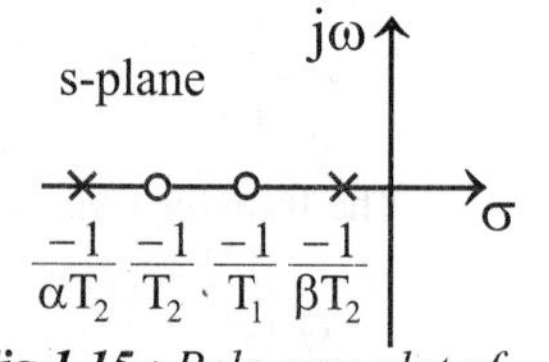

Transfer function of lag – lead compensator $\left.\right\}$ $G_c(s) = \dfrac{(s+1/T_1)}{\underbrace{(s+1/\beta T_1)}_{\text{lag section}}} \dfrac{(s+1/T_2)}{\underbrace{(s+1/\alpha T_1)}_{\text{lead section}}}$

where $\beta > 1$ **and** $0 < \alpha < 1$

Fig 1.15 : Pole-zero plot of lag-lead compensator.

.....(1.36)

REALISATION OF LAG-LEAD COMPENSATOR USING ELECTRICAL NETWORK

The lag-lead compensator can be realised by the R-C network shown in fig 1.16.

Let $E_i(s)$ = Input voltage

and $E_o(s)$ = Output voltage.

In the network shown in fig 1.16, the input voltage is applied to the series combination of $R_1 \parallel C_1$, R_2 and C_2. The output voltage, is obtained across series combination of R_2 and C_2. By voltage division rule,

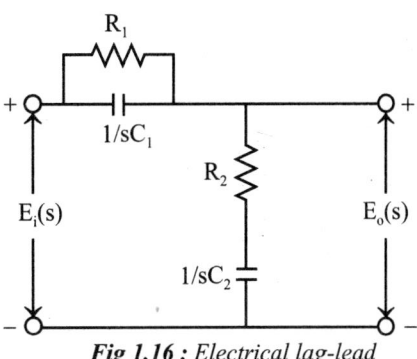

Fig 1.16 : Electrical lag-lead compensator.

$$E_o(s) = E_i(s) \frac{R_2 + \dfrac{1}{sC_2}}{\left(R_1 \parallel \dfrac{1}{sC_1}\right) + R_2 + \dfrac{1}{sC_2}}$$

$$\therefore \frac{E_o(s)}{E_i(s)} = \frac{\dfrac{sR_2C_2+1}{sC_2}}{\dfrac{R_1 \dfrac{1}{sC_1}}{R_1 + \dfrac{1}{sC_1}} + \dfrac{sR_2C_2+1}{sC_2}} = \frac{\dfrac{sR_2C_2+1}{sC_2}}{\dfrac{R_1}{sR_1C_1+1} + \dfrac{sR_2C_2+1}{sC_2}}$$

$$= \frac{\dfrac{sR_2C_2+1}{sC_2}}{\dfrac{sR_1C_2+(sR_1C_1+1)(sR_2C_2+1)}{(sR_1C_1+1)sC_2}} = \frac{(sR_1C_1+1)(sR_2C_2+1)}{sR_1C_2(sR_1C_1+1)(sR_2C_2+1)}$$

$$= \frac{R_1C_1R_2C_2\left(s+\dfrac{1}{R_1C_1}\right)\left(s+\dfrac{1}{R_2C_2}\right)}{sR_1C_2 + R_1C_1R_2C_2\left(s+\dfrac{1}{R_1C_1}\right)\left(s+\dfrac{1}{R_2C_2}\right)}$$

On dividing the numerator and denominator by $R_1C_1\,R_2C_2$ we get,

$$\frac{E_o(s)}{E_i(s)} = \frac{\left(s+\dfrac{1}{R_1C_1}\right)\left(s+\dfrac{1}{R_2C_2}\right)}{\dfrac{s}{R_2C_1} + \left(s+\dfrac{1}{R_1C_1}\right)\left(s+\dfrac{1}{R_2C_2}\right)}$$

$$= \frac{\left(s+\dfrac{1}{R_1C_1}\right)\left(s+\dfrac{1}{R_2C_2}\right)}{s^2 + s\left(\dfrac{1}{R_1C_1} + \dfrac{1}{R_2C_2} + \dfrac{1}{R_2C_1}\right) + \dfrac{1}{R_1R_2C_1C_2}} \qquad \text{.....(1.37)}$$

The transfer function of lag-lead compensator is given by,

$$G_c(s) = \frac{(s+1/T_1)(s+1/T_2)}{(s+1/\beta T_1)(s+1/\alpha T_1)}$$

$$= \frac{(s+1/T_1)(s+1/T_2)}{s^2 + s((s+1/\beta T_1)(1/\alpha T_2)) + 1/\alpha\beta T_1 T_2} \qquad \text{.....(1.38)}$$

On comparing equations (1.37) and (1.38) we get,

$$T_1 = R_1 C_1 \qquad \qquad .....(1.39)$$

$$T_2 = R_2 C_2 \qquad \qquad .....(1.40)$$

$$R_1 R_2 C_1 C_2 = \alpha\beta\, T_1 T_2 \qquad \qquad .....(1.41)$$

$$\frac{1}{R_1 C_1} + \frac{1}{R_2 C_2} + \frac{1}{R_2 C_1} = \frac{1}{\beta T_1} + \frac{1}{\alpha T_1} \qquad \qquad .....(1.42)$$

From equation (1.41) we get, $\alpha\beta = \dfrac{R_1 R_2 C_1 C_2}{T_1 T_2}$ $\qquad \qquad .....(1.43)$

From equations (1.39), (1.40) and (1.43), we can say that, $\alpha\beta = 1$ $\qquad .....(1.44)$

The equation (1.44) implies that a single lag-lead network does not allow an independent choice of α and β. (But separate lag and lead network will allow independent choice of α and β). Hence in the transfer function of electrical lag-lead compensator replace α by $1/\beta$ as shown below.

$$\therefore\ \frac{E_o(s)}{E_i(s)} = G_c(s) = \frac{(s + 1/T_1)(s + 1/T_2)}{(s + 1/\beta T_1)(s + \beta/T_2)} \qquad \qquad .....(1.45)$$

Where $\beta > 1$, $T_1 = R_1 C_1$, $T_2 = R_2 C_2$ and $\dfrac{1}{R_1 C_1} + \dfrac{1}{R_2 C_2} + \dfrac{1}{R_2 C_1} = \dfrac{1}{\beta T_1} + \dfrac{\beta}{T_2}$

FREQUENCY RESPONSE OF LAG-LEAD COMPENSATOR

Consider the transfer function of lag-lead compensator.

$$G_c(s) = \frac{(s + 1/T_1)(s + 1/T_2)}{(s + 1/\beta T_1)(s + 1/\alpha T_2)} = \alpha\beta\,\frac{(s + 1/T_1)(1 + sT_2)}{(1 + s\beta T_1)(1 + s\alpha T_2)} \qquad .....(1.46)$$

The sinusoidal transfer function of lag-lead compensator is obtained by letting s = jω in equation (1.46).

$$\therefore\ G_c(j\omega) = \alpha\beta\,\frac{(1 + j\omega T_1)(1 + j\omega T_2)}{(1 + j\omega\beta T_1)(1 + j\omega\alpha T_2)} \qquad \qquad .....(1.47)$$

For a single lag-lead compensator, $\alpha\beta = 1$. Hence from equation (1.47) we can say that the lag-lead compensator provides a dc gain of unity.

$$\therefore\ G_c(j\omega) = \frac{(1 + j\omega T_1)(1 + j\omega T_2)}{(1 + j\omega\beta T_1)(1 + j\omega\alpha T_2)} \qquad \qquad .....(1.48)$$

The sinusoidal transfer function shown in equation (1.48) has four corner frequencies and they are ω_{c1}, ω_{c2}, ω_{c3} and ω_{c4}, where $\omega_{c1} < \omega_{c2} < \omega_{c3} < \omega_{c4}$.

Here $\omega_{c1} = \dfrac{1}{\beta T_1}$; $\omega_{c2} = \dfrac{1}{T_1}$; $\omega_{c3} = \dfrac{1}{T_2}$ and $\omega_{c4} = \dfrac{1}{\alpha T_2}$

By an analysis similar to that of lag compensator the bode plot of lag-lead compensator is sketched as shown in fig 1.17.

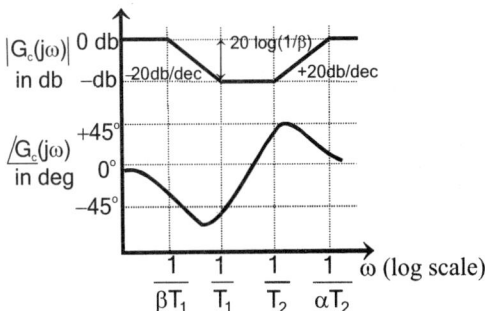

Fig 1.17 : Bode plot of lag-lead compensator.

PROCEDURE FOR DESIGN OF LAG-LEAD COMPENSATOR USING BODE PLOT

The lag-lead compensator is employed only when a large error constant and a large bandwidth are required. First design a lag section and then take $\alpha = 1/\beta$ and design a lead section. The step by step procedure for the design of lag-lead compensator is given below.

Step-1 : Determine the open loop gain K of the uncompensated system to satisfy the specified error requirement.

Step-2 : Draw the bode plot of uncompensated system.

Step-3 : From the bode plot determine the gain margin of the uncompensated system.

Let, ϕ_{gc} = Phase of G(jω) at gain crossover frequency.

and γ = Phase margin of uncompensated system.

Now, $\gamma = 180° + \phi_{gc}$

If the gain margin is not satisfactory then compensation is required.

Step-4 : Choose a new phase margin

Let, γ_d = Desired phase margin

Now, new phase margin, $\gamma_n = \gamma_d + \in$

Choose an initial value of $\in = 5°$.

Step-5 : From the bode plot, determine the new gain crossover frequency, which is the frequency corresponding to a phase margin of γ_n.

Let, ω_{gcn} = New gain crossover frequency

and ϕ_{gcn} = Phase of G(jω) at ω_{gcn}

γ_n = $180° + \phi_{gcn}$ (or) $\phi_{gcn} = \gamma_n - 180°$

In the phase plot of uncompensated system, the frequency corresponding to a phase of ϕ_{gcn} is the new gain crossover frequency ω_{gcn}.

Choose the gain crossover frequency of the lag compensator, ω_{gcl}, somewhat greater than ω_{gcn} (i.e., choose ω_{gcl} such that $\omega_{gcl} > \omega_{gcn}$).

Step-6 : Calculate β of lag compensator.

Let, $A_{gcl} = |G(j\omega)|$ in db at $\omega = \omega_{gcl}$

From the bode plot find A_{gcl}

Now, $A_{gcl} = 20 \log \beta$ (or) $\beta = 10^{(A_{gcl}/20)}$

Step-7 : Determine the transfer function of lag section

The zero of the lag compensator is placed at a frequency one-tenth of ω_{gcl}.

$\therefore$ Zero of lag compensator, $z_{c1} = 1/T_1 = \omega_{gcl}/10$

$$\text{Now, } T_1 = 10/\omega_{gcl}$$

Pole of lag compensator, $p_{c1} = 1/\beta T_1$

Transfer function of lag section $\left.\right\} G_1(s) = \dfrac{(s + 1/T_1)}{(s + 1/\beta T_1)} = \beta \dfrac{(1 + sT_1)}{(1 + s\beta T_1)}$

Step-8 : Determine the transfer function of lead section

Take, $\alpha = 1/\beta$

From the bode plot find ω_m which is the frequency at which the db gain is $-20\log(1/\sqrt{\alpha})$

Now $T_2 = \dfrac{1}{\omega_m \sqrt{\alpha}}$

Transfer function of lead section $\left.\right\} G_2(s) = \dfrac{(s + 1/T_2)}{(s + 1/\alpha T_2)} = \alpha \dfrac{(1 + sT_2)}{(1 + s\alpha T_2)}$

Step-9 : Determine the transfer function of lag-lead compensator.

Transfer function of lag-lead compensator, $G_c(s) = G_1(s) \times G_2(s)$

$$= \beta \dfrac{(1 + sT_1)}{(1 + s\beta T_1)} \times \alpha \dfrac{(1 + sT_2)}{(1 + s\alpha T_2)}$$

Since $\alpha = \dfrac{1}{\beta}$,

$$G_c(s) = \dfrac{(1 + sT_1)(1 + sT_2)}{(1 + s\beta T_1)(1 + s\alpha T_2)}$$

Step-10 : Determine the open loop transfer function of compensated system.

The lag-lead compensator is connected in series with $G(s)$ as shown in fig 1.18.

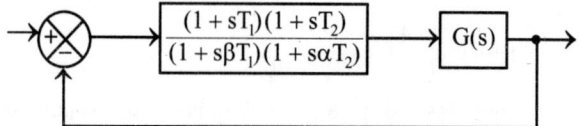

Fig 1.18 : Block diagram of lag-lead compensated system.

Open loop transfer function of compensated system $\left.\right\} G_o(s) = \dfrac{(1 + sT_1)(1 + sT_2)}{(1 + s\beta T_1)(1 + s\alpha T_2)} \times G(s)$

Step-11 : Draw the bode plot of compensated system and verify whether the specifications are satisfied or not. If the specifications are not satisfied then choose another choice of α such that, $\alpha < 1/\beta$ and repeat the steps 8 to 11.

PROCEDURE FOR DESIGN OF LAG-LEAD COMPENSATOR USING ROOT LOCUS

The lag-lead compensation is employed to improve both the transient and steady state responses of a system. First design a lead section to realize the required ζ and ω_n for the dominant closed loop poles. Then determine the error constant of lead compensated system. If it is satisfactory then only lead compensation will meet the requirement. If the error constant has to be increased then design a lag section. The step-by-step procedure for the design of lag-lead compensator is given below.

Step-1 : Determine the dominant pole, s_d

$$s_d = -\zeta\omega_n \pm \omega_n \sqrt{\zeta^2 - 1}$$

where, ζ = Damping ratio ; ω_n = Natural frequency of oscillation, rad/sec.

Step-2 : Mark the poles and zeros of open loop transfer function and the dominant pole on the s-plane. Let the dominant pole be point P.

Step-3 : Find the angle to be contributed by lead network to make the point P as a point on root locus.

Let, ϕ = Angle to be contributed by lead network to make point, P as a point on root locus.

Draw vectors from all open loop poles and zeros to point P. Measure the angle contributed by the vectors. *[For the procedure to find angle contribution by vectors refer root locus in Appendix-II].*

$$\text{Now, } \phi = \begin{pmatrix} \text{Sum of angles} \\ \text{contributed by poles} \\ \text{of uncompensated system} \end{pmatrix} - \begin{pmatrix} \text{Sum of angles} \\ \text{contributed by zeros} \\ \text{of uncompensated system} \end{pmatrix} \pm n180°$$

where n is an odd integer, so that n180° is nearest to the difference between angles contributed by poles and zeros.

Step-4 : Determine the pole and zero of the lead section.

Let point O be the origin of s-plane and point P be the dominant pole. Draw straight lines OP and AP such that AP is parallel to x-axis as shown in fig 1.19. Draw a line PC so as to bisect the angle APO [$\angle$APO] where the point C is on the real axis. With line PC as reference, draw angles BPC and CPD such that each equal to $\phi/2$. Here the points B and D are located on the real axis.

Now the point B is the location of the pole of the compensator $(-1/\alpha T_2)$ and the point D is the location of the zero of the compensator $(-1/T_2)$. Compute T_2 and α from the values of point D and B.

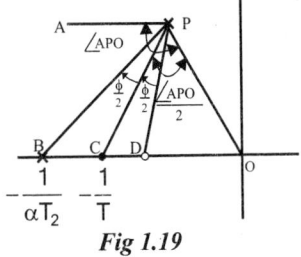

Step-5 : Determine the transfer function of lead section.

$$\left. \begin{array}{l} \text{Transfer function} \\ \text{of lead section} \end{array} \right\} G_2(s) = \frac{\left(s + \dfrac{1}{T_2}\right)}{\left(s + \dfrac{1}{\alpha T_2}\right)}$$

Fig 1.19

Step-6 : Determine the open loop gain, K.

The open loop gain K is the value of gain at $s = s_d$. The value of gain, K is determined from pole-zero plot of lead compensated system and by using the magnitude condition given below.

$$K = \frac{\text{Product of vector lengths from all poles to } s = s_d}{\text{Product of vector lengths from all zeros to } s = s_d}$$

Note : *The length of vectors should be measured to scale. For details of magnitude condition refer root locus in Appendix-II.*

Open loop transfer function of lead compensated system $\left.\right\}$ $G_{02}(s) = G_2(s) \times G(s)$

Step-7 : Determine the velocity error constant of lead compensated system.

Velocity error constant of lead compensated system $\left.\right\}$ $K_{v2} = \underset{s \to 0}{Lt} \ s \ G_{02}(s)$

If K_{v2} satisfies the requirement then only lead compensation is sufficient but if K_{v2} is less than the desired value then provide lag compensation.

Step-8 : Determine the parameter, β of lag section.

Let, K_{vd} = Desired velocity error constant ; A = Factor by which K_v is increased.

Now, $A = K_{vd}/K_{v2}$. Select β, such that $\beta > A$. [i.e., $\beta = (1.1$ to $1.2) \times A]$

Step-9 : Determine the transfer function of lag section. Choose the zero of lag section as 10% of the second pole of uncompensated system.

∴ Zero of lag section, $z_{c1} = 0.1 \times$ second pole of G(s)

Also, $z_{c1} = \dfrac{-1}{T_1}$; ∴ $T_1 = \dfrac{-1}{z_{c1}}$

Pole of lag section, $p_{c1} = \dfrac{-1}{\beta T_1}$

Transfer function of lag section, $G_1(s) = \dfrac{\left(s + \dfrac{1}{T_1}\right)}{\left(s + \dfrac{1}{\beta T_1}\right)}$

Step-10 : Determine the transfer function of lag-lead compensator and compensated system

Transfer function of lag – lead compensator $\left.\right\}$ $G_c(s) = G_1(s) \times G_2(s) = \dfrac{\left(s + \dfrac{1}{T_1}\right)\left(s + \dfrac{1}{T_2}\right)}{\left(s + \dfrac{1}{\beta T_1}\right)\left(s + \dfrac{1}{\alpha T_2}\right)}$

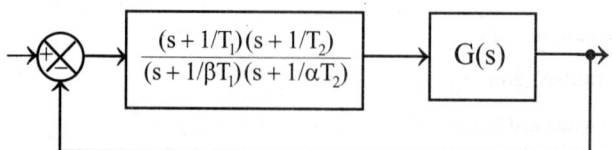

Fig 1.20 : *Block diagram of lag-lead compensated system.*

The lag-lead compensator is connected in series with G(s) as shown in fig 1.20.

$$\left.\begin{array}{l}\text{Open loop transfer function of}\\ \text{lag - lead compensated system}\end{array}\right\} G_0(s) \ = \ \frac{\left(s+\dfrac{1}{T_1}\right)\left(s+\dfrac{1}{T_2}\right)}{\left(s+\dfrac{1}{\beta T_1}\right)\left(s+\dfrac{1}{\alpha T_2}\right)}$$

Step-11 : Check the velocity error constant of compensated system, if it is satisfactory then the design is accepted, otherwise repeat the design by modifying the locations of poles and zeros of the compensator.

EXAMPLE 1.9

Consider the unity feedback system whose open loop transfer function is G(s) = K/s(s + 3) (s + 6). Design a lag-lead compensator to meet the following specifications. (i) Velocity error constant, K_v = 80. (ii) Phase margin, $\gamma \ge 35°$.

SOLUTION

Step-1 : Determine K

For unity feedback system,

Velocity error constant, $K_v \ = \ \underset{s\to 0}{\text{Lt}} \ sG(s)$

Given that, $K_v = 80$.

$$\therefore \ \underset{s\to 0}{\text{Lt}} \ sG(s) \ = \ \underset{s\to 0}{\text{Lt}} \ s\frac{K}{s(s+3)(s+6)} \ = \ 80$$

$$\frac{K}{3\times 6} \ = \ 80 \quad (\text{or}) \quad K \ = \ 80\times 3\times 6 \ = \ 1440$$

$$\therefore \ G(s) \ = \ \frac{1440}{s(s+3)(s+6)} \ = \ \frac{1440}{s\times 3(1+s/3)\times 6(1+s/6)}$$

$$= \ \frac{80}{s(1+0.33s)(1+0.167s)}$$

Step-2: Bode plot of uncompensated system.

In G(s), put $s = j\omega$

$$\therefore \ G(j\omega) \ = \ \frac{80}{j\omega(1+j0.33\omega)(1+j0.167\omega)}$$

MAGNITUDE PLOT

The corner frequencies are ω_{c1} and ω_{c2}.

Here ω_{c1} = 1/0.33 = 3 rad/sec and ω_{c2} = 1/0.167 = 6 rad/sec.

The various terms of G(jω) are listed in table-1. Also the table shows the slope contributed by each term and the change in slope at the corner frequency.

TABLE-1

Term	Corner frequency rad/sec	Slope db/dec	Change in slope db/dec
$\dfrac{80}{j\omega}$	–	–20	–
$\dfrac{1}{1+j0.33\omega}$	$\omega_{c1} = \dfrac{1}{0.33} = 3$	–20	$-20 - 20 = -40$
$\dfrac{1}{1+j0.167\omega}$	$\omega_{c2} = \dfrac{1}{0.167} = 6$	–20	$-40 - 20 = -60$

Choose a low frequency ω_l such that $\omega_l < \omega_{c1}$ and choose a high frequency ω_h such that $\omega_h > \omega_{c2}$

Let $\omega_l = 0.5$ rad/sec and $\omega_h = 20$ rad/sec.

Let $A = |G(j\omega)|$ in db

At $\omega = \omega_1$, $A = 20\log\dfrac{80}{\omega} = 20\log\dfrac{80}{0.5} = 44$ db

At $\omega = \omega_{c1}$, $A = 20\log\dfrac{80}{\omega} = 20\log\dfrac{80}{3} = 28.5 \approx 28$ db

At $\omega = \omega_{c2}$, $A = \left[\text{slope from } \omega_{c1} \text{ to } \omega_{c2} \times \log\dfrac{\omega_{c2}}{\omega_{c1}}\right] + A \text{ at } (\omega = \omega_{c1})$

$$= -40 \times \log\frac{6}{3} + 28 = 16 \text{ db}$$

At $\omega = \omega_h$, $A = \left[\text{slope from } \omega_{c2} \text{ to } \omega_h \times \log\dfrac{\omega_h}{\omega_{c2}}\right] + A \text{ at } (\omega = \omega_{c2})$

$$= -60 \times \log\frac{20}{6} + 16 = -15 \text{ db}$$

Let the points a, b, c and d be the points corresponding to frequencies ω_l, ω_{c1}, ω_{c2} and ω_h respectively on the magnitude plot. In a semilog graph sheet choose appropriate scales and fix the points a, b, c and d. Join the points by straight lines and mark the slope on the respective region. The magnitude plot is shown in fig 1.9.2.

PHASE PLOT

The phase angle of $G(j\omega)$ as a function of ω is given by

$\phi = \angle G(j\omega) = -90° - \tan^{-1}0.33\omega - \tan^{-1}0.167\omega$.

The phase angle of $G(j\omega)$ are calculated for various values of ω and listed in table-2.

TABLE-2

ω rad/sec	0.5	1.0	3.0	6	10	20
$\angle G(j\omega)$ deg	–104	–118	–161 ≈ -160	–198	–222	–244.7 ≈ -244

On the same semilog sheet take another y-axis, choose appropriate scale and draw phase plot as shown in fig 1.9.2.

Step-3 : Find phase margin of uncompensated system.

Let, ϕ_{gc} = Phase of G(jω) at gain crossover frequency

and γ = Phase margin of uncompensated system.

From the bode plot of uncompensated system we get, $\phi_{gc} = -226°$.

Now, γ = $180° + \phi_{gc} = 180° - 226° = -46°$

Step-4 : Choose a new phase margin

The desired phase margin, $\gamma_d = 35°$

The phase margin of compensated system, $\gamma_n = \gamma_d + \epsilon$

Let initial choice of $\epsilon = 5°$

$\therefore \gamma_n = \gamma_d + \epsilon = 35° + 5° = 40°$.

Step-5 : Determine new gain crossover frequency

Let, ω_{gcn} = New gain crossover frequency   and ϕ_{gcn} = Phase of G(jω) at ω_{gcn}

Now, $\gamma_n = 180° + \phi_{gcn}$, $\therefore \phi_{gcn} = \gamma_n - 180° = 40° - 180° = -140°$

From the bode plot we found that the frequency corresponding to a phase of $-140°$ is 1.8 rad/sec.

Let, ω_{gcl} = Gain crossover frequency of lag compensator.

Choose ω_{gcl} such that, $\omega_{gcl} > \omega_{gcn}$.   Let $\omega_{gcl} = 4$ rad/sec.

Step-6 : Calculate β of lag compensator

From the bode plot we found that the db magnitude at ω_{gcl} is 23 db.

$\therefore |G(j\omega)|$ in db at $(\omega = \omega_{gcl}) = A_{gcl} = 23$ db.

Also, $A_{gcl} = 20\log\beta$; $\therefore \beta = 10^{A_{gcl}/20} = 10^{23/20} = 14$

Step-7 : Determine the transfer function of lag section.

The zero of the lag compensator is placed at a frequency one-tenth of ω_{gcl}.

$\therefore$ Zero of lag compensator, $z_{c1} = \dfrac{-1}{T_1} = \dfrac{\omega_{gcl}}{10}$

$$\text{Now, } T_1 = \frac{10}{\omega_{gcl}} = \frac{10}{4} = 2.5$$

Pole of lag compensator, $p_{c1} = \dfrac{1}{\beta T_1} = \dfrac{1}{14 \times 2.5} = \dfrac{1}{35}$

Transfer function of lag section $\Big\} \quad G_1(s) = \beta \dfrac{(1+sT_1)}{(1+s\beta T_1)} = 14\dfrac{(1+2.5s)}{(1+35s)}$

Step-8 : Determine the transfer function of lead section.

Let $\alpha = 1/\beta$; $\therefore \alpha = 1/14 = 0.07$

The db gain (magnitude)$\left. \begin{array}{l} \\ \end{array} \right\}$ corresponding to ω_m $= -20 \log \dfrac{1}{\sqrt{\alpha}} = -20 \log \dfrac{1}{\sqrt{0.07}} = -11.5 \text{ db} \approx -12 \text{ db}$

From the bode plot of uncompensated system the frequency ω_m corresponding to a db pair of -12 db is found to be 17 rad/sec.

$\therefore \ \omega_m = 17 \text{ rad/ sec}$

$\therefore \ T_2 = \dfrac{1}{\omega_m \sqrt{\alpha}} = \dfrac{1}{17\sqrt{0.07}} = 0.22$

Transfer function$\left. \begin{array}{l} \\ \end{array} \right\}$ of lead section $G_2(s) = \alpha \dfrac{(1 + sT_2)}{(1 + s\alpha T_2)} = 0.07 \dfrac{(1 + 0.22s)}{(1 + 0.0154s)}$

Step-10 : Determine the transfer function of lag-lead compensator.

Transfer function$\left. \begin{array}{l} \\ \end{array} \right\}$ of lag – lead compensator $G_c(s) = G_1(s) \times G_2(s)$

$= 14 \dfrac{(1 + 2.5s)}{(1 + 35s)} \times 0.07 \dfrac{(1 + 0.22s)}{(1 + 0.0154s)}$

$= \dfrac{(1 + 2.5s)(1 + 0.22s)}{(1 + 35s)(1 + 0.0154s)}$

Step-11 : Determine open loop transfer function of compensated system.

The lag-lead compensator is connected in series with G(s) as shown in fig 1.9.1.

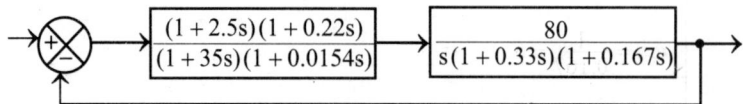

Fig 1.9.1 : *Block diagram of lag-lead compensated system.*

Open loop transfer function$\left. \begin{array}{l} \\ \end{array} \right\}$ of compensated system $G_0(s) = \dfrac{80(1 + 2.5s)(1 + 0.22s)}{s(1 + 35s)(1 + 0.0154s)(1 + 0.33s)(1 + 0.167s)}$

Step-12 : Bode plot of compensated system.

Put $s = j\omega$ in $G_0(s)$

$\therefore \ G_0(j\omega) = \dfrac{80(1 + 2.5j\omega)(1 + 0.22j\omega)}{s(1 + 35j\omega)(1 + 0.0154j\omega)(1 + 0.33j\omega)(1 + 0.167j\omega)}$

MAGNITUDE PLOT

There are six corner frequencies, which are given below.

$\omega_{c1} = \dfrac{1}{35} = 0.03 \text{ rad/ sec} \ ; \ \omega_{c2} = \dfrac{1}{2.5} = 0.4 \text{ rad/ sec} \ ; \ \omega_{c3} = \dfrac{1}{0.33} = 3 \text{ rad/ sec}$

$$\omega_{c4} = \frac{1}{0.22} = 4.5 \text{ rad/sec} \; ; \; \omega_{c5} = \frac{1}{0.167} = 6 \text{ rad/sec} \; ; \; \omega_{c6} = \frac{1}{0.0154} = 65 \text{ rad/sec}$$

The various terms of $G_0(j\omega)$ are listed in table-3. Also the table shows the slope contributed by each term and the change in slope at the corner frequency.

TABLE-3

Term	Corner frequency rad/sec	Slope db/dec	Change in slope db/dec
$\dfrac{80}{j\omega}$	—	-20	—
$\dfrac{1}{1+j35\omega}$	$\omega_{c1} = \dfrac{1}{35} = 0.03$	-20	$-20 -20 = -40$
$1+j2.5\omega$	$\omega_{c2} = \dfrac{1}{2.5} = 0.4$	$+20$	$-40 +20 = -20$
$\dfrac{1}{1+j0.33\omega}$	$\omega_{c3} = \dfrac{1}{0.33} = 3$	-20	$-20 -20 = -40$
$1+j0.22\omega$	$\omega_{c4} = \dfrac{1}{0.22} = 4.5$	$+20$	$-40 +20 = -20$
$\dfrac{1}{1+j0.167\omega}$	$\omega_{c5} = \dfrac{1}{0.167} = 6$	-20	$-20 -20 = -40$
$\dfrac{1}{1+j0.0154\omega}$	$\omega_{c6} = \dfrac{1}{0.0154} = 65$	-20	$-40 -20 = -60$

Choose a low frequency ω_l such that $\omega_l < \omega_{c1}$ and choose a high frequency ω_h such that $\omega_h > \omega_{c6}$.

Let $\omega_l = 0.01$ rad/sec and $\omega_h = 80$ rad/sec

Let $A_0 = |G_0(j\omega)|$ in db.

At $\omega = \omega_l$, $\qquad\qquad A_0 = 20 \log \dfrac{80}{0.01} = 78 \text{ db}$

At $\omega = \omega_{c1}$, $\qquad\qquad A_0 = 20 \log \dfrac{80}{0.03} = 68.5 \text{ db} \approx 68 \text{ db}$

At $\omega = \omega_{c2}$, $\qquad\qquad A_0 = -40 \times \log \dfrac{0.4}{0.03} + 68 = 23 \text{ db}$

At $\omega = \omega_{c3}$, $\qquad\qquad A_0 = -20 \times \log \dfrac{3}{0.4} + 23 = 5 \text{ db}$

At $\omega = \omega_{c4}$, $\qquad\qquad A_0 = -40 \times \log \dfrac{4.5}{3} + 5 = -2 \text{ db}$

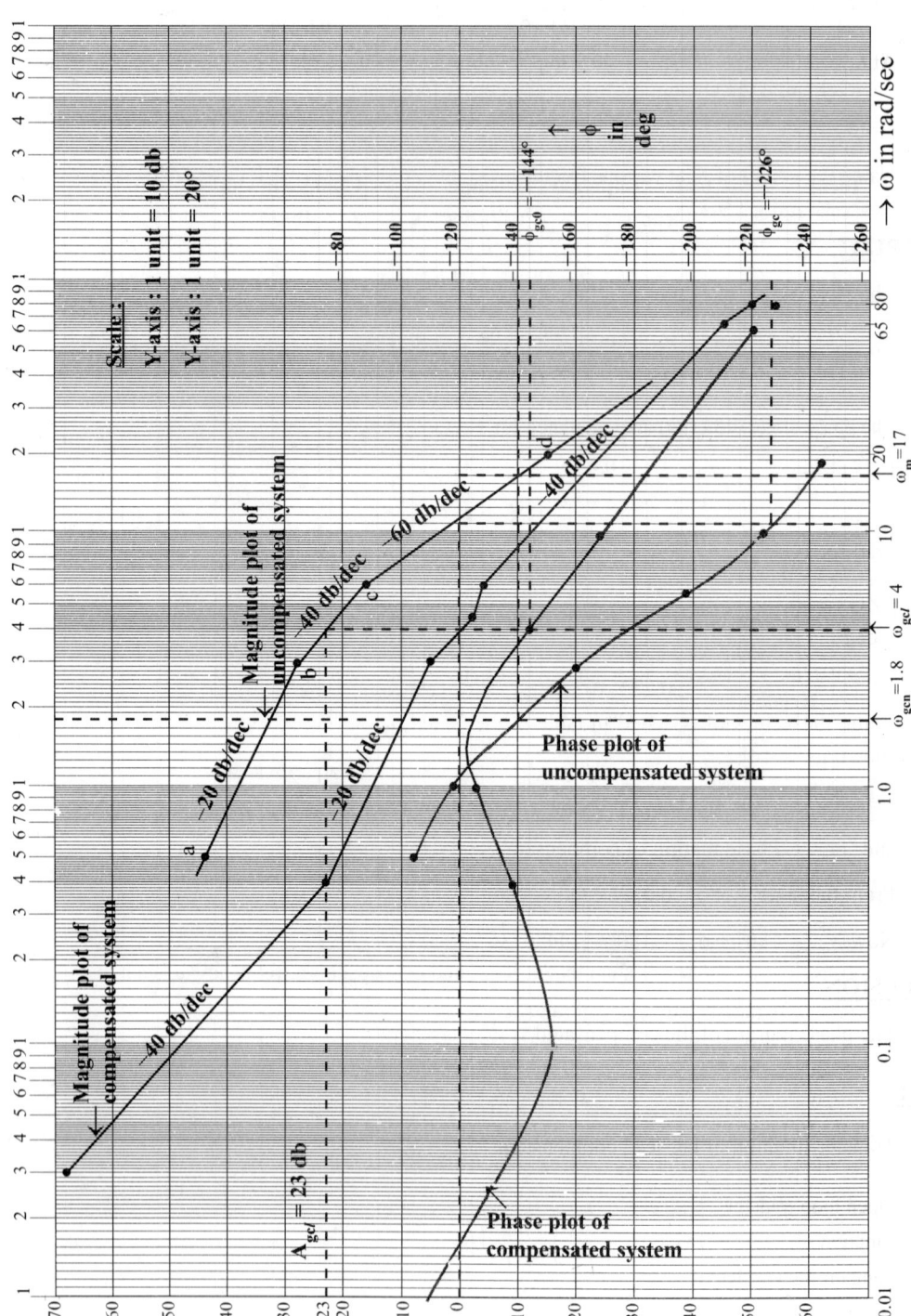

Fig 1.9.2 : Bode plot of $G(j\omega) = 80/j\omega(1 + j0.33\omega)(1 + j0.167\omega)$.

At $\omega = \omega_{c5}$, $\quad A_0 = -20 \times \log \dfrac{6}{4.5} + (-2) = -4 \text{ db}$

At $\omega = \omega_{c6}$, $\quad A_0 = -40 \times \log \dfrac{65}{6} + (-4) = -45 \text{ db}$

At $\omega = \omega_h$, $\quad A_0 = -60 \times \log \dfrac{80}{65} + (-45) = -50 \text{ db}$

Using the values of A_0 at various frequencies the magnitude plot of compensated system is drawn as shown in fig 1.9.2.

PHASE PLOT

The phase angle of $G_0(j\omega)$ as a function of ω is given by

$$\phi_0 = \angle G_0(j\omega) = \tan^{-1} 2.5\omega + \tan^{-1} 0.22\omega - 90° - \tan^{-1} 35\omega$$
$$-\tan^{-1} 0.0154\omega - \tan^{-1} 0.33\omega - \tan^{-1} 0.167\omega.$$

The phase angle of $G_0(j\omega)$ are calculated for various values of ω and listed in table-4.

TABLE-4

ω rad/sec	0.01	0.03	0.1	0.4	1	4	10	65	80
$\angle G_0(j\omega)$ deg	−108	−132	−152	−138	−126	−144	−168	−221	−228
								≈ -220	

Using the values of ϕ_0 listed in table-4, the phase plot of compensated system is sketched as shown in fig 1.9.2.

Let ϕ_{gc0} = Phase of $G_0(j\omega)$ at the gain crossover frequency of compensated system.

and γ_0 = Phase margin of compensated system.

From the bode plot of compensated system, we get, $\phi_{gc0} = -144°$

Now, $\gamma_0 = 180° + \phi_{gc0} = 180° - 144° = 36°$

CONCLUSION

The phase margin of the compensated system is satisfactory. Hence the design is acceptable.

RESULT

Transfer function of lag-lead compensator, $G_c(s) = \dfrac{(1+2.5s)(1+0.22s)}{(1+35s)(1+0.0154s)}$

Open loop transfer function of compensated system $\Bigg\} \quad G_0(s) = \dfrac{80(1+2.5s)(1+0.22s)}{s(1+35s)(1+0.0154s)(1+0.33s)(1+0.167s)}$

EXAMPLE 1.10

Design a lag-lead compensator for a system with open loop transfer function $G(s) = K/s(s+0.5)$ to satisfy the following specifications. (i) Damping ratio of dominant closed-loop poles, $\zeta = 0.5$. (ii) Undamped natural frequency of dominant closed loop poles, $\omega_n = 5$ rad/sec. (iii) Velocity error constant, $K_v = 80$ sec^{-1}.

SOLUTION

Step-1: Determine dominant pole, s_d.

Dominant pole, $s_d = -\zeta\omega_n \pm j\omega_n\sqrt{1-\zeta^2}$.

Given that, $\zeta = 0.5$ and $\omega_n = 5$ rad/sec.

$\therefore s_d = -0.5 \times 5 \pm j5 \times \sqrt{1 - 0.5^2} = -2.5 \pm j4.3$

Step-2 : Draw the pole-zero plot

The pole-zero plot of open loop transfer function is shown fig 6.10.1. Poles are represented by the symbol "x". The pole at point P is the dominant pole, s_d.

Step-3 : To find the angle to be contributed by lead network.

Let ϕ = Angle to be contributed by lead network to make point, P as a point on root locus.

Now, $\phi = \left(\begin{array}{c} \text{Sum of angles} \\ \text{contributed by poles} \\ \text{of uncompensated system} \end{array} \right) - \left(\begin{array}{c} \text{Sum of angles} \\ \text{contributed by zeros} \\ \text{of uncompensated system} \end{array} \right) \pm n180°$

From fig 1.10.1, we get,

$\left. \begin{array}{c} \text{Sum of angles contributed} \\ \text{by poles of uncompensated system} \end{array} \right\} = \theta_1 + \theta_2 = 120° + 115° = 235°$

Since there is no finite zero in uncompensated system, there is no angle contribution by zeros.

$\therefore \phi = 235° \pm n180°$

Let $n = 1$, $\therefore \phi = 235° - 180° = 55°$.

Step-4 : To find the pole and zero of the lead section

Draw a line AP parallel to x-axis as shown in fig 1.10.1. The bisector PC is drawn to bisect the angle APO. The angles CPD and BPC are constructed as shown in fig 1.10.1. Here $\angle CPD = \angle BPC = \phi/2 = 55°/2 = 27.5° \approx 27°$.

From fig 1.10.1., Pole of the lead section, $p_{c2} = -10$

Zero of the lead section, $z_{c2} = -2.65$.

We know that, $z_{c2} = -1/T_2$ $\therefore T_2 = 1/2.65 = 0.377$

We know that, $p_{c2} = -1/\alpha T_2$ $\therefore \alpha T_2 = 1/10$ (or) $\alpha = 1/(T_2 \times 10) = 0.265$.

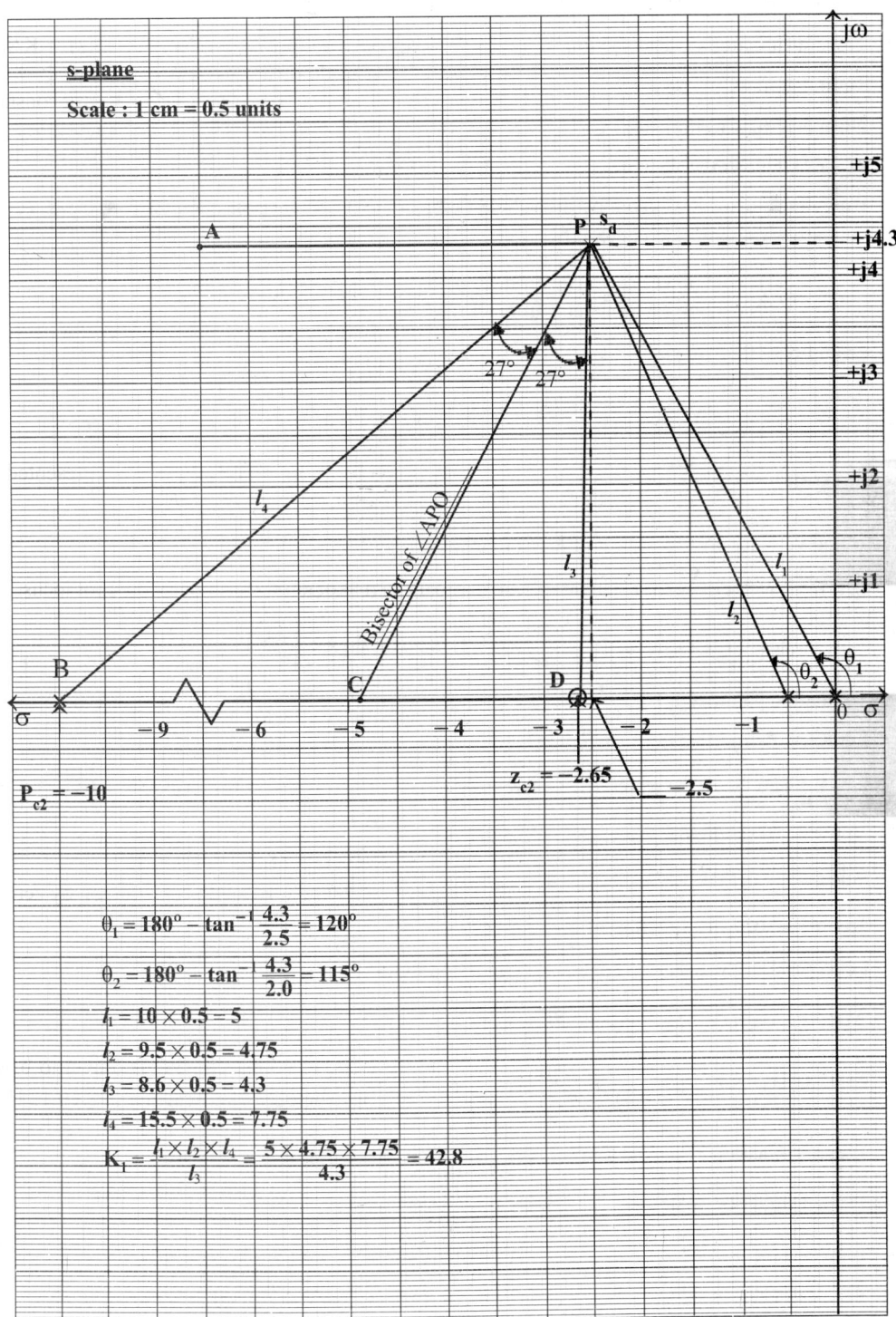

Fig 1.10.1. : Pole zero plot of lead compensated system.

The text inside the figure:

s-plane

Scale : 1 cm = 0.5 units

$\theta_1 = 180° - \tan^{-1} \dfrac{4.3}{2.5} = 120°$

$\theta_2 = 180° - \tan^{-1} \dfrac{4.3}{2.0} = 115°$

$l_1 = 10 \times 0.5 = 5$

$l_2 = 9.5 \times 0.5 = 4.75$

$l_3 = 8.6 \times 0.5 = 4.3$

$l_4 = 15.5 \times 0.5 = 7.75$

$K_1 = \dfrac{l_1 \times l_2 \times l_4}{l_3} = \dfrac{5 \times 4.75 \times 7.75}{4.3} = 42.8$

$P_{c2} = -10$

$z_{c2} = -2.65$

Step-5 : Transfer function of lead compensator

Transfer function of lead section, $G_2(s) = \dfrac{(s + 1/T_2)}{(s + 1/\alpha T_2)} = \dfrac{(s + 2.65)}{(s + 10)}$

Step-6 : To find gain, K

$\left.\begin{array}{l}\text{Open loop transfer function}\\\text{of lead compensated system}\end{array}\right\}$ $G_{02}(s) = G_2(s) \times G(s) = \dfrac{(s + 2.65)}{(s + 10)} \times \dfrac{K}{s(s + 0.5)} = \dfrac{K(s + 2.65)}{s(s + 0.5)(s + 10)}$

Here the value of K is given by the value of gain at the dominant pole s_d on the root locus. From magnitude condition K is given by,

$$K = \dfrac{\text{Product of vector lengths from all poles to } s = s_d}{\text{Product of vector lengths from all zeros to } s = s_d}$$

From fig 1.10.1, we get,

$$K = \dfrac{l_1 \times l_2 \times l_4}{l_3} = \dfrac{5 \times 4.75 \times 7.75}{4.3} = 42.8$$

$$G_{02}(s) = \dfrac{42.8(s + 2.65)}{s(s + 0.5)(s + 10)}$$

Step-7 : To find velocity error constant of lead compensated system.

Let, K_{v2} = Velocity error constant of lead compensated system.

$$\therefore\ K_{v2} = \underset{s \to 0}{\text{Lt}}\ s.\,G_{02}(s) = \underset{s \to 0}{\text{Lt}}\ s\,\dfrac{42.8(s + 2.65)}{s(s + 0.5)(s + 10)} = \dfrac{42.8 \times 2.65}{0.5 \times 10} = 22.684$$

Step-8 : To find the parameter, β

Let K_{vd} = Desired velocity error constant

A = The factor by which K_v is increased

Now, $A = K_{vd}/K_{v2} = 80/22.684 = 3.5267$

Select β, such that β > A. Let β = 4.

Step-9 : To find the transfer function of lag section.

Let zero of lag section, $z_{cl} = 0.1 \times$ second pole of $G(s) = 0.1 \times (-0.5) = -0.05$

Also, $z_{cl} = \dfrac{-1}{T_1}$; $\therefore\ T_1 = \dfrac{1}{0.05} = 20$

Pole of lag section, $p_{cl} = \dfrac{-1}{\beta T_1} = \dfrac{-1}{4 \times 20} = \dfrac{-1}{80} = -0.0125$

Transfer function of lag section, $G_1(s) = \dfrac{(s + 1/T_1)}{(s + 1/\beta T_1)} = \dfrac{(s + 0.05)}{(s + 0.0125)}$

Step-10 : Transfer function of compensated system

$\left.\begin{array}{l}\text{Transfer function of}\\\text{lag} - \text{lead compensator}\end{array}\right\}$ $G_c(s) = G_1(s) \times G_2(s) = \dfrac{(s + 0.05)(s + 2.65)}{(s + 0.0125)(s + 10)}$

The lag-lead compensator is connected in series with G(s) as shown in fig 1.10.2.

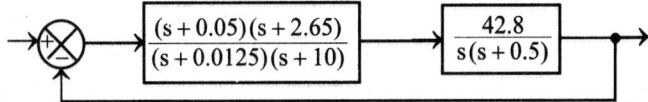

Fig 1.10.2 : Block diagram of lag-lead compensated system.

$$\left.\begin{array}{l}\text{Open loop transfer function}\\\text{of compensated system}\end{array}\right\} G_0(s) = \frac{42.8(s+0.05)(s+2.65)}{s(s+0.0125)(s+0.5)(s+10)}$$

Step-11 : Check velocity error constant of compensated system.

$$\left.\begin{array}{l}\text{Velocity error}\\\text{constant of}\\\text{compensated system}\end{array}\right\} K_{vc} = \underset{s\to 0}{\text{Lt}}\ sG_0(s) = \underset{s\to 0}{\text{Lt}}\ s\ \frac{42.8(s+0.05)(s+2.65)}{s(s+0.0125)(s+0.5)(s+10)}$$

$$= \frac{42.8 \times 0.05 \times 2.65}{0.0125 \times 0.5 \times 10} = 90.7$$

CONCLUSION

The velocity error constant of the compensated system satisfies the requirement. Hence the design is accepted.

RESULT

$$\text{Transfer function of lag} - \text{lead compensator, } G_c(s) = \frac{(s+0.05)(s+2.65)}{(s+0.0125)(s+10)}$$

$$\left.\begin{array}{l}\text{Open loop transfer function of}\\\text{lag} - \text{lead compensated system}\end{array}\right\} G_0(s) = \frac{42.8(s+0.05)(s+2.65)}{s(s+0.0125)(s+0.5)(s+10)}$$

1.5 PI, PD AND PID CONTROLLERS

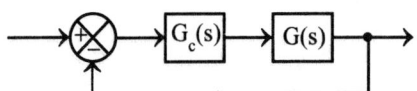

A controller with transfer function, $G_c(s)$ can be introduced in cascade with open loop transfer function, G(s) as shown in fig 1.21, to modify the transient and steady state response of the system.

Fig 1.21 : Block diagram of a system with a controller in cascade.

The different types of controllers employed in control system are the following

    1. Proportional controller (P-controller)

    2. Proportional-plus-integral controller (PI-controller)

    3. Proportional-plus-derivative controller (PD-controller)

    4. Proportional-plus-derivative-plus-integral controller (PID-controller)

The proportional controller is a device that produces an output signal, u(t), which is proportional to the input signal, e(t).

In P-controller, $u(t) \propto e(t)$

$$\therefore u(t) = K_p\, e(t) \qquad\qquad\qquad\qquad\qquad .....(1.49)$$

where, K_p = Proportional gain or constant.

On taking laplace transform of equation (6.49) we get,

$$U(s) = K_p E(s)$$

Transfer function of P-controller, $G_c(s) = \dfrac{U(s)}{E(s)} = K_P$ .....(1.50)

The proportional controller improves the steady state tracking accuracy, disturbance signal rejection and relative stability. It also decreases the sensitivity of the system to parameter variations. The proportional control is not used alone because it produces a constant steady state error.

The proportional plus integral controller (PI-controller) is a device that produces an output signal, u(t) consisting of two terms-one proportional to input signal, e(t) and the other proportional to the integral of input signal, e(t).

In PI-controller, $u(t) \propto [e(t) + \int e(t) \, dt]$

$$\therefore u(t) = K_p \, e(t) + K_i \, \int e(t) \, dt$$ .....(1.51)

where, K_p = Proportional gain

$\quad\quad K_i$ = Integral constant or gain.

On taking laplace transform of equation (1.51) with zero initial conditions we get,

$$U(s) = K_p E(s) + K_i \frac{E(s)}{s} = E(s)\left(K_p + \frac{K_i}{s}\right)$$

Transfer function of PI controller, $G_c(s) = \dfrac{U(s)}{E(s)} = K_P + \dfrac{K_i}{s}$ .....(1.52)

The PI controller reduces the steady state error. The introduction of PI controller increases the order and type number of the system by one.

The proportional plus derivative controller is a device that produces an output signal, u(t), consisting of two terms-one proportional to input signal, e(t) and the other proportional to the derivative of input signal, e(t).

In PD controller, $u(t) \propto \left[e(t) + \dfrac{d}{dt} e(t)\right]$

$$\therefore u(t) = K_p \, e(t) + K_d \frac{d}{dt} e(t)$$

where, K_p = Proportional gain

$\quad\quad K_d$ = Derivative constant or gain

On taking laplace transform of equation (1.53) with zero initial conditions we get,

$$U(s) = K_p E(s) + K_d \, sE(s) = E(s) [K_p + K_d \, s]$$

Transfer function of PD controller, $G_c(s) = \dfrac{U(s)}{E(s)} = K_P + K_d s$ .....(1.54)

The PD controller increases the damping of the system which results in reducing the peak overshoot.

The PID controller is a device which produces an output signal, u(t) consisting of three terms-one proportional to input signal, e(t), another one proportional to integral of input signal, e(t) and the third one proportional to derivative of input signal, e(t).

In PID controller, $u(t) \alpha \left[e(t) + \int e(t) \, dt + \dfrac{d}{dt} e(t) \right]$

$$\therefore \; u(t) = K_p e(t) + K_i \int e(t) \, dt + K_d \dfrac{d}{dt} e(t) \qquad\qquad .....(1.55)$$

On taking laplace transform of equation (1.55) we get,

$$U(s) = K_p E(s) + K_i \dfrac{E(s)}{s} + K_d s E(s) = E(s) \left[K_p + \dfrac{K_i}{s} + K_d s \right]$$

Transfer function of PID-controller, $G_c(s) = \dfrac{U(s)}{E(s)} = K_p + \dfrac{K_i}{s} + K_d s \qquad .....(1.56)$

The PID controller have the combined effect of all the three control actions-i.e., proportional, integral and derivative control actions. Hence the introduction of PID controller stabilises the gain, reduces the steady state error and peak overshoot of the system.

DESIGN OF PI, PD AND PID CONTROLLERS IN FREQUENCY DOMAIN

In frequency domain PI, PD and PID controllers can be designed to satisfy a specified gain margin and error constant.

Let ω_1 = Gain crossover frequency of the system with cascade PI/PD/PID controller.

A_1 = $|G(j\omega)|$ at ω_1

ϕ_1 = Phase of $G(j\omega)$ at ω_1

γ_u = Phase margin of uncompensated system at ω_1

γ_d = Desired phase margin at ω_1

θ = Angle to be contributed by PI/PD/PID controller at gain crossover frequency, ω_1 to achieve a phase margin of γ_d.

By definition of phase margin

$$\gamma_u = 180° + \phi_1$$

The desired phase margin, γ_d is given by,

$$\gamma_d = \theta + \gamma_u$$

$$\therefore \theta = \gamma_d - \gamma_u \qquad\qquad .....(1.57)$$

Transfer function of PID controller } $G_c(s) = K_p + \dfrac{K_i}{s} + K_d s = \dfrac{K_d s^2 + K_p s + K_i}{s}$

On replacing s by $j\omega$ in $G_c(s)$ we get,

$$G_c(j\omega) = \dfrac{K_d(j\omega)^2 + K_p(j\omega) + K_i}{j\omega} = \dfrac{(-K_d \omega^2 + K_i) + j K_p \omega}{j\omega} \qquad .....(1.58)$$

Let, $A_c = |G_c(j\omega)|$ at $\omega = \omega_1$

θ = Phase of $G_c(j\omega)$ at $\omega = \omega_6$.

$\cdot \; \therefore$ At $\omega = \omega_1$, $G_c(j\omega) = G_c(j\omega_1) = |G_c(j\omega_1)| \angle G_c(j\omega_1) = A_c \angle \theta \qquad .....(1.59)$

The magnitude of open loop transfer function at gain crossover frequency is unity.

$\therefore$ At $\omega = \omega_1$, $|G_c(j\omega_1).G(j\omega_1)| = 1$

$\therefore$ $A_cA_1 = 1$ (or) $A_c = \dfrac{1}{A_1}$ .....(1.60)

From equation (1.59) and (1.60) we can write,

$G_c(j\omega) = \dfrac{1}{A_1}\angle\theta$ .....(1.61)

From equation (1.58) at $\omega = \omega_1$ we get,

$G_c(j\omega_1) = \dfrac{(-K_d\omega_1^2 + K_i) + jK_p\omega_1}{j\omega_1}$ .....(1.62)

On equating, equations (1.61) and (1.62) we get,

$\dfrac{-K_d\omega_1^2 + K_i + jK_p\omega_1}{j\omega_1} = \dfrac{1}{A_1}\angle\theta$

$-K_d\omega_1^2 + K_i + jK_p\omega_1 = j\omega_1\dfrac{1}{A_1}\angle\theta$

Here, $j\omega_1 = \omega_1\angle 90°$

$-K_d\omega_1^2 + K_i + jK_p\omega_1 = \omega_1\angle 90°\dfrac{1}{A_1}\angle\theta$

$-K_d\omega_1^2 + K_i + jK_p\omega_1 = \dfrac{\omega_1}{A_1}\angle(90° + \theta)$

$-K_d\omega_1^2 + K_i + jK_p\omega_1 = \dfrac{\omega_1}{A_1}\cos(90° + \theta) + j\dfrac{\omega_1}{A_1}\sin(90° + \theta)$

$-K_d\omega_1^2 + K_i + jK_p\omega_1 = -\dfrac{\omega_1}{A_1}\sin\theta + j\dfrac{\omega_1}{A_1}\cos\theta$ .....(1.63)

On equating real parts of equation (1.63) we get,

$-K_d\omega_1^2 + K_i = -\dfrac{\omega_1}{A_1}\sin\theta$ (or) $-K_d + \dfrac{K_i}{-\omega_1^2} = -\dfrac{\omega_1\sin\theta}{-\omega_1^2 A_1}$

$\therefore$ $K_d = \dfrac{\sin\theta}{\omega_1 A_1} + \dfrac{K_i}{\omega_1^2}$ .....(1.64)

On equating imaginary parts of equation (1.63) we get,

$K_p\omega_1 = \dfrac{\omega_1}{A_1}\cos\theta$

$\therefore$ $K_p = \dfrac{\cos\theta}{A_1}$ .....(1.65)

The equations (1.64) and (1.65) are used to design cascade PD or PI or PID controller. For designing a PD controller put $K_i = 0$, in equation (6.64). Hence the constants K_p and K_d of PD controller are given by the following two equations,

$K_d = \dfrac{\sin\theta}{\omega_1 A_1}$ and $K_p = \dfrac{\cos\theta}{A_1}$

For designing a PI controller put $K_d = 0$, in equation (6.64). Hence the constants K_p and K_i of PI controller are given by the following two equations,

$$K_i = \frac{-\omega_1 \sin\theta}{A_1} \quad \text{and} \quad K_p = \frac{\cos\theta}{A_1}$$

For designing a PID controller, determine the constant K_i to satisfy the specified error constant. Then calculate K_d and K_p from equations (1.64) and (1.65).

PROCEDURE FOR DESIGN OF PD/PI/PID CONTROLLER IN FREQUENCY DOMAIN

The following procedure can be followed to design a PD/PI/PID controller when the given specifications are desired phase margin, γ_d at a gain crossover frequency, ω_6.

Step-1 : Determine the magnitude and phase of uncompensated open loop sinusoidal transfer function [i.e., $G(j\omega)$]

Let, $A_1 = |G(j\omega)|$ at $\omega = \omega_1$ and $\phi_1 = \angle G(j\omega)$ at $\omega = \omega_1$

Step-2 : Determine the phase margin of uncompensated system and the angle to be contributed by the controller to achieve the desired phase margin.

Let, γ_u = Phase margin of uncompensated system.

γ_d = Desired phase margin at ω_6.

θ = Phase angle of the controller at $\omega = \omega_6$.

Now, $\gamma_u = 180° + \phi_1$; $\theta = \gamma_d - \gamma_u$

Step-3 : Determine the transfer function of the controller.

a. PD controller

Derivative constant, $K_d = \dfrac{\sin\theta}{\omega_1 A_1}$

Proportional constant, $K_p = \dfrac{\cos\theta}{A_1}$

Transfer function of PD controller } $G_c(s) = (K_p + K_d s) = K_p\left(1 + \dfrac{K_d}{K_p} s\right)$

b. PI controller

Integral constant, $K_i = \dfrac{-\omega_1 \sin\theta}{A_1}$

Proportional constant, $K_p = \dfrac{\cos\theta}{A_1}$

Transfer function of PI controller } $G_c(s) = \left(K_p + \dfrac{K_i}{s}\right) = \dfrac{K_i\left(1 + \dfrac{K_p}{K_i} s\right)}{s}$

c. PID controller

Transfer function of PID controller } $G_c(s) = \left(K_p + K_d s + \dfrac{K_i}{s}\right) = \dfrac{K_d\left(s^2 + \dfrac{K_p}{K_d} s + \dfrac{K_i}{K_d}\right)}{s}$

Evaluate K_i such that the compensated system satisfies the error requirement. For example if the compensated system is type-1 system then, $K_v = \underset{s\to0}{Lt}\ sG_c(s)\,G(s)$ will give the value of K_i.

Derivative constant, $K_d = \dfrac{\sin\theta}{\omega_1 A_1} + \dfrac{K_i}{\omega_1^2}$

Proportional constant, $K_p = \dfrac{\cos\theta}{A_1}$

Step-4 : Determine the open loop transfer function of compensated system.

The transfer function of the controller is placed in s with G(s) as shown in fig 1.22.

Open loop transfer function of compensated system,
$G_0(s) = G_c(s) \times G(s)$

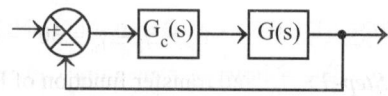

Fig 1.22 : Block diagram of system with cascade controller.

Step-5 : Verify the design by calculating phase margin of compensated system.

Let $A_0 = |G_0(j\omega)|$ at $\omega = \omega_1$

$\phi_0 = \angle G_0(j\omega)$ at $\omega = \omega_1$

γ_0 = Phase margin of compensated system.

Now, $\gamma_0 = 180° + \phi_0$

It can be observed that $A_0 \approx 1$ and γ_0 satisfies the specifications.

EXAMPLE 1.11

Consider a unity feedback system with open loop transfer function, $G(s)=5/s(s+0.5)(s+1)$. Design a PD controller so that the phase margin of the system is 30° at a frequency of 1.2 rad/sec.

SOLUTION

Step-1 : To find magnitude and phase of $G(j\omega_1)$

Given that, $G(s) = \dfrac{5}{s(s+0.5)(s+1)} = \dfrac{5}{s \times 0.5\left(1+\dfrac{s}{0.5}\right)(1+s)} = \dfrac{10}{s(1+2s)(1+s)}$

Put, $s = j\omega$ in G(s).

$\therefore G(j\omega) = \dfrac{10}{j\omega(1+j2\omega)(1+j\omega)} = \dfrac{10}{\omega\angle90°\sqrt{1+4\omega^2}\angle\tan^{-1}2\omega\sqrt{1+\omega^2}\angle\tan^{-1}\omega}$

$|G(j\omega)| = \dfrac{10}{\omega\sqrt{1+4\omega^2}\sqrt{1+\omega^2}}$

$\angle G(j\omega) = -90° - \tan^{-1}2\omega - \tan^{-1}\omega$

The gain crossover frequency of compensated system, $\omega_1 = 1.2$ rad/sec.

Let, $A_1 = |G(j\omega)|$ at $\omega = \omega_1$; $\phi_1 = \angle G(j\omega)$ at $\omega = \omega_1$

$\therefore A_1 = \dfrac{10}{1.2 \times \sqrt{1+4\times1.2^2} \times \sqrt{1+1.2^2}} = 2.052$

$\phi_1 = -90° - \tan^{-1}(2\times1.2) - \tan^{-1}(1.2) = -207.5°$

Step-2 : To find γ_u and θ

Let γ_u = Phase margin of uncompensated system

γ_d = Desired phase margin of compensated system.

θ = Phase of $G_c(j\omega)$ at $\omega = \omega_6$.

Now, $\gamma_u = 180° + \phi_1 = 180° + (-207.5°) = -27.5°$

$\theta = \gamma_d - \gamma_u = 30° - (-27.5°) = 57.5°$

Step-3 : To find transfer function of PD-controller.

Derivative constant, $K_d = \dfrac{\sin\theta}{\omega_1 A_1} = \dfrac{\sin 57.5°}{1.2 \times 2.052} = 0.343$

Proportional constant, $K_p = \dfrac{\cos\theta}{A_1} = \dfrac{\cos 57.5°}{2.052} = 0.262$

Transfer function of PD controller $\Big\}$ $G_c(s) = (K_p + K_d s) = K_p\left(1 + \dfrac{K_d}{K_p}s\right)$

$= 0.262\left(1 + \dfrac{0.343}{0.262}s\right) = 0.262(1 + 1.3s)$

Step-4 : To find open loop transfer function of compensated system.

The PD-controller is connected in cascade with G(s) as shown in figure 1.11.1

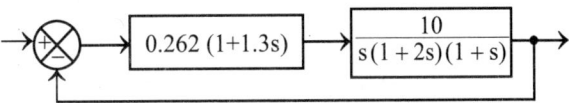

Fig 1.11.1 : Block diagram of compensated system.

Open loop transfer function of compensated system $\Big\}$ $G_0(s) = G_c(s) \times G(s)$

$= 0.262(1 + 1.3s) \times \dfrac{10}{s(1+2s)(1+s)}$

$= \dfrac{2.62(1+1.3s)}{s(1+2s)(1+s)}$

Step-5 : To verify the design

Put $s = j\omega$ in $G_0(s)$,

$\therefore G_0(j\omega) = \dfrac{2.62(1+j1.3\omega)}{j\omega(1+j2\omega)(1+j\omega)} = \dfrac{2.62\sqrt{1+1.69\omega^2}\,\angle\tan^{-1}1.3\omega}{\omega\angle 90°\,\sqrt{1+4\omega^2}\,\angle\tan^{-1}2\omega\sqrt{1+\omega^2}\,\tan^{-1}\omega}$

Let, $A_0 = |G_0(j\omega)|$ and $\phi_0 = G_0(j\omega)$

$\therefore A_0 = \dfrac{2.62\sqrt{1+1.69\omega^2}}{\omega\sqrt{1+4\omega^2}\,\sqrt{1+\omega^2}}$

$\phi_0 = \tan^{-1}1.3\omega - 90° - \tan^{-1}2\omega - \tan^{-1}\omega$

At $\omega = \omega_1$, $A_0 = A_{01} = \dfrac{2.62\sqrt{1+1.69\times 1.2^2}}{1.2\times\sqrt{1+4\times 1.2^2}\times\sqrt{1+1.2^2}} \approx 1$

At $\omega = \omega_1$, $\phi = \phi_{01} = \tan^{-1}(1.3 \times 1.2) - 90^\circ - \tan^{-1}(2 \times 1.2) - \tan^{-1}1.2 = -150^\circ$

Phase margin of compensated system, $\gamma_o = 180^\circ + \phi_{01} = 180^\circ - 150^\circ = 30^\circ$

CONCLUSION

The phase margin of the compensated system is satisfactory. Hence the design is acceptable.

RESULT

1. Transfer function of PD controller, $G_c(s) = (0.262 + 0.343s) = 0.262(1 + 1.3s)$

2. Open loop transfer function of compensated system, $G_o(s) = \dfrac{2.62(1 + 1.3s)}{s(1 + 2s)(1 + s)}$

EXAMPLE 1.12

Consider a unity feedback system with open loop transfer function, $G(s) = \dfrac{100}{(s+1)(s+2)(s+5)}$.
Design a PI controller, so that the phase margin of the system is 60° at a frequency of 0.5 rad/sec.

SOLUTION

Step-1 : To find magnitude and phase of $G(j\omega_1)$

Given that,

$$G(s) = \frac{100}{(s+1)(s+2)(s+5)} = \frac{100}{(s+1) \times 2 \times \left(1+\frac{s}{2}\right) \times 5 \times \left(1+\frac{s}{5}\right)}$$

$$= \frac{10}{(s+1)(1+0.5s)(1+0.2s)}$$

Put, $s = j\omega$ in $G(s)$.

$$\therefore \ G(j\omega) = \frac{10}{(1+j\omega)(1+j0.5\omega)(1+j0.2\omega)}$$

$$= \frac{10}{\sqrt{1+\omega^2} \angle \tan^{-1}\omega \sqrt{1+0.25\omega^2} \angle \tan^{-1}0.5\omega \sqrt{1+0.04\omega^2} \angle \tan^{-1}0.2\omega}$$

$$|G(j\omega)| = \frac{10}{\sqrt{1+\omega^2}\sqrt{1+0.25\omega}\sqrt{1+0.04\omega^2}}$$

$$\angle G(j\omega) = -\tan^{-1}\omega - \tan^{-1}0.5\omega - \tan^{-1}0.2\omega$$

The gain crossover frequency of compensated system, $\omega_1 = 0.5$ rad/sec.

Let $A_1 = |G(j\omega)|$ at $\omega = \omega_1$

$\phi_1 = \angle G(j\omega)$ at $\omega = \omega_1$

$$\therefore A_1 = \frac{10}{\sqrt{1+0.5^2} \times \sqrt{1+0.25 \times 0.5^2} \times \sqrt{1+0.04 \times 0.5^2}} = 8.63$$

$$\phi = -\tan^{-1}0.5 - \tan^{-1}(0.5 \times 0.5) - \tan^{-1}(0.2 \times 0.5) = -46^\circ$$

Step-2 : To find γ_u and θ

     Let γ_u = Phase margin of uncompensated system.

       γ_d = Desired phase margin of uncompensated system.

       θ = Phase of $G_c(j\omega)$ at $\omega = \omega_6$.

     Now, $\gamma_u = 180° + \phi_1 = 180° - 46° = 134°$

         $\theta = \gamma_d - \gamma_u = 60° - 134° = -74°$

Step-3 : To find transfer function of PI-controller.

     Integral constant, $K_i = \dfrac{-\omega_1 \sin\theta}{A_1} = \dfrac{-0.5 \times \sin(-74°)}{8.63} = 0.056$

     Proportional constant, $K_p = \dfrac{\cos\theta}{A_1} = \dfrac{\cos(-74°)}{8.63} = 0.0352$

     $\left.\begin{array}{l}\text{Transfer function}\\\text{of PI controller}\end{array}\right\}$ $G_c(s) = \left(K_p + \dfrac{K_i}{s}\right) = \left(0.032 + \dfrac{0.056}{s}\right)$

$$= \frac{0.032s + 0.056}{s} = \frac{0.056\left(\frac{0.032}{0.056}s + 1\right)}{s}$$

$$= \frac{0.056(1 + 0.57s)}{s}$$

Step-4 : To find open loop transfer function of compensated system.

     The PI-controller is connected in cascade with G(s) as shown in figure 1.12.1

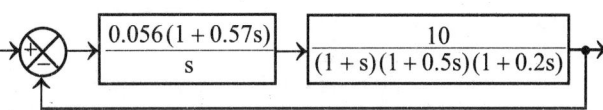

           Fig 1.12.1 : Block diagram of compensated system.

     $\left.\begin{array}{l}\text{Open loop transfer}\\\text{function of}\\\text{compensated system}\end{array}\right\}$ $G_o(s) = G_c(s) \times G(s)$

$$= \frac{0.056(1 + 0.57s)}{s} \times \frac{10}{(1 + s)(1 + 0.5s)(1 + 0.2s)}$$

$$= \frac{0.56(1 + 0.57s)}{s(1 + s)(1 + 0.5s)(1 + 0.2s)}$$

Step-5 : To verify the design

     Put $s = j\omega$ in $G_o(s)$,

     $\therefore G_o(j\omega) = \dfrac{0.56(1 + j0.57\omega)}{(j\omega)(1 + j\omega)(1 + j0.5\omega)(1 + j0.2\omega)}$

$$= \frac{0.56\sqrt{1 + (0.57\omega)^2}\,\angle\tan^{-1}0.57\omega}{\omega\angle 90°\sqrt{1 + \omega^2}\,\angle\tan^{-1}\omega\sqrt{1 + (0.5\omega)^2}\,\angle\tan^{-1}0.5\omega\sqrt{1 + (0.2\omega)^2}\,\angle\tan^{-1}0.2\omega}$$

Let, $A_0 = |G_0(j\omega)|$ and $\phi_0 = G_0(j\omega)$

$$\therefore A_0 = \frac{0.56\sqrt{1+0.3249\omega^2}}{\omega\sqrt{1+\omega^2}\sqrt{1+0.25\omega^2}\sqrt{1+0.04\omega^2}}$$

$$\phi_0 = \tan^{-1}0.57\omega - 90° - \tan^{-1}\omega - \tan^{-1}0.5\omega - \tan^{-1}0.2\omega$$

At $\omega = \omega_1$, $A_0 = A_{01} = \dfrac{0.56 \times \sqrt{1+0.3249 \times 0.5^2}}{0.5\sqrt{1+0.5^2}\sqrt{1+0.25 \times 0.5^2}\sqrt{1+0.04 \times 0.5^2}} \approx 1$

At $\omega = \omega_1$, $\phi_0 = \phi_{01} = \tan^{-1}(0.57 \times 0.5) - 90° - \tan^{-1}0.5 - \tan^{-1}(0.5 \times 0.5) - \tan^{-1}(0.2 \times 0.5)$

$$= -120°$$

Phase margin of compensated system, $\gamma_0 = 180° + \phi_{01} = 180° - 120° = 60°$

CONCLUSION

The phase margin of the compensated system, meets the given specification. Hence the design is acceptable.

RESULT

1. Transfer function of PI controller } $G_c(s) = \left(0.032 + \dfrac{0.056}{s}\right) = \dfrac{0.056(1+0.57s)}{s}$

2. Open loop transfer function of compensated system } $G_0(s) = \dfrac{0.56(1+0.57s)}{s(1+s)(1+0.5s)(1+0.2s)}$

EXAMPLE 1.13

Consider a unity feedback system with open loop transfer function, $G(s) = \dfrac{100}{(s+1)(s+2)(s+10)}$.
Design a PID controller, so that the phase margin of the system is 45° at a frequency of 4 rad/sec and the steady state error for unit ramp input is 0.1.

SOLUTION

Step-1 : To find magnitude and phase of $G(j\omega_1)$

$$G(s) = \frac{100}{(s+1)(s+2)(s+10)} = \frac{100}{(1+s)\times 2\times\left(1+\frac{s}{2}\right)\times 10\times\left(1+\frac{s}{10}\right)}$$

$$= \frac{5}{(1+s)(1+0.5s)(1+0.1s)}$$

Put, $s = j\omega$ in $G(s)$

$$\therefore G(j\omega) = \frac{5}{(1+j\omega)(1+j0.5\omega)(1+j0.1\omega)}$$

$$= \frac{5}{\sqrt{1+\omega^2}\angle\tan^{-1}\omega\sqrt{1+0.25\omega^2}\angle\tan^{-1}0.5\omega\sqrt{1+0.01\omega^2}\angle\tan^{-1}0.1\omega}$$

$$|G(j\omega)| = \frac{5}{\sqrt{1+\omega^2}\sqrt{1+0.25\omega^2}\sqrt{1+0.01\omega^2}}$$

$$\angle G(j\omega) = -\tan^{-1}\omega - \tan^{-1}0.5\omega - \tan^{-1}0.1\omega$$

The gain crossover frequency of compensated system, $\omega_1 = 4$ rad/sec.

Let, $A_1 = |G(j\omega)|$ at $\omega = \omega_1$

$\quad\quad \phi_1 = \angle G(j\omega)$ at $\omega = \omega_1$

$\therefore A_1 = \dfrac{5}{\sqrt{1+4^2} \times \sqrt{1+0.25 \times 4^2} \times \sqrt{1+0.01 \times 4^2}} = 0.5$

$\quad\quad \phi_1 = -\tan^{-1}4 - \tan^{-1}(0.5 \times 4) - \tan^{-1}(0.1 \times 4) = -161°$

Step-2 : To find γ_u and θ

Let γ_u = Phase margin of uncompensated system

$\quad\quad \gamma_d$ = Desired phase margin of compensated system.

$\quad\quad \theta$ = Phase of $G_c(j\omega)$ at $\omega = \omega_6$.

Now, $\gamma_u = 180° + \phi_1 = 180° - 161° = 19°$

$\quad\quad \theta = \gamma_d - \gamma_u = 45° - 19° = 26°$

Step-3 : To find transfer function of PID-controller.

Given that, steady state error, $e_{ss} = 0.1$ for unit ramp input.

$\therefore$ Velocity error constant, $K_v = \dfrac{1}{e_{ss}} = \dfrac{1}{0.1} = 10$

The velocity error constant of compensated system is given by,

$\quad\quad K_v = \underset{s \to 0}{\text{Lt}}\ s G_c(s)\, G(s)$

Here $G_c(s) = K_p + K_d s + \dfrac{K_i}{s} = \dfrac{K_d s^2 + K_p s + K_i}{s}$

and $G(s) = \dfrac{5}{(1+s)(1+0.5s)(1+0.1s)}$

$\therefore K_v = \underset{s \to 0}{\text{Lt}}\ s\, \dfrac{(K_d s^2 + K_p s + K_i)}{s} \times \dfrac{5}{(1+s)(1+0.5s)(1+0.1s)} = 10$

$\therefore 5K_i = 10 \quad$ (or) $\quad K_i = \dfrac{10}{5} = 2$

Derivative constant, $K_d = \dfrac{\sin\theta}{\omega_1 A_1} + \dfrac{K_i}{\omega_1^2} = \dfrac{\sin(26)°}{4 \times 0.5} + \dfrac{2}{4^2} = 0.344$

Proportial constant, $K_p = \dfrac{\cos\theta}{A_1} = \dfrac{\cos 26°}{0.5} = 1.8$

Transfer function of PID controller $\left.\vphantom{\dfrac{K_i}{s}}\right\}\ G_c(s) = \left(K_p + K_d s + \dfrac{K_i}{s}\right) = \left(1.8 + 0.344s + \dfrac{2}{s}\right) = \dfrac{0.344s^2 + 1.8s + 2}{s}$

$\quad\quad\quad = \dfrac{0.344\left(s^2 + \dfrac{1.8}{0.344}s + \dfrac{2}{0.344}\right)}{s} = \dfrac{0.344(s^2 + 5.23s + 5.81)}{s}$

Step-4 : To find open loop transfer function of compensated system.

PID-controller is connected in cascade with G(s) as shown in figure 1.13.1.

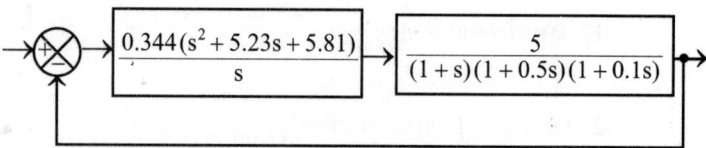

Fig 1.13.1 : Block diagram of system with cascade PID controller.

Open loop transfer function of compensated system $\left.\begin{array}{c} \\ \\ \end{array}\right\}$ $G_0(s) = G_c(s) \times G(s)$

$$= \frac{0.344(s^2 + 5.23s + 5.81)}{s} \times \frac{5}{(1+s)(1+0.5s)(1+0.1s)}$$

$$= \frac{1.72(s^2 + 5.23s + 5.81)}{s(1+s)(1+0.5s)(1+0.1s)}$$

Step-5 : To verify the design

Put $s = j\omega$ in $G_0(s)$,

$$\therefore G_0(j\omega) = \frac{1.72(-\omega^2 + j5.23\omega + 5.81)}{j\omega(1+j\omega)(1+j0.5\omega)(1+j0.1\omega)}$$

$$= \frac{1.72\sqrt{(5.81 - \omega^2)^2 + (5.23\omega)^2} \angle \tan^{-1}\dfrac{5.23}{5.81 - \omega^2}}{\omega\angle 90° \sqrt{1+\omega^2} \angle \tan^{-1}\omega \sqrt{1+(0.5\omega)^2} \angle \tan^{-1}0.5\omega \sqrt{1+(0.1\omega)^2} \angle \tan^{-1}0.1\omega}$$

Let $A_0 = |G_0(j\omega)|$ and $\phi_0 = G_0(j\omega)$

$$\therefore A_0 = \frac{1.72\sqrt{(5.81 - \omega^2)^2 + (5.23\omega)^2}}{\omega\sqrt{1+\omega^2} \sqrt{1+(0.5\omega)^2} \sqrt{1+(0.1\omega)^2}}$$

$$\phi_0 = \tan^{-1}\frac{5.23\omega}{5.81 - \omega^2} - 90° - \tan^{-1}\omega - \tan^{-1}0.5\omega - \tan^{-1}0.1\omega, \text{ for } \omega < \sqrt{5.81}$$

$$= 180° + \tan^{-1}\frac{5.23\omega}{5.81 - \omega^2} - 90° - \tan^{-1}\omega - \tan^{-1}0.5\omega - \tan^{-1}0.1\omega, \text{ for } \omega > \sqrt{5.81}$$

At $\omega = \omega_1$, $A_0 = A_{01} = \dfrac{1.72 \times \sqrt{(5.81 - 4^2)^2 + (5.23 \times 4)^2}}{4 \times \sqrt{1+4^2} \times \sqrt{1+(0.5 \times 4)^2} \times \sqrt{1+(0.1 \times 4)^2}} = 1$

At $\omega = \omega_1$, $\phi_0 = \phi_{01} = 180° + \tan^{-1}\dfrac{5.23 \times 4}{5.81 - 4^2} - 90° - \tan^{-1}4 - \tan^{-1}(0.5 \times 4) - \tan^{-1}(0.1 \times 4)$

$$= -135°$$

Phase margin of compensated system, $\gamma_0 = 180° + \phi_{01} = 180° - 135° = 45°$

CONCLUSION

The phase margin of the compensated system, meets the given specification. Hence the design is acceptable.

RESULT

1. Transfer function of PID controller $\Bigg\}$ $G_c(s) = \left(1.8 + 0.344s + \dfrac{2}{s}\right) = \dfrac{0.344(s^2 + 5.23s + 5.81)}{s}$

2. Open loop transfer function of compensated system $\Bigg\}$ $G_0(s) = \dfrac{1.72(s^2 + 5.23s + 5.81)}{s(1+s)(1+0.5s)(1+0.1s)}$

DESIGN OF PD, PI AND PID CONTROLLER USING ROOT LOCUS TECHNIQUE

In this technique a controller can be introduced in cascade with open loop system so that, it can have a pair of dominant closed loop poles, to satisfy a specified time domain specifications ζ and ω_n.

The dominant pole, $s_d = -\zeta\omega_n \pm j\omega_n\sqrt{1-\zeta^2} = D\angle\beta$ .....(1.66)

where, $D = |s_d| = \sqrt{(\zeta\omega_n)^2 + \omega_n^2(1-\zeta^2)}$ and $\beta = \angle s_d = \tan^{-1}\dfrac{\omega_n\sqrt{1-\zeta^2}}{-\zeta\omega_n}$

$\zeta =$ Damping ratio and $\omega_n =$ Natural frequency of oscillation, rad/sec.

Let, $G(s) =$ Open loop transfer function of the system.

$G_c(s) =$ Transfer function of PID controller.

∴ Open loop transfer function of compensated system $= G_c(s)\,G(s)$ .....(1.67)

For the dominant pole, s_d to be a point on root locus, the magnitude condition says that,

$G_c(s_d)\,G(s_d) + 1 = 0$ .....(1.68)

$\therefore G_c(s_d)G(s_d) = -1$ .....(1.69)

Let $G(s_d) = A_d \angle \phi_d$. ....(1.70)

where $A_d = |G(s_d)|$ and $\phi_d = \angle G(s_d)$

Also, $-1 = 1 \angle 180°$

$\therefore G_c(s_d) = \dfrac{-1}{G(s_d)} = \dfrac{1\angle 180°}{A_d\angle\phi_d} = \dfrac{1}{A_d}\angle(180° - \phi_d)$ .....(1.71)

The transfer function of PID controller is given by,

$G_c(s) = K_p + \dfrac{K_i}{s} + K_d s = \dfrac{K_p s + K_i + K_d s^2}{s} = \dfrac{K_d s^2 + K_p s + K_i}{s}$ .....(1.72)

Replace s by s_d in equation (1.72).

$\therefore G_c(s_d) = \dfrac{K_d s_d^2 + K_p s_d + K_i}{s_d}$ .....(1.73)

From equation (1.66) we know that $s_d = D\angle\beta$.

$s_d = D\angle\beta = D\cos\beta + jD\sin\beta$ .....(1.74)

$s_d^2 = (D\angle\beta)^2 = D^2\angle 2\beta = D^2\cos 2\beta + jD^2\sin 2\beta$ .....(1.75)

On substituting for s_d and s_d^2 from equations (1.74) and (1.75) in equation (1.73) we get,

$$G_c(s_d) = \frac{K_d(D^2\cos 2\beta + jD^2\sin 2\beta) + K_p(D\cos\beta + jD\sin\beta) + K_i}{D\angle\beta} \qquad .....(1.76)$$

On equating equations (6.71) and (6.76) we get,

$$\frac{1}{A_d}\angle(180° - \phi_d) = \frac{K_d(D^2\cos 2\beta + jD^2\sin 2\beta) + K_p(D\cos\beta + jD\sin\beta) + K_i}{D\angle\beta}$$

$$\frac{1}{A_d}\angle(180° - \phi_d)D\angle\beta = K_dD^2\cos 2\beta + K_pD\cos\beta + K_i + j(K_dD^2\sin 2\beta + K_pD\sin\beta)$$

$$\frac{D}{A_d}(\angle(180° - (\phi_d - \beta))) = K_dD^2\cos 2\beta + K_pD\cos\beta + K_i + j(K_dD^2\sin 2\beta + K_pD\sin\beta)$$

$$\frac{D}{A_d}\cos(180° - (\phi_d - \beta)) + j\frac{D}{A_d}\sin(180 - (\phi_d - \beta)) = K_dD^2\cos 2\beta + K_pD\cos\beta + K_i$$
$$+ j(K_dD^2\sin 2\beta + K_pD\sin\beta)$$

$$-\frac{D}{A_d}\cos(\phi_d - \beta) + j\frac{D}{A_d}\sin(\phi_d - \beta) = K_dD^2\cos 2\beta + K_pD\cos\beta + K_i$$
$$+ j(K_dD^2\sin 2\beta + K_pD\sin\beta) \qquad .....(1.77)$$

On equating real parts of equation (1.77) we get,

$$-\frac{D}{A_d}\cos(\phi_d - \beta) = K_dD^2\cos 2\beta + K_pD\cos\beta + K_i \qquad .....(1.78)$$

On equating imaginary parts of equation (1.77) we get,

$$\frac{D}{A_d}\sin(\phi_d - \beta) = K_dD^2\sin 2\beta + K_pD\sin\beta$$

$$\therefore K_pD\sin\beta = \frac{D}{A_d}\sin(\phi_d - \beta) - K_dD^2 2\sin\beta\cos\beta \qquad \boxed{\because \sin 2\beta = 2\sin\beta\cos\beta}$$

$$\therefore K_p = \frac{\sin(\phi_d - \beta)}{A_d\sin\beta} - 2K_dD\cos\beta \qquad .....(1.79)$$

On substituting for K_p from equation (1.79) in equation (1.78) we get,

$$-\frac{D}{A_d}\cos(\phi_d - \beta) = K_dD^2\cos 2\beta + \left[\frac{\sin(\phi_d - \beta)}{A_d\sin\beta} - 2K_dD\cos\beta\right]D\cos\beta + K_i$$

$$-\frac{D}{A_d}\cos(\phi_d - \beta) = K_dD^2\cos 2\beta + \frac{D}{A_d}\frac{\cos\beta}{\sin\beta}\sin(\phi_d - \beta) - 2K_dD^2\cos^2\beta + K_i$$

$$\therefore K_dD^2(\cos 2\beta - 2\cos^2\beta) = -\frac{D}{A_d}\cos(\phi_d - \beta) - \frac{D}{A_d}\frac{\cos\beta\sin(\phi_d - \beta)}{\sin\beta} - K_i$$

Put, $\cos 2\beta = 1 - 2\sin^2\beta$

$$K_dD^2(1 - 2\sin^2\beta - 2\cos^2\beta) = -\frac{D}{A_d}\frac{\sin\beta\cos(\phi_d - \beta) + \cos\beta\sin(\phi_d - \beta)}{\sin\beta} - K_i$$

$$K_dD^2(1 - 2(\sin^2\beta - \cos^2\beta)) = -\frac{D}{A_d}\frac{\sin(\beta + \phi_d - \beta)}{\sin\beta} - K_i$$

$$\therefore -K_d D^2 = -\frac{D}{A_d}\frac{\sin\phi_d}{\sin\beta} - K_i$$

$$\boxed{\text{Note}: \sin^2\theta + \cos^2\theta = 1 \\ \sin(A+B) = \sin A \cos B + \cos A \sin B}$$

$$\therefore -K_d = \frac{\sin\phi_d}{DA_d\sin\beta} - \frac{K_i}{D^2} \qquad\qquad .....(1.80)$$

On substituting for K_d from equation (1.80) in equation (1.79) we get,

$$K_p = \frac{\sin(\phi_d-\beta)}{A_d\sin\beta} - 2\left[\frac{\sin\phi_d}{DA_d\sin\beta} + \frac{K_i}{D^2}\right]D\cos\beta$$

$$K_p = \frac{\sin\phi_d\cos\beta - \cos\phi_d\sin\beta}{A_d\sin\beta} - \frac{2D\cos\beta\sin\phi_d}{DA_d\sin\beta} - \frac{2K_i D\cos\beta}{D^2}$$

$$= \frac{\sin\phi_d\cos\beta - \cos\phi_d\sin\beta - 2\cos\beta\sin\phi_d}{A_d\sin\beta} - \frac{2K_i\cos\beta}{D}$$

$$= \frac{-(\sin\beta\cos\phi_d + \cos\beta\sin\phi_d)}{A_d\sin\beta} - \frac{2K_i\cos\beta}{D}$$

$$\boxed{\sin(A+B) = \sin A\,\cos B + \cos A\,\sin B}$$

$$\therefore K_p = \frac{-\sin(\beta+\phi_d)}{A_d\sin\beta} - \frac{2K_i\cos\beta}{D} \qquad\qquad .....(1.81)$$

The equations (1.80) and (1.81) are used to design a PID controller. Here the parameter K_i is first determined such that the compensated system satisfies the specified error requirement. Then the parameters K_d and K_p are calculated using equations (1.80) and (1.81) respectively.

For designing a PD controller put $K_i = 0$ in equations (1.80) and (1.81). Hence the design equations are,

$$K_d = \frac{\sin\phi_d}{DA_d\sin\beta} \quad\text{and}\quad K_p = \frac{-\sin(\beta+\phi_d)}{A_d\sin\beta}$$

For designing a PI controller, put $K_d = 0$ in equation (1.80). Hence the design equations are,

$$K_i = \frac{-D\sin\phi_d}{A_d\sin\beta} \quad\text{and}\quad K_p = \frac{-\sin(\beta+\phi_d)}{A_d\sin\beta} - \frac{2K_i\cos\beta}{D}$$

PROCEDURE FOR DESIGN OF PD/ PI/PID CONTROLLER

The following procedure can be followed to design a PD/PI/PID controller when the specifications are damping ratio, ζ and natural frequency of oscillation, ω_n.

Step-1 : Determine the dominant pole, s_d and calculate its magnitude and phase.

Dominant pole, $s_d = -\zeta\omega_n \pm j\omega_n\sqrt{1-\zeta^2}$

Let, $D = |s_d|$ and $\beta = \angle s_d$

By considering the dominant pole at $-\zeta\omega_n + j\omega_n\sqrt{1-\zeta^2}$ we get,

$$D = \sqrt{\zeta^2\omega_n^2 + \omega_n^2(1-\zeta^2)} \qquad\text{and}\qquad \beta = \tan^{-1}\frac{\sqrt{1-\zeta^2}}{-\zeta}$$

Step-2 : Determine the magnitude and phase of G(s) at s = s_d.

Let $A_d = |G(s)|$ at s = s_d (i.e., $A_d = |G(s_d)|$)

and, $\phi_d = \angle G(s)$ at s = s_d (i.e., $\phi_d = \angle G(s_d)$)

Step-3 : Determine the transfer function of PD/PI/PID controller.

a.   **PD controller**

Derivative constant, $K_d = \dfrac{\sin \phi_d}{DA_d \sin \beta}$

Proportional constant, $K_p = \dfrac{-\sin(\beta + \phi_d)}{A_d \sin \beta}$

Transfer function of PID controller $\Big\}$ $G_c(s) = K_p + K_d s = K_d \left(s + \dfrac{K_p}{K_d} \right)$

b. **PI controller**

Integral constant, $K_i = \dfrac{-D \sin \phi_d}{A_d \sin \beta}$

Proportional constant, $K_p = \dfrac{-\sin(\beta + \phi_d)}{A_d \sin \beta} - \dfrac{2K_i \cos \beta}{D}$

Transfer function of PI controller $\Big\}$ $G_c(s) = K_p + \dfrac{K_i}{s} = \dfrac{K_p s + K_i}{s} = \dfrac{K_p (s + (K_i/K_p))}{s}$

c. **PID controller**

Transfer function of PID controller $\Big\}$ $G_c(s) = K_p + \dfrac{K_i}{s} + K_d s = \dfrac{K_p s + K_i + K_d s^2}{s}$

$$= \dfrac{K_d \left(s^2 + \dfrac{K_p}{K_d} s + \dfrac{K_i}{K_d} \right)}{s}$$

Determine K_i from the specified error constant, such that the compensated system meets the error requirement.

For example, if the system is type-0 system and velocity error constant, K_v is specified then K_i is obtained by evaluating the following expression.

$$K_v = \underset{s \to 0}{Lt} \, sG_c(s) \, G(s)$$

Calculate the parameter K_d and K_p using the following equations

Proportional constant, $K_p = \dfrac{-\sin(\beta + \phi_d)}{A_d \sin \beta} - \dfrac{2K_i \cos \beta}{D}$

Derivative constant, $K_d = \dfrac{\sin \phi_d}{DA_d \sin \beta} + \dfrac{K_i}{D^2}$

Step-4 : Verify the design

Open loop transfer function of compensated system, $G_0(s) = G_c(s)G(s)$.

The design is accepted if the root locus of compensated system pass through the dominant pole, s_d. This can be verified from the magnitude condition, which states that the point $s = s_d$ will be a point on root locus if $1+G_0(s_d) = 0$, where $G_0(s_d)$ is the value of $G_0(s)$ at $s = s_d$. It can be shown that $G_0(s_d) = -1$

EXAMPLE 1.14

Consider a unity feedback system with open loop transfer function, $G(s) = 20/s\,(s+2)\,(s+4)$. Design a PD controller so that the closed loop has a damping ratio of 0.8 and natural frequency of oscillation as 2 rad/sec.

SOLUTION

Step-1 : To find dominant pole, s_d

Dominant pole, $s_d = -\zeta\omega_n \pm j\omega_n \sqrt{1-\zeta^2}$

Given that, $\zeta = 0.8$ and $\omega_n = 2$ rad/sec.

$\therefore\ s_d = -0.8 \times 2 \pm j2 \times \sqrt{1-0.8^2} = -1.6 \pm j1.2$

Let $D = |s_d|$ and $\beta = \angle s_d$

By considering dominant pole at $(-1.6 + j1.2)$ we get,

$s_d = -1.6+j1.2 = 2\angle 143°$ $\therefore\ D = 2$ and $\beta = 143°$

Step-2 : To find magnitude and phase of G(s) at $s = s_d$.

     | *Note : Use polar to rectangular conversion* |

Let $A_d = |G(s_d)|$ and $\phi_d = \angle G(s_d)$

Given that, $G(s) = 20/s(s+2)(s+4)$

$$G(s_d) = \frac{20}{s_d(s_d+2)(s_d+4)} = \frac{20}{(-1.6+j1.2)(-1.6+j1.2+2)(-1.6+j1.2+4)}$$

$$= \frac{20}{(-1.6+j1.2)(0.4+j1.2)(2.4+j1.2)} = \frac{20}{2\angle 143° \times 1.26\angle 71° \times 2.68\angle 26°}$$

$$= \frac{20}{2 \times 1.26 \times 2.68}\angle(-143° - 71° - 26°) = 2.96\angle -240°$$

$\therefore\ A_d = 2.96$ and $\phi_d = -240°$

Step-3 : Determine the transfer function of PD controller

Derivative constant, $K_d = \dfrac{\sin\phi_d}{DA_d\sin\beta} = \dfrac{\sin(-240°)}{2 \times 2.96 \times \sin(143°)} = 0.243$

Proportional constant, $K_p = \dfrac{-\sin(\beta+\phi_d)}{A_d\sin\beta} = \dfrac{-\sin(143°-240°)}{2.96 \times \sin(143°)} = 0.557$

Transfer function of PD controller $\}$ $G_c(s) = K_p + K_d s = (0.557 + 0.243s)$

$$= 0.243\left(\frac{0.557}{0.243} + s\right)$$

$$= 0.243(2.292 + s)$$

Step-4 : Verify the design

Open loop transfer function of compensated system $\}$ $G_0(s) = G_c(s)\,G(s)$

$$= 0.243(s + 2.292)\frac{20}{s(s+2)(s+4)}$$

$$= \frac{4.86(s + 2.292)}{s(s+2)(s+4)}$$

$G_c(s_d) = 0.243(s_d + 2.292)$ $= 0.243(-1.6 + j1.2 + 2.292)$

$\qquad = 0.243(0.692 + j1.2)$ $= 0.243 \times 1.385\angle 60° = 0.337 \angle 60°$

$\therefore\ G_0(s_d) = G_c(s_d)\,G(s_d)$ $= 0.337 \angle 60° \times 2.96\angle -240°$

$\qquad\qquad\qquad = 1 \angle -180° = -1$

Since $G_0(s_d) = -1$, the root locus will pass through s_d and so the design is accepted.

RESULT

Transfer function of PD controller $\}$ $G_c(s) = (0.557 + 0.243s) = 0.243(s + 2.292)$

Open loop transfer function of compensated system $\}$ $G_0(s) = \dfrac{4.86(s + 2.292)}{s(s+2)(s+4)}$

EXAMPLE 1.15

Consider a unity feedback system with open loop transfer function, $G(s) = 4/(s + 1)\,(s + 5)$. Design a PI controller so that the closed loop has a damping ratio of 0.9 and natural frequency of oscillation as 2.5 rad /sec.

SOLUTION

Step-1 : To find dominant pole, s_d

Dominant pole, $s_d = -\zeta\omega_n \pm j\omega_n\sqrt{1 - \zeta^2}$

Given that, $\zeta = 0.9$ and $\omega_n = 2.5$ rad/sec.

$\therefore\ s_d = -0.9 \times 2.5 \pm j2.5 \times \sqrt{1 - 0.9^2} = -2.25 \pm j1.09$

Let $D = |s_d|$ and $\beta = \angle s_d$

By considering the dominant pole at $(-2.25 + j1.09)$ we get,

$$s_d = -2.25 + j1.09 = 2.5\angle 154°$$

$\therefore\ D = 2.5$ and $\beta = 154°$ | *Note : Use polar to rectangular conversion* |

Step-2 : To find magnitude and phase of G(s) at $s = s_d$.

Let $A_d = |G(s_d)|$ and $\phi_d = \angle G(s_d)$

Given that, $G(s) = 4/(s+1)(s+5)$

$$G(s_d) = \frac{4}{(s_d + 1)(s_d + 5)} = \frac{4}{(-2.25 + j1.09 + 1)(-2.25 + j1.09 + 5)}$$

$$= \frac{4}{(-1.25 + j1.09)(2.75 + j1.09)} = \frac{4}{1.66\angle 139°\ 2.96\angle 21°}$$

$$= \frac{4}{1.66 \times 2.96}\angle(-139° - 21°) = 0.81\angle - 160°$$

$\therefore\ A_d = 0.81$ and $\phi_d = -160°$ | *Note : Use polar to rectangular conversion* |

Step-3 : Determine the transfer function of PI controller

Integral constant, $K_i = \dfrac{-D\sin\phi_d}{A_d\sin\beta} = \dfrac{-2.5\sin(-160°)}{0.81\sin(154°)} = 2.4$

Proportional constant, $K_p = \dfrac{-\sin(\beta + \phi_d)}{A_d\sin\beta} - \dfrac{2K_i\cos\beta}{D}$

$$= \frac{-\sin(154° - 160°)}{0.81\sin(154°)} - \frac{2 \times 2.4\cos(154°)}{2.5} = 2.02$$

Transfer function of PI controller $\Big\}$ $G_c(s) = K_p + \dfrac{K_i}{s} = 2.02 + \dfrac{2.4}{s} = \dfrac{2.02s + 2.4}{s}$

$$= \frac{2.02\left(s + \dfrac{2.4}{2.02}\right)}{s} = \frac{2.02(s + 1.19)}{s}$$

Step-4 : Verify the design

Open loop transfer function of compensated system $\Big\}$ $G_0(s) = G_c(s)\,G(s)$

$$= \frac{2.02(s + 1.19)}{s} \times \frac{4}{(s + 1)(s + 5)}$$

$$= \frac{8.08(s + 1.19)}{s(s + 1)(s + 5)}$$

$$G_c(s_d) = \frac{2.02(s_d + 1.19)}{s_d} = \frac{2.02(-2.25 + j1.09 + 1.19)}{-2.25 + j1.09}$$

$$= \frac{2.02(-1.06 + j1.09)}{-2.25 + j1.09} = \frac{2.02 \times 1.52\angle 134°}{2.5\angle 154°} = 1.228\angle - 20°$$

$\therefore G_0(s_d) = G_c(s_d) G(s_d) = 1.228 \angle -20° \times 0.81 \angle -160° = 1 \angle -180° = -1$

Since $G_0(s_d) = -1$, the root locus will pass through s_d and so the design is accepted.

RESULT

Transfer function of PI controller $\left.\right\}$ $G_c(s) = \left(2.02 + \dfrac{2.4}{s}\right) = \dfrac{2.02(s+1.19)}{s}$

Open loop transfer function of compensated system $\left.\right\}$ $G_0(s) = \dfrac{8.80(s+1.19)}{s(s+1)(s+5)}$

EXAMPLE 1.16

Consider a unity feedback system with open loop transfer function, $G(s) = 75/(s+1)\ (s+3)\ (s+8)$ Design a PID controller to satisfy the following specifications. (a) The steady state error for unit ramp input should be less than 0.08. (b) Damping ratio $= 0.8$. (c) Natural frequency of oscillation $= 2.5$ rad/sec.

SOLUTION

Step-1 : To find the dominant pole, s_d

Dominant pole, $s_d = -\zeta\omega_n \pm j\omega_n \sqrt{1-\zeta^2}$

Given that, $\zeta = 0.8$ and $\omega_n = 2.5$ rad/sec.

$\therefore s_d = -0.8 \times 2.5 \pm j2.5 \times \sqrt{1-0.8^2} = -2 \pm j1.5$

Let $D = |s_d|$ and $\beta = \angle s_d$

By considering the dominant pole at $(-2 + j6.5)$ we get,

$s_d = -2 + j6.5 = 2.5 \angle 143°$

$\therefore D = 2.5$ and $\beta = 143°$

| Note : Use polar to rectangular conversion |

Step-2 : To find magnitude and phase of $G(s)$ at $s = s_d$.

Let $A_d = |G(s_d)|$ and $\phi_d = \angle G(s_d)$

Given that, $G(s) = 75/(s+1)(s+3)(s+8)$

$$G(s_d) = \dfrac{75}{(s_d+1)(s_d+3)(s_d+8)} = \dfrac{75}{(-2+j1.5+1)(-2+j1.5+3)(-2+j1.5+8)}$$

$$= \dfrac{75}{(-1+j1.5)(1+j1.5)(6+j1.5)} = \dfrac{75}{1.8\angle 124° \ 1.8\angle 56° \ 6.18\angle 14°}$$

$$= \dfrac{75}{1.8 \times 1.8 \times 6.18} \angle (-124° - 56° - 14°) = 3.75 \angle -194°$$

$\therefore A_d = 3.75$ and $\phi_d = -194°$

| Note : Use polar to rectangular conversion |

Step-3 : Determine the transfer function of PID controller

Given that, $e_{ss} \le 0.08$ for unit ramp input.

$\therefore$ Velocity error constant, $K_v \ge \dfrac{1}{e_{ss}} = \dfrac{1}{0.08} = 12.5$

Transfer function of PID controller $\left\} \quad G_c(s) = \left(K_p + \dfrac{K_i}{s} + K_d s \right) = \left(\dfrac{K_d s^2 + K_p s + K_i}{s} \right)$

Open loop transfer function of compensated system $\left\} \quad G_0(s) = G_c(s) G(s) \right.$

$$= \dfrac{(K_d s^2 + K_p s + K_i)}{s} \times \dfrac{75}{(s+1)(s+3)(s+8)}$$

$$= \dfrac{75(K_d s^2 + K_p s + K_i)}{s(s+1)(s+3)(s+8)}$$

Velocity error constant of compensated system $\left\} = \underset{s \to 0}{Lt} \; sG_0(s) \right.$

$$= \underset{s \to 0}{Lt} \; s \dfrac{75(K_d s^2 + K_p s + K_i)}{s(s+1)(s+3)(s+8)}$$

$$= \dfrac{75 K_i}{3 \times 8} = 3.125 \, K_i$$

But the velocity error constant of compensated system should be greater than or equal to 12.5.

$$\therefore \; 3.125 K_i = 12.5 \quad (or) \quad K_i = \dfrac{12.5}{3.125} = 4$$

Derivative constant, $K_d = \dfrac{\sin \phi_d}{D \, A_d \sin \beta} + \dfrac{K_i}{D^2}$

$$= \dfrac{\sin(-194°)}{2.5 \times 3.75 \times \sin(143°)} + \dfrac{4}{2.5^2} = 0.68$$

Proportional constant, $K_p = \dfrac{-\sin(\beta + \phi_d)}{A_d \sin \beta} - \dfrac{2K_i \cos \beta}{D}$

$$= \dfrac{-\sin(143° - 194°)}{3.75 \times \sin(143°)} - \dfrac{2 \times 4 \times \cos(143°)}{2.5} = 2.9$$

Transfer function of PID controller $\left\} \quad G_c(s) = K_p + \dfrac{K_i}{s} + K_d s = 2.9 + \dfrac{4}{s} + 0.68s = \dfrac{0.68 s^2 + 2.9 s + 4}{s} \right.$

$$= \dfrac{0.68 \left(s^2 + \dfrac{2.9}{0.68} s + \dfrac{4}{0.68} \right)}{s} = \dfrac{0.68 \, (s^2 + 4.26s + 5.88)}{s}$$

Step-4 : Verify the design

Open loop transfer function of compensated system $\left\} \quad G_0(s) = G_c(s) G(s) = \dfrac{0.68 \, (s^2 + 4.26s + 5.88)}{s} \times \dfrac{75}{(s+1)(s+3)(s+8)} \right.$

$$= \dfrac{51 \, (s^2 + 4.26s + 5.88)}{s(s+1)(s+3)(s+8)}$$

$$G_c(s_d) = \frac{0.68\,(s_d^2 + 4.26 s_d + 5.88)}{s_d}$$

$$s_d = 2.5 \angle 143° = -2 + j1.5$$

$$s_d^2 = (2.5\angle 143°) = 2.5^2 \angle 2 \times 143° = 6.25\angle 286° = 1.72 - j6$$

$$G_c(s_d) = \frac{0.68\,(1.72 - j6 + 4.26(-2+j1.5) + 5.88)}{-2+j1.5} = \frac{0.68\,(-0.92 + j0.39)}{-2+j1.5}$$

$$= \frac{0.68 \times 1\angle 157°}{2.5\angle 143°} = 0.272\angle 14°$$

$$\therefore\ G_0(s_d) = G_c(s_d)\,G(s_d) = 0.272\angle 14° \times 3.75\angle -194°$$

$$= 1\angle -180° = -1$$

Since $G_0(s_d) = -1$, the root locus will pass through s_d and so the design is accepted.

RESULT

$$\left.\begin{array}{l}\text{Transfer function}\\\text{of PID controller}\end{array}\right\}G_c(s) = \left(2.9 + \frac{4}{s} + 0.68s\right) = \frac{0.68\,(s^2 + 4.26s + 5.88)}{s}$$

$$\left.\begin{array}{l}\text{Open loop transfer function}\\\text{of compensated system}\end{array}\right\}G_0(s) = \frac{51\,(s^2 + 4.26s + 5.88)}{s(s+1)(s+3)(s+8)}$$

1.6   FEEDBACK COMPENSATION

In feedback compensation a compensating device is placed in an internal feedback path around one or more components of the forward path. It is also called minor loop feedback compensation.

The decision to use a feedback compensation is sometimes a matter of convenience, sometimes a matter of necessity and for some problems it can be shown that feedback scheme will offer better performance. For example, use of velocity feedback for damping is very effective and quite popular. The feedback compensation scheme may be necessary in one of the following situations.

1.   *Feedback compensation is provided when derivative action is needed on controlled variable in order to avoid the derivative kick on set point changes.*

2.   *In non-electrical systems, when suitable cascade device is not available, feedback compensation is employed.*

3.   *In systems subjected to frequent load disturbances, feedback compensation may be preferred as it provides greater stiffness against load disturbances.*

Besides all the factors discussed above, the available components and designer's experience and preferences influence the choice between a cascade and a feedback compensation scheme. A commonly used arrangement for feedback compensation is shown in fig 1.23.

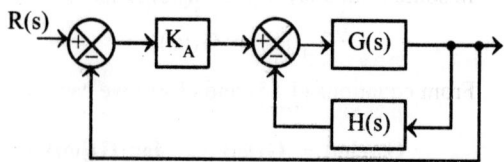

Fig 1.23 : Feedback compensation scheme.

The most popular schemes employ rate feedback or tachometer feedback in which, $H(s) = K_t s$. The disadvantage in the rate feedback is that the system velocity error constant, K_v is reduced. This undesirable effect can be eliminated by reducing the feedback signal in the region of low frequencies. Hence a high pass filter of the type shown in fig 1.24 can be connected in cascade with rate device as shown in fig 1.25.

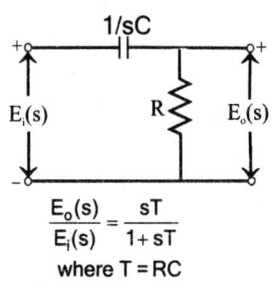

1/sC
+
+
E_i(s)
R
E_o(s)
-
E_o(s)/E_i(s) = sT/(1+sT)
where T = RC

***Fig 1.24 :** High pass filter.*

R(s)
C(s)
K_A
G(s)
sT/(1+sT)
K_t s
H(s)

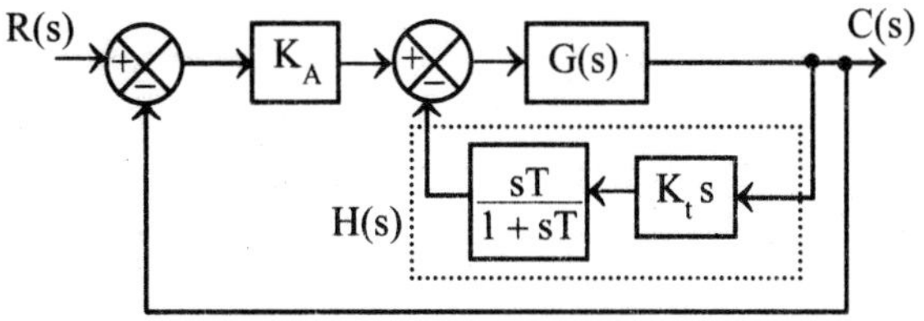

***Fig 1.25 :** Feedback compensation with high-pass filter in the feedback path.*

DESIGN OF FEEDBACK COMPENSATION SCHEME USING BODE PLOT

The minor loop feedback in fig 1.23 can be eliminated and the forward path transfer function, G(s) can be replaced by an equivalent forward path transfer function, $G_{eq}(s)$ as shown in fig 1.26.

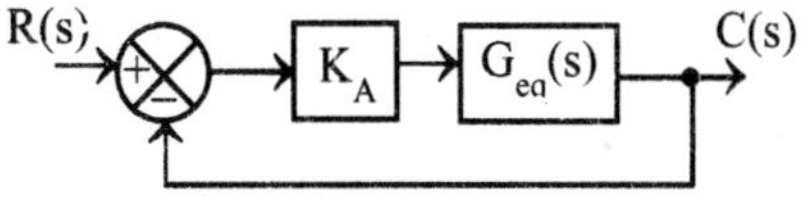

R(s)
C(s)
K_A
G_eq(s)

Fig 1.26

$$\text{Equivalent transfer function, } G_{eq}(s) = \frac{G(s)}{1+G(s)H(s)} \quad .....(1.82)$$

$$\left.\begin{array}{l}\text{Open loop transfer function} \\ \text{of compensated system}\end{array}\right\} G_0(s) = K_A G_{eq}(s) \quad .....(1.83)$$

The transfer function, $G_{eq}(s)$ and the parameter K_t and K_A can be designed in frequency domain using bode plot.

$$\text{Put } s = j\omega \text{ in Geq(s)}, \quad \therefore\; G_{eq}(j\omega) = \frac{G(j\omega)}{1+G(j\omega)H(j\omega)} \quad .....(1.84)$$

In general there will be a range of frequencies for which $|G(j\omega)|$ is large and some other range for which it is small. This is also true for $H(j\omega)$ and the ranges of frequencies will probably overlap but will not be identical.

In some frequency range $|G(j\omega)\,H(j\omega)| >> 1$ and in this range the equation(1.84) can be written as

$$G_{eq}(j\omega) \approx \frac{G(j\omega)}{G(j\omega)H(j\omega)} = \frac{1}{H(j\omega)} \quad .....(1.85)$$

In some frequency range $|G(j\omega)\,H(j\omega)| << 1$ and in this range the equation(1.84) can be written as,

$$G_{eq}(j\omega) \approx G(j\omega) \quad .....(1.86)$$

From equations (1.85) and (1.86) we can conclude that,

$$\begin{aligned} \mathbf{G_{eq}(j\omega)} &= \mathbf{G(j\omega)} && \textbf{for } |\mathbf{G(j\omega)H(j\omega)}| \le 1 \\ &= \frac{\mathbf{1}}{\mathbf{H(j\omega)}} && \textbf{for } |\mathbf{G(j\omega)H(j\omega)}| \ge 1 \end{aligned} \quad .....(1.87)$$

The approximate bode magnitude plot of G(jω) and 1/H(jω) are drawn and by graphically interpreting equation(1.87), the $G_{eq}(j\omega)$ can be determined.

The db magnitude of G(jω)H(jω) can be expressed as,

$$20\log |G(j\omega)H(j\omega)| = 20\log |G(j\omega)| - 20 \log |1/H(j\omega)| \qquad .....(1.88)$$

From equation (1.88) and bode magnitude plots of G(jω) &1/H(jω) we can obtain following conclusions.

1.   *If |G(jω)| curve is above |1/H(jω)| curve,*

      then |G(jω)H(jω)| > 1 and $G_{eq}(j\omega) \approx |1/H(j\omega)|$

2.   *If |G(jω)| curve is below |1/H(jω)| curve,*

      then |G(jω)H(jω)| < 1 and $G_{eq}(j\omega) \approx |G(j\omega)|$

3.   *At the intersection of |G(jω)| curve and |1/H(jω)| curve,|G(jω)H(jω)| = 1.*

PROCEDURE FOR DESIGN USING BODE PLOT

Step-1 : Draw the magnitude plot of G(jω) on semilog graph sheet.

Step-2 : Draw the magnitude plot of 1/H(jω) on the same graph sheet and using the same scales. The feedback compensator consists of a rate device and a high pass filter in cascade.

$$\left. \begin{array}{l} \text{Transfer function of} \\ \text{feedback compensator} \end{array} \right\} H(s) = K_t s \frac{sT}{1+sT} = \frac{K_t Ts^2}{1+sT}$$

$$\text{Put } s = j\omega, \quad \therefore \ H(j\omega) = K_i T \frac{(j\omega)^2}{1+j\omega T}$$

$$\text{Now,} \qquad \frac{1}{H(j\omega)} = \frac{1}{K_t T} \frac{1+j\omega T}{(j\omega)^2}$$

By taking an initial value of $K_t T = 1$ and assuming suitable value for T, draw the bode plot of 1/H(jω). (Typical values of T is in the range of 1 to 2 sec).

Step-3 : Determine the desired gain crossover frequency of feedback compensated system.

Let, ω_{gc} = Desired gain crossover frequency.

       γ_d = Desired (or specified) phase margin.

Choose a new value of phase margin, γ_n such that, $\gamma_n = \gamma_d + \epsilon$

Let initial choice of ϵ be 10° to 25°

Now, $\gamma_n = 180° + \phi_{hgc}$

where, $\phi_{hgc} = \angle(1/H(j\omega))$ at $\omega = \omega_{gc}$

From the transfer function of 1/H(jω) we get,

$$\phi_{hgc} = \tan^{-1}\omega_{gc} T - 180°$$

$$\therefore \ \gamma_n = 180° + \tan^{-1}\omega_{gc} T - 180° \quad \text{(or)} \quad \omega_{gc} T = \tan \gamma_n \qquad \therefore \ \omega_{gc} = \frac{\tan \gamma_n}{T}$$

Step-4 : Determine the parameter K_t

Now shift $|1/H(j\omega)|$ curve vertically up or down such that it crosses $|G(j\omega)|$ curve at a frequency atleast 4 times greater than ω_{gc}. Let the $|1/H(j\omega)|$ curve be shifted by $\pm x$ db ("+" for upward shift and "−" for downward shift).

Now, $20\log\dfrac{1}{K_t T} = \pm x$

$\log\dfrac{1}{K_t T} = \pm\dfrac{x}{20}$

$\therefore \dfrac{1}{K_t T} = 10^{\pm x/20}$ (or) $K_t = \dfrac{1}{T\times 10^{\pm x/20}}$

Step-5 : Determine nthe transfer function, $G_{eq}(j\omega)$ from the bode plots of $G(j\omega)$ and $1/H(j\omega)$.

A typical magnitude plot of $G(j\omega)$ and $1/H(j\omega)$ with $K_tT = 1$ are shown in fig 1.27. The magnitude plot of $1/H(j\omega)$ is shifted to cross $|G(j\omega)|$ curve as mentioned in step-4. Let the shifted plot of $1/H(j\omega)$ intersects, the $|G(j\omega)|$ curve as shown in fig 1.28 and the frequencies corresponding to crossing points be ω_{c1} and ω_{c3}.

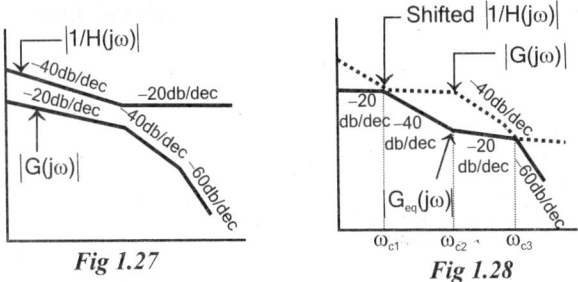

Fig 1.27   *Fig 1.28*

In the frequency range, ω_{c3} to ∞, the magnitude plot of $G_{eq}(j\omega)$ is given by portion of $|G(j\omega)|$ curve in the range, ω_{c3} to ∞.

In the frequency range, 0 to ω_{c1}, the magnitude plot of $G_{eq}(j\omega)$ is given by portion of $|G(j\omega)|$ curve in the range, 0 to ω_{c1}.

In the frequency range, ω_{c1} to ω_{c3}, the magnitude plot of $G_{eq}(j\omega)$ is given by portion of shifted $|1/H(j\omega)|$ curve in the range, ω_{c1} to ω_{c3}.

The bode magnitude plot of $G_{eq}(j\omega)$ is shown as a bold curve in fig 1.28.

From the bode magnitude plot of $G_{eq}(j\omega)$ the transfer function $G_{eq}(j\omega)$ can be determined. The example plot shown in fig 1.28 can be interpreted as follows,

1. *The slope in the starting portion of the curve is −20 db/dec and it is due to an integral factor $K/j\omega$ where K is given by G(jω) when ω = 0.*

2. *At ω = ω_{c1}, the slope changes from −20 db/dec to −40 db/dec. This is due to a first order factor $1/(1+j\omega T_1)$, where $T_1 = 1/\omega_{c1}$.*

3. *At ω = ω_{c2}, the slope changes from −40 db/dec to −20 db/dec. This is due to a first order factor $(1+j\omega T_2)$, where $T_2 = 1/\omega_{c2}$.*

4. At $\omega = \omega_{c3}$, the slope changes from -20 db/dec to -60 db/dec. This is due to a first order factor $1/(1+j\omega T_3)^2$, where $T_3 = 1/\omega_{c3}$. Hence transfer function, $G_{eq}(j\omega)$ can be written as,

$$G_{eq}(jw) = \frac{K(1+j\omega T_2)}{j\omega(1+j\omega T_1)(1+j\omega T_3)^2}$$

Put, $j\omega = s$, $\therefore$ $G_{eq}(s) = \dfrac{K(1+sT_2)}{s(1+sT_1)(1+sT_3)^2}$

Step-6 : Determine the parameter, K_A.

From the bode magnitude plot of $G_{eq}(j\omega)$ find the vertical shift to make ω_{gc} as the gain crossover frequency of $G_{eq}(j\omega)$, (i.e., at $\omega = \omega_{gc}$, $|G_{eq}(j\omega)|$ in db = 0). Let the vertical shift be $\pm$ x db (+ for upward shift and – for downward shift).

Now, $20\log K_A = \pm x$

$$\log K_A = \frac{\pm x}{20} \qquad \text{(or)} \qquad K_A = 10^{\pm x/20}$$

Step-7 : Verify the design.

Determine the velocity error constant, K_v of the feedback compensated system.

where, $K_v = \underset{s \to 0}{\mathrm{Lt}}\ s K_A G_{eq}(s)$

If the velocity error constant is satisfactory then the design is acceptable. Otherwise repeat the design with another suitable value of T or $\in$.

DESIGN OF FEEDBACK COMPENSATION SCHEME USING ROOT LOCUS TECHNIQUE

In time domain the feedback compensation scheme can be designed using root locus technique. The most popular feedback scheme used in this design technique is shown in fig 1.29. It employs a rate device or a tachometer feedback in a minor loop around G(s).

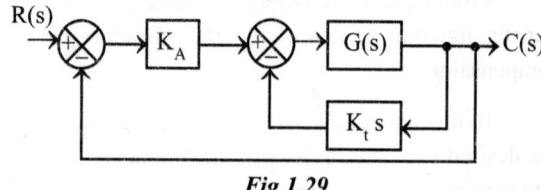

Fig 1.29

The minor loop around G(s) can be eliminated and an equivalent forward transfer function can be obtained as shown below.

$$G_{eq}(s) = \frac{G(s)}{1+G(s)K_t s} \qquad\qquad .....(1.89)$$

Let, $G(s) = \dfrac{K}{s(s+1/T_1)(s+1/T_2)}$ $\qquad\qquad .....(1.90)$

From equation(1.89) and (1.90) we can write,

$$G_{eq}(s) = \frac{\dfrac{K}{s(s+1/T_1)(s+1/T_2)}}{1+\dfrac{K}{s(s+1/T_1)(s+1/T_2)}K_t s}$$

$$= \frac{K}{s(s+1/T_1)(s+1/T_2)+KK_t s} \qquad\qquad .....(1.91)$$

Closed loop transfer function, $\dfrac{C(s)}{R(s)} = \dfrac{K_A G_{eq}(s)}{1 + K_A G_{eq}(s)}$ $\qquad$ .....(1.92)

On substituting for $G_{eq}(s)$ from equation(1.91) in equation(1.92) we get,

$$\frac{C(s)}{R(s)} = \frac{K_A \dfrac{K}{s(s + 1/T_1)(s + 1/T_2) + KK_t s}}{1 + K_A \dfrac{K}{s(s + 1/T_1)(s + 1/T_2)KK_t s}} = \frac{K_A K}{s(s + 1/T_1)(s + 1/T_2) + KK_t s + K_A K} \qquad .....(1.93)$$

The denominator polynomial of closed loop transfer function, C(s)/R(s) is the characteristic equation. Hence from equation(1.93) the characteristic equation is,

$$s(s + 1/T_1)(s + 1/T_2) + KK_t s + K_A K = 0 \qquad .....(1.94)$$

The equation(1.94) can be rearranged in the form of $1 + G(s)H(s) = 0$ as shown below.

On dividing equation(6.94) by $s(s + 1/T_1)(s + 1/T_2)$, we get,

$$1 + \frac{KK_t(s + K_A/K_t)}{s(s + 1/T_1)(s + 1/T_2)} = 0 \qquad .....(1.95)$$

From equation(1.95) we get a new loop transfer function $[G(s)H(s)]_{new}$,

where, $[G(s)H(s)]_{new} = \dfrac{KK_t(s + K_A/K_t)}{s(s + 1/T_1)(s + 1/T_2)}$ $\qquad$ .....(1.96)

From equation(1.96) we can conclude that the effect of feedback is to introduce a zero in the loop transfer function. Hence the design procedure includes a search for a suitable location for zero of the compensator.

In this design, the root locus is made to pass through a pair of dominant poles so that the system has desired transient response. The dominant poles are determined from the time domain specifications. The zero as defined by equation(1.96) is located on the real axis such that the angle condition is satisfied at the dominant pole, s_d.

PROCEDURE FOR DESIGN USING ROOT LOCUS TECHNIQUE

Step-1 : Determine the dominant pole, s_d.

Dominanat pole, $s_d = -\zeta\omega_n + j\omega_n\sqrt{1 - \zeta^2}$

where, ζ = Damping ratio

ω_n = Natural frequency of oscillation, rad/sec

Step-2 : Draw the pole-zero plot of G(s)

In an ordinary graph sheet, take suitable scales and mark the poles by "x" and zeros by "o". Also mark the dominant pole, s_d. Let point P be the dominant pole.

Step-3 : Determine the new loop transfer function.

Choose a compensation scheme as shown in fig 1.29. Determine the closed loop transfer function C(s)/R(s). The new loop transfer function is obtained by reconstructing the characteristic equation of C(s)/R(s) in the form $1 + [G(s)H(s)]_{new}$. In the new loop transfer function, there will be an additional term, $(s + K_A / K_t)$ which contributes a zero, $z_c = -K_A / K_t$.

Step-4 : Locate the zero introduced by the feedback compensator

Let, ϕ = Angle contributed by zero of compensator to make point, P as a point on root locus.

Now, $\phi = \begin{pmatrix} \text{Sum of angles contributed} \\ \text{by poles of G(s)} \end{pmatrix} - \begin{pmatrix} \text{Sum of angles contributed} \\ \text{by zeros of G(s)} \end{pmatrix} \pm n180°$

where n is an odd integer.

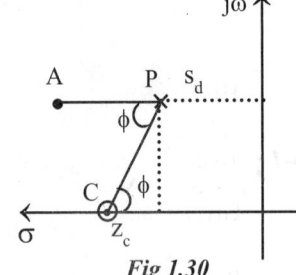

With line AP as reference draw a line PC such that $\angle APC = \phi$, as shown in fig 1.30. This line will intersect the real axis at point C. The value corresponding to point C is the zero of the compensator, z_c.

Step-5 : Determine the parameter K_A and K_t.

From $[G(s)H(s)]_{new}$ we get,

$z_c = -K_A / K_t$

Choose a suitable value for K_t and calculate K_A.

Fig 1.30

Step-6 : Determine the gain K of G(s)

Let, $[G(s)H(s)]_{new} = \dfrac{KK_t(s + K_A/K_t)}{s(s + 1/T_1)(s + 1/T_2)}$ $\boxed{\textit{Note : Vector lengths should be measured to scale.}}$

The value of $K\, K_t$ can be obtained from the magnitude condition.

$\therefore K\, K_t = \dfrac{\text{Product of vector lengths from poles to } s = s_d}{\text{Product of vector lengths from zeros to } s = s_d}$

Step-7 : Verify the design.

Check the velocity error constant, K_v of the new loop transfer function.

$K_v = \underset{s \to 0}{\text{Lt}}\ s[G(s)H(s)]_{new}$

If the velocity error constant is satisfactory then the design is accepted otherwise this procedure calls for modifications in time domain specifications.

EXAMPLE 1.17

Design a feedback compensation scheme for a unity feedback system with open loop transfer function, G(s) = 3/s(s+1) to satisfy the following specifications. (1) phase margin of system should be atleast 45°. (2) Velocity error constant, $K_v \ge 20$.

SOLUTION

The feedback compensation scheme is shown in fig 1.17.1. The feedback transfer function, H(s) consists of a rate device in cascade with a high pass filter. The minor-loop feedback can be eliminated and replaced by an equivalent transfer function, $G_{eq}(s)$ as shown in fig 1.17.2.

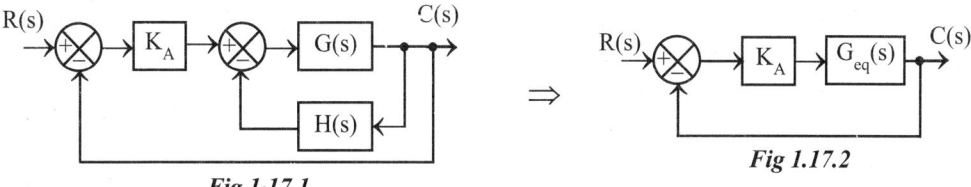

Fig 1.17.1   *Fig 1.17.2*

$$H(s) = K_t s \frac{sT}{1+sT} = \frac{K_t Ts^2}{(1+sT)} \quad \text{and} \quad G_{eq}(s) = \frac{G(s)}{1+G(s)H(s)}$$

Step-1 : To draw the magnitude plot of $G(j\omega)$.

     Given that, $G(s) = \dfrac{3}{s(s+1)}$

     Put $s = j\omega$, $\therefore\ G(j\omega) = \dfrac{3}{j\omega(1+j\omega)}$

     The corner frequency is, $\omega_c = 1$ rad/sec.

     The various terms of $G(j\omega)$ are listed in table-1. Also the table shows the slope contributed by each term and the change in slope at the corner frequency.

TABLE-1

Term	Corner frequency rad/sec	Slope db/dec	Change in slope db/dec
$\dfrac{3}{j\omega}$	—	-20	—
$\dfrac{1}{1+j\omega}$	$\omega_c = 1$	-20	$-20 -20 = -40$

Choose a low frequency, ω_l such that $\omega_l < \omega_{cl}$ and choose a high frequency, ω_h such that $\omega_h > \omega_c$.

Let $\omega_l = 0.5$ rad/sec and $\omega_h = 10$ rad/sec.

Let $A = |G(j\omega)|$ in db

At $\omega = \omega_l$, $A = 20 \log \dfrac{3}{\omega} = 20 \log \dfrac{3}{0.5} = 15.5 \approx 16$ db

At $\omega = \omega_c$, $A = 20 \log \dfrac{3}{\omega} = 20 \log \dfrac{3}{1} = 9.5$ db ≈ 10 db

At $\omega = \omega_h$, $A = \left[\text{slope from } \omega_c \text{ to } \omega_h \times \log \dfrac{\omega_h}{\omega_c}\right] + A \text{ at } (\omega = \omega_c) = -40 \times \log \dfrac{10}{1} + 10 = -30$ db

Let the points a, b and c be the points corresponding to frequencies ω_l, ω_c and ω_h respectively on the magnitude plot. In a semilog graph sheet choose appropriate scales and fix the points a, b and c. Join the points by straight lines and mark the slope on the respective region. The magnitude plot is shown in fig 1.17.3.

Step-2 : Draw the magnitude plot of $1/H(j\omega)$.

$$\left.\begin{array}{l}\text{Transfer function of}\\\text{feedback compensator}\end{array}\right\} H(s) = \frac{K_t Ts^2}{1+sT}$$

Put $s = j\omega$, $\therefore H(j\omega) = \dfrac{K_t T(j\omega)^2}{(1+j\omega T)}$

Now, $\dfrac{1}{H(j\omega)} = \dfrac{1}{K_t T} \cdot \dfrac{1+j\omega T}{(j\omega)^2}$

Let $K_t T = 1$, $\therefore \dfrac{1}{H(j\omega)} = \dfrac{1+j\omega T}{(j\omega)^2}$

Let $T = 1.25 \text{ sec}$, $\therefore \dfrac{1}{H(j\omega)} = \dfrac{1+j1.25\omega}{(j\omega)^2}$

The corner frequency, $\omega_{ch} = \dfrac{1}{1.25} = 0.8 \text{ rad/sec}$

The various terms of $1/H(j\omega)$ are listed in table-2.

TABLE-2

Term	Corner frequency rad/sec	Slope db/dec	Change in slope db/dec
$\dfrac{1}{(j\omega)^2}$	–	–40	–
$1 + j1.25\omega$	$\omega_{ch} = \dfrac{1}{1.25} = 0.8$	+20	$-40 + 20 = -20$

Choose a low frequency, ω_{lh} such that $\omega_{lh} < \omega_{ch}$ and choose a high frequency, ω_{hh} such that $\omega_{hh} > \omega_{ch}$.

Let $\omega_{lh} = 0.1 \text{ rad/sec}$ and $\omega_{hh} = 2.0 \text{ rad/sec}$.

Let $H = |1/H(j\omega)|$ in db

At $\omega = \omega_{lh}$, $H = 20 \log \dfrac{1}{\omega^2} = 20 \log \dfrac{1}{0.1^2} = 40 \text{ db}$

At $\omega = \omega_{ch}$, $H = 20 \log \dfrac{1}{\omega^2} = 20 \log \dfrac{1}{0.8^2} = 4 \text{ db}$

At $\omega = \omega_{hh}$, $H = \left[\text{slope from } \omega_{ch} \text{ to } \omega_{ch} \times \log \dfrac{\omega_{hh}}{\omega_{ch}} \right] + H \text{ at } (\omega = \omega_{ch})$

$\qquad\qquad = -20 \times \log \dfrac{2}{0.8} + 4 = -4 \text{ db}$

Let the points d, e and f be the points corresponding to frequencies ω_{lh}, ω_{ch} and ω_{hh} respectively on the magnitude plot. In the same semilog graph sheet and using the same scales draw the magnitude plot as shown in fig 1.17.3.

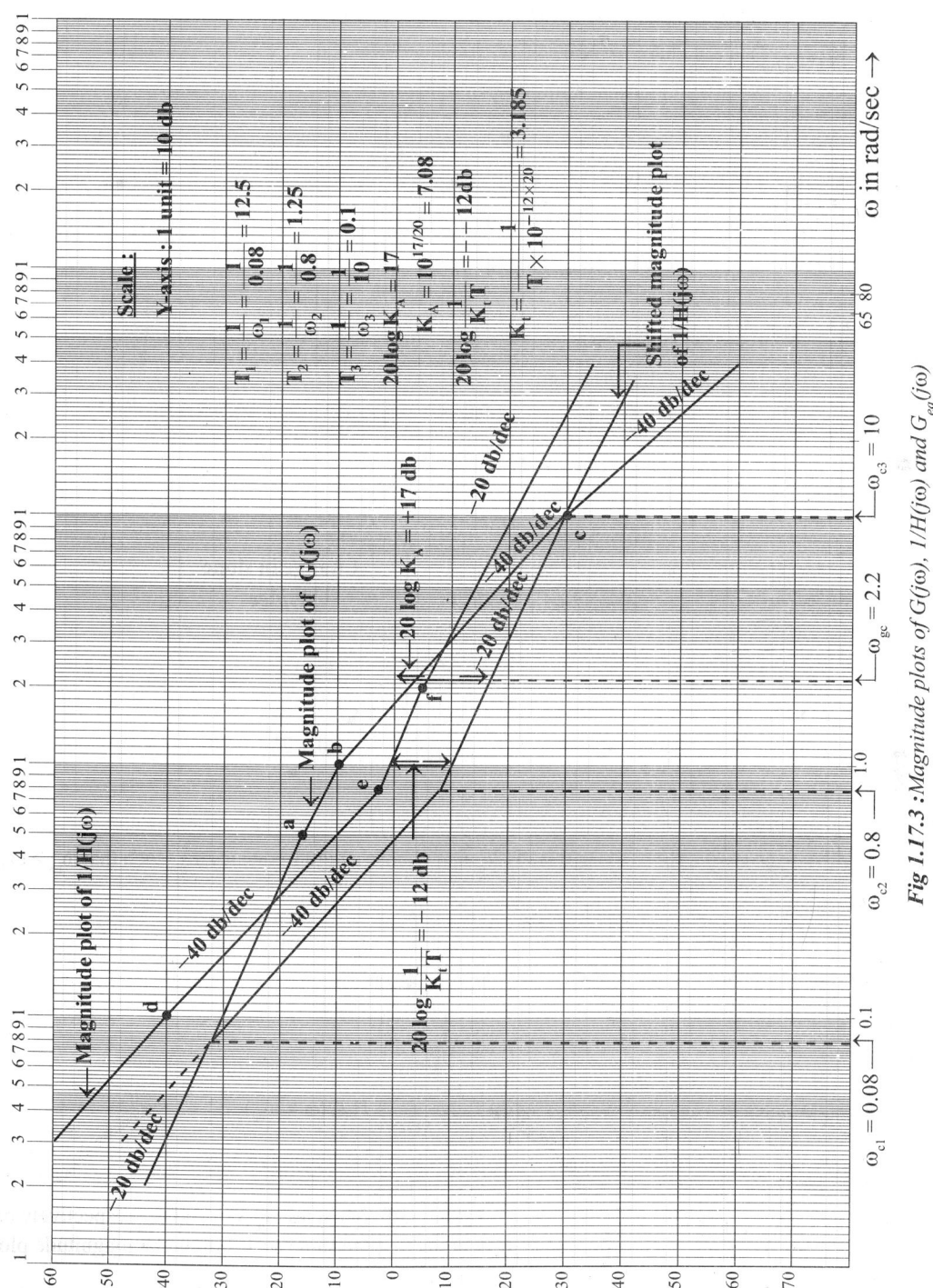

Fig 1.17.3 : Magnitude plots of $G(j\omega)$, $1/H(j\omega)$ and $G_{eq}(j\omega)$

Step-3 : Determine the gain cross-over frequency.

Let ω_{gc} = Desired gain crossover frequency

γ_d = Desired phase margin

Choose a new value of phase margin, γ_n such that, $\gamma_n = \gamma_d + \epsilon$.

Let, $\epsilon = 25°$, $\therefore \gamma_n = \gamma_d + \epsilon = 45° + 25° = 70°$.

Let, $\phi_{hgc} = \angle(1/H(j\omega))$ at $\omega = \omega_{gc}$

From the transfer function of $1/H(j\omega)$ we get,

$$\phi_{hgc} = \tan^{-1}\omega_{gc}T - 180° = \tan^{-1}1.25\omega_{gc} - 180°$$

Here, $\gamma_n = 180° + \phi_{hgc}$

$$\therefore 70° = 180° + \tan^{-1}1.25\omega_{gc} - 180°$$

$$1.25\omega_{gc} = \tan 70° \quad \text{(or)} \quad \omega_{gc} = \frac{\tan 70°}{1.25} = 2.2 \text{ rad/sec}$$

Step-4 : Determine the parameter, K_t

The $|1/H(j\omega)|$ curve is shifted down to cross $|G(j\omega)|$ curve at $\omega = 10$ rad/sec. (here $4\,\omega_{gc} = 4 \times 2.2 = 8.8$, hence frequency corresponding to crossing point is chosen to be 10 rad/sec).

From the bode plots we found that $|1/H(j\omega)|$ curve is shifted by 12 db downwards.

$$\therefore 20 \log \frac{1}{K_t T} = -12 \text{ db}$$

$$\log \frac{1}{K_t T} = \frac{-12}{20}$$

$$\frac{1}{K_t T} = 10^{-12/20} = 0.2512$$

$$\therefore K_t = \frac{1}{T \times 0.2512} = \frac{1}{1.25 \times 0.2512} = 3.185$$

Step-5 : Determine the new transfer function, $G_{eq}(j\omega)$

From the bode plots it is observed that the shifted $|1/H(j\omega)|$ curve crosses the $|G(j\omega)|$ curve at $\omega = \omega_{c1} = 0.08$ rad/sec and $\omega = \omega_{c3} = 10$ rad/sec.

In the frequency range, $0 < \omega < \omega_{c1}$, $|G_{eq}(j\omega)| = |G(j\omega)|$

In the frequency range, $\omega_{c1} < \omega < \omega_{c2}$, $|G_{eq}(j\omega)| = |1/H(j\omega)|$ (Shifted curve)

In the frequency range, $\omega_{c2} < \omega < \infty$, $|G_{eq}(j\omega)| = |G(j\omega)|$

The bode magnitude plot of $G_{eq}(j\omega)$ is shown as a bold curve in fig 1.17.3.

The bode magnitude plot of $G_{eq}(j\omega)$ is interpreted as follows,

1.   The slope in the starting portion of the curve is -20 db/dec and it is due to an integral factor $K/j\omega$, where K is given by $G(j\omega)$ when $\omega = 0$,

$$|G(j\omega)|_{\omega = 0} = 3, \qquad \therefore K = 3$$

2. At $\omega = \omega_{c1}$, the slope changes from -20 db/dec to -40 db/dec. This is due to a first order factor $1/(1 + j\omega T_1)$ where, $T_1 = 1/\omega_{c1} = 1/0.08 = 12.5$ sec.

3. At $\omega = \omega_{c2}$, the slope changes from -40 db/dec to -20 db/dec. This is due to a first order factor $(1 + j\omega T_2)$ where, $T_2 = 1/\omega_{c2} = 1/0.8 = 1.25$ sec.

4. At $\omega = \omega_{c3}$, the slope changes from -20 db/dec to -40 db/dec. This is due to a first order factor, $1/(1 + j\omega T_3)$ where, $T_3 = 1/\omega_{c3} = 1/10 = 0.1$ sec.

Hence the transfer function, $G_{eq}(j\omega)$ can be written as,

$$G_{eq}(j\omega) = \frac{K(1 + j\omega T_2)}{j\omega(1 + j\omega T_1)(1 + j\omega T_3)} = \frac{3(1 + j1.25\omega)}{j\omega(1 + j12.5\omega)(1 + j0.1\omega)}$$

Put $j\omega = s$, $\therefore G_{eq}(s) = \dfrac{3(1 + 1.25s)}{s(1 + 12.5s)(1 + 0.1s)}$

Step-6 : Determine the parameter, K_A

From the bode plot it is observed that $|G_{eq}(j\omega)|$ curve should be shifted 17 db upwards to make $\omega_{gc} = 2.2$ rad/sec as the gain crossover frequency of $G_{eq}(j\omega)$ (i.e., to make $|G_{eq}(j\omega)|$ in db = 0 at $\omega = \omega_{gc} = 2.2$ rad/sec).

Now, $20 \log K_A = 17$ db

$$\log K_A = \frac{17}{20} \quad \text{(or)} \quad K_A = 10^{17/20} = 7.08$$

Step-7 : Verify the design

$$\left.\begin{array}{l}\text{Open loop transfer function}\\\text{of feedback compensated system}\end{array}\right\} = K_A G_{eq}(s)$$

$$= 7.08 \times \frac{3(1 + 1.25s)}{s(1 + 12.5s)(1 + 0.1s)}$$

$$= \frac{21.24(1 + 1.25s)}{s(1 + 12.5s)(1 + 0.1s)}$$

Velocity error constant, $K_v = \underset{s \to 0}{\text{Lt}} \ s \ K_A G_{eq}(s)$

$$= \underset{s \to 0}{\text{Lt}} \ s \ \frac{21.24(1 + 1.25s)}{s(1 + 12.5s)(1 + 0.1s)} = 21.24$$

CONCLUSION

Since the velocity error constant of the compensated system satisfies the requirement, the design is accepted.

RESULT

$K_A = 7.08$, $\qquad K_t = 3.185$ $\qquad$ and $\qquad$ $T = 6.25$ sec.

$$\left.\begin{array}{l}\text{Open loop transfer function}\\\text{of feedback compensated system}\end{array}\right\} = \frac{21.24(1 + 1.25s)}{s(1 + 12.5s)(1 + 0.1s)}$$

EXAMPLE 1.18

Consider a unity feedback system with open loop transfer function, $G(s) = K/s(s+2)(s+5)$. Design a feedback compensator to satisfy the following specifications.

(a) Maximum overshoot, $M_p \leq 12\%$, (b) Settling time, $t_s \leq 3$ sec for 2% error and (c) Velocity error constant, $K_v \geq 4$.

SOLUTION

Step-1 : To find dominant pole, s_d

Given that $M_p \leq 12 \%$

Let $M_p = 10\%$, $\qquad \therefore M_p = e^{-\zeta\pi/\sqrt{1-\zeta^2}} \times 100 = 10\%$

$$\therefore e^{-\zeta\pi/\sqrt{1-\zeta^2}} = 0.1$$

On taking natural log we get, $-\zeta\pi/\sqrt{1-\zeta^2} = In(0.1) = -2.3$

On squaring we get, $\dfrac{\zeta^2\pi^2}{1-\zeta^2} = 5.29$

$$\zeta^2\pi^2 = 5.29 - \zeta^2 5.29$$

$$\zeta^2\pi^2 + 5.29\zeta^2 = 5.29$$

$$\zeta = \sqrt{\dfrac{5.29}{\pi^2 + 5.29}} = 0.59 \approx 0.6$$

Given that, $t_s \leq 3$ sec for 2% error. Let, $t_s = 3$ sec for 2% error.

Hence for 2% error, $e^{-\zeta\omega_n t_s} = \dfrac{2}{100}$

On taking natural log, we get,

$$-\zeta\omega_n t_s = In(0.02) = -4$$

$$\therefore \omega_n = \dfrac{-4}{-\zeta t_s} = \dfrac{4}{0.6 \times 3} = 2.22 \text{ rad/sec}$$

Dominant pole, $s_d = -\zeta\omega_n + j\omega_n\sqrt{1-\zeta^2}$

$$= -(0.6 \times 2.22) + j2.22\sqrt{1-0.6^2} = -1.3 + j1.8$$

Step-2 : Pole-zero plot of G(s).

The poles of G(s) are located at s = 0, –2 and –5. The poles are marked by "x" as shown in fig 1.18.1. The dominant pole, s_d is also marked in fig 1.18.1. and it is denoted by point P.

Step-3 : Determine the new transfer function.

Let us choose a feedback compensation scheme as shown in fig 1.18.2.

Given that, $G(s) = \dfrac{K}{s(s+2)(s+5)}$

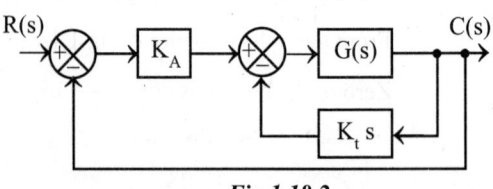

Fig 1.18.2

$$G_{eq}(s) = \frac{G(s)}{1 + G(s)K_t s} = \frac{\dfrac{K}{s(s+2)(s+5)}}{1 + \dfrac{K}{s(s+2)(s+5)}K_t s} = \frac{K}{s(s+2)(s+5) + K\,K_t s}$$

Closed loop transfer function $\Bigg\}$ $\dfrac{C(s)}{R(s)} = \dfrac{K_A G_{eq}(s)}{1 + K_A G_{eq}(s)}$

$$= \frac{K_A \dfrac{K}{s(s+2)(s+5) + K\,K_t s}}{1 + K_A \dfrac{K}{s(s+2)(s+5) + K\,K_t s}} = \frac{K\,K_A}{s(s+2)(s+5) + K\,K_t s + K\,K_A}$$

The characteristic equation is, $s(s+2)(s+5) + K K_t s + K K_A = 0$ .....(1.18.1)

On dividing the equation (1.18.1) by, $s(s+2)(s+5)$ we get,

$$1 + \frac{K K_t (s + K_A/K_t)}{s(s+2)(s+5)} = 0 \qquad\qquad .....(1.18.2)$$

The equation (1.18.2) is in the form of $1 + G(s)H(s) = 0$

$$\therefore [G(s)H(s)]_{new} = \frac{K K_t (s + K_A/K_t)}{s(s+2)(s+5)} \qquad\qquad .....(1.18.3)$$

From the equation (1.18.3) we get,

Zero introduced by the feedback compensator, $z_c = -K_A/K_t$ .....(1.18.4)

Step-4 : To locate the zero of the compensator

Let ϕ = Angle contributed by the zero, z_c to make point P as a point on root locus.

Now, $\phi = \begin{pmatrix} \text{Sum of angles contributed by} \\ \text{poles of uncompensated system} \end{pmatrix} \pm n180°$

From pole-zero plot of fig 1.18.1, we get,

$\left.\begin{matrix}\text{Sum of angles contributed} \\ \text{by poles of uncompensated system}\end{matrix}\right\} = \theta_1 + \theta_2 + \theta_3 = 126° + 69° + 26° = 221°$

Let $n = 1$, $\therefore \phi = 221° - 180° = 41°$.

With line AP as reference draw a line PC such that $\angle APC = \phi$ as shown in fig 1.18.1. This line intersects the real axis at point C.

From the pole-zero plot of fig 1.18.1, we get,

Zero of the compensator, $z_c = -3.35$ .....(1.18.5)

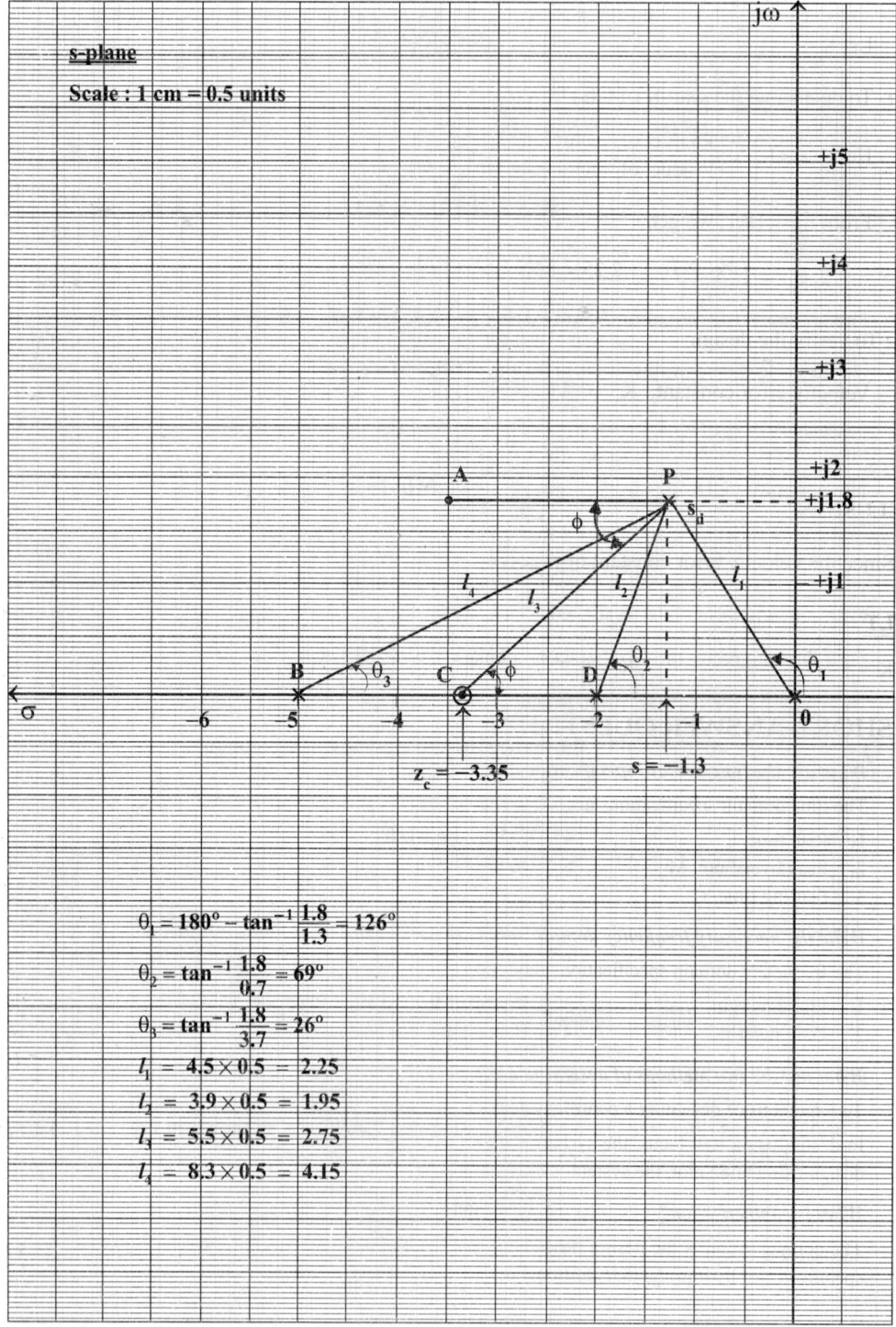

s-plane

Scale : 1 cm = 0.5 units

$z_c = -3.35$

$s = -1.3$

$\theta_1 = 180° - \tan^{-1}\dfrac{1.8}{1.3} = 126°$

$\theta_2 = \tan^{-1}\dfrac{1.8}{0.7} = 69°$

$\theta_3 = \tan^{-1}\dfrac{1.8}{3.7} = 26°$

$l_1 = 4.5 \times 0.5 = 2.25$

$l_2 = 3.9 \times 0.5 = 1.95$

$l_3 = 5.5 \times 0.5 = 2.75$

$l_4 = 8.3 \times 0.5 = 4.15$

Fig 1.18.1. : Pole zero plot

Step-5 : To find K_A and K_t

On equating equations (1.18.4) and (1.18.5) we get, $K_A/K_t = 3.35$

Let $K_t = 1.2$, $\therefore K_A = K_t \times 3.35 = 1.2 \times 3.35 = 4.02$

Step-6 : To determine the gain K of G(s)

Here, $K\,K_t = \dfrac{\text{Product of vector length from poles to } s = s_d}{\text{Product of vector length from zeros to } s = s_d}$

From fig 1.18.1, we get,

$$K\,K_t = \frac{l_1 \times l_2 \times l_4}{l_3} = \frac{2.25 \times 1.95 \times 4.15}{2.75} = 6.62 \qquad \therefore K = \frac{6.62}{K_t} = \frac{6.62}{1.2} = 5.5$$

Step-7 : To verify the design

$$\text{Velocity error constant, } K_v = \underset{s \to 0}{\text{Lt}}\, s\big[G(s)H(s)\big]_{new}$$

$$= \underset{s \to 0}{\text{Lt}}\, s\, \frac{K\,K_t(s + K_A/K_t)}{s(s+1)(s+5)} = \frac{K\,K_A}{1 \times 5} = \frac{5.5 \times 4.02}{5} = 4.422$$

CONCLUSION

Since the velocity error constant of the compensated system is satisfactory, the design is accepted.

RESULT

$K = 5.5$; $K_t = 1.2$; $K_A = 4.02$

1.7   SHORT-ANSWER QUESTIONS

Q1.1   *What are the time domain specifications needed to design a control system?*

The time domain specifications needed to design a control system are

1. Rise time, t_r 4. Damping ratio, ζ
2. Peak overshoot, M_p 5. Natural frequency of oscillation, ω_n.
3. Settling time, t_s

Q1.2   *Write the necessary frequency domain specifications for design of a control system.*

The frequency domain specifications required to design a control system are,

1. Phase margin   3. Resonant peak
2. Gain margin   4. Bandwidth

Q1.3   *What are the two methods of designing a control system?*

Two methods of designing a control system are design using root locus and design using bode plot.

In design using root locus, the system is designed to satisfy the specified time domain specifications. In design using bode plot, the system is designed to satisfy the specified frequency domain specifications.

Q1.4   *What is compensation?*

The compensation is the design procedure in which the system behaviour is altered to meet the desired specifications, by introducing additional device called compensator.

Q1.5   *What is compensator? What are the different types of compensator?*

A device inserted into the system for the purpose of satisfying the specifications is called compensator.

The different types of compensators are lag compensator, lead compensator and lag-lead compensator.

Q1.6   *What are the two types of compensation schemes?*

The two types of compensation schemes employed in control system are series compensation and feedback or parallel compensation.

Q1.7   *What is series compensation?*

The series compensation is a design procedure in which a compensator is introduced in series with plant to alter the system behaviour and to provide satisfactory performance (i.e., to meet the desired specifications). The block diagram of series compensation scheme is shown in fig Q1.7.

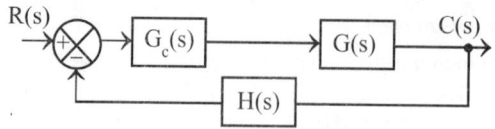

Fig Q1.7 : Series compensation.

$G_c(s)$ = Transfer function of series compensator

$G(s)$ = Open loop transfer function of the plant.

$H(s)$ = Feedback path transfer function.

Q1.8   *What is feedback compensation?*

The feedback compensation is a design procedure in which a compensator is introduced in the feedback path so as to meet the desired specifications. It is also called parallel compensation. The block diagram of feedback compensation scheme is shown in fig Q1.8.

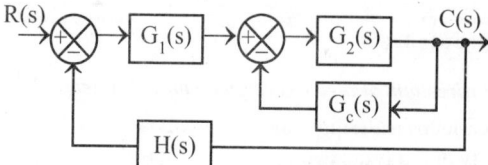

Fig Q1.8 : Feedback compensation

$G_c(s)$ = Transfer function of feedback compensator

$H(s)$ = Feedback path transfer function.

$G_1(s), G_2(s)$ = Open loop transfer function of the components of the plant.

Q1.9   *What are the factors to be considered for choosing series or shunt/feedback compensation?*

The choice between series, shunt or feedback compensation depends on the following

1.   Nature of signals in the system.
2.   Power levels at various points.
3.   Components available.
4.   Designer's experience.
5.   Economic considerations.

Q1.10   *When lag/lead/lag-lead compensation is employed?*

Lag compensation is employed for a stable system for improvement in steady state performance.

Lead compensation is employed for stable/unstable system for improvement in transient-state performance.

Lag-lead compensation is employed for stable/unstable system for improvement in both steady-state and transient state performance.

Q1.11   *Why compensation is necessary in feedback control system?*

In feedback control systems compensation is required in the following situations.

1. When the system is absolutely unstable, then compensation is required to stabilize the system and also to meet the desired performance.

2. When the system is stable, compensation is provided to obtain the desired performance.

Q1.12   *Discuss the effect of adding a pole to open loop transfer function of a system.*

The addition of a pole to open loop transfer function of a system will reduce the steady-state error. The closer the pole to origin lesser will be the steady-state error. Thus the steady-state performance of the system is improved.

Also the addition of pole will increase the order of the system, which inturn makes the system less stable than the original system.

Q1.13   *Discuss the effect of adding a zero to open loop transfer function of a system.*

The addition of a zero to open loop transfer function of a system will improve the transient response. The addition of zero reduces the rise time. If the zero is introduced close to origin then the peak overshoot will be larger. If the zero is introduced far away from the origin in the left half of s-plane then the effect of zero on the transient response will be negligible.

Q1.14   *How root loci are modified when a zero is added to open loop transfer function?*

The addition of a zero to open loop transfer function will pull the root locus to the left which make the system more stable and reduce the settling time.

Q1.15   *How the root loci are modified when a pole is added to open loop transfer function of the system?*

The addition of a pole to the open-loop transfer function has the effect of pulling the root locus to the right, which reduce the relative stability of the system and increase the settling time.

Q1.16   *How control system design is carried using root locus?*

In design using root locus the transient response specifications are translated into desired locations for a pair of dominant closed loop poles.

In order to satisfy the performance specifications, the root loci should pass through these points. Hence a compensator is introduced in open loop transfer function which will reshape the root loci and force them to pass through the points where the dominant closed loop poles are located.

Q1.17   *What is the advantage in design using root locus.*

The advantage in design using root locus technique is that the information about closed loop transient response and frequency response are directly obtained from the pole-zero configuration of the system in the s-plane.

Q1.18   *What are the informations that can be obtained from frequency response plots?*

The low frequency region of frequency response plot provides information regarding the steady-state performance and high frequency region provides information regarding the transient performance of the system. The mid-frequency region provides information regarding relative stability.

Q1.19   *What are the advantages and disadvantages in frequency domain design.*

The advantages of frequency domain design are the following.

1. The effect of disturbances, sensor noise and plant uncertainities are easy to visualize and asses in frequency domain.

2. The experimental information can be used for design purposes.

The disadvantage of frequency response design is that it gives the information on closed loop system's transient response indirectly.

Q1.20   *What is lag compensation?*

The lag compensation is a design procedure in which a lag compensator is introduced in the system so as to meet the desired specifications.

Q1.21   *What is lag compensator? Give an example.*

A compensator having the characteristics of a lag network is called lag compensator. If a sinusoidal signal is applied to a lag compensator, then in steady state the output will have a phase lag with respect to input. An electrical lag compensator can be realised by a R-C network. The R-C network shown in fig Q1.21 is an example of electrical lag compensator.

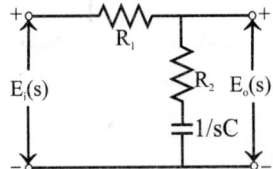

$E_i(s)$ R_1 R_2 $E_o(s)$ $1/sC$

Fig Q1.21 : Lag compensator.

Q1.22   *Write the transfer function of lag compensator and draw its pole-zero plot.*

$$\left.\begin{array}{l}\text{Transfer function of}\\ \text{lag compensator}\end{array}\right\} g_c(s) = \frac{s + \dfrac{1}{T}}{s + \dfrac{1}{\beta T}}$$

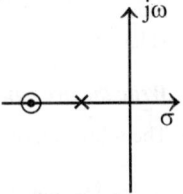

The lag compensator has a pole at $s = -1/\beta T$ and a zero at $s = -1/T$. Since $\beta > 1$ and $T > 0$, the pole of lag compensator is nearer to origin. The pole-zero plot of lag compensator is shown in fig Q1.22.

Fig Q 1.22 : Pole-zero plot of lag compensator

Q1.23   *What are the characteristics of lag compensation? When lag compensation is employed?*

The lag compensation improves the steady state performance, reduce the bandwidth and increases the rise time. (The increase in rise time results in slower transient response). If the pole introduced by the compensator is not cancelled by a zero in the system then the lag compensator increases the order of the system by one.

When a system is stable and does not satisfy the steady-state performance specifications then lag compensation can be employed so that the system is redesigned to satisfy the steady-state requirements.

Q1.24   *Draw the bode plot of lag compensator.*

Let, $G_c(j\omega)$ = Sinusoidal transfer function of lag compensator.

The approximate magnitude plot and phase plot at $G_c(j\omega)$ are shown in fig Q1.24.

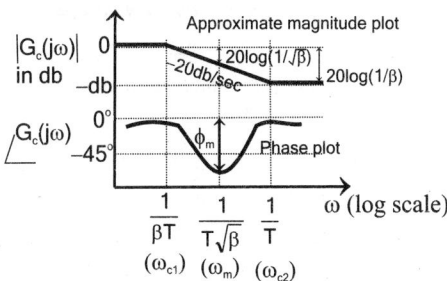

Fig Q1.24 : Bode plot of lag compensator.

Q1.25   **When maximum phase lag occurs in lag compensator? Give the expressions for maximum lag angle and the corresponding frequency.**

The maximum phase lag occurs at the geometric mean of two corner frequencies of the lag compensator.

Maximum phase lag angle, $\phi_m = \tan^{-1}\dfrac{(1-\beta)}{2\sqrt{\beta}}$

Frequency corresponding to maximum phase lag angle $\Bigg\}$ $\omega_m = \sqrt{\omega_{c1}\omega_{c2}} = \sqrt{\dfrac{1}{T}\dfrac{1}{\beta T}} = \dfrac{1}{T\sqrt{\beta}}$

Q1.26   **What is the relation between ϕ_m and β in lag compensator?**

In lag compensator the ϕ_m and β are related by the expressions,

$$\phi_m = \tan^{-1}\left(\dfrac{1-\beta}{2\sqrt{\beta}}\right) \qquad \text{or} \qquad \beta = \dfrac{1-\sin\phi_m}{1+\sin\phi_m}$$

Since $\beta > 1$, from the above expressions we can conclude that, larger the value of β the larger will be the value of ϕ_m.

Q1.27   **Write the two equations that relates β and ϕ_m of lag compensator.**

The following two equations relates ϕ_m and β of lag compensator.

$$\phi_m = \tan^{-1}\dfrac{1-\beta}{2\sqrt{\beta}} \quad ; \quad \beta = \dfrac{1-\sin\phi_m}{1+\sin\phi_m}$$

Q1.28   **What is lead compensation?**

The lead compensation is a design procedure in which a lead compensator is introduced in the system so as to meet the desired specifications.

Q1.29   **What is lead compensator? Given an example.**

A compensator having the characteristic of a lead network is called a lead compensator. If a sinusoidal signal is applied to a lead compensator, then in steady state the output will have a phase lead with respect to input. An electrical lead compensator can be realised by a R-C network. The R-C network shown in fig Q6.29. is an example of electrical lead compensator.

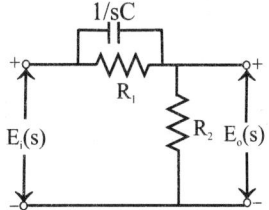

Fig Q1.29 : Electrical lead compensator.

Q1.30   *Write the transfer function of lead compensator and draw its pole-zero plot.*

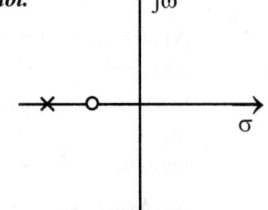

Transfer function of lead compensator, $g_c(s) = \dfrac{s + \dfrac{1}{T}}{s + \dfrac{1}{\alpha T}}$

The lead compensator has a pole at $s = -1/\alpha T$ and a zero at $s = -1/T$. Since $\alpha < 1$ and $T > 0$, the zero of lead compensator is nearer to origin. The pole-zero plot of lead compensator is shown in fig Q1.30.

Fig Q1.30 : Pole-zero plot of lead compensator

Q1.31   *What are the characteristics of lead compensation? When lead compensation is employed?*

The lead compensation increases the bandwidth and improves the speed of response. It also reduces the peak overshoot. If the pole introduced by the compensator is not cancelled by a zero in the system, then lead compensation increases the order of the system by one. When the given system is stable/unstable and requires improvement in transient state response then lead compensation is employed.

Q1.32   *Draw the bode plot of lead compensator.*

Let, $G_c(j\omega)$ = Sinusoidal transfer function of lead compensator. The approximate magnitude plot and phase plot of $G_c(j\omega)$ are shown in fig Q1.32.

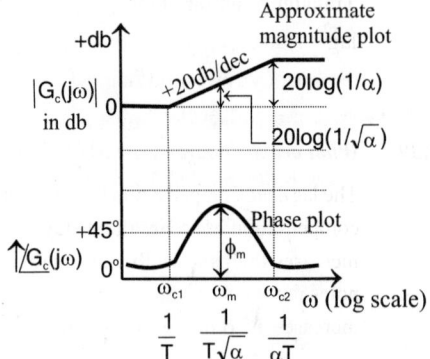

Q1.33   *What is the relation between ϕ_m and α in lead compensator?*

In lead compensator the ϕ_m and α are related by the expression,

$$\phi_m = \tan^{-1}\left(\frac{1-\alpha}{2\sqrt{\alpha}}\right) \quad \text{or} \quad \alpha = \frac{1 - \sin\phi_m}{1 + \sin\phi_m}$$

Since $\alpha < 1$ from the above expressions we can conclude that, smaller the value of α the larger will be the value of ϕ_m.

Fig Q1.32 : Bode plot of lead compensator.

Q1.34   *When maximum phase lead occurs in lead compensator? Give the expressions for maximum lead angle and the corresponding frequency.*

The maximum phase lead occurs at the geometric mean of two corner frequencies of the lead compensator.

Maximum phase lead angle, $\phi_m = \tan^{-1}\dfrac{(1-\alpha)}{2\sqrt{\alpha}}$

Frequency corresponding to maximum phase lead angle $\Bigg\}$ $\omega_m = \sqrt{\omega_{c1}\omega_{c2}} = \sqrt{\dfrac{1}{T}\dfrac{1}{\alpha T}} = \dfrac{1}{T\sqrt{\alpha}}$

Q1.35   *Write the two equations that relates α and ϕ_m of lead compensator*

$$\dot{\phi}_m = \tan^{-1}\left(\frac{1-\alpha}{2\sqrt{\alpha}}\right) \quad ; \quad \alpha \; \frac{1 - \sin\phi_m}{1 + \sin\phi_m}$$

Q1.36   *What is lag-lead compensation?*

The lag-lead compensation is a design procedure in which a lag-lead compensator is introduced in the system so as to meet the desired specifications.

Q1.37   **What is lag-lead compensator? Given an example.**

A compensator having the characteristics of lag-lead network
is called lag-lead compensator. If a sinusoidal signal is applied
to a lag-lead compensator then the output will have both phase
lag and lead with respect to input, but in different frequency
regions.

An electrical lag-lead compensator can be realised by a   R-C
network. The R-C network shown in fig Q1.37 is an example
of electrical lag-lead compensator.

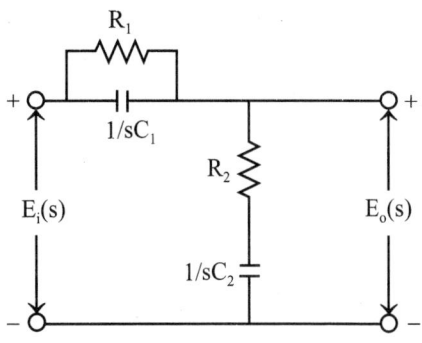

Fig Q1.37 : Electrical lag-lead compensator.

Q1.38   **Write the transfer function of lag-lead compensator and draw its pole-zero plot.**

$$\left.\begin{array}{l}\text{Transfer function of}\\ \text{lag} - \text{lead compensator}\end{array}\right\} G_c(s) = \underbrace{\frac{(s + 1/T_1)}{(s + 1/\beta T_1)}}_{\text{lag section}} \underbrace{\frac{(s + 1/T_2)}{(s + 1/\alpha T_2)}}_{\text{lead section}}$$

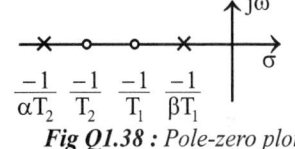

Fig Q1.38 : Pole-zero plot of lag-lead compensator

Q1.39   **What are the characteristics of lag-lead compensation? When lag-lead compensation is employed?**

The lag-lead compensation has the characteristics of both lag compensation and lead compensation. The lag
compensation improves the steady state performance and decreases the bandwidth. The lead compensation
increases the bandwidth and improves the speed of response. It also reduces the peak overshoot. If the
poles introduced by the compensator is not cancelled by zeros in the system then the lag-lead compensator
increases the order of the system by two.

The lag-lead compensation is employed when improvements in both steady-state and transient response
are required.

Q1.40   **Draw the bode plot of lag-lead compensator.**

Let, $G_c(j\omega)$ = Sinusoidal transfer function of lag-lead compensator.

The approximate magnitude plot and phase plot of $G_c(j\omega)$ are shown in fig Q1.40.

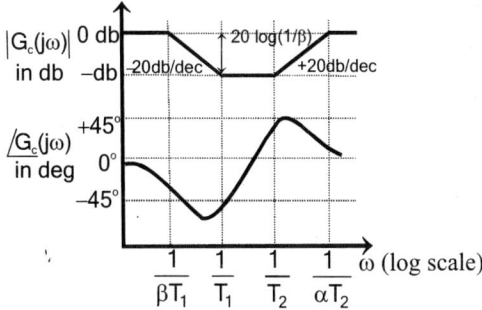

Fig Q1.40 : Bode plot of lag-lead compensator.

Q1.41   What is P-controller and what are its characteristics?

The proportional controller is a device that produces an output signal which is proportional to the input signal.

The proportional controller improves the steady state tracking accuracy, disturbance signal rejection and relative stability. It also decreases the sensitivity of the system to parameter variations.

Q1.42.   What is PI-controller and what are its effect on system performance?

The PI-controller is a device that produces an output signal consisting of two terms-one proportional to input signal and the other proportional to the integral of input signal.

The introduction of PI-controller in the system reduces the steady state error and increases the order and type number of the system by one.

Q1.43   Write the transfer function of PI-controller.

$$\left.\begin{array}{l}\text{Transfer function}\\\text{of PI}-\text{controller}\end{array}\right\} G_c(s) = K_p + \frac{K_i}{s} = \frac{K_p(s + K_i/K_p)}{s}$$

Q1.44   What is PD-controller and what are its effect on system performance?

The PD-controller is a device that produces an output signal consisting of two terms-one proportional to input signal and the other proportional to the derivative of input signal.

The PD-controller increases the damping of the system which results in reducing the peak overshoot.

Q1.45   Write the transfer function of PD-controller.

$$\text{Transfer function of PD}-\text{controller, } G_c(s) = K_p + K_d s = K_d(s + K_p/K_d)$$

Q1.46   What is PID controller and what are its effect on system performance?

The PID controller is a device which produces an output signal consisting of three terms-one proportional to input signal, another one proportional to integral of input signal and the third one proportional to derivative of input signal.

The PID controller stabilises the gain, reduces the steady state error and peak overshoot of the system.

Q1.47   Write the transfer function of PID controller.

$$\left.\begin{array}{l}\text{Transfer function}\\\text{of PID controller}\end{array}\right\} G_c(s) = K_p + \frac{K_i}{s} + K_d s = \frac{K_d s^2 + K_p s + K_i}{s}$$

Q1.48   What is feedback compensation?

The feedback compensation is a design procedure in which a compensator is placed in an internal feedback path around one or more components of the forward path so as to meet the desired specifications.

Q1.49   Draw the block diagram of a feedback compensation scheme.

The block diagram of a popular feedback scheme employed in control systems is shown in fig Q1.49.

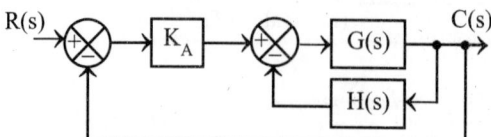

Fig Q1.49 : Feedback compensation scheme

Here, H(s) = Transfer function of feedback compensator

K_A = A parameter to adjust the velocity error constant of the system.

Q1.50   What is the disadvantage in rate feedback and how it is eliminated?

The disadvantage in the rate feedback is that the system velocity error constant, K_v is reduced. This undesirable effect can be eliminated by reducing the feedback signal in the low frequencies by introducing a high pass filter in cascade with rate device as shown in fig Q1.50.

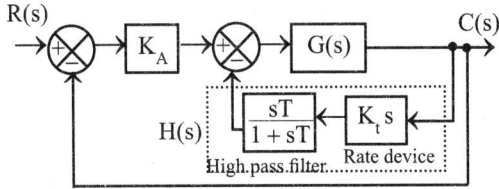

Fig Q1.50 : *Feedback compensation with high-pass filter in cascade with rate device*

1.8   EXERCISES

I. FILL IN THE BLANKS

1.   A device inserted into the system for the purpose of satisfying the specifications is called ....

2.   In .... compensation the compensator is placed in series with the ....

3.   The lag compensator will reduce the .... of the system which will result in slower .... response.

4.   The addition of a .... to the open loop transfer function has the effect of pulling the root locus to the ....

5.   The addition of a .... to the open loop transfer function has the effect of pulling the root locus to the ....

6.   In lead compensation the conflict between dominance condition and noise elimination can be avoided by locating by pole at .... times the value of zero location.

7.   The .... frequency region of bode plot provides information regarding .... performance of the system.

8.   The .... frequency region of bode plot provides information regarding .... performance of the system.

9.   The frequency corresponding to maximum phase lag, ω_m = ...

10. The lead compensator increase the .... and reduce the ....

11. In lead compensator if α is too low then the .... at the output will be very low.

12. Typical choices of α in lead compensator and β in lag compensator are .... and .... respectively.

13. The frequency corresponding to maximum phase lead, ω_m = ...

14. In lag-lead compensator, the phase lag occurs in .... region and phase lead occurs in .... region.

15. The transfer function of lag-lead compensator has two real .... and ....

16. The propotional control is not used alone because it produces a constant ....

17. The PI controller .... the steady state error.

18. The PD controller increse the .... of the system which results in reducing the ....

19. In system subjected to frequent load disturbances, ... compenasation may be preferred.

20. In rate feedback a .... is connected in cascade with rate device in order to eliminate the reduction in velocity error constant (K_v).

ANSWERS

1. compensator	11. signal / noise ratio
2. series ; plant	12. 0.1 and 10
3. bandwidth ; transient	13. $1/T\sqrt{\beta}$
4. pole, right	14. low frequency; high frequency
5. zero, left	15. poles; zeros
6. 3 to 10	16. steady state error
7. low, transient state	17. reduce
8. high, steady state	18. damping; peak overshot
9. $1/T\sqrt{\beta}$	19. feedback
10. bandwidth; peak overshoot	20. high pass filter

II. State whether the following statements are TRUE/FALSE

1. Lag compensator increases the bandwidth of the system

2. Lag compensator will shift the gain crossover frequency to a lower frequency point.

3. In lag compensator the maximum phase lag occurs at the geometric mean of its two corner frequencies.

4. Lag compensation is provided to make an unstable system as a stable system.

5. Lead compensation reduce the bandwidth of the system.

6. Lead compensation improves the transient response.

7. In lead compensator the maximum phase lead occurs at the geometric mean of its two corner frequencies.

8. The lead compensation increse the rise time and lag compensation decrease the rise time.

9. The lag compensator is a low pass filter and lead compensator is a high pass filter.

10. The lag-lead compensator is employed for improvements in both transient and steady-state response.

11. The lag-lead compensation will increase the order of the system by one, unless cancellation of poles and zeros occurs in the compensated system.

12. In a single lag-lead compensator, $\alpha\beta = 1$.

13. The propotional controller stabilizes the gain.

15. The PD controller increase peak overshoot.

16. The PID controller does not stablises the gain.

17. In rate feedback the velocity error constant is increased.

18. In design using root locus technique, the desired transient response is achieved by making the root locus of the system to pass through a pair of dominant poles.

19. The required steady state performance is usually specified in terms of error constants.

20. The transient response specifications can be translated into desired locations for a pair of dominant closed loop poles.

ANSWERS

1. False	6. True	11. False	16. False
2. True	7. True	12. True	17. False
3. True	8. False	13. False	18. True
4. False	9. True	14. True	19. True
5. False	10. True	15. False	20. True

III. UNSOLVED PROBLEM

E1.1 Design a phase lag network for a system having $G(s) = K/s(1+0.2s)^2$ to have a phase margin of 30°.

E1.2 The open loop transfer function of a certain unity feedback control system is given by $G(s) = K/s(s+1)$. It is desired to have the velocity error constant, $K_v = 10$ and the phase margin to be atleast 60°. Design a phase lag series compensator.

E1.3 The controlled plant of a unity feedback system is $G(s) = K/s(s+5)$. It is specified that velocity error constant of the system be equal to 15, while the damping ratio is 0.6 and velocity error is less than 0.25 rad per unit ramp input. Design a suitable lag compensator.

E1.4 Consider the unity feedback system with an open loop transfer function $G(s) = K/s(s+2)^2$. Design a suitable lead compensator so that phase margin is atleast 50° and velocity error constant is 20 s⁻⁶.

E1.5 The open loop transfer function of certain unity feedback control system is given by $G(s) = K/s(0.1s+1)$ $(0.2s+1)$. It is desired to have the phase margin to be atleast 30°. Design a suitable phase lead series compensator.

E1.6 A unity feedback system with open loop transfer function $G(s) = K/s^2(s+1.5)$ is to be lead compensated to satisfy the following specifications.
    (i)   Damping ratio = 0.45
    (ii)   Undamped natural frequency, $\omega_n = 2.2$ rad/sec
    (ii)   Velocity error constant, $K_v = 30$.

E1.7 Consider a unity feedback control system whose forward transfer function is $G(s) = K/s(s+2)(s+8)$. Design a lag-lead compensator so that $K_v = 80$ s^{-1} and dominant closed loop poles are located at $-2 \pm j2\sqrt{3}$.

E1.8 The open loop transfer function of uncompensated system is $G(s) = K/s(s+1)(s+4)$. Design a lag-lead compensator to meet the following specifications.

  (i)   Velocity error constant ≥ 5

  (ii)   Damping ratio $= 0.4$.

E1.9 Consider a unity feedback system with open loop transfer function, $G(s) = K/s(2s+1)(0.5s+1)$. Design a suitable lag-lead compensator to meet the following specifications.

  (i) $K_v = 30$ (ii) Phase margin ≥ 50

E1.10 Consider a unity feedback system with open loop transfer function, $G(s) = 1/s(s+1)$. Design a PD controller so that the phase margin of the system is $30°$ at a frequency of 2 rad/sec.

E1.11 A unity feedback system has an open loop transfer function as $G(s) = 50/(s+3)(s+1)$. Design a PI controller so that phase margin of the system is $35°$ at a frequency of 1.2 rad/sec.

E1.12 Consider a unity feedback system with open loop transfer function, $G(s) = 20/(s+0.5)(s+2)(s+4)$. Design a PID controller so that the phase margin of the system is $30°$ at a frequency of 2 rad/sec and steady state error for unit ramp input is 0.1.

E1.13 Consider a unity feedback system with open loop transfer function $G(s) = 10/s(s+4)$. The dominant poles are $-2 \pm j\sqrt{6}$. Design a suitable PD controller.

E1.14 Consider a unity feedback system with open loop transfer function $G(s) = 10/(s+1)(s+2)$. Design a PI controller so that the closed loop has damping ratio of 0.707 and natural frequency of oscillation as 1.2 rad/sec.

E1.15 Consider a unity feedback system with $G(s) = 50/(s+2)(s+10)$. Design a PID controller to satisfy the following specifications.

  (i) $K_v \geq 2$

  (ii)   Damping ratio $= 0.6$

  (iii)   Natural frequency of oscillations $= 2$ rad/sec.

E1.16 Design a feedback compensation scheme for a unity feedback system with open loop transfer function $G(s) = 1/s^2(s+5)$ to satisfy the following specifications. (i) phase margin of the system should be atleast $50°$. (ii) Velocity error constant, $K_v \geq 20$.

E1.17 Consider a unity feedback system with open loop transfer function, $G(s) = K/s(s+1)(s+4)$. Design a feedback compensator to satisfy the following specifications.

  (a)   Maximum overshoot, $M_p \leq 12\%$

  (b)   Settling time, $t_s = 10$s

  (c)   Velocity error constant, $K_v = 10$.

Chapter 2

NonLinear Systems

2.1 INTRODUCTION TO NONLINEAR SYSTEMS

The nonlinear systems are systems which does not obey the principle of superposition. The linear systems are systems which satisfy the principle of superposition.

The principle of superposition implies that if a system has responses $y_1(t)$ and $y_2(t)$ to any two inputs $x_1(t)$ and $x_2(t)$ respectively then the system response to the linear combination of these inputs $\alpha_1 x_1(t) + \alpha_2 x_2(t)$ is given by the linear combination of the individual outputs, i.e, $\alpha_1 y_1(t) + \alpha_2 y_2(t)$, where α_1 and α_2 are constants.

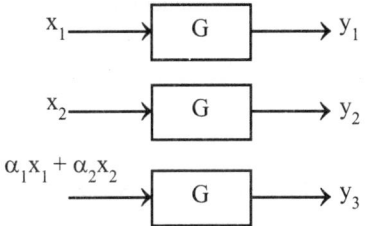

To satisfy the principle of superposition, $y_3 = \alpha_1 y_1 + \alpha_2 y_2$

Example of linear system : $y = ax + b\dfrac{dx}{dt}$

Example of nonlinear system : $y = ax^2 + e^{bx}$

EXAMPLE 2.1

a) The response of a system is, $y = ax + b\dfrac{dx}{dt}$, Test whether the system is linear or nonlinear.

b) The response of a system is, $y = ax^2 + e^{bx}$. Test whether the system is linear or nonlinear.

SOLUTION

a) Let x_1 and x_2 be the two inputs to the system and y_1 and y_2 be their responses, respectively.

Given that $y = ax + b\dfrac{dx}{dt}$

When x = x_1,   y = y_1, $\therefore$ $y_1 = ax_1 + b\dfrac{dx_1}{dt}$

When x = x_2,   y = y_2, $\therefore$ $y_2 = ax_2 + b\dfrac{dx_2}{dt}$

Consider a linear combination of inputs $\alpha_1 x_1 + \alpha_2 x_2$ and let the response of the system for this linear combination of inputs be y_3.

When x = $\alpha_1 x_1 + \alpha_2 x_2$,   y = y_3

$\therefore$ $y_3 = a(\alpha_1 x_1 + \alpha_2 x_2) + b\dfrac{d}{dt}(\alpha_1 x_1 + \alpha_2 x_2)$

$= \alpha_1 ax_1 + \alpha_2 ax_2 + \alpha_1 b\dfrac{dx_1}{dt} + \alpha_2 b\dfrac{dx_2}{dt}$

Consider the same linear combination of output, $\alpha_1 y_1 + \alpha_2 y_2$.

$\alpha_1 y_1 + \alpha_2 y_2 = \alpha_1\left[ax_1 + b\dfrac{dx_1}{dt}\right] + \alpha_2\left[ax_2 + b\dfrac{dx_2}{dt}\right]$

$= \alpha_1 ax_1 + \alpha_2 ax_2 + \alpha_1 b\dfrac{dx_1}{dt} + \alpha_2 b\dfrac{dx_2}{dt}$

It is observed that $y_3 = \alpha_1 y_1 + \alpha_2 y_2$. Hence the system is linear.

---

b) Let x_1 and x_2 be two inputs to the system and y_1 and y_2 be their responses respectively.

Given that y = $ax^2 + e^{bx}$

When x = x_1,   y = y_1, $\therefore$ $y_1 = ax_1^2 + e^{bx_1}$

When x = x_2,   y = y_2, $\therefore$ $y_1 = ax_2^2 + e^{bx_2}$

Consider a linear combination of inputs $\alpha_1 x_1 + \alpha_2 x_2$ and let the response of the system for this linear combination of inputs be y_3.

When x = $\alpha_1 x_1 + \alpha_2 x_2$, y = y_3

$\therefore$ $y_3 = a(\alpha_1 x_1 + \alpha_2 x_2)^2 + e^{b(\alpha_1 x_1 + \alpha_2 x_2)}$

$= a(\alpha_1^2 x_1^2 + \alpha_2^2 x_2^2 + 2\alpha_1 x_1 \alpha_2 x_2) + e^{\alpha_1 bx_1}.e^{\alpha_2 bx_2}$

$= a\alpha_1^2 x_1^2 + a\alpha_2^2 x_2^2 + 2a\alpha_1 \alpha_2 x_1 x_2 + e^{\alpha_1 bx_1}.e^{\alpha_2 bx_2}$

Consider the same linear combination of output , $\alpha_1 y_1 + \alpha_2 y_2$

$\alpha_1 y_1 + \alpha_2 y_2 = \alpha_1\left[ax_1^2 + e^{bx_1}\right] + \alpha_2\left[ax_2^2 + e^{bx_2}\right]$

$= a\alpha_1 x_1^2 + \alpha_1 e^{bx_1} + a\alpha_2 x_2^2 + \alpha_2 e^{bx_2}$

It is observed that $y_3 \neq \alpha_1 y_1 + \alpha_2 y_2$. Hence the system is nonlinear.

In all practical engineering systems, there will be always some nonlinearity due to friction, inertia, stiffness, backlash, hysteresis, saturation and dead-zone. The effect of the nonlinear components can be avoided by restricting the operation of the component over a narrow limited range. Moreover most of the automatic control systems operate within a narrow range, e.g. the speed controller of an electric drive for constant speed operation of 1500 rpm will be required to operate between 1450 to 1550 rpm. Similarly, automatic voltage controller will be operating within ± 5% of the specified voltage. Thus the characteristics of components may be considered as linear over this limited range.

Further, some components behave linearly over its working range, e.g., a spring when loaded, gets extended. As the load is being increased the load-displacement curve is linear within the working range. However, when the load is increased beyond the maximum of the working range, the spring material starts to yield and it becomes permanently deformed. It can be concluded that the spring behaves linearly over its working range and beyond this range it is nonlinear.

Although nonlinearities in systems may generally be due to imperfections of a physical device, some times we deliberately introduce nonlinear device or operate the linear devices in nonlinear regions with a view to improve system performance.

The characteristics of non-linear system are given below:

1. *The response of nonlinear system to a particular test signal is no guide to their behaviour to other inputs, since the principle of superposition does not holds good for nonlinear systems.*

2. *The nonlinear system response may be highly sensitive to input amplitude. The stability study of nonlinear systems requires the information about the type and amplitude of the anticipated inputs, initial conditions, etc., in addition to the usual requirement of the mathematical model.*

3. *The nonlinear systems may exhibit limit cycles which are self sustained oscillations of fixed frequency and amplitude.*

4. *The nonlinear systems may have jump resonance in the frequency response.*

5. *The output of a nonlinear system will have harmonics and sub-harmonics when excited by sinusoidal signals.*

6. *The nonlinear systems will exhibit phenomena like frequency entrainment and asynchronous quenching.*

BEHAVIOUR OF NONLINEAR SYSTEMS

In nonlinear systems, the response (output) depends on the magnitude and type of input signal. The principle of superposition will not hold good for nonlinear systems. The nonlinear systems may exhibit various phenomena like jump resonance, subharmonic oscillation, limit cycles, frequency entrainment and asynchronous quenching. The various phenomena that occur in nonlinear system are explained in this section.

Frequency-amplitude dependence

The frequency-amplitude dependence is one of the most fundamental characteristics of the oscillations of nonlinear systems. The frequency-amplitude dependence can be best studied by considering the mechanical system shown in fig 2.1 in which the spring is nonlinear. The differential equation governing the dynamics of the system may be written as

$$M\ddot{x} + B\dot{x} + K\,x + K'\,x^3 = 0 \qquad .....(2.1)$$

where $Kx + K'x^3$ = Opposing force due to nonlinear spring.

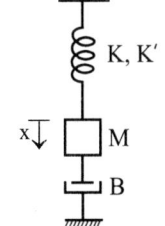

Fig 2.1 : Mechanical system with nonlinear spring.

The parameters M, B and K are positive constants. The parameter K' may be positive or negative. If K' is positive, the spring is called *hard spring* and if K' is negative the spring is called *soft spring.* The equation (2.1) is nonlinear differential equation and it is also called *Duffing's equation.*

When the system of fig 2.1 has non-zero initial conditions, the free response (i.e., solution of equation 2.1) is damped oscillatory. The frequency of free oscillations depends on the amplitude of oscillations. When K' < 0 (soft spring) the frequency decreases with decreasing amplitude. When K' > 0 (hard spring) the frequency increases with decreasing amplitude. When K' = 0 (corresponding to linear system) the frequency remains unchanged as the amplitude of free oscillation decreases. The frequency-amplitude dependence characteristic of nonlinear mechanical system of fig 2.1 is shown in fig 2.2.

Jump resonance

In the frequency response of nonlinear systems, the amplitude of the response (output) may jump from one point to another for increasing or decreasing values of frequency, ω. This phenomenon is called jump resonance and it can be observed in the frequency response of the system shown in fig 2.1., when it is subjected to sinusoidal input.

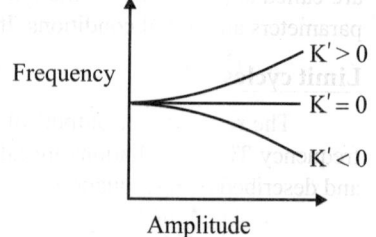

Fig 2.2 : Amplitude Vs frequency curves for free oscillations in the system described by equation 2.1.

Let the mechanical system of fig 2.1, be subjected to an input of type Acosωt. Now the differential equation governing the mechanical system is

$$M\ddot{x} + B\dot{x} + K x + K'x^3 = A \cos \omega t \qquad .....(2.2)$$

Let X be the amplitude of the response or output of the system. In frequency response studies, the amplitude, A of the input is held constant, while its ω is varied and the amplitude, X of the output is observed. The frequency response curve is plotted between X and ω. The frequency response curves of the mechanical system of fig 2.1 are shown in fig 2.3a and 2.3b for hard and soft springs respectively.

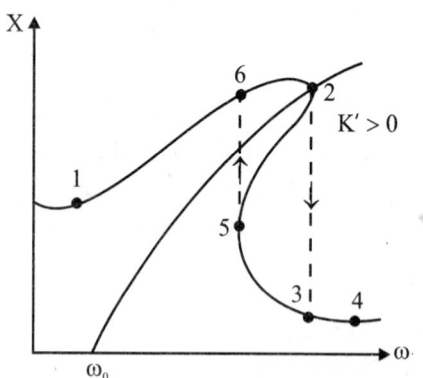

a: Mechanical system with hard spring

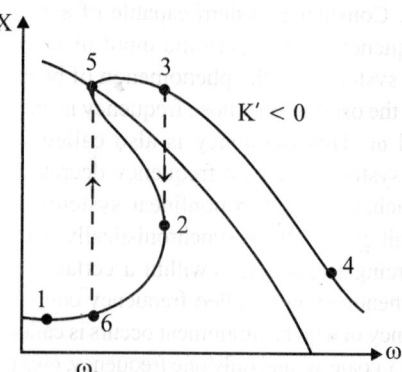

b: Mechanical system with soft spring

Fig 2.3 : Frequency response curves showing jump resonance.

In the frequency response curve shown in fig 2.3a and b, as the frequency ω is increased, the amplitude X increases, until point-2 is reached. A further increase in frequency will cause a jump from point-2 to point-3. This phenomenon is called jump resonance. As the frequency is increased further, the amplitude X follows the curve from point-3 towards point-4.

When the frequency is reduced starting from a high value corresponding to point-4, the amplitude X slowly increases through point-3, until point-5 is reached. A further decrease in ω will cause another jump from point-5 to point-6. This phenomenon is called jump resonance. After this jump, the amplitude X decreases with ω and follows the curve from point-6 towards point-1.

For jump resonance to take place, it is necessary that the damping term be small and the amplitude of the forcing function be large enough to drive the system into a region of appreciably nonlinear operation.

Subharmonic oscillations

When a nonlinear system is excited by a sinusoidal signal, the response or output will have steady-state oscillation whose frequency is an integral submultiple of the forcing frequency. These oscillations are called subharmonic oscillations. The generation of subharmonic oscillations depends on the system parameters and initial conditions. It also depends on amplitude and frequency of the forcing functions.

Limit cycles

The response (or output) of nonlinear systems may exhibit oscillations with fixed amplitude and frequency. These oscillations are called limit cycles. Consider a mechanical system with nonlinear damping and described by the equation,

$$M\ddot{x} + B(1 - x^2)\dot{x} + K x = 0 \qquad .....(2.3)$$

where M, B and K are positive constants. The equation (2.3) is called the van der pol equation. For small values of x the damping will be negative which implies the stored energy in the damper is fed to the system. For large values of x the damping is positive which implies that it absorbs energy from the system. Thus, it can be expected that such a system may exhibit a sustained oscillation. Since the system explained above is not a forced system, this oscillation is called a self-excited oscillation or zero input limit cycle.

Frequency entrainment

The phenomena of frequency entrainment is observed in the frequency response of nonlinear systems that exhibit limit cycles. Consider a system capable of exhibiting a limit cycle of frequency ω_l. If a periodic input of frequency ω is applied to this system then the phenomenon of beats is observed. [The beat is the oscillation whose frequency is the difference between ω_l and ω. This frequency is also called beat frequency]. In linear systems, the beat frequency decreases indefinitely as ω approaches ω_l. But in nonlinear systems, the frequency ω_l of the limit cycle falls in synchronistically with or is entrained by the forcing frequency, ω within a certain band of frequencies. This phenomenon is called frequency entrainment. The band of

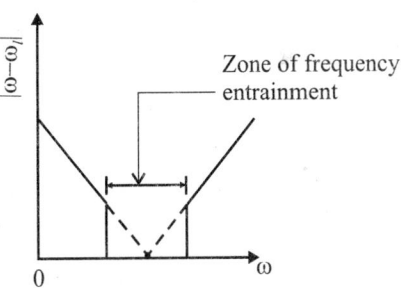

Fig 2.4 : $|\omega-\omega_l|$ Vs ω curve showing the zone of frequency entrainment.

frequency in which entrainment occurs is called the zone of frequency entrainment. In this zone, the frequencies ω and ω_l coalesce and only one frequency, ω exists. The relationship between $|\omega - \omega_l|$ and ω is shown in figure 2.4.

Asynchronous quenching

In a nonlinear system that exhibits a limit cycle of frequency ω_l, it is possible to quench (stop or eliminate) the limit cycle oscillation by forcing the system of a frequency ω_q, where ω_q and ω_l are not related to each other. The phenomenon is called signal stabilization or asynchronous quenching.

INVESTIGATION OF NONLINEAR SYSTEMS

For analysis, the nonlinear system can be approximated by a linear model in the entire operating region. The nonlinear systems can be piecewise approximated. Each piece can be analysed by a differential equation governing the systems.

The two popular methods of analysing nonlinear systems are phase-plane method and describing function method.

The phase plane method is basically a graphical method from which information about transient behaviour and stability is easily obtained by constructing phase trajectories. This method is restricted to second order systems. Higher order systems may first be approximated by their second-order equivalent for investigation by the phase plane method.

The Describing function method is based on harmonic linearization. Here the input to nonlinear component is sinusoidal and depending upon the filtering properties of the linear part of the overall system, the output is adequately represented by the fundamental frequency term in fourier series.

The phase-plane and describing function methods use complimentary approximations. The phase-plane method retains, the nonlinearity as such and uses the second-order approximation of a higher-order linear part, while on the other hand, the describing function method retains the linear part and harmonically linearizes the nonlinearity.

COMMON PHYSICAL NONLINEARITIES

The nonlinearites can be classified as incidental and intentional.

The incidental nonlinearities are those which are inherently present in the system. Common examples of incidental nonlinearities are saturation, dead-zone, coulomb friction, stiction, backlash, etc.

The intentional nonlinearities are those which are deliberately inserted in the system to modify system characteristics. The most common example of this type of nonlinearity is a relay.

SATURATION : In this type of nonlinearity the output is proportional to input for a limited range of input signals. When the input exceeds this range, the output tends to become nearly constant as shown in figure 2.5.

All devices when driven by sufficiently large signals, exhibit the phenomenon of saturation due to limitations of their physical capabilities. Saturation in the output of electronic, rotating and flow (hydraulic and pneumatic) amplifiers, speed and torque saturation in electric and hydraulic motors, saturation in the output of sensors for measuring position, velocity, temperature, etc., are the well known examples.

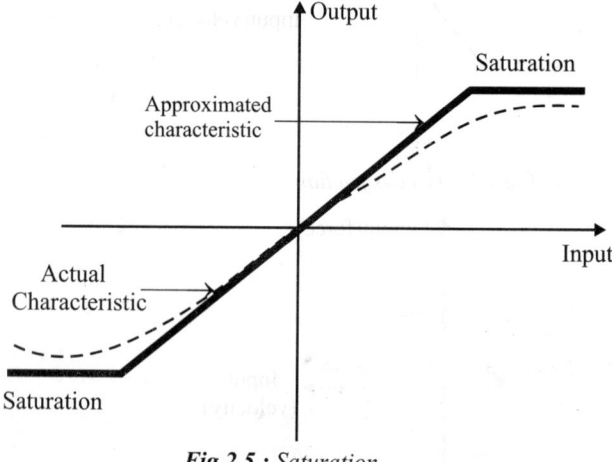

DEADZONE : The deadzone is the region in which the output is zero for a given input. Many physical devices do not respond to small signals, i.e., if the input amplitude

Fig 2.5 : Saturation.

is less than some small value, there will be no output. The region in which the output is zero is called deadzone. When the input is increased beyond this deadzone value, the output will be linear.

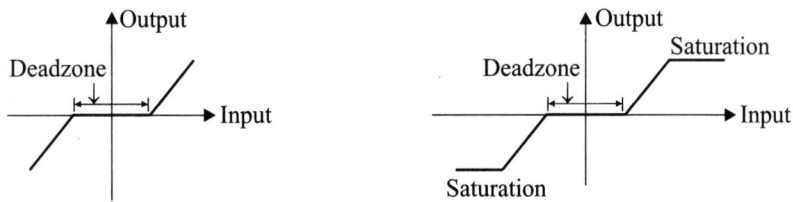

Fig 2.6 : Dead zone nonlinearity.   *Fig 2.7 : Dead zone and saturation nonlinearity.*

The figure 2.6 shows the deadzone nonlinearity and the fig 2.7 shows the combination of dead zone and saturation nonlinearity.

FRICTION : Friction exists in any system when there is relative motion between contacting surfaces. The different types of friction are viscous friction, coulomb friction and stiction.

The viscous friction is linear in nature and the frictional force is directly proportional to relative velocity of the sliding surfaces.

The coulomb friction and stiction are nonlinear frictions. The coulomb friction offers a constant retarding force only when the motion is initiated. Due to interlocking of surface irregularities, more force is required to move an object from rest than to maintain it in motion. Hence the force of stiction is always greater than that of coulomb friction.

In actual practice, the stiction force gradually decreases with velocity and changes over to coulomb friction at reasonably low velocities as shown in fig 2.10. The composite characteristics of various frictions are shown in fig 2.8 to 2.11.

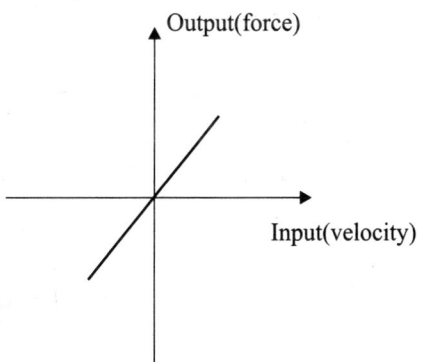

Fig 2.8 : Viscous friction.

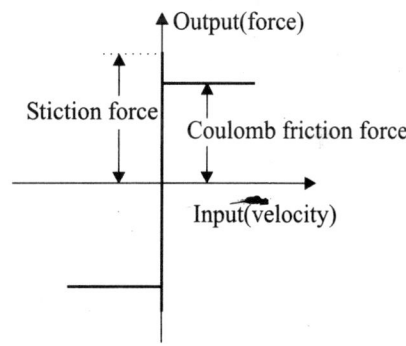

Fig 2.9 : Ideal stiction and coulomb friction.

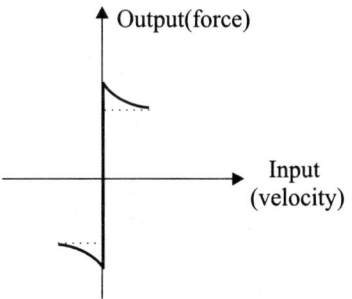

Fig 2.10 : Actual stiction and coulomb friction.

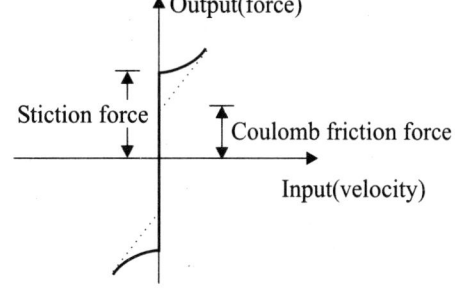

Fig 2.11 : Stiction, coulomb friction and viscous friction.

2.2   DESCRIBING FUNCTION

Consider the block diagram of the nonlinear system shown in figure 2.12.

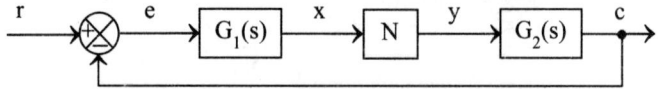

Fig 2.12 : A nonlinear system.

In the above system the blocks $G_1(s)$ and $G_2(s)$ represents linear elements and the block N represent nonlinear element.

Let $x = X \sin\omega t$ be the input to nonlinear element. Now the output y of the nonlinear element will be in general a nonsinusoidal periodic function. The fourier series representation of the output y can be expressed as (by assuming that the nonlinearity does not generate subharmonics).

$$y = A_0 + A_1 \sin\omega t + B_1 \cos\omega t + A_2 \sin2\omega t + B_2 \cos2\omega t + \dots \qquad \dots(2.4)$$

If the nonlinearity is symmetrical the average value of y is zero and hence the output y is given by

$$y = A_1 \sin\omega t + B_1 \cos\omega t + A_2 \sin2\omega t + B_2 \cos2\omega t + \dots \qquad \dots(2.5)$$

In the absence of an external input (i.e, when $r = 0$)the output y of the nonlinearity N is fedback to its input through the linear elements $G_2(s)$ and $G_1(s)$ in tandem. If $G_1(s)G_2(s)$ has low-pass characteristics, then all the harmonics of y are filtered, so that the input x to the nonlinear element N is mainly contributed by the fundamental component of y and hence x remains sinusoidal. Under such conditions the harmonics of the output are neglected and the fundamental component of y alone considered for the purpose of analysis.

$$\therefore y = y_1 = A_1 \sin\omega t + B_1 \cos\omega t = Y_1 \angle \phi_1 = Y_1 \sin(\omega t + \phi_1) \qquad \dots(2.6)$$

where, $Y_1 = \sqrt{A_1^2 + B_1^2}$ $\qquad\qquad\qquad\qquad\qquad\qquad\qquad\qquad\qquad\qquad\qquad\dots(2.7)$

and $\phi_1 = \tan^{-1} \dfrac{B_1}{A_1}$ $\qquad\qquad\qquad\qquad\qquad\qquad\qquad\qquad\qquad\qquad\dots(2.8)$

Y_1 =   Amplitude of the fundamental harmonic component of the output.

ϕ_1 =   Phase shift of the fundamental harmonic component of the output with respect to the input.

The coefficients A_1 and B_1 of the fourier series are given by

$$A_1 = \frac{2}{2\pi} \int_0^{2\pi} y \sin\omega t\, d(\omega t) \qquad\qquad\qquad\qquad\qquad\qquad\dots(2.9)$$

$$B_1 = \frac{2}{2\pi} \int_0^{2\pi} y \cos\omega t\, d(\omega t) \qquad\qquad\qquad\qquad\qquad\qquad\dots(2.10)$$

When the input, x to the nonlinearity is sinusoidal (i.e., $x = X\sin\omega t$) the describing function of the nonlinearity is defined as,

$$K_N(X,\omega) = \frac{Y_1}{X} \angle \phi_1 \qquad\qquad\qquad\qquad\qquad\qquad\qquad\dots(2.11)$$

The nonlinear element N in the system can be replaced by the describing function as shown in figure 2.13.

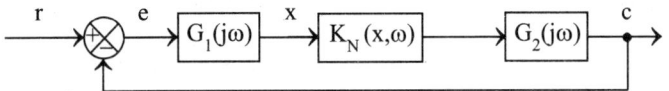

Fig 2.13 : Nonlinear system with nonlinearity replaced by describing function

If the nonlinearity is replaced by a describing function then all linear theory frequency domain techniques can be used for the analysis of the system. The describing functions are used only for stability analysis and it is not directly applied to the optimization of system design. The describing function is a frequency domain approach and no general correlation is possible between time and frequency responses.

2.3 DESCRIBING FUNCTION OF DEAD-ZONE AND SATURATION NONLINEARITY

The input and the output relationship of nonlinearity with dead-zone and saturation is shown in figure 2.14.

The dead-zone region is from $x = -D/2$ to $+D/2$ and in this region the output is zero. The input-output relation is linear for $x = \pm D/2$ to $\pm S$ and when the input, $x > S$, the output reaches a saturated value of $\pm K(S-D/2)$.

The output equation for the linear region can be obtained from the general equation of straight line as shown below.

The equation of straight line is , $y = mx + c$ .....(2.12)

In the linear region, when $x = D/2$, $y = 0$. On substituting this values of x and y in equation (2.12) we get,

$0 = mD/2 + c$ .....(2.13)

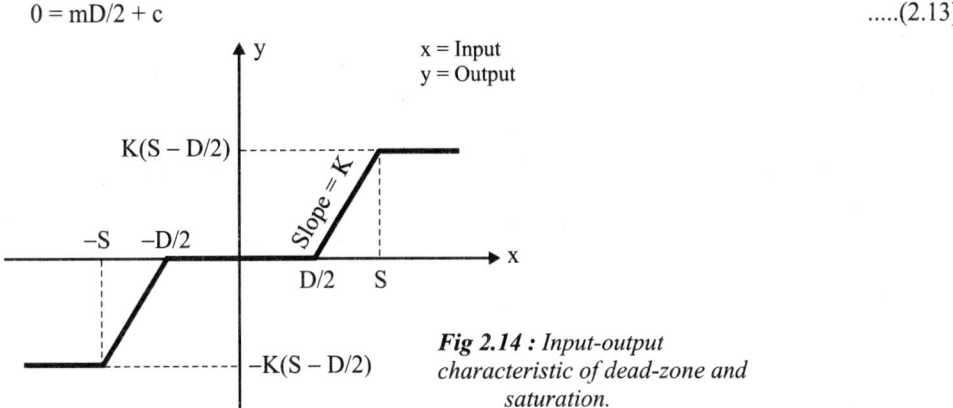

Fig 2.14 : Input-output characteristic of dead-zone and saturation.

In the linear region, when $x = S$, $y = K(S-D/2)$. On substituting this values of x and y in equation (2.12) we get,

$K(S-D/2) = mS + c$ .....(2.14)

Equ(2.14) - equ(2.13) yields, $K\left(S - \dfrac{D}{2}\right) = mS + c - m\dfrac{D}{2} - c$

$$K\left(S - \frac{D}{2}\right) = m\left(S - \frac{D}{2}\right)$$

$$\therefore m = K \qquad\qquad\qquad .....(2.15)$$

Put m = K in eqn(2.13), $\therefore 0 = K\dfrac{D}{2} + c$ (or) $c = -K\dfrac{D}{2}$ .....(2.16)

From equations (2.12), (2.15) and (2.16) the output equation for the linear region can be written as,

$$y = mx + c = Kx - K\frac{D}{2} = K\left(x - \frac{D}{2}\right) \qquad\qquad .....(2.17)$$

The response or output of the nonlinearity when the input is sinusoidal signal (x = X sinωt) is shown in fig 2.15.

The input x is sinusoidal , $\therefore$ x = X sinωt .....(2.18)

where X = Maximum value of input.

In fig 2.15, when, ωt = α,   x = D/2

Hence from equ (2.18) we get

   D/2 = X sin α

   sin α = D/2X

$$\therefore \ \alpha = \sin^{-1}\frac{D}{2X} \qquad\qquad .....(2.19)$$

In fig 2.15, when, ωt = β,   x = S

Hence from equ (2.18) we get

   S = X sin β

   sin β = S/X

$$\therefore \ \beta = \sin^{-1}\frac{S}{X} \qquad\qquad .....(2.20)$$

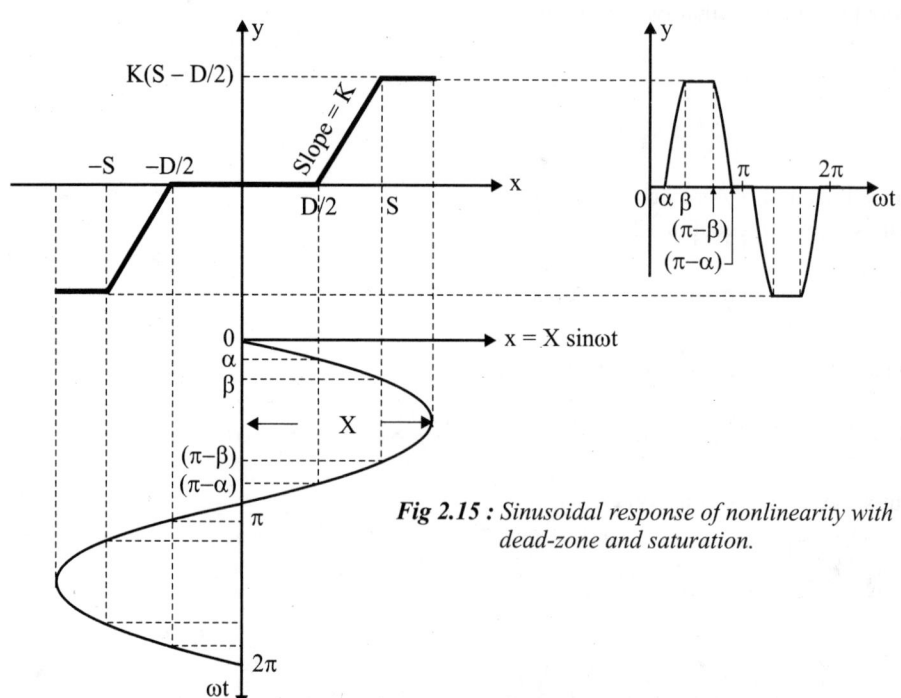

Fig 2.15 : *Sinusoidal response of nonlinearity with dead-zone and saturation.*

The output y of the nonlinearity can be divided into five regions in a period of π and the output equation for the five regions are given below.

$$y = \begin{cases} 0 & ; \ 0 \le \omega t \le \alpha \\ K\left(x - \dfrac{D}{2}\right) & ; \ \alpha \le \omega t \le \beta \\ K\left(S - \dfrac{D}{2}\right) & ; \ \beta \le \omega t \le (\pi - \beta) \\ K\left(x - \dfrac{D}{2}\right) & ; \ (\pi - \beta) \le \omega t \le (\pi - \alpha) \\ 0 & ; \ (\pi - \alpha) \le \omega t \le \pi \end{cases} \qquad \text{.....(2.21)}$$

Let Y_1 = Amplitude of the fundamental harmonic component of the output.

ϕ_1 = Phase shift of the fundamental harmonic component of the output with respect to the input.

The describing function is given by

$$K_N(X, \omega) = \frac{Y_1}{X} \angle \phi_1$$

where, $Y_1 = \sqrt{A_1^2 + B_1^2}$ and $\phi_1 = \tan^{-1} \dfrac{B_1}{A_1}$

$$A_1 = \frac{2}{2\pi} \int_0^{2\pi} y \sin \omega t \, d(\omega t) \quad \text{and} \quad B_1 = \frac{2}{2\pi} \int_0^{2\pi} y \cos \omega t \, d(\omega t)$$

Here the output has half wave and quarter wave symmetries

$$\therefore \ B_1 = 0$$

and $A_1 = \dfrac{2}{\pi/2} \displaystyle\int_0^{\pi/2} y \sin \omega t \, d(\omega t)$ $\qquad$ .....(2.22)

Since the output, y is given by different expressions in the range 0 to $\pi/2$, the equation (2.22) can be written as shown in equation (2.23).

$$A_1 = \frac{4}{\pi} \int_0^{\alpha} y \sin \omega t \, d(\omega t) + \frac{4}{\pi} \int_{\alpha}^{\beta} y \sin \omega t \, d(\omega t) + \frac{4}{\pi} \int_{\beta}^{\frac{\pi}{2}} y \sin \omega t \, d(\omega t) \qquad \text{.....(2.23)}$$

On substituting the values of y in the range 0 to $\pi/2$ from equation (2.21) in equation (2.23) we get,

$$A_1 = \frac{4}{\pi} \int_{\alpha}^{\beta} K\left(x - \frac{D}{2}\right) \sin \omega t \, d(\omega t) + \frac{4}{\pi} \int_{\beta}^{\frac{\pi}{2}} K\left(S - \frac{D}{2}\right) \sin \omega t \, d(\omega t) \qquad \text{.....(2.24)}$$

Put $x = X \sin \omega t$ in equation (2.24),

$$\therefore \ A_1 = \frac{4K}{\pi} \left[\int_{\alpha}^{\beta} \left(X \sin \omega t - \frac{D}{2}\right) \sin \omega t \, d(\omega t) + \int_{\beta}^{\frac{\pi}{2}} \left(S - \frac{D}{2}\right) \sin \omega t \, d(\omega t) \right]$$

$$= \frac{4K}{\pi} \left[\int_{\alpha}^{\beta} X \sin^2 \omega t \, d(\omega t) - \int_{\alpha}^{\beta} \frac{D}{2} \sin \omega t \, d(\omega t) + \int_{\beta}^{\frac{\pi}{2}} \left(S - \frac{D}{2}\right) \sin \omega t \, d(\omega t) \right]$$

$$= \frac{4K}{\pi}\left[\frac{X}{2}\int_\alpha^\beta (1-\cos 2\omega t)\, d(\omega t) - \frac{D}{2}\int_\alpha^\beta \sin \omega t\, d(\omega t) + \left(S - \frac{D}{2}\right)\int_\beta^{\frac{\pi}{2}} \sin \omega t\, d(\omega t)\right]$$

$$= \frac{4K}{\pi}\left[\frac{X}{2}\left[\omega t - \frac{\sin 2\omega t}{2}\right]_\alpha^\beta - \frac{D}{2}\left[-\cos \omega t\right]_\alpha^\beta + \left(S - \frac{D}{2}\right)\left[-\cos \omega t\right]_\beta^{\frac{\pi}{2}}\right]$$

$$\boxed{\left(\because \cos \frac{\pi}{2} = 0\right)}$$

$$= \frac{4K}{\pi}\left[\frac{X}{2}\left(\beta - \sin \frac{2\beta}{2} - \alpha + \frac{\sin 2\alpha}{2}\right) - \frac{D}{2}(-\cos \beta + \cos \alpha) + + \left(S - \frac{D}{2}\right)\left(-\cos \frac{\pi}{2} + \cos \beta\right)\right]$$

$$= \frac{4K}{\pi}\left[\frac{X}{2}\left(\beta - \alpha - \frac{\sin 2\beta}{2} + \frac{\sin 2\alpha}{2}\right) + \frac{D}{2}\cos \alpha - \frac{D}{2}\cos \beta + S\cos \beta - \frac{D}{2}\cos \beta\right] \quad .....(2.25)$$

We know that, $\sin \alpha = \frac{D}{2X}$ (or) $\frac{D}{2} = X \sin \alpha$ $\qquad .....(2.26)$

Also, $\sin \beta = \frac{S}{X}$ (or) $S = X \sin \beta$ $\qquad .....(2.27)$

On substituting for D/2 and S from equations (2.26) and (2.27) in equation (2.25) we get,

$$A_1 = \frac{4K}{\pi}\left[\frac{X}{2}\left(\beta - \alpha - \frac{\sin 2\beta}{2} + \frac{\sin 2\alpha}{2}\right) - X\sin \alpha \cos \alpha + X\sin \beta \cos \beta\right]$$

$$= \frac{4KX}{\pi}\left[\frac{\beta}{2} - \frac{\alpha}{2} - \frac{\sin 2\beta}{4} + \frac{\sin 2\alpha}{4} - \frac{\sin 2\alpha}{2} + \frac{\sin 2\beta}{2}\right]$$

$$= \frac{4KX}{\pi}\left[\frac{1}{2}(\beta - \alpha) + \frac{\sin 2\beta}{4} - \frac{\sin 2\alpha}{4}\right]$$

$$= \frac{2KX}{\pi}\left[(\beta - \alpha) + \frac{\sin 2\beta}{2} - \frac{\sin 2\alpha}{2}\right]$$

$$= \frac{KX}{\pi}\left[2(\beta - \alpha) + \sin 2\beta - \sin 2\alpha\right] \quad .....(2.28)$$

$$Y_1 = \sqrt{A_1^2 + B_1^2} = \sqrt{A_1^2 + 0} = A_1$$

$$\therefore\ Y_1 = A_1 = \frac{KX}{\pi}\left[2(\beta - \alpha) + \sin 2\beta - \sin 2\alpha\right] \quad .....(2.29)$$

$$\phi_1 = \tan^{-1}\frac{B_1}{A_1} = \tan^{-1}0 = 0 \quad .....(2.30)$$

The describing function $K_N(X,\omega) = \frac{Y_1}{X}\angle \phi_1$ $\qquad .....(2.31)$

On substituting for Y_1 and ϕ_1 from equations (2.29) and (2.30) in equation (2.31) we get

$$K_N(X,\omega) = \frac{Y_1}{X}\angle \phi_1 = \frac{K}{\pi}\left[2(\beta - \alpha) + \sin 2\beta - \sin 2\alpha\right]\angle 0^0 \quad .....(2.32)$$

Depending on the maximum value of input, X, the describing function of equation (2.32) can be written as,

If $X < \frac{D}{2}$, then $\alpha = \beta = \frac{\pi}{2}$ and $K_N(X,\omega) = 0$ $\qquad .....(2.33)$

If $\frac{D}{2} < X < S$, then $\beta = \frac{\pi}{2}$ and $K_N(X,\omega) = K\left[1 - \frac{2}{\pi}(\alpha + \sin \alpha \cos \alpha)\right]$ $\qquad .....(2.34)$

If $X > S$, $K_N(X,\omega) = \frac{K}{\pi}\left[2(\beta - \alpha) + \sin 2\beta - \sin 2\alpha\right]$ $\qquad .....(2.35)$

2.4 DESCRIBING FUNCTION OF SATURATION NONLINEARITY

The input-output relationship0 of saturation nonlinearity is shown in fig 2.16.

The input-output relation is linear for x = 0 to S. When the input x > S, the output reaches a saturated value of KS.

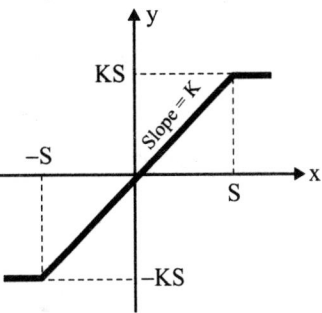

The response of the nonlinearity when the input is sinusoidal signal (x = Xsin ωt) is shown in fig 2.17.

The input x is sinusoidal,

$$\therefore x = X\sin \omega t \qquad .....(2.36)$$

where X is the maximum value of input.

In fig (2.17), when ωt = β, x = S.

Fig 2.16: Input-output characteristic of saturation nonlinearity.

Hence equation (2.36) can be written as, $S = X \sin\beta$ .....(2.37)

$$\therefore \sin\beta = \frac{S}{X} \quad \text{(or)} \quad \beta = \sin^{-1}\left(\frac{S}{X}\right) \qquad .....(2.38)$$

The output y of the nonlinearity can be divided into three regions in a period of π. The output equation for the three regions are given by equation (2.39).

$$y = \begin{cases} Kx & ; \ 0 \le \omega t \le \beta \\ KS & ; \ \beta \le \omega t \le (\pi - \beta) \\ Kx & ; \ (\pi - \beta) \le \omega t \le \pi \end{cases} \qquad .....(2.39)$$

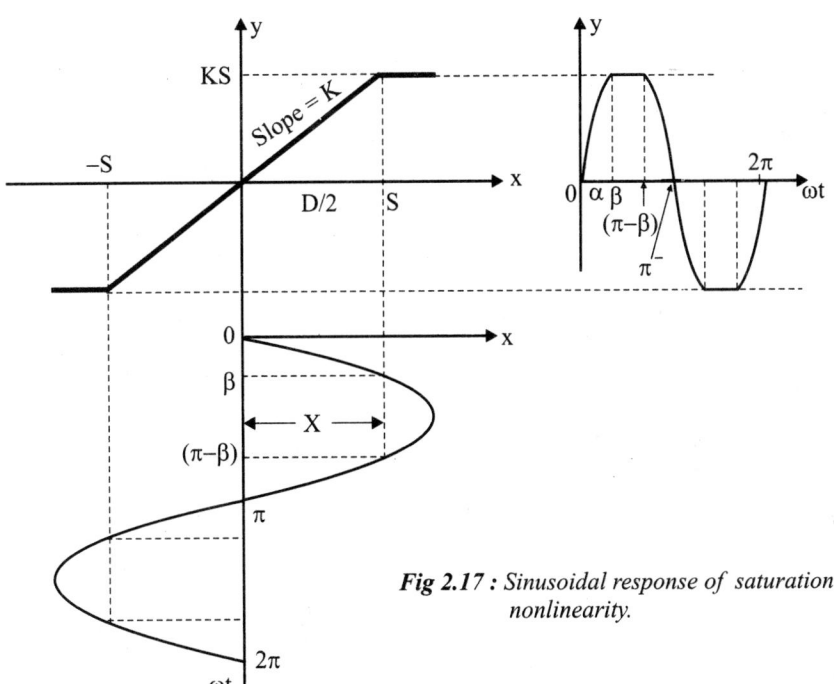

Fig 2.17 : Sinusoidal response of saturation nonlinearity.

Let Y_1 = Amplitude of the fundamental harmonic component of the output.

ϕ_1 = Phase shift of the fundamental harmonic component of the output with respect to the input.

The describing function is given by, $K_N(X,\omega) = (Y_1/X) \angle \phi_1$

where $Y_1 = \sqrt{A_1^2 + B_1^2}$ and $\phi_1 = \tan^{-1}(B_1/A_1)$

The output y has half wave and quarter wave symmetries

$$\therefore B_1 = 0 \quad \text{and} \quad A_1 = \frac{2}{\pi/2} \int_0^{\frac{\pi}{2}} y \sin \omega t \, d(\omega t) \qquad .....(2.40)$$

The output, y is given by two different expressions in the period 0 to $\pi/2$. Hence equation (2.40) can be written as shown in equation (2.41).

$$A_1 = \frac{2}{\pi/2} \int_0^{\beta} y \sin \omega t \, d(\omega t) + \frac{4}{\pi} \int_{\beta}^{\frac{\pi}{2}} y \sin \omega t \, d(\omega t) \qquad .....(2.41)$$

On substituting the values of y from equation (2.39) in equation (2.41) we get,

$$A_1 = \frac{4}{\pi} \int_0^{\beta} Kx \sin \omega t \, d(\omega t) + \frac{4}{\pi} \int_{\beta}^{\frac{\pi}{2}} KS \sin \omega t \, d(\omega t)$$

On substituting $x = X \sin \omega t$, we get,

$$A_1 = \frac{4K}{\pi} \int_0^{\beta} X \sin \omega t \times \sin \omega t \, d(\omega t) + \frac{4KS}{\pi} \int_{\beta}^{\frac{\pi}{2}} \sin \omega t \, d(\omega t)$$

$$= \frac{4KX}{\pi} \int_0^{\beta} \sin^2 \omega t \, d(\omega t) + \frac{4KS}{\pi} \int_{\beta}^{\frac{\pi}{2}} \sin \omega t \, d(\omega t)$$

$$= \frac{4KX}{\pi} \int_0^{\beta} \frac{1 - \cos 2\omega t}{2} \, d(\omega t) + \frac{4KS}{\pi} \int_{\beta}^{\frac{\pi}{2}} \sin \omega t \, d(\omega t)$$

$$= \frac{2KX}{\pi} \left[\omega t - \frac{\sin 2\omega t}{2} \right]_0^{\beta} + \frac{4KS}{\pi} \left[-\cos \omega t \right]_{\beta}^{\frac{\pi}{2}}$$

$$= \frac{2KX}{\pi} \left[\beta - \frac{\sin 2\beta}{2} \right] + \frac{4KS}{\pi} \left[-\cos \frac{\pi}{2} + \cos \beta \right]$$

$$= \frac{2KX}{\pi} \left[\beta - \frac{\sin 2\beta}{2} \right] + \frac{4KS}{\pi} \cos \beta \qquad .....(2.42)$$

On substituting for S, (i.e, $S = X \sin \beta$) from equation (2.37) in equation (2.42) we get,

$$A_1 = \frac{2KX}{\pi} \left[\beta - \frac{\sin 2\beta}{2} \right] + \frac{4K}{\pi} X \sin \beta \cos \beta$$

$$= \frac{2KX}{\pi} \left[\beta - \frac{2 \sin 2\beta \cos \beta}{2} \right] + \frac{4KX}{\pi} \sin \beta \cos \beta$$

$$= \frac{2KX}{\pi} \left[\beta - \sin \beta \cos \beta + 2 \sin \beta \cos \beta \right]$$

$$= \frac{2KX}{\pi} \left[\beta + \sin \beta \cos \beta \right] \qquad .....(2.43)$$

$$Y_1 = \sqrt{A_1^2 + B_1^2} = \sqrt{A_1^2 + 0} = A_1 = \frac{2KX}{\pi}[\beta + \sin\beta\cos\beta] \qquad .....(2.44)$$

$$\phi_1 = \tan^{-1}\frac{B_1}{A_1} = \tan^{-1}0 = 0 \qquad .....(2.45)$$

The describing function, $K_N(X,\omega) = \frac{Y_1}{X} \angle \phi_1$ .....(2.46)

Using equations (2.44) and (2.45), the describing function of equation (2.46) can be written as,

$$K_N(X,\omega) = \frac{Y_1}{X} \angle \phi_1 = \frac{2K}{\pi}[\beta + \sin\beta\cos\beta] \angle 0^\circ \qquad .....(2.47)$$

Depending on the maximum value of input X, the describing function can be written as,

If $X < S$, then $\beta = \frac{\pi}{2}$, $K_N(X,\omega) = K$ .....(2.48)

If $X > S$, $K_N(X,\omega) = \frac{2K}{\pi}[\beta + \sin\beta\cos\beta]$ .....(2.49)

The equation (2.49) can be expressed in another form as shown below.

From equation (2.37) we get, $S = X \sin\beta$, $\therefore \sin\beta = \frac{S}{X}$ .....(2.50)

On constructing right angle triangle with unity hypotenuse as shown in fig 2.18, $\cos\beta$ can be evaluated. From fig 2.18 we get.

$$adj = \sqrt{1 - \left(\frac{S}{X}\right)^2} \quad \therefore \cos\beta = \frac{adj}{hyp} = \sqrt{1 - \left(\frac{S}{X}\right)^2} \qquad .....(2.51)$$

In the describing function of equation (2.49), substitute for β, $\sin\beta$ and $\cos\beta$ from equations (2.38), (2.50) and (2.51)

$$\therefore K_N(X,\omega) = \frac{2K}{\pi}\left[\sin^{-1}\left(\frac{S}{X}\right) + \left(\frac{S}{X}\right)\sqrt{1 - \left(\frac{S}{X}\right)^2}\right] \text{ for } X > S \qquad .....(2.52)$$

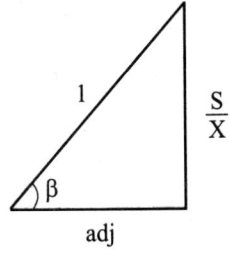

adj

Fig 2.18

2.5   DESCRIBING FUNCTION OF DEAD-ZONE NONLINEARITY

The input-output relationship of dead-zone nonlinearity is shown in figure 2.19 The output is zero, when the input is less than D/2. The input-output relationship is linear when the input is greater than D/2. The response of the nonlinearity when input is sinusoidal signal (x = X sinωt) is shown in fig 2.20.

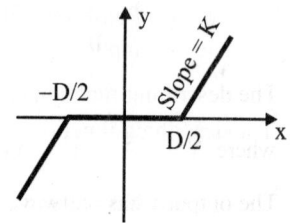

Fig 2.19 : Input-Output characteristic of dead-zone nonlinearity.

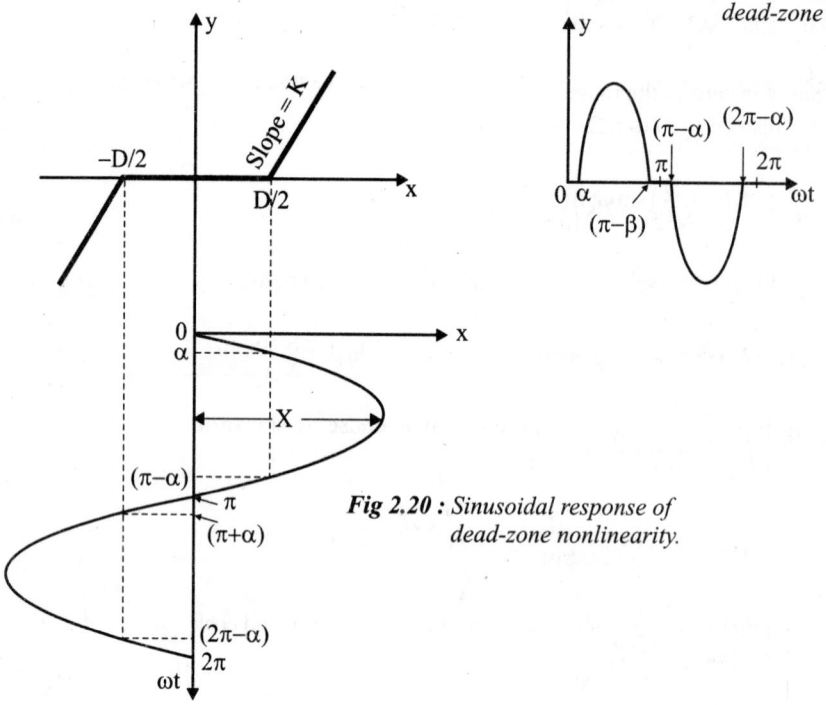

Fig 2.20 : Sinusoidal response of dead-zone nonlinearity.

The input x is sinusoidal,   ∴ x = X sinωt   .....(2.53)

where X is the maximum value of input.

In fig 2.20, when ωt = α, x = D/2,

Hence when ωt = α, the equation (2.53) can be written as, D/2 = X sin α   .....(2.54)

$$\therefore \sin \alpha = \frac{D}{2X}$$ .....(2.55)

$$\text{and } \alpha = \sin^{-1} \frac{D}{2X}$$ .....(2.56)

The output y can be divided into three regions in a period of π. The output equation for the three regions are given by equation (2.57).

$$y = \begin{cases} 0 & ; \ 0 \le \omega t \le \alpha \\ K\left(x - \dfrac{D}{2}\right) & ; \ \alpha \le \omega t \le (\pi - \alpha) \\ 0 & ; \ (\pi - \alpha) \le \omega t \le \pi \end{cases}$$ .....(2.57)

Let Y_1 = Amplitude of the fundamental harmonic component of the output.

ϕ_1 = Phase shift of the fundamental harmonic component of the output with respect to the input.

The describing function is given by, $K_N(X,\omega) = (Y_1/X) \angle \phi_1$

where $Y_1 = \sqrt{A_1^2 + B_1^2}$ and $\phi_1 = \tan^{-1}(B_1/A_1)$

The output y has halfwave and quarter wave symmetries.

$$\therefore \ B_1 = 0 \quad \text{and} \quad A_1 = \frac{2}{\pi/2} \int_0^{\frac{\pi}{2}} y \sin \omega t \, d(\omega t) \qquad \dots(2.57)$$

Since the output, y is zero in the range, $0 \le \omega t \le \alpha$, the limits of integration in equation (2.58) can be changed to, α to $\pi/2$ instead of, 0 to $\pi/2$.

$$\therefore \ A_1 = \frac{4}{\pi} \int_\alpha^{\frac{\pi}{2}} K\left(x - \frac{D}{2}\right) \sin \omega t \, d(\omega t) \qquad \dots(2.58)$$

Put $x = X \sin \omega t$, in equation (2.59)

$$\therefore \ A_1 = \frac{4K}{\pi}\left[\int_\alpha^{\pi/2} \left(X \sin \omega t - \frac{D}{2}\right) \sin \omega t \, d(\omega t)\right]$$

$$= \frac{4K}{\pi}\left[\int_\alpha^{\pi/2} X \sin^2 \omega t \, d(\omega t) - \frac{D}{2} \int_\alpha^{\pi/2} \sin \omega t \, d(\omega t)\right]$$

$$= \frac{4K}{\pi}\left[\frac{X}{2} \int_\alpha^{\pi/2} (1 - \cos 2\omega t) \, d(\omega t) - \frac{D}{2} \int_\alpha^{\pi/2} \sin \omega t \, d(\omega t)\right]$$

$$= \frac{4K}{\pi}\left[\frac{X}{2}\left[\omega t - \frac{\sin 2\omega t}{2}\right]_\alpha^{\frac{\pi}{2}} - \frac{D}{2}\left[-\cos \omega t\right]_\alpha^{\frac{\pi}{2}}\right]$$

$$= \frac{4K}{\pi}\left[\frac{X}{2}\left[\frac{\pi}{2} - \frac{\sin \pi}{2} - \alpha + \frac{\sin 2\alpha}{2}\right]_\alpha^{\frac{\pi}{2}} - \frac{D}{2}\left(-\cos \frac{\pi}{2} + \cos \alpha\right)\right]$$

$$= \frac{4K}{\pi}\left[\frac{X}{2}\left(\frac{\pi}{2} - \alpha + \frac{\sin 2\alpha}{2}\right) - \frac{D}{2}(\cos \alpha)\right] \qquad \dots(2.60)$$

From equ (2.55) we get, $\sin \alpha = \frac{D}{2X}$ $\quad \therefore \ D = 2X \sin \alpha$ $\qquad \dots(2.61)$

On substituting for D from equation (2.61) in equation (2.60) we get,

$$A_1 = \frac{4K}{\pi}\left[\frac{X}{2}\left(\frac{\pi}{2} - \alpha + \frac{\sin 2\alpha}{2}\right) - X \sin \alpha \, \cos \alpha\right] = \frac{4KX}{\pi}\left[\frac{\pi}{4} - \frac{\alpha}{2} + \frac{2 \sin \alpha \cos \alpha}{4} - \sin \alpha \cos \alpha\right]$$

$$= \frac{4KX}{\pi}\left[\frac{\pi}{4} - \frac{\alpha}{2} - \frac{\sin \alpha \cos \alpha}{2}\right] = KX\left[\frac{4}{\pi} \times \frac{\pi}{4} - \frac{4}{\pi} \times \frac{1}{2}(\alpha + \sin \alpha \cos \alpha)\right]$$

$$= KX\left[1 - \frac{2}{\pi}(\alpha + \sin \alpha \cos \alpha)\right] \qquad \dots(2.62)$$

$$Y_1 = \sqrt{A_1^2 + B_1^2} = \sqrt{A_1^2 + 0} = A_1 = KX\left[1 - \frac{2}{\pi}(\alpha + \sin\alpha\cos\alpha)\right] \qquad .....(2.63)$$

$$\phi_1 = \tan^{-1}\frac{B_1}{A_1} = \tan^{-1}0 = 0 \qquad .....(2.64)$$

The describing function, $K_N(X,\omega) = \dfrac{Y_1}{X} \angle \phi_1$ $\qquad .....(2.65)$

Using equations (2.63) and (2.64) the describing function of equation (2.65) can be written as,

$$K_N(X,\omega) = K\left[1 - \frac{2}{\pi}(\alpha + \sin\alpha\cos\alpha)\right]\angle 0^0 \qquad .....(2.66)$$

Depending on the maximum value of input X, the describing function can be written as,

$$\text{If } X < \frac{D}{2}, \qquad K_N(X,\omega) = 0 \qquad .....(2.67)$$

$$\text{If } X > \frac{D}{2}, \qquad K_N(X,\omega) = K\left[1 - \frac{2}{\pi}(\alpha + \sin\alpha\cos\alpha)\right] \qquad .....(2.68)$$

The equation (2.68) can be expressed in another form as shown below.

From equation (2.55), we get, $\sin\alpha = D/2X$

On constructing right angle triangle with unity hypotenuse as shown in fig 2.21 $\cos\alpha$ can be evaluated.

From fig 2.21, we get,

$$\text{adj} = \sqrt{1 - \left(\frac{D}{2X}\right)^2} \quad \therefore \ \cos\alpha = \frac{\text{adj}}{\text{hyp}} = \sqrt{1 - \left(\frac{D}{DX}\right)^2} \qquad .....(2.69)$$

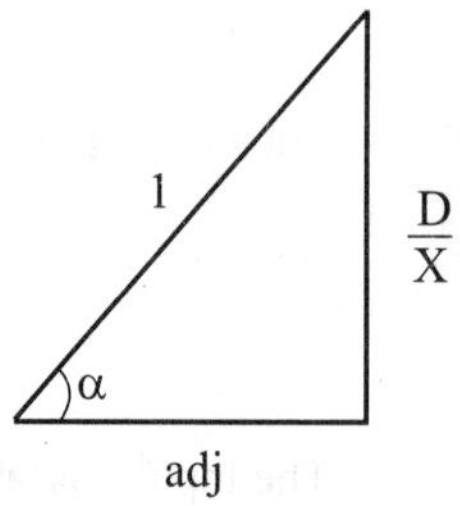

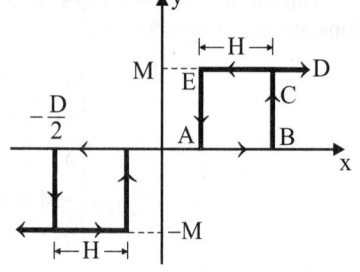

Fig 2.21

In the describing function of equation (2.68), substitute for α, $\sin\alpha$ and $\cos\alpha$ from equations (2.56) (2.55) and (2.69) respectively.

$$\therefore \ K_N(X,\omega) = K\left[1 - \frac{2}{\pi}\left(\sin^{-1}\left(\frac{D}{2X}\right) + \left(\frac{D}{2X}\right)\sqrt{1 - \left(\frac{D}{2X}\right)^2}\right)\right] \text{ for } X > \frac{D}{2} \qquad .....(2.70)$$

2.6 DESCRIBING FUNCTION OF RELAY WITH DEAD-ZONE AND HYSTERESIS

The input and the output relationship of a relay with dead-zone and hysteresis is shown in fig 2.22.

Due to dead-zone the relay will respond only after a definite value of input. Due to hysteresis the output follows a different paths for increasing and decreasing values of input. When the input x is increased from zero, the output follows the path ABCD and when the input is decreased from a maximum value, the output follows the path DCEA.

For increasing values of input, the output is zero when x<(D/2) and the output is M when x>(D/2). For decreasing values of input the output is M when x>(D/2−H) and output is zero when x < (D/2−H).

The response or output of the relay when the input is sinusoidal signal (x = Xsinωt) is shown in fig 2.23.

Fig 2.22 : Input Output characteristic of relay with dead-zone and hysteresis.

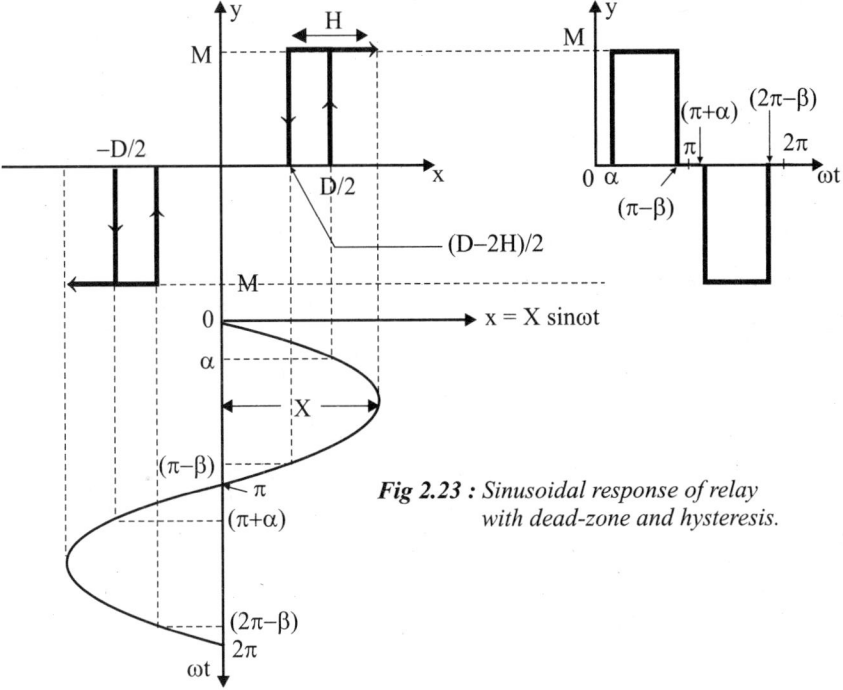

Fig 2.23 : *Sinusoidal response of relay with dead-zone and hysteresis.*

The input x is sinusoidal , $\therefore$ x = Xsinωt   .....(2.71)

where X = maximum value of input

In fig 2.23, when ωt = α, x = D/2

Hence equation (2.71) can be written as

$$D/2 = X \sin\alpha$$

$$\therefore \sin\alpha = D/2X \qquad .....(2.72)$$

$$\text{and } \alpha = \sin^{-1}\frac{D}{2X} \qquad .....(2.73)$$

In fig 2.23, when ωt = π − β, x = (D/2) − H

Hence equation (2.71) can be written as

$$D/2 - H = X \sin(\pi - \beta)$$

$$D/2 - H = X \sin\beta$$

$$\sin\beta = \frac{1}{X}\left(\frac{D}{2} - H\right) \qquad .....(2.74)$$

$$\beta = \sin^{-1}\left(\frac{1}{X}\left(\frac{D}{2} - H\right)\right) \quad . \qquad .....(2.75)$$

The output can be divided into five regions in a period of 2π and the output equation for the five regions are given by equation (2.76).

$$y = \begin{cases} 0 & ; \ 0 \le \omega t \le \alpha \\ M & ; \ \alpha \le \omega t \le (\pi - \beta) \\ 0 & ; \ (\pi - \beta) \le \omega t \le (2\pi - \beta) \\ -M & ; \ (\pi + \alpha) \le \omega t \le (2\pi - \beta) \\ 0 & ; \ (2\pi - \beta) \le \omega t \le 2\pi \end{cases} \qquad .....(2.76)$$

Let Y_1 = Amplitude of the fundamental harmonic component of the output.

ϕ_1 = Phase shift of the fundamental harmonic component of the output with respect to the input.

The describing function is given by, $K_N(X,\omega) = (Y_1/X) \angle \phi_1$

where $Y_1 = \sqrt{A_1^2 + B_1^2}$ and $\phi_1 = \tan^{-1}(B_1/A_1)$

$$A_1 = \frac{2}{\pi}\int_0^{\pi} y \sin\omega t \, d(\omega t) = \frac{2}{\pi}\int_{\alpha}^{\pi-\beta} M \sin\omega t \, d(\omega t)$$

Note : $\cos(\pi - \beta) = -\cos\beta$

$$= \frac{2M}{\pi}[-\cos\omega t]_{\alpha}^{\pi-\beta} = \frac{2M}{\pi}[-\cos(\pi-\beta) + \cos\alpha]$$

$$= \frac{2M}{\pi}[\cos\alpha + \cos\beta] \quad .....(2.77)$$

From equation (2.72) we get, $\sin\alpha = D/2X$

On constructing right angle triangle with unity hypotenuse as shown in fig 2.24, $\cos\alpha$ can be evaluated

$$\text{adj} = \sqrt{1-(D/2X)^2} \qquad \therefore \; \cos\alpha = \frac{\text{adj}}{\text{hyp}} = \sqrt{1-(D/2X)^2} \quad .....(2.78)$$

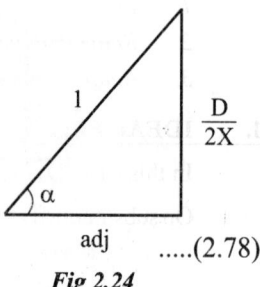

Fig 2.24

From equation (2.74) we get $\sin\beta = \left(\frac{D}{2X} - \frac{H}{X}\right)$.

On constructing right angle triangle with unity hypotenuse as shown in fig(2.25), cosβ can be evaluated

$$\text{adj} = \sqrt{1-\left(\frac{D}{2X}-\frac{H}{X}\right)^2}$$

$$\cos\beta = \frac{\text{adj}}{\text{hyp}} = \sqrt{1-\left(\frac{D}{2X}-\frac{H}{X}\right)^2} \quad .....(2.79)$$

Fig 2.25

On substituting for cos α and cos β from equations (2.78) and (2.79) in equation (2.77) we get,

$$A_1 = \frac{2M}{\pi}\left[\sqrt{1-\left(\frac{D}{2X}\right)^2} + \sqrt{1-\left(\frac{D}{2X}-\frac{H}{X}\right)^2}\right] \quad .....(2.80)$$

$$B_1 = \frac{2}{\pi}\int_0^{\pi} y \cos\omega t \, d(\omega t) = \frac{2}{\pi}\int_{\alpha}^{\pi-\beta} M \cos\omega t \, d(\omega t)$$

Note : $\sin(\pi - \beta) = \sin\beta$

$$= \frac{2M}{\pi}[\sin\omega t]_{\alpha}^{\pi-\beta} = \frac{2M}{\pi}[\sin(\pi-\beta) - \sin\alpha] = \frac{2M}{\pi}(\sin\beta - \sin\alpha)$$

On substituting for sin α and sin β from equation (2.72) and equation (2.74) we get,

$$B_1 = \frac{2M}{\pi}\left[\frac{D}{2X} - \frac{H}{X} - \frac{D}{2X}\right] = \frac{2M}{\pi}\left(\frac{-H}{X}\right) = \frac{2M}{\pi}\left(-\frac{H}{X}\right) \quad .....(2.81)$$

$$\therefore \; Y_1 = \sqrt{A_1^2 + B_1^2} = [A_1^2 + B_1^2]^{\frac{1}{2}}$$

$$Y_1 = \left[\frac{4M^2}{\pi^2}\left\{\sqrt{1-\left(\frac{D}{2X}\right)^2} + \sqrt{1-\left(\frac{D}{2X}-\frac{H}{X}\right)^2}\right\}^2 + \frac{4M^2}{\pi^2}\left(\frac{H^2}{X^2}\right)\right]^{\frac{1}{2}} \quad .....(2.82)$$

$$\phi_1 = \tan^{-1}\frac{B_1}{A_1} = \tan^{-1}\left(\frac{\frac{2M}{\pi}\left(-\frac{H}{X}\right)}{\frac{2M}{\pi}\sqrt{1-\left(\frac{D}{2X}\right)^2} + \sqrt{1-\left(\frac{D}{2X}-\frac{H}{X}\right)^2}}\right) \quad .....(2.83)$$

The describing function of the relay with dead-zone and hystersis is given by

$$K_N(X,\omega) = \frac{Y_1}{X}\angle\phi_1 \quad .....(2.84)$$

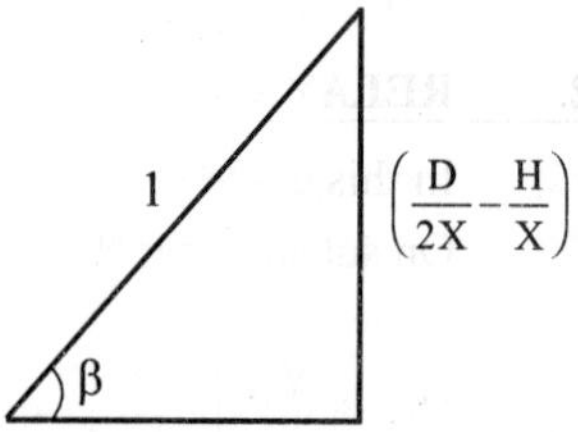

Where Y_1 is given by equation (2.82) and ϕ_1 is given by equation (2.83).

From the equation (2.84), the describing functions of the following three cases of relay can be obtained.

1. *Ideal relay.*
2. *Relay with dead-zone.*
3. *Relay with hysteresis.*

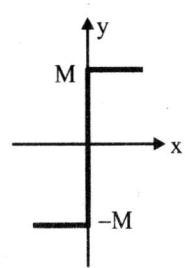

1. IDEAL RELAY

In this case, $D = H = 0$,

On substituting $D = H = 0$, in equation (2.82) and equation (2.83) we get,

$$Y_1 = \frac{2M}{\pi} \quad \text{and} \quad \phi_1 = 0$$

Fig 2.26 : Input-Output characteristics of ideal relay.

Hence the describing function of the ideal relay is given by,

$$K_N(X, \omega) = \frac{Y_1}{X} \angle \phi_1 = \frac{2M}{\pi X} \qquad \qquad .....(2.85)$$

2. RELAY WITH DEAD-ZONE

In this case $H = 0$

On substituting $H = 0$, in equation (2.82) and (2.83) we get,

$$Y_1 = \left[\frac{4M^2}{\pi^2}\left\{2\sqrt{1-\left(\frac{D}{2X}\right)^2}\right\}^2\right]^{\frac{1}{2}}$$

$$= \frac{4M}{\pi}\sqrt{1-\left(\frac{D}{2X}\right)^2}$$

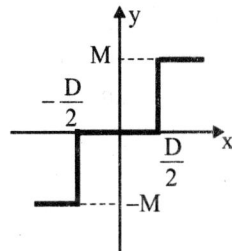

Fig 2.27 : Input-Output characteristics of relay with dead-zone.

$$\phi_1 = 0$$

Hence the describing function of relay with dead-zone is given by

$$K_N(X,\omega) = \frac{Y_1}{X} \angle \phi_1 = \begin{cases} 0 & ; X < \frac{D}{2} \\ \frac{4M}{\pi X}\sqrt{1-\left(\frac{D}{2X}\right)^2} & ; X > \frac{D}{2} \end{cases} \qquad .....(2.86)$$

3. RELAY WITH HYSTERESIS

In this case $D = H$

On substituting $D = H$ in equation (2.82) we get,

$$Y_1 = \left[\frac{4M^2}{\pi^2}\left\{\sqrt{1-\left(\frac{H}{2X}\right)^2}+\sqrt{1-\left(\frac{-H}{2X}\right)^2}\right\}^2 + \frac{4M^2}{\pi^2}\left(\frac{H^2}{X^2}\right)\right]^{\frac{1}{2}}$$

$$= \frac{2M}{\pi}\left[4\left(1-\frac{H^2}{4X^2}\right)+\left(\frac{H^2}{X^2}\right)\right]^{\frac{1}{2}}$$

$$= \frac{2M}{\pi}\left[4 - \frac{H^2}{X^2} + \frac{H^2}{X^2}\right]^{\frac{1}{2}}$$

$$= \frac{4M}{\pi}$$

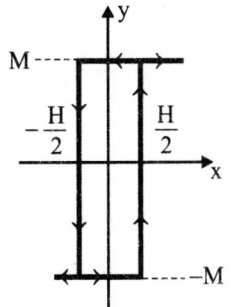

Fig 2.28 : Input-Output characteristics of relay with hystersis.

$$.....(2.87)$$

On substituting D = H in equation (2.83) we get,

$$\phi_1 = \tan^{-1} \frac{\dfrac{2M}{\pi}\left(-\dfrac{H}{X}\right)}{\dfrac{2M}{\pi}\left[\sqrt{1-\left(\dfrac{H}{2X}\right)^2}+\sqrt{1-\left(\dfrac{-H}{2X}\right)^2}\right]} = \tan^{-1}\frac{-\dfrac{H}{X}}{2\sqrt{1-\dfrac{H^2}{4X^2}}}$$

$$\phi_1 = -\tan^{-1}\frac{\dfrac{H}{2X}}{\sqrt{1-\dfrac{H^2}{4X^2}}}$$

$$\therefore \; -\phi_1 = \tan^{-1}\frac{\dfrac{H}{2X}}{\sqrt{1-\dfrac{H^2}{4X^2}}}$$

$$\tan(-\phi_1) = \frac{\dfrac{H}{2X}}{\sqrt{1-\dfrac{H^2}{4X^2}}} \qquad\qquad .....(2.87)$$

Using the numerator and denominator of equation (2.88) as two sides, we can construct a right angle triangle as shown in fig 2.29.

From fig 2.29 we get, $\sin(-\phi_1) = \dfrac{H}{2X}$

$$\therefore \; -\phi_1 = \sin^{-1}\frac{H}{2X}$$

$$\text{(or) } \phi_1 = \sin^{-1}\frac{H}{2X} \qquad\qquad .....(2.89)$$

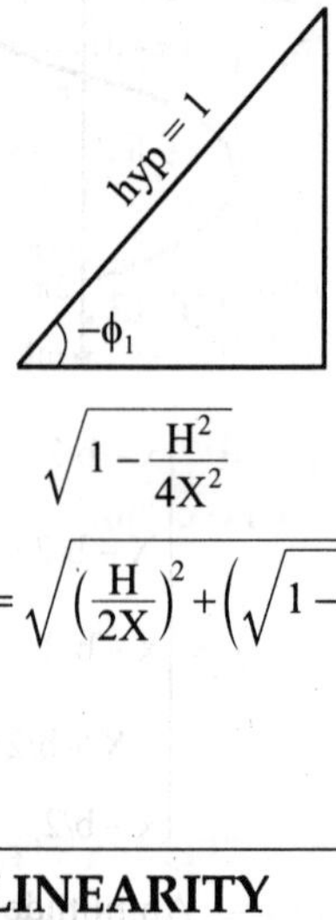

Using equations(2.87) and (2.89), the describing function of relay with hystersis can be written as,

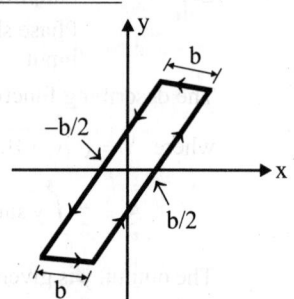

hyp = 1

$\dfrac{H}{2X}$

$-\phi_1$

Fig 2.29.

$$\sqrt{1-\frac{H^2}{4X^2}}$$

$$\text{hyp} = \sqrt{\left(\frac{H}{2X}\right)^2+\left(\sqrt{1-\frac{H^2}{4X^2}}\right)^2} = 1$$

$$K_N(X,\omega) = \frac{Y_1}{X}\angle\,\phi_1 = \begin{cases} 0 & ; \; X < \dfrac{H}{2} \\[2ex] \dfrac{4M}{\pi X}\angle\left(-\sin^{-1}\dfrac{H}{2X}\right) & ; \; X > \dfrac{H}{2} \end{cases} \qquad .....(2.90)$$

2.7 DESCRIBING FUNCTION OF BACKLASH NONLINEARITY

The input-output relationship of Backlash nonlinearity is shown in fig 2.30.

The response of the nonlinearity when the input is sinusoidal signal (x = Xsinωt) is shown in fig 2.31.

In fig 2.31, when ωt = (π−β),   x = X − b

On substituting this value of x and ωt in the input signal, x = X sinωt, we get,

X − b = X sin(π−β)

X − b = X sinβ

Fig 2.30 : *Input-Output characteristic of backlash nonlinearity.*

$$\therefore \sin\beta = \frac{X-b}{X} = 1 - \frac{b}{X} \qquad\qquad .....(2.91)$$

$$\text{and } \beta = \sin^{-1}\left(1 - \frac{b}{X}\right) \qquad\qquad .....(2.92)$$

The output can be divided into five regions in a period of 2π and the output equation for the five regions are given by equation (2.93).

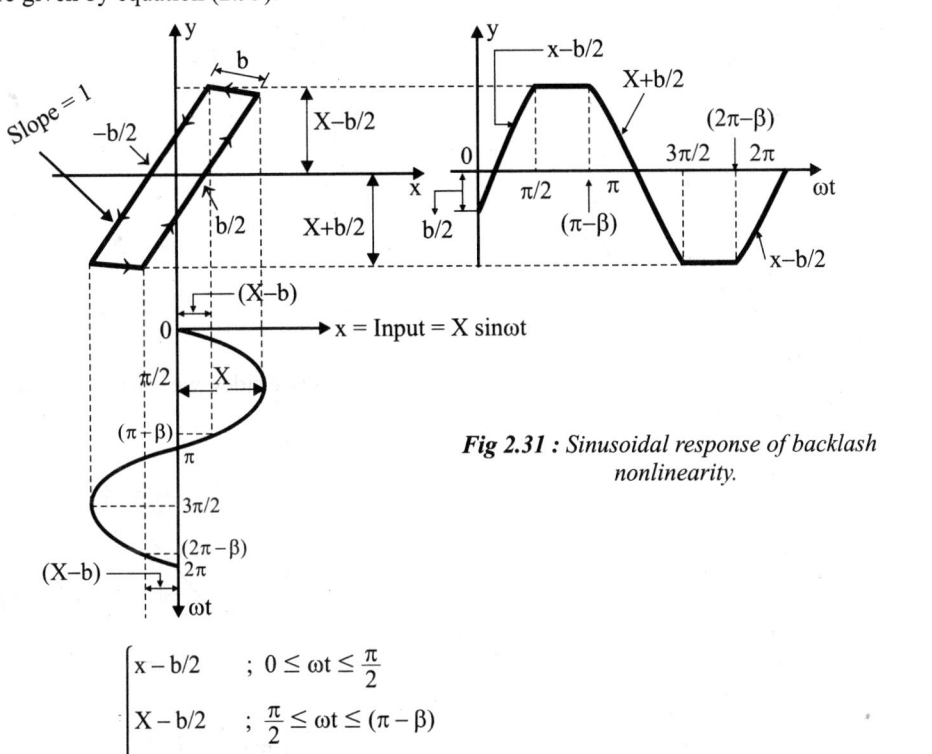

Fig 2.31 : Sinusoidal response of backlash nonlinearity.

$$y = \begin{cases} x - b/2 & ; \ 0 \le \omega t \le \frac{\pi}{2} \\ X - b/2 & ; \ \frac{\pi}{2} \le \omega t \le (\pi - \beta) \\ x + b/2 & ; \ (\pi - \beta) \le \omega t \le 3\frac{\pi}{2} \\ -X + b/2 & ; \ 3\frac{\pi}{2} \le \omega t \le (2\pi - \beta) \\ x - b/2 & ; \ (2\pi - \beta) \le \omega t \le 2\pi \end{cases} \qquad .....(2.93)$$

Let Y_1 = Amplitude of the fundamental harmonic component of the output.

ϕ_1 = Phase shift of the fundamental harmonic component of the output with respect to the input.

The describing function is given by, $K_N(X,\omega) = (Y_1/X)\angle\phi_1$

where $Y_1 = \sqrt{A_1^2 + B_1^2}$ and $\phi_1 = \tan^{-1}(B_1/A_1)$

$$A_1 = \frac{2}{\pi}\int_0^\pi y \sin\omega t \ d(\omega t) \qquad\qquad .....(2.94)$$

The output, y is given by three different equations in range 0 to π, hence equ(2.94) can be written as

$$A_1 = \frac{2}{\pi}\int_0^{\frac{\pi}{2}}\left(x - \frac{b}{2}\right)\sin\omega t \ d(\omega t) + \frac{2}{\pi}\int_{\frac{\pi}{2}}^{\pi-\beta}\left(X - \frac{b}{2}\right)\sin\omega t \ d(\omega t) + \frac{2}{\pi}\int_{\pi-\beta}^{\pi}\left(x + \frac{b}{2}\right)\sin\omega t \ d(\omega t) \quad .....(2.95)$$

Put x = X sinωt in equation (2.95)

$$A_1 = \frac{2}{\pi}\int_0^{\frac{\pi}{2}}\left(X\sin\omega t - \frac{b}{2}\right)\sin\omega t\, d(\omega t) + \frac{2}{\pi}\int_{\frac{\pi}{2}}^{\pi-\beta}\left(X - \frac{b}{2}\right)\sin\omega t\, d(\omega t)$$

$$+ \frac{2}{\pi}\int_{\pi-\beta}^{\pi}\left(X\sin\omega t + \frac{b}{2}\right)\sin\omega t\, d(\omega t)$$

$$= \frac{2X}{\pi}\int_0^{\frac{\pi}{2}}\sin^2\omega t\, d(\omega t) - \frac{b}{\pi}\int_0^{\frac{\pi}{2}}\sin\omega t\, d(\omega t) + \frac{2\left(X - \frac{b}{2}\right)}{\pi}\int_{\frac{\pi}{2}}^{\pi-\beta}\sin\omega t\, d(\omega t)$$

$$+ \frac{2X}{\pi}\int_{\pi-\beta}^{\pi}\sin^2\omega t\, d(\omega t) + \frac{b}{\pi}\int_{\pi-\beta}^{\pi}\sin\omega t\, d(\omega t) \qquad\qquad .....(2.96)$$

Put $\sin^2\omega t = \dfrac{1-\cos 2\omega t}{2}$ in equation (2.96)

$$\therefore\ A_1 = \frac{X}{\pi}\int_0^{\frac{\pi}{2}}(1-\cos 2\omega t)\, d(\omega t) - \frac{b}{\pi}\int_0^{\frac{\pi}{2}}\sin\omega t\, d(\omega t) + \frac{2\left(X-\frac{b}{2}\right)}{\pi}\int_{\frac{\pi}{2}}^{\pi-\beta}\sin\omega t\, d(\omega t)$$

$$+ \frac{X}{\pi}\int_{\pi-\beta}^{\pi}(1-\cos 2\omega t)\, d(\omega t) + \frac{b}{\pi}\int_{\pi-\beta}^{\pi}\sin\omega t\, d(\omega t)$$

$$= \frac{X}{\pi}\left[\omega t - \frac{\sin 2\omega t}{2}\right]_0^{\frac{\pi}{2}} - \frac{b}{\pi}\left[-\cos\omega t\right]_0^{\frac{\pi}{2}} + \frac{2\left(X-\frac{b}{2}\right)}{\pi}\left[-\cos\omega t\right]_{\frac{\pi}{2}}^{\pi-\beta}$$

$$+ \frac{X}{\pi}\left[\omega t - \frac{\sin 2\omega t}{2}\right]_{\pi-\beta}^{\pi} + \frac{b}{\pi}\left[-\cos\omega t\right]_{\pi-\beta}^{\pi}$$

$$= \frac{X}{\pi}\left(\frac{\pi}{2}\right) - \frac{b}{\pi}(1) + \frac{2\left(X-\frac{b}{2}\right)}{\pi}\left(-\cos(\pi-\beta) + \cos\frac{\pi}{2}\right)$$

$$+ \frac{X}{\pi}\left[\pi - \frac{\sin 2\pi}{2} - (\pi-\beta) + \frac{\sin 2(\pi-\beta)}{2}\right] + \frac{b}{\pi}[-\cos\pi + \cos(\pi-\beta)]$$

$$= \frac{X}{2} - \frac{b}{\pi} + \frac{2}{\pi}\left(X - \frac{b}{2}\right)\cos\beta + \frac{X}{\pi}\left(\beta - \frac{\sin 2\beta}{2}\right) + \frac{b}{\pi}(1-\cos\beta)$$

$$= \frac{X}{2} - \frac{b}{\pi} + \frac{2}{\pi}\left(X - \frac{b}{2}\right)\cos\beta + \frac{X\beta}{\pi} - \frac{X}{2\pi}\sin 2\beta + \frac{b}{\pi} - \frac{b}{\pi}\cos\beta$$

$$= \frac{X}{2} - \frac{X\beta}{\pi} + \frac{2}{\pi}\left(X - \frac{b}{2} - \frac{b}{2}\right)\cos\beta - \frac{X}{2\pi}\sin 2\beta$$

$$= \frac{X}{2} + \frac{X\beta}{\pi} + \frac{2X}{\pi}\left(1 - \frac{b}{X}\right)\cos\beta - \frac{X}{2\pi}\sin 2\beta \qquad\qquad .....(2.97)$$

On substituting for (1−b/X) from equation (2.91) in equation (2.97) we get,

$$A_1 = \frac{X}{2} + \frac{X\beta}{\pi} + \frac{2X}{\pi}\sin\beta\cos\beta - \frac{X}{2\pi}\sin 2\beta = \frac{X}{2} + \frac{X\beta}{\pi} + \frac{X}{\pi}\sin 2\beta - \frac{X}{2\pi}\sin 2\beta$$

$$= \frac{X}{2} + \frac{X\beta}{\pi} + \frac{X}{2\pi} \sin 2\beta = \frac{X}{\pi}\left(\frac{\pi}{2} + \beta + \frac{1}{2}\sin 2\beta\right)$$

.....(2.98)

$$B_1 = \frac{2}{\pi} \int_0^\pi y \cos \omega t \, d(\omega t)$$

.....(2.99)

The output, y is given by three different equations in the range 0 to π, hence equation (2.99) can be expressed as,

$$B_1 = \frac{2}{\pi} \int_0^{\frac{\pi}{2}} \left(x - \frac{b}{2}\right) \cos \omega t \, d(\omega t) + \frac{2}{\pi} \int_{\frac{\pi}{2}}^{\pi-\beta} \left(X - \frac{b}{2}\right) \cos \omega t \, d(\omega t)$$

.....(2.99)

$$+ \frac{2}{\pi} \int_{\pi-\beta}^{\pi} \left(x + \frac{b}{2}\right) \cos \omega t \, d(\omega t)$$

Put x = X sinωt in equation (2.100)

$$\therefore B_1 = \frac{2}{\pi} \int_0^{\frac{\pi}{2}} \left(X\sin \omega t - \frac{b}{2}\right) \cos \omega t \, d(\omega t) + \frac{2}{\pi} \int_{\frac{\pi}{2}}^{\pi-\beta} \left(X - \frac{b}{2}\right) \cos \omega t \, d(\omega t)$$

.....(2.100)

$$+ \frac{2}{\pi} \int_{\pi-\beta}^{\pi} \left(x + \frac{b}{2}\right) \cos \omega t \, d(\omega t)$$

Put x = X sinωt in equ(2.100)

$$\therefore B_1 = \frac{2}{\pi} \int_0^{\frac{\pi}{2}} \left(X\sin \omega t - \frac{b}{2}\right) \cos \omega t \, d(\omega t) + \frac{2}{\pi} \int_{\frac{\pi}{2}}^{\pi-\beta} \left(X - \frac{b}{2}\right) \cos \omega t \, d(\omega t)$$

$$+ \frac{2}{\pi} \int_{\pi-\beta}^{\pi} \left(X\sin \omega t + \frac{b}{2}\right) \cos \omega t \, d(\omega t)$$

$$= \frac{X}{\pi} \int_0^{\frac{\pi}{2}} 2\sin \omega t \cos \omega t \, d(\omega t) - \frac{b}{\pi} \int_0^{\frac{\pi}{2}} \cos \omega t \, d(\omega t) + \frac{2}{\pi}\left(X - \frac{b}{2}\right) \int_{\frac{\pi}{2}}^{\pi-\beta} \cos \omega t \, d(\omega t)$$

$$+ \frac{X}{\pi} \int_{\pi-\beta}^{\pi} 2\sin \omega t \cos \omega t \, d(\omega t) + \frac{b}{\pi} \int_{\pi-\beta}^{\pi} \cos \omega t \, d(\omega t)$$

$$B_1 = \frac{X}{\pi} \int_0^{\frac{\pi}{2}} \sin 2\omega t \, d(\omega t) - \frac{b}{\pi} \int_0^{\frac{\pi}{2}} \cos \omega t \, d(\omega t) + \frac{2}{\pi}\left(X - \frac{b}{2}\right) \int_{\frac{\pi}{2}}^{\pi-\beta} \cos \omega t \, d(\omega t)$$

$$+ \frac{X}{\pi} \int_{\pi-\beta}^{\pi} \sin 2\omega t \, d(\omega t) + \frac{b}{\pi} \int_{\pi-\beta}^{\pi} \cos \omega t \, d(\omega t)$$

$$= \frac{X}{\pi}\left[-\frac{\cos 2\omega t}{2}\right]_0^{\frac{\pi}{2}} - \frac{b}{\pi}\left[\sin \omega t\right]_0^{\frac{\pi}{2}} + \frac{2}{\pi}\left(X - \frac{b}{2}\right)\left[\sin \omega t\right]_{\frac{\pi}{2}}^{\pi-\beta}$$

$$+ \frac{X}{\pi}\left[-\frac{\cos 2\omega t}{2}\right]_{\pi-\beta}^{\pi} + \frac{b}{\pi}\left[\sin \omega t\right]_{\pi-\beta}^{\pi}$$

$$= \frac{X}{\pi}\left[-\frac{\cos\pi}{2} + \frac{\cos 0}{2}\right] - \frac{b}{\pi}\left[\sin\frac{\pi}{2} - 0\right] + \frac{2}{\pi}\left(X - \frac{b}{2}\right)\left[\sin(\pi - \beta) - \sin\frac{\pi}{2}\right]$$

$$+ \frac{X}{\pi}\left[-\frac{\cos 2\pi}{2} + \frac{\cos 2(\pi - \beta)}{2}\right] + \frac{b}{\pi}[\sin\pi - \sin(\pi - \beta)]$$

$$= \frac{X}{\pi}\left(\frac{1}{2} + \frac{1}{2}\right) - \frac{b}{\pi}(1 - 0) + \frac{2}{\pi}\left(X - \frac{b}{2}\right)(\sin\beta - 1) + \frac{X}{\pi}\left(-\frac{1}{2} + \frac{\cos 2\beta}{2}\right) + \frac{b}{\pi}(0 - \sin\beta)$$

$$= \frac{X}{\pi} - \frac{b}{\pi} + \frac{2X}{\pi}\sin\beta - \frac{b}{\pi}\sin\beta - \frac{2X}{\pi} + \frac{b}{\pi} - \frac{X}{2\pi} + \frac{X}{2\pi}\cos 2\beta - \frac{b}{\pi}\sin\beta$$

$$= \frac{X}{\pi}\left(1 - 2 - \frac{1}{2}\right) + \frac{2X}{\pi}\sin\beta - \frac{2b}{\pi}\sin\beta + \frac{X}{2\pi}\cos 2\beta$$

$$= -\frac{3X}{2\pi} + \frac{2X}{\pi}\sin\beta\left(1 - \frac{b}{X}\right) + \frac{X}{2\pi}\cos 2\beta \qquad\qquad .....(2.101)$$

Since $(1-b/X) = \sin\beta$ and $\cos 2\beta = (1 - 2\sin^2\beta)$, the equation (2.98) can be written as

$$B_1 = -\frac{3X}{2\pi} + \frac{2X}{\pi}\sin\beta(\sin\beta) + \frac{X}{2\pi}(1 - 2\sin^2\beta)$$

$$= -\frac{3X}{2\pi} + \frac{2X}{\pi}\sin^2\beta + \frac{X}{2\pi} - \frac{X}{\pi}\sin^2\beta = -\frac{X}{\pi} + \frac{X}{\pi}\sin^2\beta$$

> **Note:**
> $\sin^2\beta + \cos^2\beta = 1$
> $\therefore \sin^2\beta = 1 - \cos^2\beta$

$$= -\frac{X}{\pi} + \frac{X}{\pi}(1 - \cos^2\beta) = -\frac{X}{\pi} + \frac{X}{\pi} - \frac{X}{\pi}\cos^2\beta$$

$$= -\frac{X}{\pi}\cos^2\beta \qquad\qquad .....(2.102)$$

$$Y_1 = \sqrt{A_1^2 + B_1^2}$$

$$= \sqrt{\frac{X^2}{\pi^2}\left(\frac{\pi}{2} + \beta + \frac{1}{2}\sin 2\beta\right)^2 + \frac{X^2}{\pi^2}\cos^4\beta}$$

$$\therefore Y_1 = \frac{X}{\pi}\left[\left(\frac{\pi}{2} + \beta + \frac{1}{2}\sin 2\beta\right)^2 + \cos^4\beta\right]^{\frac{1}{2}} \qquad\qquad .....(2.103)$$

$$\phi_1 = \tan^{-1}\frac{B_1}{A_1} = \tan^{-1}\frac{-\frac{X}{\pi}\cos^2\beta}{\frac{X}{\pi}\left(\frac{\pi}{2} + \beta + \frac{1}{2}\sin 2\beta\right)} \qquad\qquad .....(2.104)$$

The describing function of backlash nonlinearity is given by,

$$K_N(X, \omega) = \frac{Y_1}{X} \angle \phi_1 \qquad\qquad .....(2.105)$$

where Y_1 is given by equation (2.103) and ϕ_1 is given by equation (2.104).

2.8 DESCRIBING FUNCTION ANALYSIS OF NONLINEAR SYSTEMS

The describing functions of nonlinear elements can be used for stability analysis of nonlinear control systems. Also it is used to predict the sustained oscillations or limit cycles in the output of the system.

Consider a unity feedback system shown in fig 2.32 in which the nonlinearity is represented by its describing function, $K_N(X, \omega)$ or K_N.

Note : *The describing function, $K_N(X, \omega)$ can be denoted by K_N.*

Let $C(j\omega)/R(j\omega)$ be the closed loop sinusoidal transfer function of the system shown in fig 2.32.

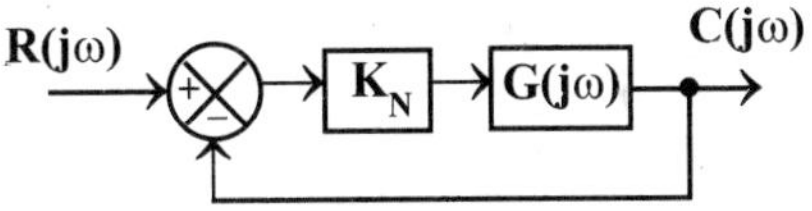

$$\therefore \frac{C(j\omega)}{R(j\omega)} = \frac{K_N G(j\omega)}{1 + K_N G(j\omega)} \qquad\qquad .....(2.106)$$

Fig 2.32

The characteristic equation of the system is obtained by equating the denominator of equation (2.106) to zero.

Hence the characteristic equation is given by,

$$1 + K_N\, G(j\omega) = 0 \qquad\qquad .....(2.107)$$

The Nyquist stability criterion can also be extended to the stability analysis of nonlinear systems. According to the Nyquist stability criterion the system will exhibit sustained oscillations or limit cycles when,

$$K_N\, G(j\omega) = -1 \qquad\qquad .....(2.108)$$

The equation(2.108) implies that the sustained oscillations or limit cycles will occur if $K_N\, G(j\omega)$ locus pass through the critical point, $-1 + j0$, in the complex plane.

The equation (2.108) can be modified as shown below

$$G(j\omega) = -\frac{1}{K_N} \qquad\qquad .....(2.109)$$

The equation(2.109) implies that the critical point, $-1 + j0$ becomes the critical locus which is the locus of $-1/K_N$. Hence the intersection point of $G(j\omega)$ locus and $-1/K_N$ locus will give the amplitude and frequency of limit cycles.

In the stability analysis, let us assume that the linear part of the system is stable. To determine the stability of the system due to nonlinearity sketch the $-1/K_N$ locus and $G(j\omega)$ locus (polar plot of $G(j\omega)$) in complex plane. (Use either a polar graph sheet or ordinary graph sheet) and from the sketches the following conclusions can be obtained.

1. *If the $-1/K_N$ locus is not enclosed by the $G(j\omega)$ locus then the system is stable or there is no limit cycle at steady state.*

2. *If the $-1/K_N$ locus is enclosed by the $G(j\omega)$ locus then the system is unstable.*

3. *If the $-1/K_N$ locus and the $G(j\omega)$ locus intersect, then the system output may exhibit a sustained oscillation or a limit cycle. The amplitude of the limit cycle is given by the value of $-1/K_N$ locus at the intersection point. The frequency of the limit cycle is given by the frequency of $G(j\omega)$ corresponding to the intersection point.*

CONCEPT OF ENCLOSURE

In a complex plane the $-1/K_N$ locus is said to be enclosed by $G(j\omega)$ locus if it lies in the region to the right of an observer travelling through $G(j\omega)$ locus in the direction of increasing ω, as shown in fig 2.33.

In a complex plane $-1/K_N$ locus is not enclosed by $G(j\omega)$ if it lies in the region to the left of an observer travelling through $G(j\omega)$ locus in the direction of increasing ω, as shown in fig 2.34.

If the $-1/K_N$ locus and $G(j\omega)$ locus intersect as shown in fig 2.35, then for an observer travelling through $G(j\omega)$ locus in the direction of increasing ω, the region on the right is unstable region and the region on the left is stable region.

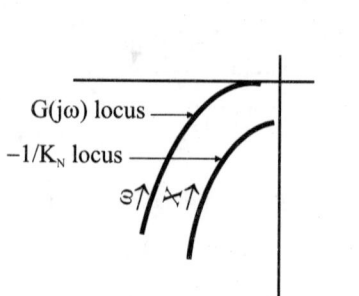

Fig 2.33 : *Figure showing enclosure of* $-1/K_N$ *locus by* $G(j\omega)$ *locus.*

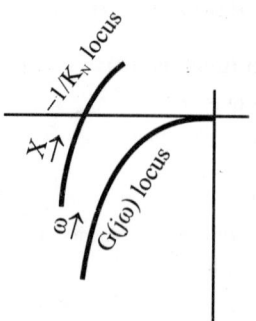

Fig 2.34 : *Figure showing nonenclosure of* $-1/K_N$ *locus by* $G(j\omega)$ *locus.*

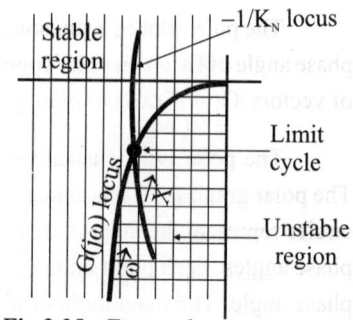

Fig 2.35 : *Figure showing intersection of* $-1/K_N$ *locus by* $G(j\omega)$ *locus.*

STABLE AND UNSTABLE LIMIT CYCLES

The $-1/K_N$ locus may intersect $G(j\omega)$ locus at one or more points. There exists a limit cycle at every intersecting point. These limit cycles can be either stable or unstable limit cycles, as shown in fig 2.36.

If $-1/K_N$ locus travels in unstable region and it intersect $G(j\omega)$ locus to enter stable region then the limit cycle corresponding to that intersection point is stable limit cycle.

If $-1/K_N$ locus travels in stable region and it intersect $G(j\omega)$ locus to enter unstable region then the limit cycle corresponding to that intersection point is unstable limit cycle.

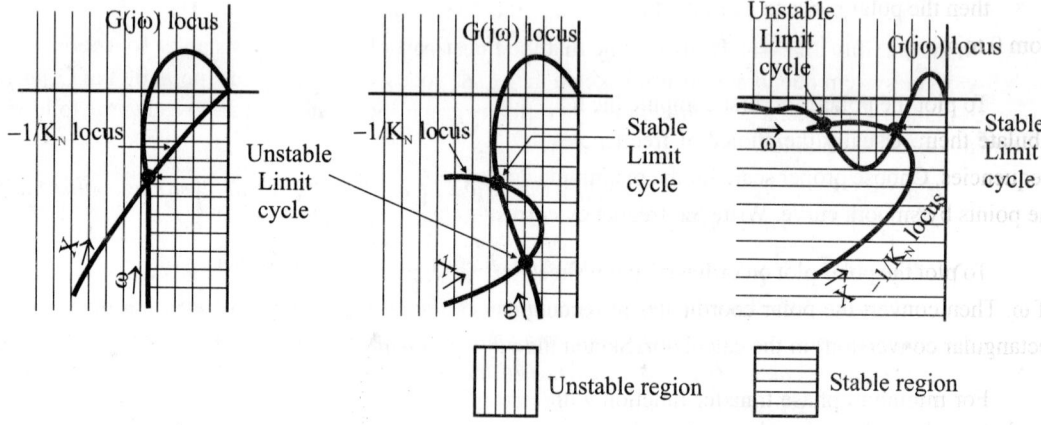

Fig 2.36 : *Stable and unstable limit cycles.*

Note : *The concept of enclosure can be extended to dp-phase angle plane (i.e., to Nichols plot) and it is same as that of complex plane.*

2.9   REVIEW OF POLAR PLOT AND NICHOLS PLOT POLAR PLOT

The polar plot of a sinusoidal transfer function, $G(j\omega)$ is a plot of the magnitude of $G(j\omega)$ versus the phase angle of $G(j\omega)$ on polar coordinates as ω is varied from zero to infinity. Thus the polar plot is the locus of vectors $|G(j\omega)|\ \angle G(j\omega)$ as ω is varied from zero to infinity. The polar plot is also called Nyquist plot.

The polar plot is usually plotted on a polar graph sheet. The polar graph sheet has concentric circles and radial lines. The circles represent the magnitude and the radial lines represent the phase angles. Each point on the polar graph has a magnitude and phase angle. The magnitude of a point is given by the value of the circle passing through that point and the phase angle is given by the radial line passing through that point. In polar graph sheet a positive phase angle is measured in anticlockwise from the reference axis (0°) and a negative angle is measured clockwise from the reference axis (0°).

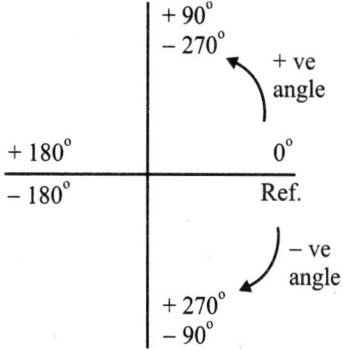

Fig 2.37 : Polar graph.

Alternatively, if $G(j\omega)$ can be expressed in rectangular coordinates as,

$$G(j\omega) = G_R(j\omega) + jG_I(j\omega)$$

where, $G_R(j\omega)$ = Real part of $G(j\omega)$

and $G_I(j\omega)$ = Imaginary part of $G(j\omega)$,

then the polar plot can be plotted in ordinary graph sheet between $G_R(j\omega)$ and $G_I(j\omega)$ as ω is varied from 0 to ∞.

To plot the polar plot, first compute the magnitude and phase of $G(j\omega)$ for various values of ω and tabulate them. Usually the choice of frequencies are corner frequencies and frequencies around corner frequencies. Choose proper scale for the magnitude circles. Fix all the points on polar graph sheet and join the points by smooth curve. Write the frequency corresponding to each point of the plot.

To plot the polar plot on ordinary graph sheet, compute the magnitude and phase for various values of ω. Then convert the polar coordinates to rectangular coordinates using $\mathbf{P \rightarrow R}$ conversion (polar to rectangular conversion) in the calculator. Sketch the polar plot using rectangular coordinates.

For minimum phase transfer function with only poles, the type number of the system determines at what quadrant the polar plot starts and the order of the system determines at what quadrant the polar plot ends.

Note : *The minimum phase systems are systems with all poles and zeros on the left half of s-plane*

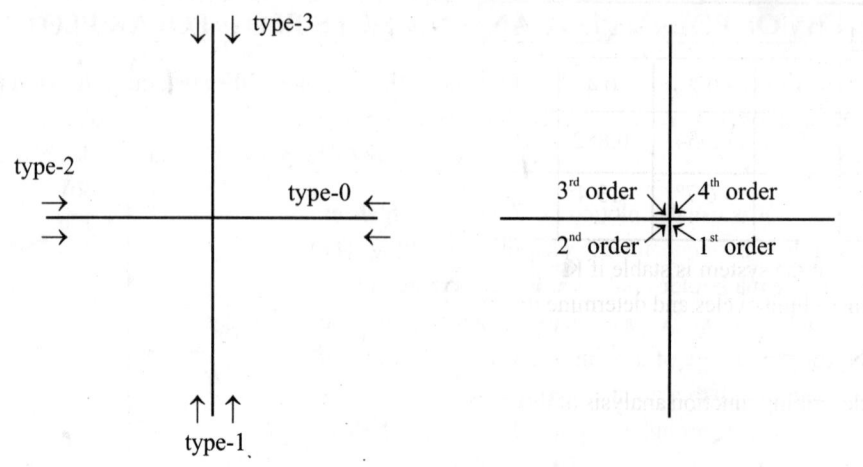

Fig 2.38 : Start of polar plot.
Fig 2.39 : End of polar plot.

NICHOLS PLOT

The **Nichols plot** is a frequency response plot of the open loop transfer function of a system. The Nichols plot is a graph between magnitude of $G(j\omega)$ in db and the phase of $G(j\omega)$ in degree, plotted on a ordinary graph sheet.

To plot the Nichols plot, first compute the magnitude of $G(j\omega)$ in db and phase of $G(j\omega)$ in deg for various values of ω and tabulate them. Usually the choice of frequencies are corner frequencies. Choose appropriate scales for magnitude on y-axis and phase on x-axis. Fix all the points on ordinary graph sheet and join the points by smooth curve. Write the frequency corresponding to each point of the plot.

In another method, first the Bode plot of $G(j\omega)$ is sketched. From the Bode plot the magnitude and phase for various values of frequency, ω are noted and tabulated. Using these values the Nichols plot is sketched as explained earlier.

In a system if the zero frequency gain K is varied then the magnitude of the transfer function alone will vary and there will not be any change in phase. This results in vertical shift of Nichols plot up or down. The constant K adds 20log K to every point of the plot. If 20log K is positive then the plot shifts upwards and if it is negative the plot shifts downwards.

EXAMPLE 2.2

A servo system used for positioning a load has backlash characteristics as shown in fig 2.2.1. The block diagram of the system is shown in fig 2.2.2. The magnitude and phase of the describing function of backlash nonlinearity for various values of b/X are listed in table-2.2.1, where X = Maximum value of input sinusoidal signal to the nonlinearity.

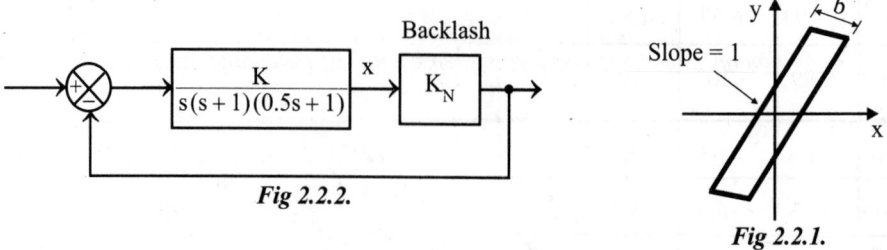

Fig 2.2.2.
Fig 2.2.1.

TABLE-2.2.1

b/X	0	0.2	0.4	1.0	1.4	1.6	1.8	1.9	2.0
$|K_N|$	1	0.954	0.882	0.592	0.367	0.248	0.125	0.064	0
$\angle K_N$	0	$-6.7°$	$-13.4°$	$-32.5°$	$-46.6°$	$-55.2°$	$-66°$	$-69.8°$	$-90°$

Show that the system is stable if K = 1. Also show that limit cycle exists when K = 2. Investigate the stability of these limit cycles and determine their frequency and b/X.

SOLUTION

The describing function analysis of the system can be carried using either polar plot or using Nichols plot.

METHOD-1 : USING POLAR PLOT

Polar plot of G(jω) when K = 1

Given that, $G(s) = \dfrac{K}{s(1+s)(1+0.5s)}$

Let K = 1 and s = jω

$\therefore G(j\omega) = \dfrac{1}{j\omega(1+j\omega)(1+j0.5\omega)}$

$= \dfrac{1}{\omega\angle 90° \sqrt{1+\omega^2} \angle \tan^{-1}\omega \sqrt{1+0.25\omega^2} \angle \tan^{-1}0.5\omega}$

$|G(j\omega)| = \dfrac{1}{\omega\sqrt{1+\omega^2}\sqrt{1+0.25\omega^2}}$

$G(j\omega) = -90° - \tan^{-1}\omega - \tan^{-1}0.5\omega$

The magnitude and phase of G(jω) are calculated for various values of ω and tabulated in table-2.2.2. Using polar to rectangular conversion the polar coordinates are converted to rectangular coordinates and listed in table-2.2.2. The polar plot of G(jω) when K = 1, is drawn in an ordinary graph sheet, as shown in fig 2.2.3.

TABLE-2.2.2

ω rad/sec	0.1	0.15	0.2	0.25	0.5	0.75	1.0	1.25
$|G(j\omega)|$	9.94	6.57	4.88	3.85	1.74	1.0	0.63	0.42
$\angle G(j\omega)$ deg	-99	-103	-107	-111	-131	-147	-162	-173
$G_R(j\omega)$	-1.6	-1.5	-1.4	-1.4	-1.1	-0.8	-0.6	-0.4
$G_I(j\omega)$	-9.8	-6.4	-4.7	-3.6	-1.3	-0.5	-0.2	-0.05

Polar plot of G(jω) when K = 2

The magnitude of G(jω) when K = 2 is given by

$$|G(j\omega)| = \frac{2}{\omega\sqrt{1+\omega^2}\sqrt{1+0.25\omega^2}}$$

(The phase of G(jω) will not change due to a change in the value of K)

The magnitude and phase of G(jω) and the real part and imaginary part of G(jω) when K = 2 are calculated for various values of ω and listed in table-2.2.3. The polar plot of G(jω) when K = 2, is drawn on the same graph sheet using the same scales as shown in fig 2.2.3.

TABLE-2.2.3

ω rad/sec	0.2	0.25	0.3	0.5	0.75	1.0	1.25
$|G(j\omega)|$	9.76	7.7	6.31	3.48	2.0	1.26	0.84
$\angle G(j\omega)$ deg	−107	−111	−115	−131	−147	−162	−173
$G_R(j\omega)$	−2.9	−2.8	−2.7	−2.3	−1.7	−1.2	−0.8
$G_I(j\omega)$	−9.3	−7.2	−5.7	−2.6	−1.1	−0.4	−0.1

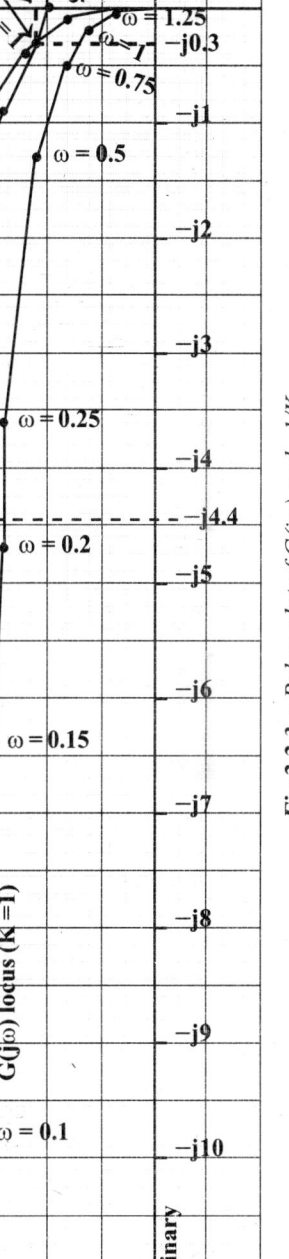

Fig 2.2.3. : Polar plot of G(jω) and −1/K_N

Polar plot of $-1/K_N$

The function $-1/K_N$ can be written as,

$$-1/K_N = -1 \times \frac{1}{K_N} = 1\angle -180 \times \frac{1}{|K_N|\angle K_N}$$

$$\therefore |-1/K_N| = \frac{1}{K_N} \quad \text{and} \quad \angle(-1/K_N) = -180° - \angle K_N$$

The values of $|K_N|$ and $\angle K_N$ are given in the problem, in table-2.2.1., for various values of b/X. Using the values in table-2.2.1, the $|-1/K_N|$ and $\angle(-1/K_N)$ are calculated for various values of b/X and listed in table-2.2.4. Then the real part and imaginary part of $-1/K_N$ are calculated using polar to rectangular convertion and listed in table-2.2.4. The locus of $-1/K_N$ is sketched using rectangular coordinates in the same graph sheet as shown in fig 2.2.3.

TABLE-2.2.4.

b/X	0	0.2	0.4	1.0	1.4	1.6	1.8	1.9	2.0
$|K_N|$	1	0.954	0.882	0.592	0.367	0.248	0.125	0.064	0
$\angle K_N$	0	−6.7°	−13.4°	−32.5°	−46.6°	−55.2°	−66°	−69.8°	−90°
$|-1/K_N|$	1	1.05	1.13	1.69	2.72	4.03	8.0	15.63	∞
$\angle(-1/K_N)$	−180°	−173°	−166°	−148°	−133°	−125°	−114°	−110°	−90°
Real part of $-1/K_N$	−1.0	−1.04	−1.1	−1.4	−1.9	−2.3	−3.3	−5.3	0
Ima. part of $-1/K_N$	0	−0.1	−0.3	−0.9	−2.0	−3.3	−7.3	−14.7	∞

STABILITY ANALYSIS

Case (i), K = 1

When K = 1, the G(jω) locus does not enclose the $-1/K_N$ locus, hence the system is stable.

Case (ii), K = 2

When K = 2, the G(jω) locus, intersects the $-1/K_N$ locus at two points. From the polar plots, it is observed that at one intersection point, unstable limit cycle exists and at another intersection point stable limit cycle exists.

From fig 2.2.3,

Coordinates corresponding to unstable limit cycle = −2.6 − j4.4 = 5.11 ∠−120°

Let ω_{l1} =   Frequency corresponding to unstable limit cycle.

and b/X_1 =   The value of b/X corresponding to unstable limit cycle

Now at $\omega \;=\; \omega_{l1}$, $G(j\omega) = 5.11 \angle -120°$

$\therefore$ At $\omega = \omega_{l1}$, $\angle G(j\omega) = -120°$

By equating the expression for $\angle G(j\omega)$ to $-120°$, the frequency ω_{l1} can be determined.

We know that, $\angle G(j\omega) = -90° - \tan^{-1} \omega - \tan^{-1} 0.5\omega$

At $\omega = \omega_{l1}$, $-90° - \tan^{-1} \omega_{l1} - \tan^{-1} 0.5\omega_{l1} = -120°$

$\therefore$ $90° + \tan^{-1} \omega_{l1} + \tan^{-1} 0.5\omega_{l1} = 120°$

$\tan^{-1} \omega_{l1} + \tan^{-1} 0.5\omega_{l1} = 120° - 90° = 30°$

On taking tan on either side we get,

$\tan(\tan^{-1} \omega_{l1} + \tan^{-1} 0.5\omega_{l1}) = \tan 30°$

$\dfrac{\tan(\tan^{-1}\omega_{l1}) + \tan(\tan^{-1}0.5\omega_{l1})}{1 - \tan(\tan^{-1}\omega_{l1}) \times \tan(\tan^{-1}0.5\omega_{l1})} = 0.577$

> **Note :** $\tan(A + B) = \dfrac{\tan A + \tan B}{1 - \tan A \tan B}$

$\dfrac{\omega_{l1} + 0.5\omega_{l1}}{1 - \omega_{l1} \times 0.5\omega_{l1}} = 0.577$

$\therefore 1.5\omega_{l1} = 0.577 - 0.2885\omega_{l1}^2$

$\therefore 0.2885\omega_{l1}^2 + 1.5\omega_{l1} - 0.577 = 0$

On dividing by 0.2885 we get,

$\omega_{l1}^2 + \dfrac{1.5}{0.2885}\,\omega_{l1} - \dfrac{0.577}{0.2885} = 0$

$\omega_{l1}^2 + 5.2\,\omega_{l1} - 2 = 0$

$\therefore \omega_{l1} = \dfrac{-5.2 \pm \sqrt{5.2^2 - 4 \times (-2)}}{2} = \dfrac{-5.2 \pm 5.92}{2}$

On taking only positive root we get,

$\omega_{l1} = \dfrac{-5.2 + 5.92}{2} = \mathbf{0.36 \; rad/sec}$

Also, at $\omega = \omega_{l1}$, $-1/K_N = 5.11 \angle -120°$

But $-1/K_N = 1\angle -180° \times \dfrac{1}{K_N}$, $\quad \therefore 1\angle -180° \times \dfrac{1}{K_N} = 5.11\angle -120°$

$\therefore K_N = \dfrac{1\angle -180°}{5.11\angle -120°} = \dfrac{1}{5.11}\angle -60° = 0.196\angle -60°$

Hence at $\omega = \omega_{l1}$, $|K_N| = 0.196$ and $\angle K_N = -60°$.

From the describing function of backlash nonlinearity we get,

$\angle K_N = \tan^{-1}\!\left(\dfrac{-\cos^2\beta}{(\pi/2) + \beta + (1/2)\sin 2\beta}\right)$

At $\omega = \omega_{l1}$, $\angle K_N = -60°$, $b/X \to b/X_1$ and $\beta \to \beta_1$

$$\therefore \tan^{-1}\left(\frac{-\cos^2\beta_1}{(\pi/2)+\beta_1+(1/2)\sin 2\beta_1}\right)=-60°$$

$$\frac{-\cos^2\beta_1}{(\pi/2)+\beta_1(1/2)\sin 2\beta_1}=\tan(-60°)$$

$$\therefore (\pi/2)+\beta_1+(1/2)\sin 2\beta_1=\frac{-\cos^2\beta_1}{\tan(-60°)}=0.577\cos^2\beta_1$$

From the describing function of backlash nonlinearity we get,

$$|K_N|=\frac{1}{\pi}[((\pi/2)+\beta+(1/2)\sin 2\beta)^2+\cos^4\beta]^{\frac{1}{2}}$$

At $\omega=\omega_{I1}$, $|K_N|=0.196°$, $b/X \rightarrow b/X_1$ and $\beta \rightarrow \beta_1$

$$\therefore \frac{1}{\pi}[((\pi/2)+\beta_1+\frac{1}{2}\sin 2\beta_1)^2+\cos^4\beta_1]^{\frac{1}{2}}=0.196$$

On substituting $((\pi/2)+\beta_1+(1/2)\sin 2\beta_1)=0.577\cos^2\beta_1$ and then squaring we get,

$$(0.577\cos^2\beta_1)^2+\cos^4\beta_1=(0.196\pi)^2$$

$$0.333\cos^4\beta_1+\cos^4\beta_1=0.379$$

$$1.333\cos^4\beta_1=0.379$$

$$\therefore \cos\beta_1=\left(\frac{0.379}{1.333}\right)^{\frac{1}{4}}=0.73 \quad ; \quad \beta_1=\cos^{-1}(0.73)=43.1°$$

We know that, $\beta=\sin^{-1}(1-b/X)$

$$\therefore \beta_1=\sin^{-1}(1-b/X_1) \quad \text{(or)} \quad b/X_1=1-\sin\beta_1=1-\sin 43.1°=0.316$$

From fig 2.2.3.

Coordinates corresponding to stable limit cycle = $-1.1-j0.3=1.14 \angle-165°$

Let ω_{I2} =   Frequency corresponding to stable limit cycle.

and b/X_2 =   The value of b/X corresponding to stable limit cycle

Now at ω = ω_{I2}, $G(j\omega)=1.14 \angle-165°$

$\therefore$ At $\omega=\dot\omega_{I2}$, $\angle G(j\omega)=-165°$

By equating the expression for $\angle G(j\omega)$ to $-165°$, the frequency ω_{I2} can be determined.

We know that, $\angle G(j\omega)=-90°-\tan^{-1}\omega-\tan^{-1}0.5\omega$

At $\omega=\omega_{I2}$, $-90°-\tan^{-1}\omega_{I2}-\tan^{-1}0.5\omega_{I2}=-165°$

$\therefore \tan^{-1}\omega_{I2}+\tan^{-1}0.5\omega_{I2}=-165°-90°=75°$

On taking tan on either side we get,

$$\tan(\tan^{-1}\omega_{I2}+\tan^{-1}0.5\omega_{I2})=\tan 75°$$

$$\frac{\tan(\tan^{-1}\omega_{I2})+\tan(\tan^{-1}0.5\omega_{I2})}{1-\tan(\tan^{-1}\omega_{I2})\times\tan(\tan^{-1}0.5\omega_{I2})}=3.732$$

$$\frac{\omega_{I2}+0.5\omega_{I2}}{1-\omega_{I2}\times 0.5\omega_{I2}}=3.732$$

$$\therefore 1.5\omega_{I2}=3.732-1.886\omega_{I2}^2$$

$$\therefore \omega_{12}^2 + \frac{1.5}{1.866}\omega_{12} - \frac{3.732}{1.866} = 0 \quad ; \quad \omega_{12}^2 + 0.8\omega_{12} - 2 = 0$$

$$\therefore \omega_{12} = \frac{-0.8 \pm \sqrt{0.8^2 - 4 \times (-2)}}{2} = \frac{-0.8 \pm 2.94}{2}$$

On taking only positive root we get,

$$\omega_{12} = \frac{-0.8 + 2.94}{2} = 1.07 \text{ rad/sec}$$

Also, at $\omega = \omega_{12}$, $\quad -1/K_N = 1.14\angle -165°$

But $\quad -1/K_N = 1\angle -180° \times \dfrac{1}{K_N}$, $\quad \therefore 1\angle 180° \times \dfrac{1}{K_N} = 1.14\angle -165°$

$$\therefore K_N = \frac{1\angle -180°}{1.14\angle -165°} = \frac{1}{1.14}\angle -15° = 0.877\angle -15°$$

Hence at $\omega = \omega_{12}$, $|K_N| = 0.877$ and $\angle K_N = -15°$.

From the describing function of backlash nonlinearity we get,

$$\angle K_N = \tan^{-1}\left(\frac{-\cos^2\beta}{(\pi/2) + \beta + (1/2)\sin 2\beta}\right)$$

At $\omega = \omega_{12}$, $\angle K_N = -15°$, $b/X \to b/X_2$ and $\beta \to \beta_2$

$$\therefore \tan^{-1}\left(\frac{-\cos^2\beta_2}{(\pi/2) + \beta_2 + (1/2)\sin 2\beta_2}\right) = -15°$$

$$\frac{-\cos^2\beta_2}{(\pi/2) + \beta_2 + (1/2)\sin 2\beta_2} = \tan(-15°)$$

$$\therefore (\pi/2) + \beta_2 + (1/2)\sin 2\beta_2 = \frac{-\cos^2\beta_2}{\tan(-15°)} = 3.732\cos^2\beta_2$$

From the describing function of backlash nonlinearity we get,

$$|K_N| = \frac{1}{\pi}\left[((\pi/2) + \beta + (1/2)\sin 2\beta)^2 + \cos^4\beta\right]^{\frac{1}{2}}$$

At, $\omega = \omega_{12}$, $|K_N| = 0.877$, $b/X \to b/X_2$ and $\beta \to \beta_2$

$$\therefore \frac{1}{\pi}\left[((\pi/2) + \beta_2 + \frac{1}{2}\sin 2\beta_2)^2 + \cos^4\beta_2\right]^{\frac{1}{2}} = 0.877$$

On substituting and then squaring we get,

$$(3.732\cos^2\beta_2)^2 + \cos^4\beta_2 = (0.877\pi)^2$$

$$13.928\cos^4\beta_2 = \cos^4\beta_2 = 7.59$$

$$14.928\cos^4\beta_2 = 7.59$$

$$\therefore \cos\beta_2 = \left(\frac{7.59}{14.928}\right)^{\frac{1}{4}} = 0.844 \quad ; \quad \therefore \beta_2 = \cos^{-1}(0.844) = 32.4°$$

We know that, $\beta = \sin^{-1}(1 - b/X)$

$$\therefore \beta_2 = \sin^{-1}(1 - b/X_2) \quad \text{(or)} \quad \mathbf{b/X_2 = 1 - \sin\beta_2 = 1 - \sin 32.4° = 0.464}$$

RESULT

1. The unstable limit cycle exist when $b/X = 0.316$ and the frequency of oscillation is 0.36 rad/sec.
2. The stable limit cycle exist when $b/X = 0.464$ and the frequency of oscillation is 1.07 rad/sec.

METHOD - 2 : USING NICHOLS PLOT

Nichols plot of G(jω) when K = 1

Given that, $G(s) = K/s(1+s)(1+0.5s)$

Let $K = 1$ and put $s = j\omega$

$$\therefore G(j\omega) = \frac{1}{j\omega(1+j\omega)(1+j0.5\omega)} = \frac{1}{\omega \angle 90° \sqrt{1+\omega^2} \angle \tan^{-1}\omega \sqrt{1+0.25\omega^2} \angle \tan^{-1}0.5\omega}$$

$$\therefore |G(j\omega)| = \frac{1}{\omega\sqrt{1+\omega^2}\sqrt{1+0.25\omega^2}}$$

$$|G(j\omega)|_{in\ db} = 20\log\left[\frac{1}{\omega\sqrt{1+\omega^2}\sqrt{1+0.25\omega^2}}\right]$$

$$\angle G(j\omega) = -90° - \tan^{-1}\omega - \tan^{-1}0.5\omega$$

The magnitude of G(jω) in db and phase of G(jω) are calculated for various values of ω and tabulated in table-2.2.5. The Nichols plot of G(jω) is sketched in an ordinary graph sheet as shown in fig 2.2.4.

TABLE-2.2.5.

ω rad/sec	0.1	0.15	0.2	0.25	0.5	0.75	1.0	1.25
|G(jω)| db	19.9	16.4	13.8	11.7	4.8	0	−4	−7.5
∠G(jω) deg	−99	−103	−107	−111	−131	−147	−162	−173

Nichols plot of G(jω) when K = 2

When K=2, the magnitude of G(jω) increases by an amount, $20\log K = 20\log 2 = 6$ db. The phase of G(jω) is not altered.

The increase in magnitude is independent of frequency. Hence G(jω) locus when $K = 2$ is obtained by shifting the locus of G(jω) when $K = 1$, by 6 db upwards as shown in fig 2.2.4.

Nichols plot of −1/K_N

The function $-1/K_N$ can be written as,

$$-1/K_N = -1 \times \frac{1}{K_N} = 1\angle -180° \times \frac{1}{|K_N|\angle K_N} = \frac{1}{|K_N|}\angle(-180° - \angle K_N)$$

$$\therefore |-1/K_N| = \frac{1}{|K_N|} \qquad |-1/K_N|_{in\ db} = 20\log\frac{1}{|K_N|}$$

$$\angle(-1/K_N) = -180° - \angle K_N$$

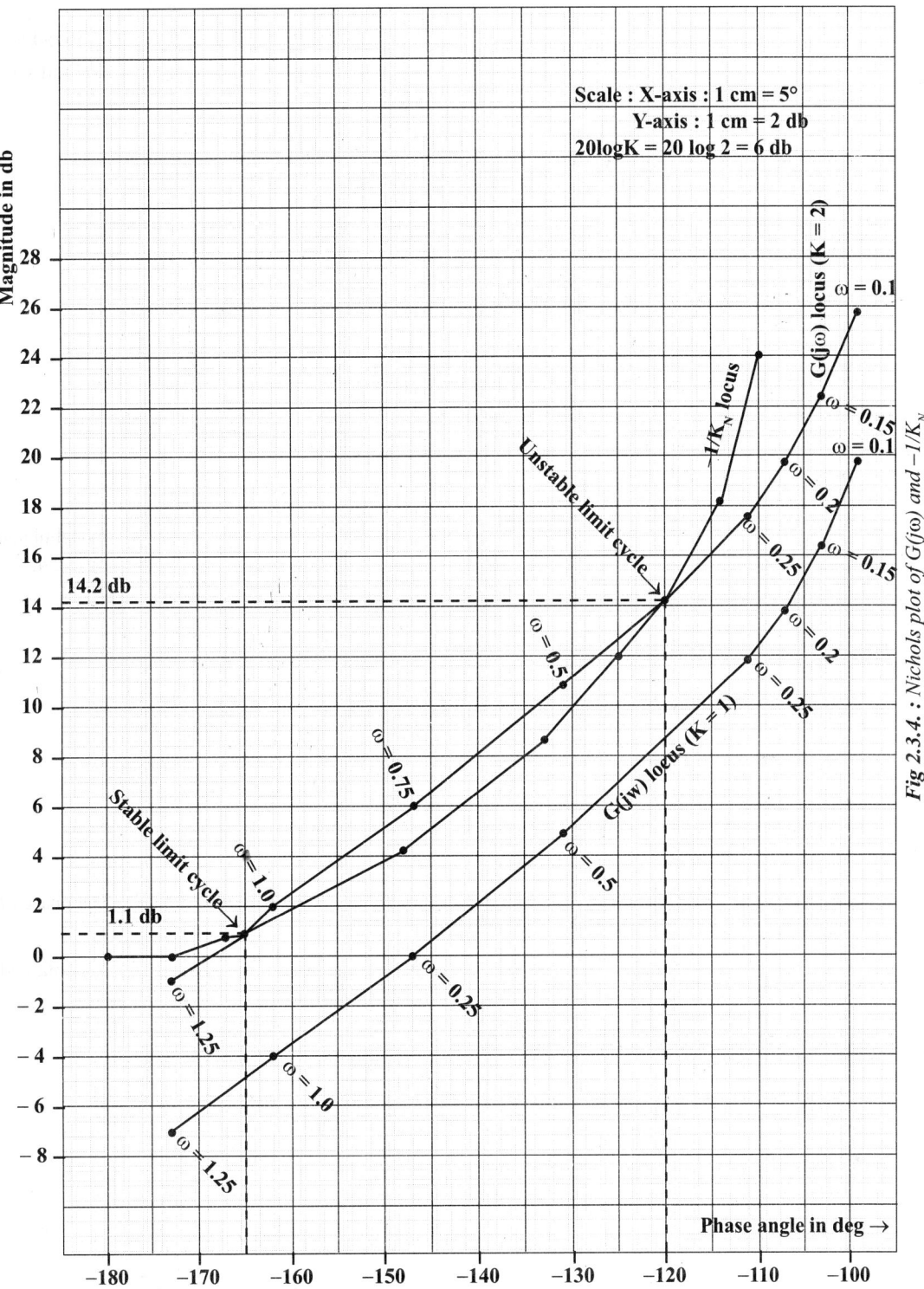

Scale : X-axis : 1 cm = 5°
Y-axis : 1 cm = 2 db
$20 \log K = 20 \log 2 = 6$ db

Fig 2.3.4. : Nichols plot of $G(j\omega)$ and $-1/K_N$

The magnitude and phase of the describing function of backlash, K_N is listed in the problem in table-2.2.1 for various values of b/X. Using the values of $|K_N|$ and $\angle K_N$ given in table-2.2.1, the values of $|-1/K_N|$ in db and $\angle(-1/K_N)$ are calculated for various values of b/X and listed in table-2.2.6. Using these values the locus of $-1/K_N$ is sketched as shown in fig 2.2.4.

TABLE-2.2.6.

b/X	0	0.2	0.4	1.0	1.4	1.6	1.8	1.9	2.0
$|K_N|$	1	0.954	0.882	0.592	0.367	0.248	0.125	0.064	0
$\angle K_N$	0	−6.7°	−13.4°	−32.5°	−46.6°	−55.2°	−66°	−69.8°	−90°
$|-1/K_N|$ in db	0	0.4	1.0	4.6	8.7	12.1	18.1	23.9	∞
$\angle(-1/K_N)$ in deg	−180°	−173°	−166°	−148°	−133°	−125°	−114°	−110°	−90°

STABILITY ANALYSIS

Case (i) when K = 1

From the Nichols plots it is observed that when K = 1, G(jω) locus does not enclose −1/K_N locus. Hence the system is stable.

Case (ii) when K = 2

From the Nichols plots it is observed that when K = 2, G(jω) locus, intersects −1/K_N locus at two points. At one intersection point unstable limit cycle exists and at another intersection point stable limit cycle exist.

The coordinates corresponding to unstable limit cycle $\Big\} = (14.2 \text{ db}, -120°) = 10^{14.2/20} \angle -120° = 5.1\angle -120°$

The coordinates corresponding to stable limit cycle $\Big\} = (1.1 \text{ db}, -165°) = 10^{1.1/20} \angle -165° = 1.14\angle -165°$

Note : It is observed that the coordinates corresponding to limit cycles are same as that obtained from polar plot, hence by an analysis similar to that of method-1. We can determine the frequency and b/X corresponding to limit cycles.

RESULT

1. The unstable limit cycle exist when b/X = 0.316 and the frequency of oscillation is 0.36 rad/sec.

2. The stable limit cycle exist when b/X = 0.464 and the frequency of oscillation is 1.07 rad/sec.

EXAMPLE 2.3

Consider a unity feedback system shown in fig 2.3.1 having a saturating amplifier with gain K. Determine the maximum value of K for the system to stay stable. What would be the frequency and nature of limit cycle for a gain of K = 2.5?

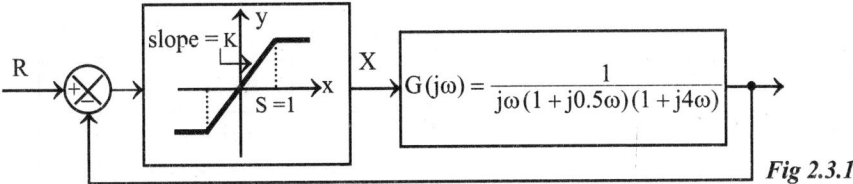

Fig 2.3.1.

SOLUTION

The stability of the system can be analysed using polar plot. The gain, K of the saturating amplifier can be attached to G(jω) and amplifier is considered to be an unity gain amplifier.

Polar plot of G(jω) when K = 1

Here, $G(j\omega) = \dfrac{K}{j\omega(1+j0.5\omega)(1+j4\omega)}$

Let, $K = 1$, $\therefore G(j\omega) = \dfrac{1}{j\omega(1+j0.5\omega)(1+j4\omega)}$

$$= \dfrac{1}{\omega\angle 90° \sqrt{1+0.25\omega^2}\angle\tan^{-1}0.5\omega \sqrt{1+16\omega^2}\angle\tan^{-1}4\omega}$$

$\therefore |G(j\omega)| = \dfrac{1}{\omega\sqrt{1+0.25\omega^2}\sqrt{1+16\omega^2}}$

$\angle G(j\omega) = -90° - \tan^{-1}0.5\omega - \tan^{-1}4\omega$

The magnitude and phase of G(jω) are calculated for various values of ω and listed in table-2.3.1. Using polar to rectangular conversion the real part and imaginary part of G(jω) are determined and listed in table-2.3.1. The polar plot of G(jω) is sketched in an ordinary graph sheet as shown in fig 2.3.2.

TABLE-2.3.1

ω rad/sec	0.4	0.5	0.6	0.8	1.0	1.2
|G(jω)|	1.299	0.868	0.614	0.346	0.216	0.145
∠G(jω)	−159°	−167°	−174°	−184°	−192°	−199°
$G_R(j\omega)$	−1.21	−0.85	−0.61	−0.35	−0.21	−0.14
$G_I(j\omega)$	−0.47	−0.2	−0.06	0.02	0.04	0.05

Polar plot of G(jω) when K = 2.5

When K = 2.5, $|G(j\omega)| = \dfrac{2.5}{\omega\sqrt{1+0.5\omega^2}\sqrt{1+16\omega^2}}$

The phase of G(jω) is not altered by the term, K. The magnitude and phase of G(jω) when K = 2.5 are calculated for various values of ω and listed in table-2.3.2. Using polar to rectangular conversion the real part and imaginary part of G(jω) when K = 2.5 are determined and listed in table-2.3.2. The polar plot of G(jω) when K = 2.5 is sketched in the same graph sheet using the same scales, as shown in fig 2.3.2.

TABLE-2.3.2

ω rad/sec	0.6	0.65	0.75	0.8	1.0	1.2
$\|G(j\omega)\|$	1.535	1.313	0.987	0.865	0.54	0.363
$\angle G(j\omega)$	−174	−177	−182	−184	−192	−199
$G_R(j\omega)$	−1.52	−1.31	−0.99	−0.87	−0.53	−0.34
$G_I(j\omega)$	−0.16	−0.07	0.03	0.06	0.11	0.12

Polar plot of −1/K_N

The function $-1/K_N$ can be expressed as,

$$\frac{-1}{K_N} = -1 \times \frac{1}{K_N} = 1\angle -180° \times \frac{1}{K_N}$$

We know that the describing function (K_N) of saturation nonlinearity is given by

$$K_N = \begin{cases} 1 & ; \text{ when } X < S \qquad (\because K = 1) \\ \dfrac{2K}{\pi}(\beta + \sin\beta\cos\beta)\angle 0° & ; \text{ when } X > S \end{cases}$$

where, $\beta = \sin^{-1}(S/X)$

and   X = Maximum value of input sinusoidal signal

Here, K = 1 and S = 1

$$\therefore -1/K_N = \begin{cases} 1\angle -180° & ; \text{ when } X < 1 \\ \dfrac{\pi}{2(\beta + \sin\beta\cos\beta)}\angle -180° & ; \text{ when } X > 1 \end{cases}$$

where, $\beta = \sin^{-1}(1/X)$

From the equation of $-1/K_N$ we can say that, the locus of $-1/K_N$ starts at $1\angle -180°$ (i.e., $-1+j0$) and travels along the negative real axis for increasing values of X as shown in fig 2.3.2. The locus of $-1/K_N$ is shown as a bold line on the negative real axis.

STABILITY ANALYSIS

Case (i) when K = 1

When K = 1, the G(jω) locus does not encloses the $-1/K_N$ locus, hence the system is stable.

Case(ii) To find maximum value of K for stability

When K is increased the G(jω) locus shifts upwards. For a particular value of K, the G(jω) locus crosses the starting point (i.e., $-1+j0$) of $-1/K_N$ locus and this value of K is the limiting value of K for stability.

If G(jω) crosses negative real axis at $-1+j0$, then, $G(j\omega) = -1 = 1\angle -180°$

$$\therefore |G(j\omega)| = 1 \text{ and } \angle G(j\omega) = -180°$$

Let, ω_{fl} = Frequency when G(jω) = -1

$$\therefore \text{At } \omega = \omega_{fl}, \angle G(j\omega) = -90° - \tan^{-1} 0.5\omega_{fl} - \tan^{-1} 4\omega_{fl} = -180°$$

$$\therefore \tan^{-1} 0.5\omega_{fl} + \tan^{-1} 4\omega_{fl} = 90°$$

On taking tan on either side we get,

$$\tan (\tan^{-1} 0.5\omega_{fl} + \tan^{-1} 4\omega_{fl}) = \tan 90°$$

$$\frac{\tan (\tan^{-1} 0.5\omega_{fl}) + \tan (\tan^{-1} 4\omega_{fl})}{1 - \tan (\tan^{-1} 0.5\omega_{fl}) \times \tan (\tan^{-1} 4\omega_{fl})} = \tan 90°$$

$$\frac{0.5\omega_{fl} + 4\omega_{fl}}{1 - 0.5\omega_{fl} \times 4\omega_{fl}} = \infty$$

For the above equation to be infinity, the denominator should be zero.

$$\therefore 1 - 2\omega_{fl}^2 = 0 \quad ; \quad \omega_{fl}^2 = 1/2 \quad \text{(or)} \quad \omega_{fl} = \frac{1}{\sqrt{2}} \text{ rad/ sec}$$

at $\omega = \omega_{fl}$, $|G(j\omega)| = 1$

$$\therefore \frac{K}{\omega_{fl}\sqrt{1+0.25\omega_{fl}^2}\sqrt{1+16\omega_{fl}^2}} = 1 \quad \text{(or)} \quad K = \omega_{fl}\sqrt{1+0.25\omega_{fl}^2}\sqrt{1+16\omega_{fl}^2}$$

$$K = \frac{1}{\sqrt{2}}\sqrt{(1+0.25 \times 0.5)(1+16 \times 0.5)} = 2.25$$

Therefore the system remains stable if, K < 2.25

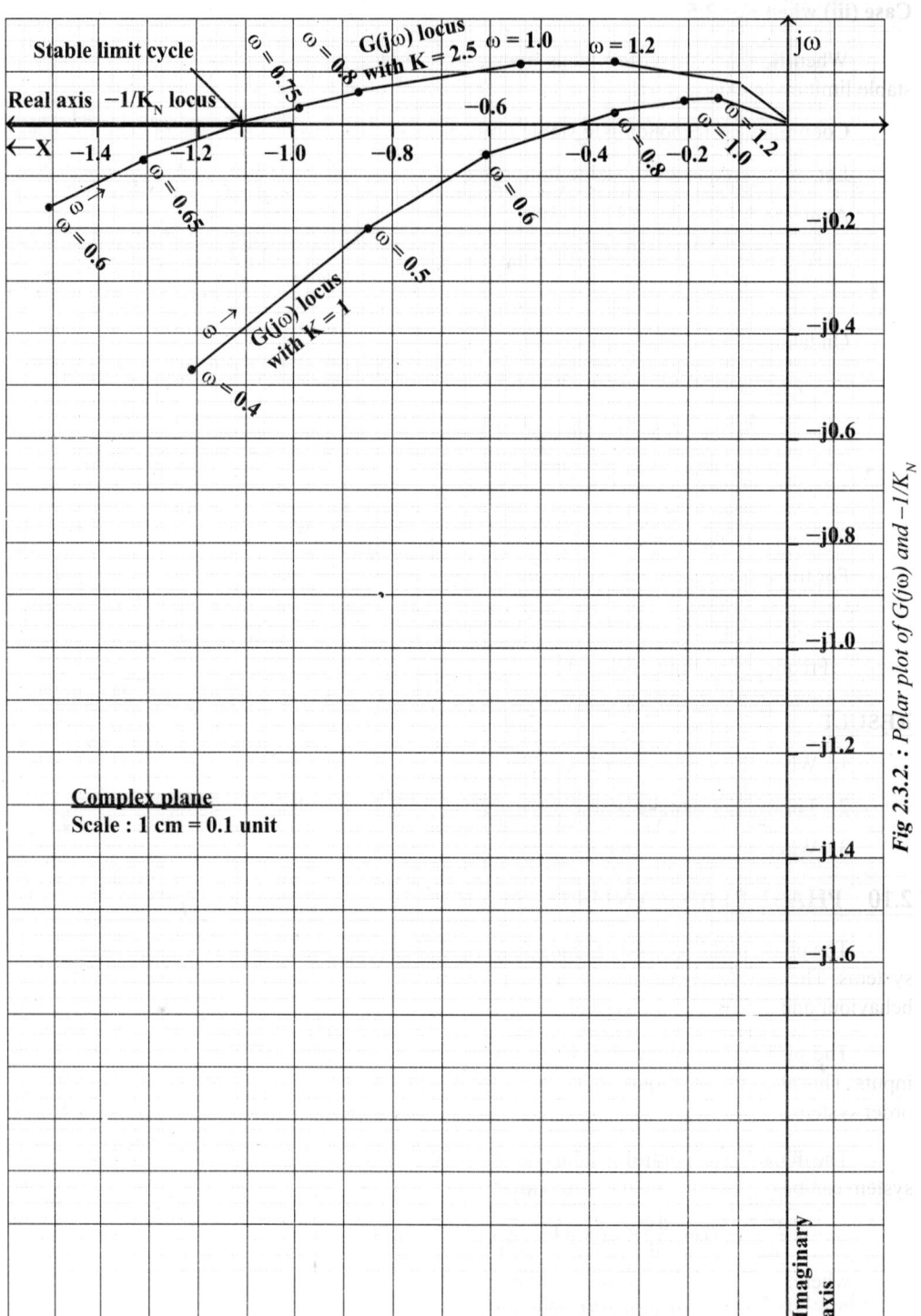

Stable limit cycle

$G(j\omega)$ locus with K = 2.5

Real axis $-1/K_N$ locus

$\omega = 0.75$

$\omega = 0.8$

$\omega = 1.0$

$\omega = 1.2$

-0.6

$j\omega$

$\leftarrow$ X -1.4 -1.2 -1.0 -0.8 -0.4 -0.2

$\omega = 0.65$

$\omega = 0.6$

$\omega = 0.8$

$\omega = 1.0$

$\omega = 1.2$

$\omega = 0.6$

$-j0.2$

$\omega = 0.5$

$G(j\omega)$ locus with K = 1

$-j0.4$

$\omega = 0.4$

$-j0.6$

$-j0.8$

$-j1.0$

$-j1.2$

Complex plane
Scale : 1 cm = 0.1 unit

$-j1.4$

$-j1.6$

Imaginary axis

Fig 2.3.2. : Polar plot of $G(j\omega)$ and $-1/K_N$

Case (iii) when K = 2.5

When K = 2.5, the $G(j\omega)$ locus intersects, $-1/K_N$ locus at $-1.11+j0$. At the intersection point a stable limit cycle exists.

Coordinate corresponding to stable limit cycle $= -1.11+j0 = 1.11\angle-180°$

Let, ω_{l2} = Frequency of stable limit cycle

At $\omega = \omega_{l2}$, $G(j\omega) = 1.11\angle-180°$

$\therefore$ At $\omega = \omega_{l2}$, $\angle G(j\omega) = -90° - \tan^{-1} 0.5\omega_{l2} - \tan^{-1} 4\omega_{l2} = -180°$

$\therefore \tan^{-1} 0.5\omega_{l2} + \tan^{-1} 4\omega_{l2} = 90°$

On taking tan on either side we get,

$$\tan (\tan^{-1} 0.5\omega_{l2} + \tan^{-1} 4\omega_{l2}) = \tan 90°$$

$$\frac{\tan (\tan^{-1} 0.5\omega_{l2}) + \tan (\tan^{-1} 4\omega_{l2})}{1 - \tan (\tan^{-1} 0.5\omega_{l2}) \times \tan (\tan^{-1} 4\omega_{l2})} = \tan 90°$$

$$\frac{0.5\omega_{l2} + 4\omega_{l2}}{1 - 0.5\omega_{l2} \times 4\omega_{l2}} = \infty$$

For the above equation to be infinity, the denominator should be zero.

$$\therefore 1 - 2\omega_{l2}^2 = 0 \quad \text{(or)} \quad \omega_{l2}^2 = 1/2 \text{ rad/ sec} \quad \text{(or)} \quad \omega_{l2} = 1/\sqrt{2} \text{ rad/ sec}$$

$\therefore$ Frequency of limit cycle $= 1/\sqrt{2} = 0.707$ rad/sec

RESULT

1. When K = 1, the system is stable

2. The system remains stable if K < 2.25

3. When K = 2.5, a stable limit cycle occurs, whose frequency of oscillation is 0.707 rad/sec.

2.10 PHASE PLANE AND PHASE TRAJECTORIES

The phase plane method of analysis is a graphical method for the analysis of linear and nonlinear systems. The analysis is carried by constructing phase trajectories. It gives an idea about the transient behaviour and stability of the system.

The phase plane analysis is usually restricted to second order systems excited by step or ramp inputs. This analysis technique can be extended to a higher order system if it is approximated as a second order system.

The dynamics of control systems can be represented by differential equations. A second order linear system can be represented by the differential equation

$$\frac{d^2x}{dt^2} + 2\zeta\omega_n \frac{dx}{dt} + \omega_n^2 x = 0 \qquad \qquad .....(2.110)$$

where,   x   =   One of the system variable (e.g. displacement in mechanical system, current in electrical system, etc.,)

         ζ =   Damping ratio

         ω_n =   Natural frequency of oscillation.

The state of the second order system represented by equation (2.110) can be described by choosing two state variables.

> **Note :** *Refer chapter 4 for state, state variables and state space modelling using phase variables.*

In state space modelling using phase variables we choose one of the system variable and its derivatives as state variables. Let x_1 and x_2 be the state variables of the second order system.

Here $x_1 = x$ and $x_2 = dx/dt$ .....(2.111)

On substituting the state variables in equation (2.110) we get,

$$\ddot{x}_2 + 2\zeta\omega_n + \dot{x}_2 + \omega_n^2 x_1 = 0 \qquad .....(2.112)$$

The state equations of the system are obtained from equations (2.111) and (2.112). The state equations are,

$$\dot{x}_1 = x_2 \qquad .....(2.113)$$

$$\dot{x}_2 = -\omega_n^2 x_1 - 2\zeta\omega_n x_2 \qquad .....(2.114)$$

For linear systems the state equations are a set of first order linear differential equations and solutions of state equations can be easily obtained by integration. But for nonlinear systems, the state equations are a set of first-order nonlinear differential equations and solving the nonlinear differential equations will not be an easy task. Hence for nonlinear systems the phase plane method of analysis will be an useful tool.

The coordinate plane with the state variables x_1 and x_2 as two axes is called the phase plane. (i.e., in phase plane, x_1 is represented in x-axis and x_2 in y-axis).

The curve describing the state point (x_1, x_2) in the phase-plane with time as running parameter is called phase trajectory [i.e., the locus of state point (x_1, x_2) in phase plane is called phase trajectory]. A trajectory can be constructed in phase-plane for each set of initial conditions. Hence a family of trajectories can be constructed for a system in a phase plane and such a family of trajectories is called a phase portrait. Typical phase trajectories of a second order system for different values of damping is shown in fig 2.40 and the phase portrait of the system for critical damping is shown in fig 2.41.

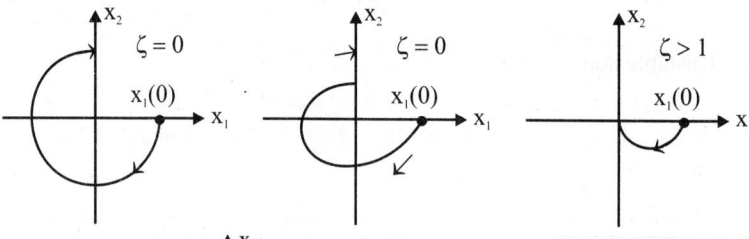

Fig 2.40 : Phase trajectories of a second order system for different values of ζ and with initial condition $x_1(0)$.

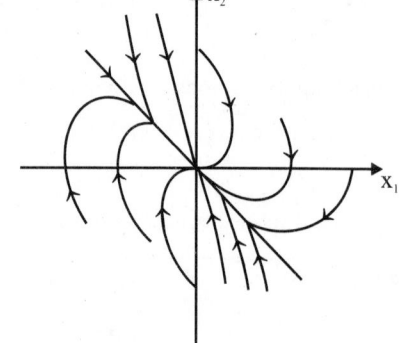

Fig 2.41 : Phase portrait for a critically damped system.

SINGULAR POINTS

A point in phase-plane at which the derivative of all state variables are zero is called singular point. It is also called equilibrium point. If the system is placed at a singular or equilibrium point, it will continue to lie there if left undisturbed (i.e., the derivatives of all the phase variables are zero and so the system state remains unchanged).

The singular points are classified as Nodal point, Saddle point, Focus point and Centre or Vortex point depending on the eigen values of the system matrix, A. The table 2.1 shows the phase portrait of systems with various types of singular points.

Note :   1.   *Refer chapter-5 for eigen values and chapter-4 for system matrix.*

           2.   *A second order system has two eigen values and phase plane analysis is performed only for second order systems.*

TABLE 2.1

Eigen Values of system matrix	Type of singular point	Phase portrait of the system with singular point at origin
Distinct, real and the two eigen values are negative	Stable node	
Distinct, real and the two eigen values are positive	Unstable node	
Distinct, real, one eigen value is positive and the other is negative	Saddle point	

TABLE 2.1 : CONTINUED .....

Eigen Values of system matrix	Type of singular point	Phase portrait of the system with singular point at origin
Complex conjugate with negative real part	Stable focus	
Complex conjugate with positive real part	Unstable focus	
Purely imaginary and conjugate	Centre or Vortex point	

The state equations of a system are formed by taking the first derivatives of state variables as functions of state variables and inputs. If the inputs are constants then the state equation will be in the form

$$\dot{X} = F(X) \qquad\qquad .....(2.115)$$

A system represented by equation (2.115) is called an autonomous system. The linearised model of the system represented by equation (2.115) may be written as

$$\dot{X} = AX \qquad\qquad .....(2.116)$$

Let, X_e = States corresponding to singular point or equilibrium state.

At equilibrium state, $\dot{X} = 0$

$$\therefore A X_e = 0 \qquad\qquad .....(2.117)$$

Since the determinant of A is non-zero, $X_e = 0$. will be the only solution for equation (2.117). Therefore we can conclude that if all the eigen values of the system are non-zero then the origin is the only singular point.

In general, if X_e is a singular point, then it is convenient to shift the origin of coordinates to X_e. To achieve this, we define new phase variables as

$$\tilde{X} = X - X_e \qquad\qquad .....(2.118)$$

The system in terms of new phase variables is represented as

$$\dot{\widetilde{X}} = F(\widetilde{X}) \qquad\qquad .....(2.119)$$

with the equilibrium or singular point lying at $\widetilde{X} = 0$.

STABILITY ANALYSIS OF NONLINEAR SYSTEMS USING PHASE TRAJECTORIES

For linear time invariant systems, the concept of stability can be defined as follows,

1. *When the input is zero, the system is stable for arbitrary initial conditions if the resulting trajectory tends towards the equilibrium state.*

2. *When the system is excited by a bounded input, the system is stable if the system output is bounded.*

In nonlinear systems the concept of stability is not clear-cut. There are many types of stability definitions in the literature.

The linear autonomous systems has only one equilibrium state. The behaviour of linear system about the equilibrium state completely determines the qualitative behaviour in the entire state-plane.

In nonlinear systems there may be multiple equilibrium state. The behaviour of nonlinear system about the equilibrium point may be different for small deviations and large deviations about the equilibrium point. In nonlinear systems with multiple equilibrium states, the system trajectories may move away from one equilibrium point and tend to other as time progresses. Hence in nonlinear systems, stability is discussed relative to the equilibrium state and the general stability of a system cannot be defined.

Consider an autonomous system described by the state equation, $\dot{X} = F(X)$. Let us assume that the system has one equilibrium point and the origin of phase plane is the equilibrium point. For this system, the following definitions of stability are proposed.

1. *The autonomous system defined by equation $\dot{X} = F(X)$ is stable at the origin, if for every initial state $X(t_o)$ which is sufficiently close to origin, $X(t)$ remains near the origin for all t.*

2. *The autonomous system defined by equation $\dot{X} = F(X)$ is asymptotically stable if $X(t)$ approaches the origin as $t \to \infty$.*

3. *The autonomous system defined by equation $\dot{X} = F(X)$ is asymptotically stable in the large if it is asymptotically stable for every initial state regardless of how near or far it is from the origin.*

LIMIT CYCLES IN PHASE-PORTRAIT

The limit cycles are oscillations of fixed amplitude and period. The existence of limit cycles in nonlinear systems can be predicted from closed trajectories in the phase-portrait.

In linear systems, when oscillations occur, the resulting trajectories will be closed curves as shown in fig 2.42. The amplitude of the oscillations is not fixed. It changes with the size of the initial conditions. Slight changes in system parameters will destroy the oscillations.

In nonlinear systems, there can be limit cycles (oscillations) that are independent of the size of initial conditions as shown in fig 2.43. These limit cycles are usually less sensitive to system parameter variations. Limit cycles of fixed amplitude and period can be sustained over a finite range of system parameters. The limit cycle is stable if the paths in its neighbourhood converge towards the limit cycle as shown in fig 2.43a. The limit cycle is unstable, if the paths in the neighbourhood of a limit cycle diverge away from it, as shown in fig 2.43b.

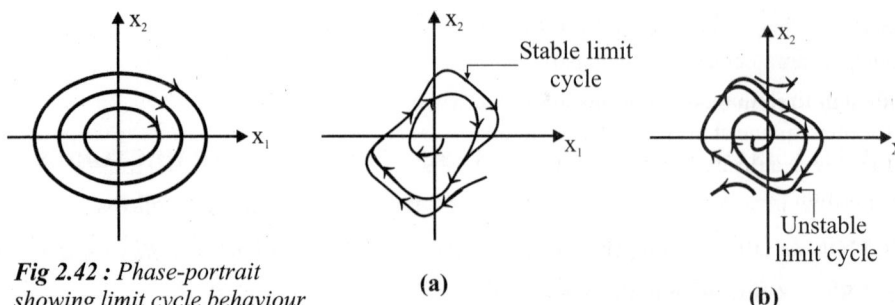

Fig 2.42 : *Phase-portrait showing limit cycle behaviour in linear system.*

(a)

(b)

Fig 2.43 : *Phase-portrait showing limit cycle behaviour in nonlinear system.*

CONSTRUCTION OF PHASE TRAJECTORIES

The state equations of a second order autonomous system are,

$$\dot{x}_1 = f_1(x_1, x_2) \qquad\qquad .....(2.120)$$

$$\dot{x}_2 = f_2(x_1, x_2) \qquad\qquad .....(2.121)$$

where x_1 and x_2 are the state variables of the system.

On dividing equ(2.121) by equ(2.120) we get,

$$\frac{\dot{x}_2}{\dot{x}_1} = \frac{f_2(x_1, x_2)}{f_1(x_1, x_2)} \qquad\qquad .....(2.122)$$

since $\dot{x}_2 = dx_2/dt$ and $\dot{x}_1 = dx_1/dt$, the equ (2.122) can be written as,

$$\frac{dx_2}{dx_1} = \frac{f_2(x_1, x_2)}{f_1(x_1, x_2)} \qquad\qquad .....(2.123)$$

The equation (2.123) defines the slope of phase trajectory at every point in the phase-plane, except at singular points. At singular points, the slope of the phase trajectory is indeterminate.

The phase trajectory can be constructed using the slope equation (i.e., 2.123) either analytically or graphically. In analytical method of construction, the equation (2.123) is integrated and the resulting equation is used to construct phase trajectories for a given set of initial conditions. In many cases, it is not possible to perform integration of the slope equation (i.e., equation 2.123). Hence a number of graphical methods have been developed for construction of phase trajectories and the two popular methods are isocline method and delta method.

CONSTRUCTION OF PHASE TRAJECTORY BY ANALYTICAL METHOD

The analytical method is used if the differential equations describing the system can be approximated by piecewise linear differential equations. *[i.e., the equations are linearized for small regions].* The slope equation dx_2/dx_1 is formed from the state equations. Then the slope equation is splitted into sections of linear equations. Each section of linear slope equation is directly integrated to get the solution of state equations. Each section of solution equation is used to construct a section of phase trajectories for various sets of initial conditions.

The solution of slope equation will be a function of x_1 and x_2. For a given set of initial conditions, assume different values of x_1 and calculate x_2 for each value of x_1 using the solution of slope equation. The values of x_1 and x_2 are tabulated. In a ordinary graph sheet, take x_1 and x_2 axis, choose appropriate scales and mark the phase points (x_1 and x_2). Join all the points by a smooth curve. For each set of initial conditions one phase trajectory can be constructed using the procedure described above.

Eventhough time domain solutions are obtained by direct integration, the construction of phase-trajectories (or phase plane analysis) will help in investigating system behaviour and the design of system parameters to achieve a desired response. Also the existence of limit cycles is sharply brought into focus by the phase-portrait.

CONSTRUCTION OF PHASE TRAJECTORY BY ISOCLINE METHOD

Let, S = Slope at any point in the phase-plane.

From equation (2.123) we get,

$$S = \frac{dx_2}{dx_1} = \frac{f_2(x_1, x_2)}{f_1(x_1, x_2)} \qquad\qquad .....(2.124)$$

Let, S_1 = Slope at a point on phase trajectory-1.

From equation (2.124) we get,

$$f_2(x_1, x_2) = S_1 \times f_1(x_1, x_2) \qquad\qquad .....(2.125)$$

The equation (2.125) defines the locus of all such points in phase-plane at which the slope of the phase-trajectory is S_1. A locus passing through the points of same slope in phase-plane is called isocline.

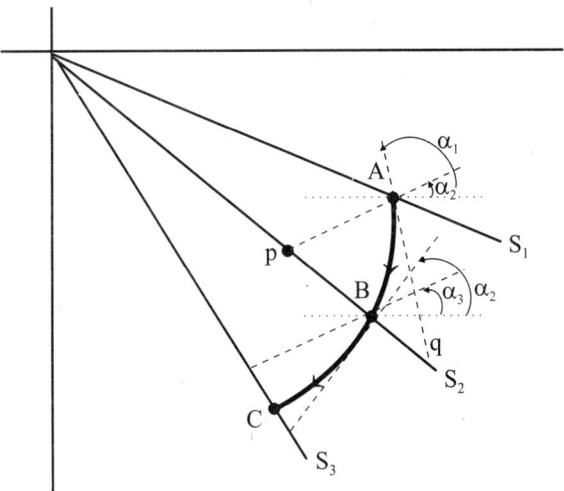

$$\alpha_1 = \tan^{-1}(S_1)$$
$$\alpha_2 = \tan^{-1}(S_2)$$
$$\alpha_3 = \tan^{-1}(S_3)$$

Fig 2.44 : Construction of phase trajectory by isocline method.

The slope of a phase trajectory at the crossing point of an isocline will be the slope of corresponding isocline. A typical plot of isoclines for various values of slope, S are shown in fig 2.44. Using these isoclines the phase trajectories can be constructed as explained below.

The phase trajectory start at a point corresponding to initial conditions. (For each set of initial conditions one phase trajectory can be constructed).

Let, S_1, S_2, S_3, etc., be the slopes associated with isoclines 1, 2, 3, etc.,

Let, $\alpha_1 = \tan^{-1}(S_1)$; $\alpha_2 = \tan^{-1}(S_2)$; $\alpha_3 = \tan^{-1}(S_3)$; etc.,

Note : If a straight line is drawn at an angle α from a point, then the slope of the line at that point is $\tan \alpha$.

In fig 2.43, let point A on isocline-1 be the point corresponding to a set of initial conditions. The phase-trajectory will leave the point A at a slope S_1. When the trajectory reaches the isocline-2, the slope changes to S_2.

Draw two lines from point A one at a slope of S_1 (i.e., at angle of $\alpha_1 = \tan^{-1}(S_1)$) and the other at a slope of S_2 (i.e., at angle of $\alpha_2 = \tan^{-1}(S_2)$). Let these two lines meet the isocline-2 at p and q. Now we can say that the trajectory would cross the isocline-2 at a point midway between p and q. Mark the point B on the isocline-2 approximately midway between p and q.

The constructional procedure is now repeated at B to find the crossing point C on the isocline-3. By similar procedure the crossing points on the isolines are determined. A smooth curve drawn through the crossing points gives the phase-trajectory starting at point A.

The accuracy of the trajectory is closely related to the spacing of the isoclines. The phase trajectory will be more accurate if large number of isoclines are used which are very close to each other. It should be noted that using a set of isoclines, any number of trajectories can be constructed.

CONSTRUCTION OF PHASE TRAJECTORIES BY DELTA METHOD

Consider a second order linear or nonlinear system represented by the equation,

$$\ddot{x} = f(x, \dot{x}, t) = 0 \qquad \qquad .....(2.126)$$

The equation (2.126) can be converted to the form shown below,

$$\ddot{x} + \omega_n^2 [x + \delta(x, \dot{x}, t)] = 0 \qquad \qquad .....(2.127)$$

In equation (2.127), δ is a function of x, and t, but for short intervals, the changes in these variables are negligible. Hence for a short interval, δ is considered as a constant.

$$\therefore \ddot{x} + \omega_n^2 (x + \delta) = 0 \qquad \qquad .....(2.128)$$

Let us choose the state variables x_1 and x_2 as,

$$x_1 = x \qquad \qquad .....(2.129)$$

$$x_2 = \dot{x}/\omega_n \qquad \qquad .....(2.130)$$

On differentiating equation (2.129) we get, $\dot{x}_1 = \dot{x}$

But $\dot{x} = \omega_n x_2, \qquad \therefore \dot{x}_1 = \omega_n x_2 \qquad \qquad .....(2.131)$

On differentiating equation (2.130) we get,

$$\dot{x}_2 = \ddot{x}/\omega_n \quad (or) \quad \ddot{x} = \omega_n \dot{x}_2$$

On substituting $\ddot{x} = \omega_n \dot{x}_2$ and $x = x_1$ in equation (2.128) we get,

$$\omega_n \dot{x}_2 + \omega_n^2 (x_1 + \delta) = 0 \qquad \qquad .....(2.132)$$

$$\therefore \dot{x}_2 = -\omega_n (x_1 + \delta)$$

The state equations are

$$\dot{x}_1 = \omega_n x_2$$

$$\dot{x}_2 = -\omega_n (x_1 + \delta)$$

From the state equations the slope equation over short interval can be written as,

$$\frac{\dot{x}_2}{\dot{x}_1} = \frac{dx_2}{dx_1} = \frac{-(x_1+\delta)}{x_2} \qquad\qquad .....(2.133)$$

Using the slope equation(2.133), a short segment of the trajectory can be drawn from the knowledge of δ at any point A on the trajectory, as explained below.

Let point A be a point on phase trajectory with coordinates (x_1, x_2) as shown in fig 2.45 a and b, (usually the point A will be the starting point of the trajectory obtained from the initial conditions).

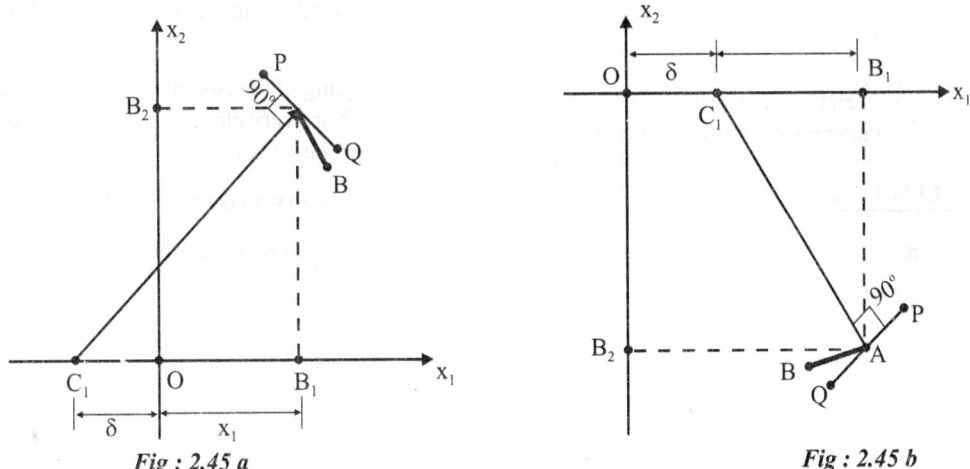

Fig : 2.45 a   Fig : 2.45 b

In fig 2.45. a and b, $OB_1 = x_1$ and $OB_2 = x_2$.

Calculate δ by substituting the value of coordinates of point A in equation (2.133). The value of δ is either positive or negative.

If δ is positive then mark point C_1 on the negative side of x_1-axis, such that length $OC_1 = \delta$ as shown in fig 2.45a. Now the coordinates of point C_1 is $(-\delta, 0)$.

If δ is negative, then mark point C_1 on the positive side of x_1-axis, such that length $OC_1 = |\delta|$ as shown in fig 2.45b. Now the coordinates of point C_1 is $(|\delta|, 0)$

Through point A, draw two perpendicular lines C_1A and PQ. Now the line PQ will have a slope of $-(x_1 + \delta)/x_2$, which is also the slope of the trajectory at point A. Hence draw a small circular arc AB with C_1 as centre and C_1A as radius. The point B is arbitrarily located on the phase plane such that, the arc AB is very small. This small circular arc will be a section of phase-trajectory with slope, $-(x_1 + \delta)/x_2$.

Using the coordinates of point B, a new value of $\delta = \delta_2$ can be calculated and by a similar procedure, the next section BC of the phase trajectory can be determined. Thus repeating the same procedure at every new point, (i.e., point B, C, D, etc.,) the complete phase trajectory is obtained.

Procedure to draw phase trajectory by delta method

1. *The phase trajectory starts at point A, corresponding to initial conditions, $[x_1(0), x_2(0)]$. Calculate δ_1 using the initial conditions, (i.e., using the coordinates of point A). (Refer fig 2.6.1. of example 2.6.).*

2. *The coordinates of point C_1 is $(-\delta_1, 0)$ if δ_1 is positive or $(|\delta_1|, 0)$ if δ_1 is negative. Mark the point C_1 on x_1 axis. With point C_1 as centre and $C_1 A$ as radius draw a circular arc AB. The point B (new point)is an arbitrary point on phase plane. Accurate phase trajectory can be constructed if we draw very short segment to locate a new point.*

Note : A compass can be used to draw the circular arc segments.

3. *Using the coordinates of point B, calculate δ_2. The coordinates of point C_2 is $(-\delta_2, 0)$ if δ_2 is positive or $(|\delta_2|, 0)$ if δ_2 is negative. Mark the point C_2 on x_1 axis. With C_2 as centre and $C_2 B$ as radius, draw a circular arc BC, to locate another new point, point-C.*

4. *Similar procedure is repeated at point C to locate point D and so on. The circular arc segments through the points A, B, C, D, etc., gives the phase trajectory.*

The delta method results in time saving when a single or a few phase trajectories are required. Using delta method, the phase trajectory for any system with step or ramp input can be conveniently drawn.

EXAMPLE 2.4

Consider a system with an ideal relay as shown in fig 2.4.1. Determine the singular point. Construct phase trajectories, corresponding to initial conditions, (i) $c(0) = 2$, $\dot{c}(0) = 1$ and (ii) $c(0) = 2$, $\dot{c}(0) = 1.5$. Take $r = 2$ volts and $M = 1.2$ volts

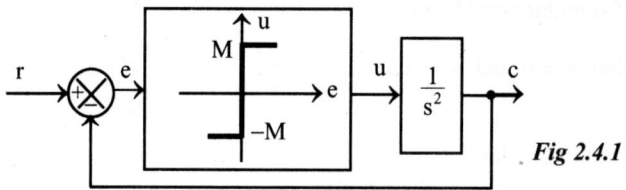

Fig 2.4.1

SOLUTION

Consider the linear portion of the system shown in fig 2.4.2.

From fig 2.4.2, we get

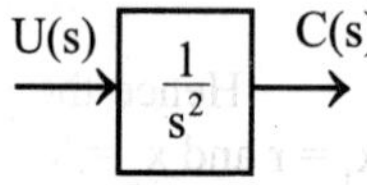

$$C(s) = \frac{1}{s^2} U(s)$$

$$\therefore s^2 C(s) = U(s) \qquad .....(2.4.1)$$

On taking inverse Laplace transform of equation (2.4.1) we get,

$$\ddot{c}(t) = u(t) \qquad .....(2.4.2)$$

Let us choose c(t) and as state variables x_1 and x_2

$$\therefore x_1 = c(t) \qquad .....(2.4.3)$$

$$x_2 = \dot{c}(t) \qquad .....(2.4.4)$$

On differentiating equations (2.4.3) and (2.4.4) we get,

$$\dot{x}_1(t) = \dot{c}(t) \qquad\qquad \dot{x}_2(t) = \ddot{c}(t)$$

$$\therefore \dot{x}_1(t) = x_2 \qquad .....(2.4.5) \qquad \therefore \dot{x}_2 = u \qquad .....(2.4.6)$$

U(s) $\boxed{\dfrac{1}{s^2}}$ C(s)

Fig : 2.4.2

The state equations are given by equations (2.4.5) and (2.4.6)

Consider the nonlinear part of the system shown in fig 2.4.3.

From fig 2.4.3. we get,

$$e = r - c$$

But $c = x_1$

$$\therefore e = r - x_1 \qquad \qquad .....(2.4.7)$$

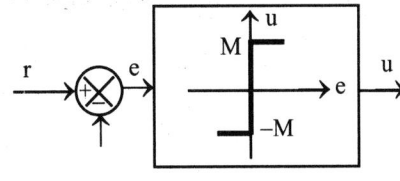

Fig : 2.4.3

When $x_1 > r$, e is negative, $\therefore u = -M$

When $x_1 < r$, e is positive, $\therefore u = +M$

On substituting the values of u in the state equation, we get

$$\dot{x}_1 = x_2$$

$$\dot{x}_2 = M \quad ; \quad \text{for } x_1 < r$$
$$\phantom{\dot{x}_2} = -M \ ; \quad \text{for } x_1 > r \qquad \qquad .....(2.4.8)$$

TO FIND SINGULAR POINT

The singular point is determined by substituting $\dot{x}_1 = 0$ and $\dot{x}_2 = 0$ in state equations.

Let the coordinates of singular point be (x_1^0, x_2^0).

In equation (2.4.5), when $\dot{x}_1 = 0$ and $\dot{x}_2 = x_2^0$. On substituting $\dot{x}_1 = 0$ and $\dot{x}_2 = x_2^0$ in equation (2.4.5) we get, $x_2^0 = 0$.

In equation (2.4.6) when , $x_2^0 = 0$, $u = 0$.

From fig 2.4.3, when $u = 0$, $e = 0$

In equation (2.4.7), when $e = 0$, $x_1 = x_1^0$

$$\therefore r - x_1^0 = 0 \quad \text{(or)} \quad x_1^0 = r$$

Hence the coordinates corresponding to singular point is $(x_1^0, x_2^0) = (r, 0)$ [i.e., at singular point $x_1 = r$ and $x_2 = 0$].

TO CONSTRUCT PHASE TRAJECTORY

From equations (2.4.5) and (2.4.8), the slope equation of the phase trajectory can be written as,

$$\frac{\dot{x}_2}{\dot{x}_1} = \frac{dx_2}{dx_1} = \frac{M}{x_2} \quad ; \quad \text{for } x_1 < r$$

$$\phantom{\frac{\dot{x}_2}{\dot{x}_1}} = -\frac{M}{x_2} \quad ; \quad \text{for } x_1 > r$$

In general we can say,

$$\frac{dx_2}{dx_1} = \pm \frac{M}{x_2} \quad ; \quad \text{" + " sign for } x_1 < r \qquad \qquad .....(2.4.9)$$
$$\phantom{\frac{dx_2}{dx_1} = \pm \frac{M}{x_2} \quad ;} \quad \text{" – " sign for } x_1 > r$$

To find the state of the system at time, t, the equation (2.4.9) can be integrated from, $t = 0$ to t.

At, $t = 0$, $x_1 = x_1(0)$ and $x_2 = x_2(0)$

At, $t = t$, $x_1 = x_1(t)$ and $x_2 = x_2(t)$

Note : *The state variables x_1 and x_2 describe the state of the system at any time t.*

$$\therefore \int_{x_2(0)}^{x_2} x_2 \, dx_2 = \int_{x_1(0)}^{x_1} \pm M \, dx_1$$

$$\left[\frac{x_2^2}{2}\right]_{x_2(0)}^{x_2} = \pm M [x_1]_{x_1(0)}^{x_1}$$

$$\frac{x_2^2}{2} - \frac{x_2^2(0)}{2} = \pm M[x_1 - x_1(0)]$$

$$\therefore x_2^2 = \pm 2M[x_1 - x_1(0)] + x_2^2(0) \qquad\qquad .....(2.4.10)$$

when $x_1 > r$, the equation (2.4.10) can be written as,

$$x_2{}^2 = -2M\,[x_1 - x_1(0)] + x_2{}^2(0)$$

$$\therefore x_2{}^2 = -2Mx_1 + 2Mx_1(0) + x_2{}^2(0) \qquad\qquad .....(2.4.11)$$

when $x_1 < r$, the equation (2.4.10) can be written as,

$$x_2{}^2 = 2M\,[x_1 - x_1(0)] + x_2{}^2(0)$$

$$\therefore x_2{}^2 = 2M\,x_1 - 2M\,x_1(0) + x_2{}^2(0) \qquad\qquad .....(2.4.12)$$

TRAJECTORY FOR FIRST SET OF INITIAL CONDITIONS

Given that $c(0) = 2$ and $\dot{c}(0) = 1$

$$\therefore x_1(0) = 2 \text{ and } x_2(0) = 1$$

On substituting these initial conditions and the value of M (i.e., M = 1.2) in equations(2.4.11) and (2.4.12) we get,

$$x_2 = \pm \sqrt{-2.4x_1 + 5.8} \quad ; \quad \text{for } x_1 > r \qquad\qquad .....(2.4.13)$$

$$x_2 = \pm \sqrt{2.4x_1 - 3.8} \quad ; \quad \text{for } x_1 < r \qquad\qquad .....(2.4.14)$$

Using equations (2.4.13) and (2.4.14) the values of x_2 are calculated by assuming various values of x_1 and tabulated in table 2.4.1. The values of x_1 are assumed such that we get real values of x_2. Using the values of x_1 and x_2 in table-2.4.1, the phase trajectory is constructed as shown in fig 2.4.4.

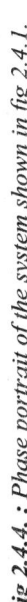

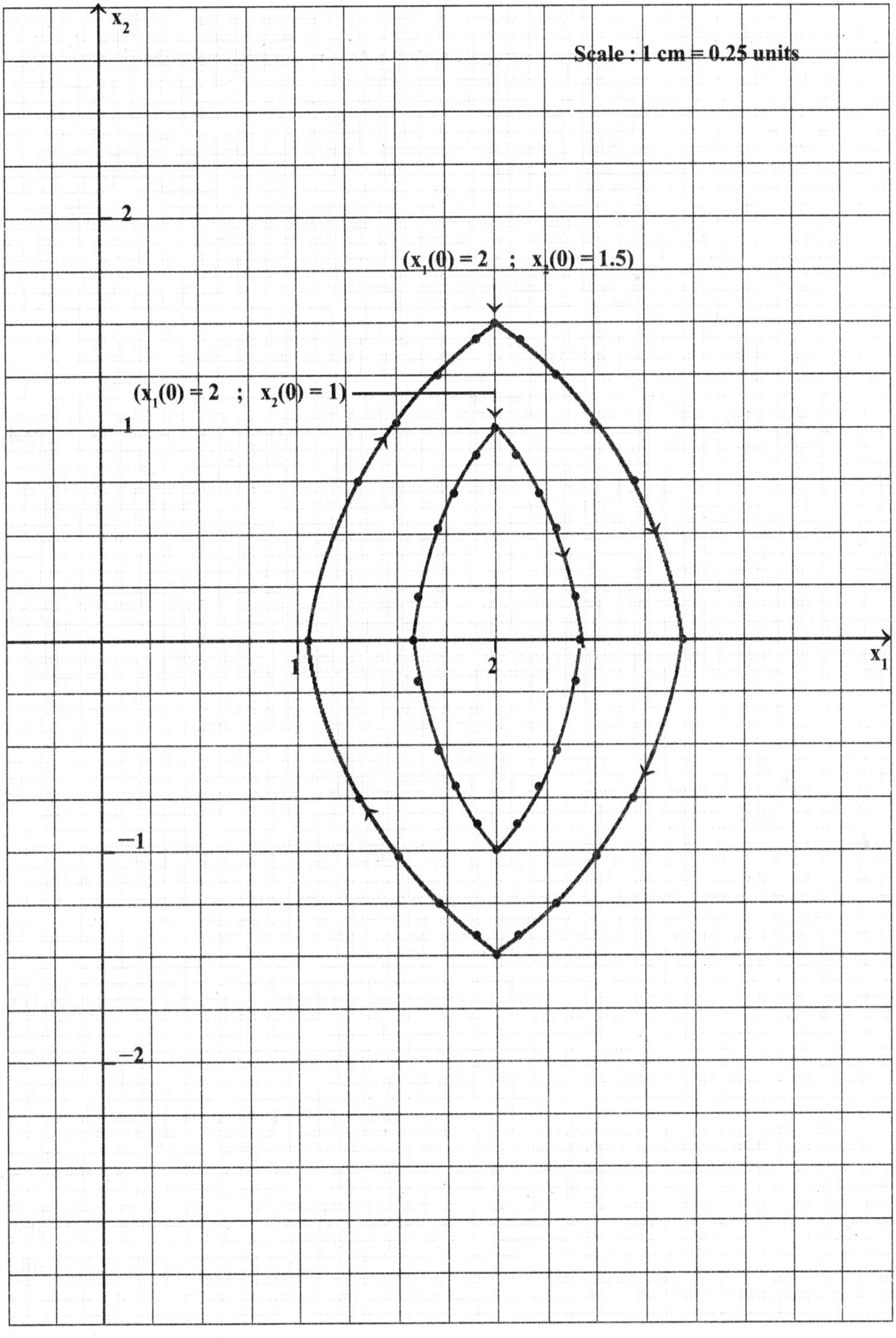
x_2
Scale : 1 cm = 0.25 units
2
$(x_1(0) = 2\ ;\ x_2(0) = 1.5)$
$(x_1(0) = 2\ ;\ x_2(0) = 1)$
1
1
2
x_1
−1
−2

Fig 2.4.4. : Phase portrait of the system shown in fig 2.4.1.

TABLE-2.4.1.

$x_1 > r$			$x_1 < r$		
x_1	x_2		x_1	x_2	
2	1	−1	1.5834	0	0
2.1	0.87	−0.87	1.6	0.2	−0.2
2.2	0.72	−0.72	1.7	0.53	−0.53
2.3	0.53	−0.53	1.8	0.72	−0.72
2.4	0.2	−0.2	1.9	0.87	−0.87
2.4167	0	0	2	1	−1

Trajectory for second set of initial conditions

Given that, $c(0) = 2$ and $\dot{c}(0) = 1.5$

$$\therefore x_1(0) = 2 \text{ and } x_2(0) = 1.5$$

On substituting these initial conditions and the value of M (i.e, M = 1.2) in equations (2.4.11) and (2.4.12) we get,

$$x_2 = \pm\sqrt{-2.4x_1 + 7.05} \quad ; \quad \text{for } x_1 > r \qquad \qquad .....(2.4.15)$$

$$x_2 = \pm\sqrt{2.4x_1 - 2.55} \quad ; \quad \text{for } x_1 < r \qquad \qquad .....(2.4.16)$$

Using equations (2.4.15) and (2.4.16) the values of x_2 are calculated by assuming various values of x_1 and tabulated in table 2.4.2. The values of x_1 are assumed such that we get real values of x_2. Using the values of x_1 and x_2 in table-2.4.2, the phase trajectory is constructed as shown in fig 2.4.4.

TABLE-2.4.2.

$x_1 > r$			$x_1 < r$		
x_1	x_2		x_1	x_2	
2	1.5	−1.5	1.0625	0	0
2.1	1.42	−1.42	1.3	0.75	−0.75
2.3	1.24	−1.24	1.5	1.02	−1.02
2.5	1.02	−1.02	1.7	1.24	−1.24
2.7	0.75	−0.75	1.9	1.42	−1.42
2.9375	0	0	2	1.5	−1.5

CONCLUSION

From fig 2.4.4, it is observed that the phase trajectory for each set of initial condition is a closed curve. Hence limit cycle exist in the system.

EXAMPLE 2.5

A linear second order servo is described by the equation

$$\ddot{e} + 2\zeta\omega_n\dot{e} + \omega_n^2 e = 0$$

where $\zeta = 0.15$, $\omega_n = 1$ rad/sec, $e(0) = 1.5$ and .

Determine the singular point. Construct the phase trajectory, using the method of isoclines.

SOLUTION

Let x_1 and x_2 be the state variables of the system and they are related to the system variable, e as shown below.

$$x_1 = e \qquad\qquad .....(2.5.1)$$

$$x_2 = \dot{e} \qquad\qquad .....(2.5.2)$$

On differentiating equation (2.5.1) we get,

$$\dot{x}_1 = \dot{e}_1 = x_2 \qquad\qquad .....(2.5.3)$$

On differentiating equation (2.5.2) we get,

$$\dot{x}_2 = \ddot{e} \qquad\qquad .....(2.5.4)$$

Given that, $\ddot{e} + 2\zeta\omega_n\dot{e} + \omega_n^2 e = 0$ $\qquad\qquad .....(2.5.5)$

On substituting equations (2.5.1), (2.5.2) and (2.5.4) in equation (2.5.5) we get,

$$\dot{x}_2 + 2\zeta\omega_n x_2 + \omega_n^2 x_1 = 0 \qquad\qquad .....(2.5.6)$$

$$\therefore \dot{x}_2 = -\omega_n^2 x_1 - 2\zeta\omega_n x_2$$

The state equations of the system are given by equations (2.5.3) and (2.5.6)

$$\dot{x}_1 = x_2$$

$$\dot{x}_2 = -\omega_n^2 x_1 - 2\zeta\omega_n x_2$$

The singular point is obtained from state equations by putting $\dot{x}_1 = 0$ and $\dot{x}_2 = 0$

Let the coordinates of singular point in phase plane $= (x_1^0, x_2^0)$

On substituting $\dot{x}_1 = 0$ and $x_2 = x_2^0$ in equation (2.5.3) we get, $x_2^0 = 0$.

On substituting $x_2^0 = 0$, $x_1 = x_1^0$ and $x_2 = x_2^0$ in equation (2.5.6) we get,

$$0 = -\omega_n^2 x_1^0 - 2\zeta\omega_n^2 x_2^0$$

But $x_2^0 = 0$, $\therefore\ 0 = -\omega_n^2 x_1^0$ (or) $x_1^0 = 0$

Therefore, the coordinates of singular point are (0, 0) and so the origin is the singular point.
The slope of the phase trajectory is given by

$$S = \frac{dx_2/dt}{dx_1/dt} = \frac{\dot{x}_2}{\dot{x}_1} \qquad\qquad .....(2.5.7)$$

On substituting for $\dot{x}_1$ and $\dot{x}_2$ from equations (2.5.3) and (2.5.6) in equation (2.5.7) we get,

$$S = -\frac{(\omega_n^2 x_1 + 2\zeta\omega_n x_2)}{x_2} \qquad\qquad .....(2.5.8)$$

Put $\zeta = 0.15$ and $\omega_n = 1$, in equation (2.5.8)

$$S = -\frac{(x_1 + 2 \times 0.15 x_2)}{x_2} = \frac{-(x_1 + 0.3 x_2)}{x_2}$$

$$= \frac{-x_1}{x_2} - \frac{0.3 x_2}{x_2} = \frac{-x_1}{x_2} - 0.3$$

$$\therefore \quad \frac{x_1}{x_2} = -0.3 - S \quad \text{(or)} \quad \frac{x_2}{x_1} = \frac{1}{-0.3 - S}$$

$$\therefore \quad x_2 = \frac{x_1}{-0.3 - S} \qquad \qquad \qquad .....(2.5.9)$$

From equation (2.5.9) we can conclude that the isoclines are straight lines. For each value of S we can draw one isocline.

Using equation (2.5.9), the coordinates (x_1, x_2) in the phase plane for various slopes can be calculated. Since there are three variables. Let us assume two variables and calculate the third variable.

Let us choose values of S as $-2, -1.0, -0.5, 0, 0.5, 1.0$ and 2.0

For each value of S, choose two values of x_1 and calculate x_2 using equation (2.5.9). The values of S, x_1 and x_2 are tabulated. The slope angle, α is calculated for each value of S, using the expression, $\alpha = \tan^{-1}(S)$ and tabulated in table-2.5.1.

TABLE-2.5.1.

S	-2.0		-1.0		-0.5		0		0.5		1.0		2.0	
α	$-63°$		$-45°$		$-27°$		0		27°		45°		63°	
	x_1	x_2	x_1	x_2	x_1	x_2	x_1	x_2	x_1	x_2	x_1	x_2	x_1	x_2
	1.0	0.6	1.0	1.4	0.25	1.25	0.25	-0.8	1.0	-1.25	1.0	-0.77	1.0	-0.43
	2.0	1.2	1.5	2.1	0.5	2.5	0.75	-2.5	2.0	-2.5	2.0	-1.54	2.0	-0.86

The isoclines corresponding to each slope is drawn using the coordinates listed in table-2.5.1. and they are shown in fig 2.5.1.

The phase trajectory starts at point A on x_1 axis, (i.e., given initial condition is (1.5, 0)). The phase trajectory is shown in fig 2.5.1 and the construction of the trajectory is explained below.

The slope of the trajectory when it crosses x_1 axis is infinite. Hence draw a line at an angle, $\tan^{-1} \alpha = 90°$ with respect to x_1 axis and passing through point A. Let this line meet the isocline corresponding to S = 2.0 at point q. Then draw a line at an angle $\tan^{-1} 2.0 = 63°$ with respect to x_1 axis and passing through point A. Let this line meet the isocline corresponding to S = 2.0 at point p. Now the phase trajectory will pass through point B on the isoclinecorresponding to S_2. The point B lies at the middle of the segment pq. Draw a smooth curve between Aand B, which is a section of phase trajectory.

In general, if F is the crossing point of phase trajectory with isocline-a corresponding to a slope of S_a and if the next isocline is isocline-b corresponding to a slope of S_b. Then draw two lines through point F, one at an angle of $\tan^{-1} S_a$ and the other at $\tan^{-1} S_b$. [The angles are marked from point F with respect to a (horizontal) line parallel to x_1 axis. Positive angles are measured in anticlockwise direction and negative angles are measured in clockwise direction]. These two lines will meet the isocline-b at points d and e. Now the crossing point of phase trajectory is fixed at the middle of d and e on the isocline-b.

Thus crossing point of phase trajectory on each isocline is determined. The complete phase trajectory is obtained by drawing a smooth curve through all the crossing points, as shown in fig 2.5.1.

RESULT

1.   The singular point lies at origin.
2.   From fig 2.5.1, it is observed that the phase trajectory spiral towards the origin, hence the type of singular point is stable focus.

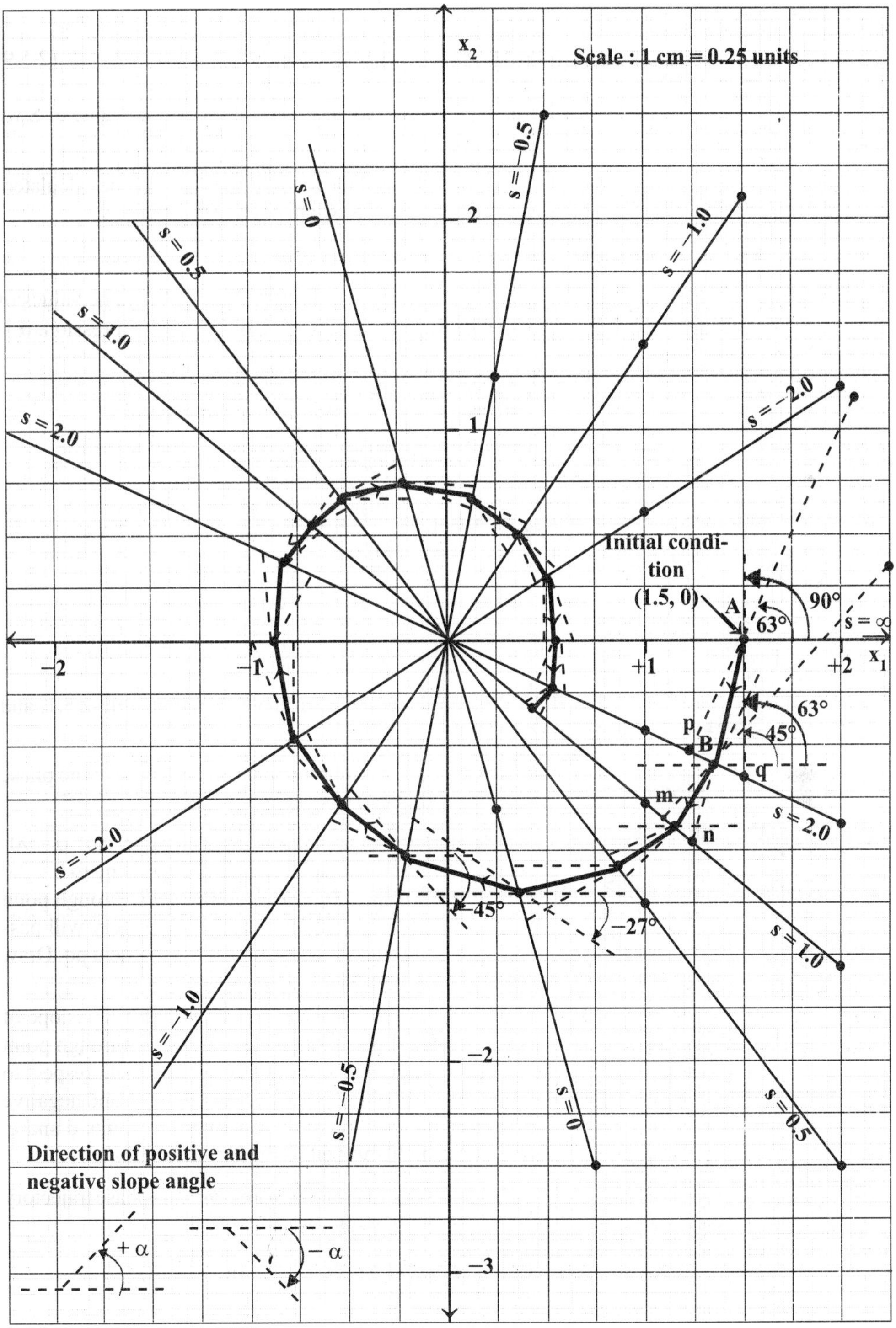

Fig 2.5.1. : Phase trajectory of second order system described by the equation,

EXAMPLE 2.6

Construct a phase trajectory by delta method for a nonlinear system represented by the differential equation, $\ddot{x} + 4\,|\dot{x}|\,\dot{x} + 4x = 0$ .Choose the initial conditions as $x(0) = 1.0$ and $\dot{x}(0) = 0$.

SOLUTION

Given that, $\ddot{x} + 4\,|\dot{x}|\,\dot{x} + 4x = 0$ .....(2.6.1)

To construct phase trajectory by delta method, the equation (2.6.1) can be converted to the form,

$$\ddot{x} + \omega_n^2(x + \delta) = 0 \qquad\qquad .....(2.6.2)$$

Hence, $\ddot{x} + 4\,|\dot{x}|\,\dot{x} + 4x = 0 = \ddot{x} + 4\big[x + |\dot{x}|\,\dot{x}\,\big] = 0$ .....(2.6.3)

On comparing equations (2.6.2) and (2.6.3) we get,

$$\delta = |\dot{x}|\,\dot{x} \qquad\qquad .....(2.6.4)$$

and $\omega_n^2 = 4$, $\therefore \omega_n = 2\ \text{rad/sec}$

Let us choose the state variables x_1 and x_2 such that,

$$x_1 = x \qquad\qquad .....(2.6.5)$$

and $x_2 = \dfrac{\dot{x}}{\omega_n} = \dfrac{\dot{x}}{2}$ .....(2.6.6)

From equation (2.6.6) we get, $\dot{x} = 2x_2$. On substituting $\dot{x} = 2x_2$ in equation (2.6.4) we get,

$$\delta = |\dot{x}|\,\dot{x} = |2x_2|\,2x_2 = 4\,|x_2|\,x_2 \qquad\qquad .....(2.6.7)$$

The given initial conditions are $x_1(0) = 1.0$ and $x_2(0) = 0$.

Let point A be the starting point of the phase trajectory.

$\therefore$ Coordinates of point $A = (x_1(0),\ x_2(0)) = (1.0,\ 0)$

Let $\delta_1 =$ Value of delta at point A.

Since, $x_2(0) = 0$, $\delta_1 = 0$

Let $C_1 =$ Centre of the circular arc AB.

The point C_1 is marked at $-\delta_1$ on x_1-axis. Since $\delta_1 = 0$, the point C_1 lies at origin.

With C_1 as centre and $C_1 A$ as radius, draw a circular arc AB as shown in fig 2.7.1. (Draw the circular arc using compass). The segment AB is a section of phase trajectory.

From the graph the value of x_2, corresponding to point B is $x_2 = -0.1875$. (Here the value of x_2 alone is needed to calculate δ).

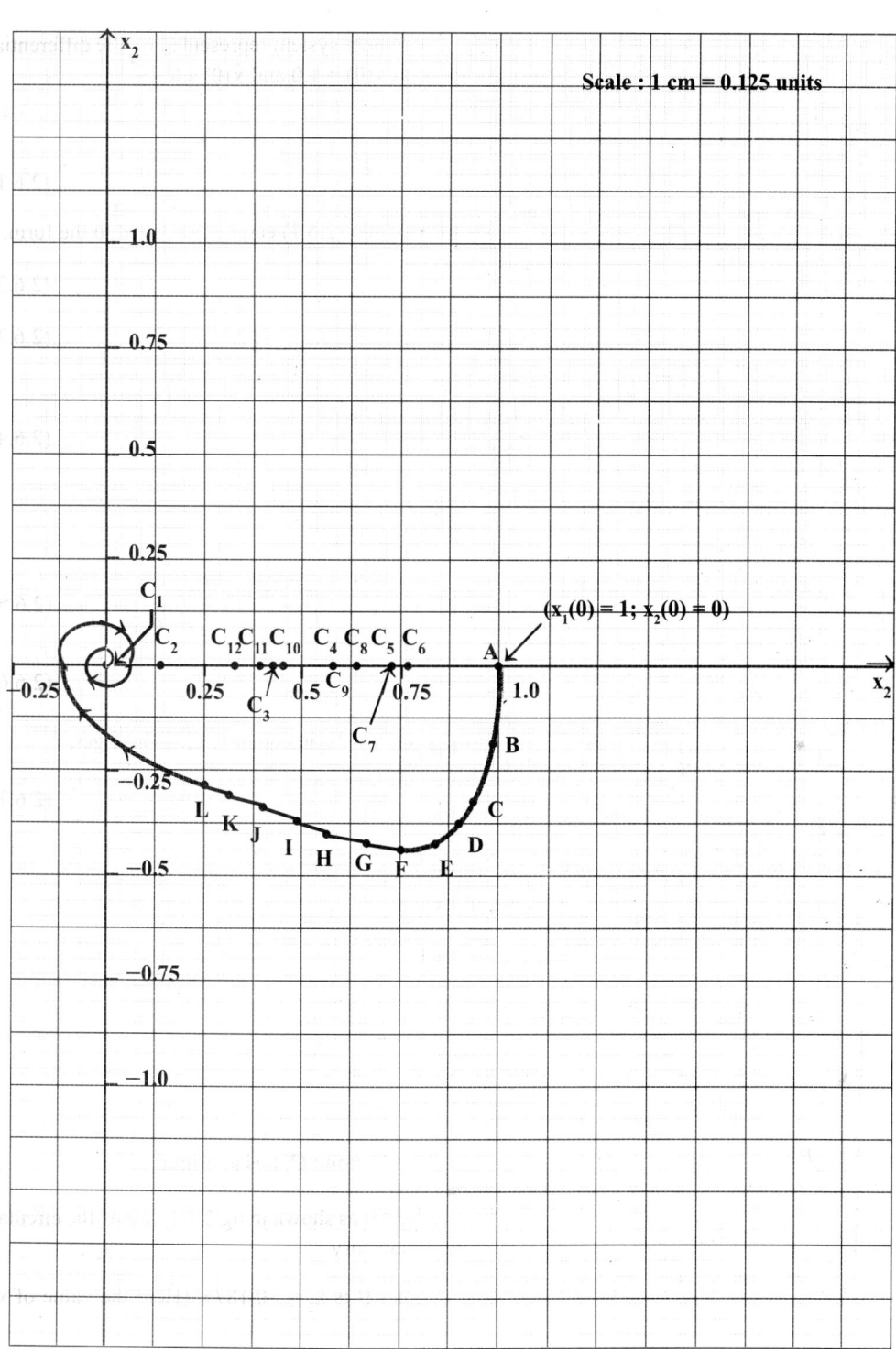

Fig 2.7.1. : Phase trajectory by delta method.

The above procedure is repeated at point B to determine the segment BC of phase trajectory shown in fig 2.7.1. Similarly the segment CD, DE, EF, FG, GH, HI, IJ, JK, and KL shown in fig 2.6.1 are determined. The values of x_2, δ and centre of the circular arc segments $(C_1, C_2, C_3, \ldots\ldots)$ are tabulated in table-2.6.1.

The complete phase trajectory shown in fig 2.7.1. is determined by repeating the above procedure again and again.

TABLE-2.6.1.

x_2		δ		$-\delta$	
0	(A)	0	(δ_1)	0	(C_1)
−0.1875	(B)	−0.140625	(δ_2)	0.140625	(C_2)
−0.325	(C)	−0.4225	(δ_3)	0.4225	(C_3)
−0.375	(D)	−0.5625	(δ_4)	0.5625	(C_4)
−0.425	(E)	−0.7225	(δ_5)	0.7225	(C_5)
−0.4375	(F)	−0.7656	(δ_6)	0.7656	(C_6)
−0.4	(G)	−0.7225	(δ_7)	0.7225	(C_7)
−0.375	(H)	−0.64	(δ_8)	0.64	(C_8)
−0.3375	(I)	−0.5625	(δ_9)	0.5625	(C_9)
−0.3125	(J)	−0.4556	(δ_{10})	0.4556	(C_{10})
−0.2875	(K)	−0.3906	(δ_{11})	0.3906	(C_{11})
		−0.3306	(δ_{12})	0.3306	(C_{12})

2.11   SHORT- ANSWER QUESTIONS

Q2.1   *What are linear and nonlinear systems? Give examples.*

The linear systems are systems which obeys the principle of superposition.The systems which does not satisfy superposition principle are called nonlinear systems.

Example of linear system   : $y = ax + b\dfrac{dx}{dt}$

Example of nonlinear system   : $y = ax^2 + e^{bx}$

Q2.2   *How nonlinearities are introduced in the systems?*

The nonlinearities are introduced in the system due to friction, inertia, stiffness, backlash, hysteresis, saturation and dead-zone of the components used in the systems.

Q2.3   *How the nonlinearities are classified? Give examples.*

The nonlinearities are classified as incidental and intentional.

The incidental nonlinearities are those which are inherently present in the system. Common examples of incidental nonlinearities are saturation, dead-zone, coulomb friction, stiction, backlash, etc.

The intentional nonlinearities are those which are deliberately inserted in the system to modify the system characteristics. The most common example of this type of nonlinearity is a relay.

Q2.4   *What is the purpose of intentionally introducing nonlinearities into the system?*

The purpose of intentionally introducing nonlinearities into the system is to improve the system performance and/or to simplify the construction of the system.

Note : The intentional nonlinear elements may improve the system performance under certain specified operating conditions, but in general they will degrade the system performance under other operating conditions.

Q2.5   *What are the methods available for the analysis of nonlinear system?*

The two popular methods of analysing nonlinear systems are phase plane method and describing function method.

Q2.6   *What is the difference between phase plane and describing function methods of analysis?*

The phase plane and describing function methods use complimentary approximations. The phase plane method retains the nonlinearity as such and uses the second order approximation of a higher-order linear part. On the other hand, the describing function method retains the linear part and harmonically linearizes the nonlinearity. The describing function analysis is frequency domain approach and phase plane analysis is time domain approach.

Q2.7   *State the limitations of analysing nonlinear systems by describing function and phase plane methods.*

1. These methods are useful only for stability analysis and to study the behaviour of the system but cannot to used for system design.
2. The phase plane method of analysis is useful only for second order systems with constant parameters and constant or zero input (i.e., second order autonomous system).
3. In describing function analysis, the accuracy of the information obtained is heavily dependent on the filtering property of the linear part of the system. This analysis does not give any useful information about transient response of the system.

Q2.8   *Write any two properties of nonlinear systems.*

1. The nonlinear systems may have jump resonance in the frequency response.
2. The output of a nonlinear system will have harmonics and sub-harmonics when excited by sinusoidal signals.

Q2.9   *Write the Duffing's equation?*

The Duffing's equation is given by

$$M\ddot{x} + nB(1 - x^2)\dot{x} = Kx + K'x^3 = 0$$

where K' = Positive for hard spring

           K' = Negative for soft spring

Q2.10   *Write the van der pol's equation for nonlinear damping.*

The Van der pol's equation for nonlinear damping is

$$M\ddot{x} + B(1 - x^2)\dot{x} + Kx = 0$$

Q2.11   *What is jump resonance?*

In the frequency response of nonlinear systems, the amplitude of the response (output) may jump from one point to another for increasing or decreasing values of ω, as shown in fig Q2.11 a & b. This phenomenon is called jump resonance.

In the frequency response curve shown in fig Q2.11 a & b, as the frequency, ω is increased, the amplitude X increases, until point 2 is reached. A further increase in frequency ω will cause a jump from point 2 to point 3. As the frequency ω is increased further, the amplitude X follows the curve from point 3 towards point 4.

If the frequency is reduced starting from a high value corresponding to point 4, then the amplitude X slowly increases through point 3, until point 5 is reached. A further decrease in ω will cause another jump from point-5 to point-6. After this jump, the amplitude X decreases with ω and follows the curve from point 6 towards point 1.

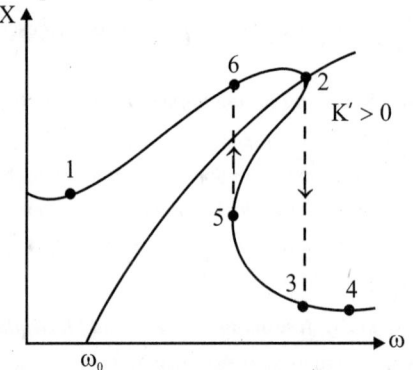 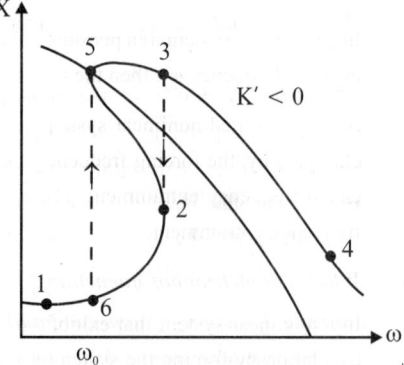

a: Mechanical system with hard spring   *b: Mechanical system with soft spring*

Fig Q2.11 : Frequency response curves showing jump resonance.

Q2.12   **Sketch the amplitude versus frequency curve for free oscillation in the system described by Duffing's equation for hard and soft springs.**

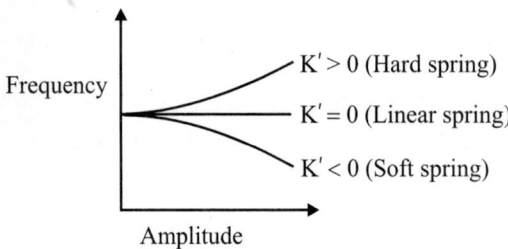

Fig Q2.12 : Amplitude Vs frequency curves for free oscillations in the system described by equation 2.1.

Q2.13   **What is meant by frequency-amplitude dependence in nonlinear system?**

In nonlinear systems the output or response may have oscillations. The frequency of this oscillations depends on the amplitude of the input signal and/or magnitude of initial conditions. This phenomenon is referred to as frequency-amplitude dependence in nonlinear systems.

Q2.14   **Write a short note on subharmonic oscillations.**

The subharmonic oscillations are nonlinear steady-state oscillations whose frequency is an integral submultiple of the forcing frequency. The generation of subharmonic oscillation depend on the system parameter and initial conditions, as well as on the amplitude and frequency of forcing function (i.e. input).

Q2.15   **What are limit cycles?**

The limit cycles are oscillations of the response (or output) of nonlinear systems with fixed amplitude and frequency. If these oscillations or limit cycles exists when there is no input then they are called zero input limit cycles.

Q2.16   *Distinguish between subharmonic and self-excited oscillations.*

1. The subharmonic oscillations are developed in a nonlinear system when it is excited by a sinusoidal input, whereas the self-excited oscillations are developed when there is no input.

2. The frequency of subharmonic oscillations depends on the frequency andamplitude of input signal but the frequency of self-excited oscillations are independent of the size of initial conditions.

Q2.17   *What is frequency entrainment?*

In a nonlinear system, if a periodic input of frequency, ω is applied to a system capable of exhibiting a limit cycle of frequency ω_1, then the well-known phenomenon of beats is observed.

In a self excited nonlinear system, it is found experimentally that the frequency ω_1 of the limit cycle is entrained by, the forcing frequency, ω, within a certain band of frequencies. This phenomenon is usually called frequency entrainment. The band of frequency in which entrainment occurs is called the zone of frequency entrainment.

Q2.18   *What is asynchronous quenching?*

In a nonlinear system that exhibits a limit cycle of frequency ω_1, it is possible to quench the limit-cycle oscillation by forcing the system at a frequency ω_q, where ω_q and ω_1 are not related to each other. This phenomenon is called asynchronous quenching or signal stabilization.

Q2.19   *What is saturation? Give an example.*

In saturation nonlinearity the output is proportional to input for a limited range of input signals. When the input exceeds this range, the output tends to become nearly constant as shown in fig Q2.19.

Saturation in the output of electronic, rotating and flow amplifiers, speed and torque saturation in electric and hydraulic motors are examples of saturation.

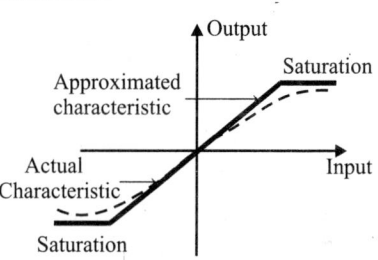

Fig Q2.19 : Saturation nonlinearity.

Q2.20   *What is dead-zone?*

The dead-zone is the region in which the output is zero for a given input. When the input is increased beyond this dead-zone value, the output will be linear.(fig.Q2.20)

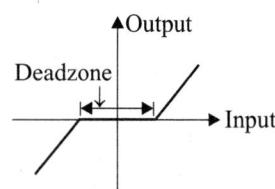

Fig Q2.20 : Dead zone nonlinearity.

Q2.21   *What are the different types of friction?*

The different types of friction are viscous friction, coulomb friction and stiction. The viscous friction is linear in nature. The coulomb friction and stiction are nonlinear frictions.

Q2.22   *What is hysteresis and backlash?*

The hysteresis is a phenomenon in which the output follows a different path for increasing and decreasing values of input. In fig Q2.22a, when the input x is increased from a minimum value, the output follows the path ABCD and when the input is decreased from a maximum value, the output follows the path DCEA.

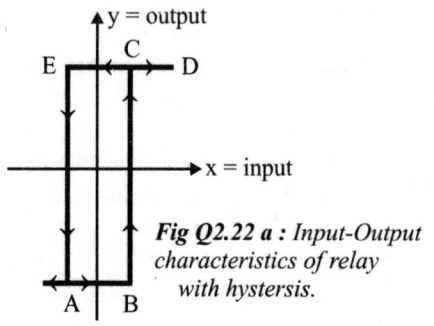

Fig Q2.22 a : Input-Output characteristics of relay with hystersis.

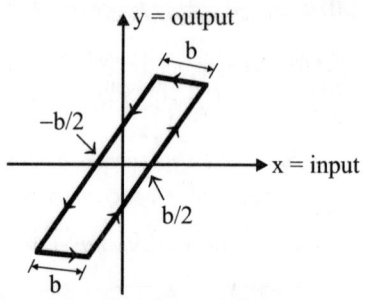

Fig Q2.22 b : Input-Output characteristic of backlash nonlinearity.

The backlash nonlinearity is a type of hysteresis in mechanical gear trains and linkages. The backlash characteristic is shown in fig Q2.22b. This is due to the play between the teeth of the drive gear and that of driven gear.

Q2.23   *What is describing function?*

When the input, x to the nonlinearity is a sinusoidal signal (i.e., x = X sinωt), the describing function of the nonlinearity is defined as,

Describing function, $K_N(X,\omega) = \dfrac{Y_1}{X} \angle \phi_1$

where, Y_1 =   Amplitude of the fundamental harmonic component of the output.

ϕ_1 =   Phase shift of the fundamental harmonic component of the output with respect to the input.

X   =   Maximum value of input signal.

ω   =   Angular frequency of input signal.

Q2.24   *Write the describing function of dead-zone and saturation nonlinearity.*

When the input, x = X sinωt, the describing function of dead-zone and saturation nonlinearity whose input-output characteristics are shown in fig Q2.24 is given by

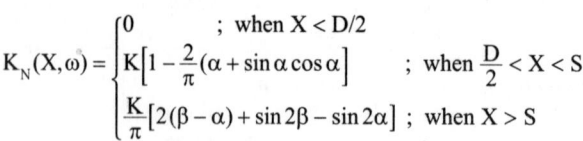

Fig Q2.24 : Input-output characteristic of dead-zone and saturation.

$$K_N(X,\omega) = \begin{cases} 0 & ; \text{ when } X < D/2 \\ K\left[1 - \dfrac{2}{\pi}(\alpha + \sin\alpha\cos\alpha)\right] & ; \text{ when } \dfrac{D}{2} < X < S \\ \dfrac{K}{\pi}[2(\beta-\alpha) + \sin 2\beta - \sin 2\alpha] & ; \text{ when } X > S \end{cases}$$

where, $\alpha = \sin^{-1} D/2X$ and $\beta = \sin^{-1} S/X$

Q2.25   *Write the describing function of saturation nonlinearity.*

When the input, x = X sinωt, the describing function of saturation nonlinearity whose input-output characteristics are shown in fig Q2.25 is given by

$$K_N(X,\omega) = \begin{cases} K & ; \text{ when } X < S \\ \dfrac{2K}{\pi}[\beta + \sin\beta\cos\beta] & ; \text{ when } X > S \end{cases}$$

where, $\beta = \sin^{-1} S/X$

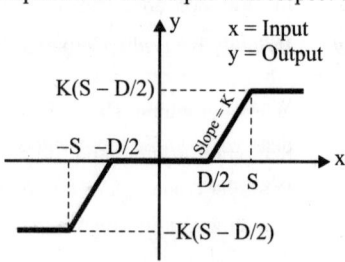
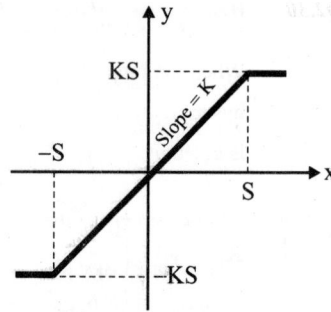

Fig Q2.25 : Input-output characteristic of saturation nonlinearity.

Q2.26 *Write the describing function for dead-zone nonlinearity.*

When the input, $x = X \sin\omega t$, the describing function of dead-zone nonlinearity whose input-output characteristics are shown in fig Q2.26 is given by

$$K_N(X,\omega) = \begin{cases} 0 & ; \text{ when } X < \dfrac{D}{2} \\ K\left[1 - \dfrac{2}{\pi}(\alpha + \sin\alpha\cos\alpha)\right] & ; \text{ when } X > \dfrac{D}{2} \end{cases}$$

where, $\alpha = \sin^{-1} D/2X$

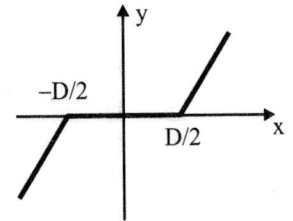

Fig Q2.26 : Input-Output characteristic of dead-zone nonlinearity.

Q2.27 *Draw the input-output characteristic of a relay with dead-zone and hysteresis.*

The input-output characteristics of a relay with dead-zone and hysteresis is shown in fig Q2.27.

Q2.28 *Write the describing function of ideal relay.*

When the input, $x = X \sin\omega t$, the describing function of ideal relay whose input-output characteristics is shown in fig Q2.28 is given by

$$K_N(X,\omega) = \frac{2M}{\pi X}$$

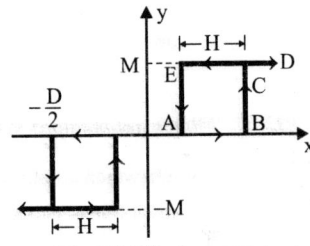

Fig Q2.27 : Input Output characteristic of relay with dead-zone and hysteresis.

Q2.29 *Write the describing function of relay with dead- zone.*

When the input, $x = X \sin\omega t$, the describing function of relay with dead-zone whose input-output characteristics is shown in fig Q2.29 is given by

$$K_N(X,\omega) = \begin{cases} 0 & ; X < \dfrac{D}{2} \\ \dfrac{4M}{\pi X}\sqrt{1 - \left(\dfrac{D}{2X}\right)^2} & ; X > \dfrac{D}{2} \end{cases}$$

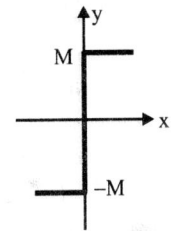

Fig Q2.28 : Input-Output characteristics of ideal relay.

Q2.30 *Write the describing function of relay with hysteresis.*

When the input, $x = X \sin\omega t$, the describing function of relay with hysteresis whose input-output characteristics is shown in fig Q2.30 is given by

$$K_N(X,\omega) = \begin{cases} 0 & ; \text{ when } X < \dfrac{H}{2} \\ \dfrac{4M}{\pi X}\angle\left(-\sin^{-1}\dfrac{H}{2X}\right) & ; \text{ when } X > \dfrac{H}{2} \end{cases}$$

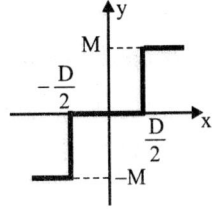

Fig Q2.29 : Input-Output characteristics of relay with dead-zone.

Q2.31 *Write the describing function of backlash nonlinearity.*

When the input, $x = X \sin\omega t$, the describing function of backlash nonlinearity whose input-output characteristic is shown in fig Q2.31 is given by

$$K_N(X, \omega) = \frac{Y_1}{X} \angle \phi_1$$

where, $Y_1 = \frac{X}{\pi}[((\pi/2) + \beta + (1/2)\sin 2\beta)^2 + \cos^4\beta]^{\frac{1}{2}}$

$$\phi_1 = \tan^{-1}\left(\frac{-\cos^2\beta}{(\pi/2) + \beta + (1/2)\sin 2\beta}\right) \quad \text{and} \quad \beta = \sin^{-1}\left(1 - \frac{b}{X}\right)$$

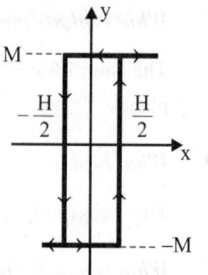

Fig Q2.30 : Input-Output characteristics of relay with hystersis.

Q2.32 *What is autonomous system?*

A system which is both free (or unforced or zero input or constant input) and time-invariant is called an autonomous system. The state equations of an autonomous system will be functions of state variables alone. i.e., $\dot{X} = F(X)$.

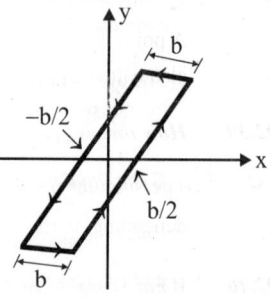

Q2.33 *State the stability criterion for nonlinear systems, when the nonlinearity is replaced by the decribing function K_N.*

Fig Q2.31 : Input-Output characteristic of backlash nonlinearity.

Nyquist stability criterion, when applied to nonlinear systems states that, the system will exhibit sustained oscillations or limit cycles when, $K_N G(j\omega) = -1$. This equation implies that limit cycle will occur if $K_N G(j\omega)$ locus pass through the critical point $-1 + j0$.

The equation $K_N G(j\omega) = -1$ can be modified such that $G(j\omega) = -1/K_N$, which implies that the critical point becomes critical locus $-1/K_N$. Hence limit cycle exists at the intersection point of $G(j\omega)$ locus and $-1/K_N$ locus.

Q2.34 *In describing function, analysis how the stability of nonlinear system is determined.*

In the stability analysis, let us assume that the linear part of the system is stable. To determine the stability of the system due to nonlinearity sketch the $-1/K_N$ locus and $G(j\omega)$ locus (polar plot of $G(j\omega)$) in complex plane, (Use either a polar graph sheet or ordinary graph sheet) and from the sketches the following conclusions can be obtained.

1. If the $-1/K_N$ locus is not enclosed by the $G(j\omega)$ locus then the system is stable or there is no limit cycle at steady state.

2. If the $-1/K_N$ locus is enclosed by the $G(j\omega)$ locus then the system is unstable.

3. If the $-1/K_N$ locus and the $G(j\omega)$ locus intersect, then the system output may exhibit a sustained oscillation or a limit cycle. The amplitude of the limit cycle is given by the value of $-1/K_N$ locus at the intersection point. The frequency of the limit cycle is given by the frequency of $G(j\omega)$ corresponding to the intersection point.

Q2.35 ***What is phase plane?***

The coordinate plane with the state variables x_1 and x_2 as two axes is called the phase plane (i.e., in phase plane x_1 is represented in x-axis, and x_2 in y-axis).

Q2.36 ***What is phase trajectory?***

The locus of the state point (x_1, x_2) in phase plane with time as running parameter is called phase trajectory.

Q2.37 ***What is phase portrait?***

A family of phase trajectories corresponding to various sets of initial conditions is called a phase portrait.

Q2.38 ***What is singular point?***

A point in phase-plane at which the derivatives of all state variables are zero is called singular point. It is also called equilibrium point.

Q2.39 ***How the singular points are classified?***

The singular points are classified as Nodal point, Saddle point, Focus point and Centre or Vortex point depending on the eigen values of the system matrix.

Q2.40 ***What is stable node? Draw the phase portrait of a stable node.***

If the system matrix has real and distinct roots then the singular point at origin is called stable node. The phase portrait of the system is shown in fig Q2.40.

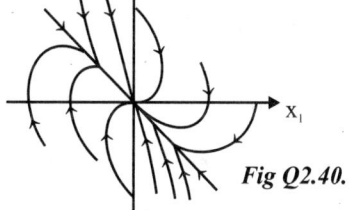

Fig Q2.40.

Q2.41 ***What is the difference in stability analysis of linear and nonlinear systems?***

In linear system the stability of the system in the entire phase-plane can be judged from the behaviour of the system at equilibrium state (i.e., at singular point), because the linear systems has only one equilibrium state.

In nonlinear systems there may be multiple equilibrium states (singular points). The behaviour of nonlinear system about the equilibrium point may be different for small deviations and large deviations about the equilibrium point. Hence in nonlinear systems, stability is discussed relative to equilibrium state and the general stability of a system cannot be defined.

Q2.42 ***Define the stability of a nonlinear system at origin.***

The autonomous system defined by equation $\dot{X} = F(X)$ is stable at the origin, if for every initial state $X(t_0)$ which is sufficiently close to origin, $X(t)$ remains near the origin for all t.

Q2.43 ***Define asymptotic stability.***

The autonomous system defined by equation $\dot{X} = F(X)$ is asymptotically stable if $X(t)$ approaches the origin as $t \to \infty$.

Q2.44   What is stable-in-the large?

The autonomous system defined by equation $\dot{X} = F(X)$ is asymptotically stable-in-the large if it is asymptotically stable for every initial state regardless of how near or far it is from the origin.

Q2.45   How limit cycles are determined from phase portrait?

If the phase portrait has closed trajectories then limit cycles exists in the system.

Q2.46   How will you determine the stable and unstable limit cycles using phase portrait?

The limit cycle is stable if the path in the neighbourhood of a closed trajectory converge towards it as shown in fig Q2.46a. The limit cycle is unstable if the path in the neighbourhood of a closed trajectory diverge away from it as shown in fig Q2.46b.

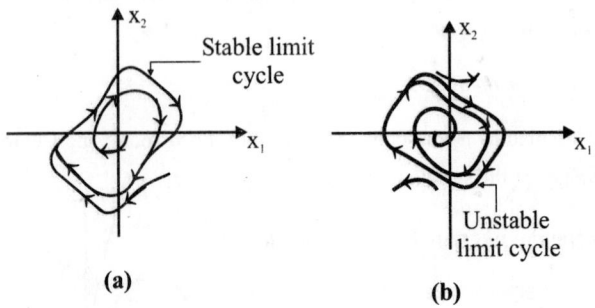

Fig Q2.46 : *Phase-portrait showing limit cycle behaviour in nonlinear system.*

Q2.47   Write the slope equation of phase trajectories.

The slope equation of phase trajectories is given by

$$\frac{dx_2}{dx_1} = \frac{f_2(x_1, x_2)}{f_1(x_1, x_2)}$$

Q2.48   What the methods available for constructing phase trajectories?

The methods available for constructing phase trajectories are analytical method, isocline method and delta method.

Q2.49   How the phase trajectory is constructed in analytical method?

In analytical method the slope equation is directly integrated to get solution equation. The various points of phase trajectory are directly calculated using solution equation. A smooth curve through these points will give the phase trajectory.

Q2.50   What is isocline?

A locus passing through the points of same slope in phase-plane is called isocline.

2.12 EXERCISES

I. FILL IN THE BLANKS

1. The nonlinear systems are systems which does not obey the ....

2. The limit cycle is a phenomenon observed in ....

3. The .... are oscillations of fixed frequency and amplitude.

4. The .... are those which are inherently present in the system.

5. The .... are those which are deliberately inserted in the system to modify its characteristics.

6. The .... is the region in which the output is zero for a given input.

7. The coulomb friction and stiction are .... frictions.

8. In .... nonlinearity the output remains constant for increasing or decreasing values of input.

9. The .... is a phenomenon in which the output follows a different path for increasing or decreasing values of input.

10. The locus of $-1/K_N$ is called .... (Note : K_N is describing function).

11. When the $-1/K_N$ locus is not enclosed by $G(j\omega)$ locus the system is ....

12. When the $-1/K_N$ locus is enclosed by $G(j\omega)$ locus the system is ....

13. The .... exists at the intersection point of $-1/K_N$ locus and $G(j\omega)$ locus.

14. The coordinate plane with the state variables x_1 and x_2 as two axes is called ....

15. The locus of the state point in .... with time as running parameter is called ....

16. A family of .... corresponding to various sets of initial conditions is called a ....

17. A point in .... at which the derivaties of all state variables are zero is called ....

18. A locus passing through the points of same slope in phase plane is called ....

19. When the eigen values are distinct, real and negative then the singular point is called ....

20. When the eigen values are complex conjugate with negative real part then the singular point is called ....

ANSWERS

1.	principle of superposition	11.	stable
2.	nonlinear system	12.	unstable
3.	limit cycles	13.	limit cycle
4.	incidental nonlinearities	14.	phase plane
5.	intentional nonlinearities	15.	phase plane, phase trajectory
6.	dead zone	16.	phase trajectories, phase portrait.
7.	nonlinear	17.	phase plane, singular point
8.	saturation	18.	isocline
9.	hysteresis	19.	stable node
10.	critical locus	20.	stable focus.

II.   State whether the following statements are TRUE/FALSE

1.   The systems which satisfy the principle of superposition are called nonlinear systems.

2.   The nonlinear system response may be highly sensitive to input amplitude.

3.   In nonlinear systems the response for any input can be judged from the response for standard inputs.

4.   The nonlinear systems will exhibit phenomena like jump resonance, subharmonic oscillations and limit cycles.

5.   The phase-plane analysis can be performed only for second order systems.

6.   In describing function analysis it is assumed that the output is adequately represented by the fundamental term in fourier series of the output.

7.   The hysteresis is a phenomenon in which the output follows the same path for increasing and decreasing values of input.

8.   The locus of the describing function, K_N is called critical locus.

9.   In describing function analysis, the limit cycles exists at the intersection points of critical locus and $G(j\omega)$ locus.

10.   If the critical locus is enclosed by the $G(j\omega)$ locus then the system is stable.

11.   In a complex plane the critical locus is said to be enclosed by $G(j\omega)$ locus if it lies in the region to the left of an observer travelling through $G(j\omega)$ locus in the direction of increasing ω.

12.   The starting point of a phase trajectory is the initial condition.

13.   At equilibrium point the derivative of all state variables are nonzero.

14.   When the eigen values are complex conjugate with positive real part then the singular point is stable focus.

15.   If all the eigen values of a system are nonzero then the origin is the only singular point.

16.   The linear autonomous systems has only one equilibrium state.

17.   At singular points the slope of the phase trajectory is indeterminate.

18.   In nonlinear systems stability is discussed relative to equilibrium state.

19.   The locus of the state point in phase plane with frequency as running parameter is called phase trajectory.

20.   The limit cycle is unstable if the path in the neighbourhood of a closed trajectory converge towards it.

ANSWERS

1.	False	6.	True	11.	False	16.	True
2.	True	7.	False	12.	True	17.	True
3.	False	8.	False	13.	False	18.	True
4.	True	9.	True	14.	False	19.	False
5.	True	10.	False	15.	True	20.	False

III. UNSOLVED PROBLEMS

E2.1 The block diagram of a system with saturation nonlinearity is shown in fig E2.1. Investigate the stability of the system by describing function method.

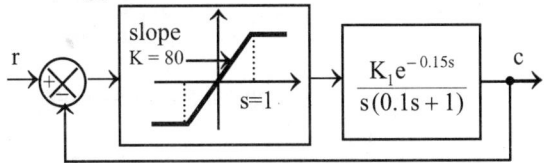

Fig E2.1.

E2.2 Derive the describing function of the element whose input-output characteristic is shown in fig E2.2.

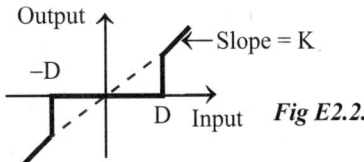

Fig E2.2.

E2.3 The block diagram of a system with hysteresis is shown in figE2.3.Using describing function method, determine whether limit cycle exists in the system. If limit cycles exists then, determine their amplitude and frequency.

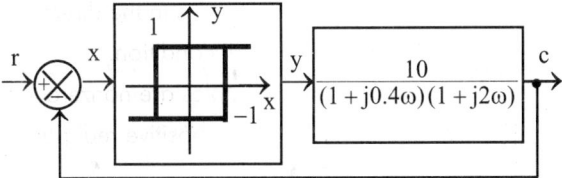

Fig E2.3.

E2.4 A system has a nonlinear element, with describing function, $K_N = (1/X) \angle -45°$ in cascade with, $G(j\omega) = 10\sqrt{2}/j\omega(1 + j0.5\omega)$. Determine the limit cycle of the system.

E2.5 Using the describing function analysis, prove that no limit cycle exists in the system shown in fig E2.5. Find the range of values of the dead-zone of the on-off controller for which limit cycle is predicted?

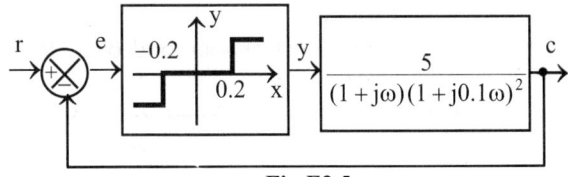

Fig E2.5.

E2.6 A second order servo containing a relay with dead-zone and hysteresis is shown in fig E2.6. Construct the phase trajectory of the system with initial conditions e(0) = 0.65 and $\dot{e}(0) = 0$.

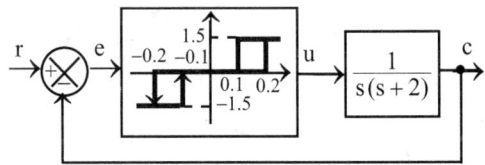

Fig E2.6.

E2.7   Construct phase trajectory for the system described by the equation,

$$\frac{dx_2}{dx_1} = \frac{4x_1 + 3x_2}{x_1 + x_2}. \text{ Comment on the stability of the system.}$$

E2.8   Draw the phase trajectory of the system described by the equation $\ddot{x} + \dot{x} + x^2 = 0$. *Comment on the stability of the system.*

E2.9   Determine the type of singularity for each of the following differential equations. Also locate the singular points on the phase plane.

(a) $\ddot{x} + 3\dot{x} + 2x = 0$ (b) $\ddot{x} + 5\dot{x} + 6x = 6$ (c) $\ddot{x} - 8\dot{x} + 17x = 34$

Note : Type of singular points can be determined from eigen values.

E2.10   Consider a nonlinear system with dead-zone shown in fig E2.10. Construct phase trajectories and show that the closed-loop system is stable with equilibrium zone defined by $-1 \le x_1 \le 1$; $\dot{x} = 0$.

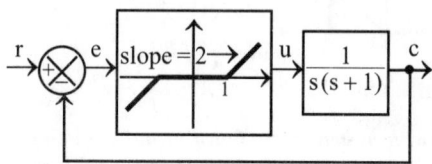

Fig E2.10

E2.11   A position control system comprises of a dc servomotor, potentiometer, error detector, a relay amplifier and a tachogenerator coupled to the motor shaft. The differential equation governing this system is (i) Reaction torque = $\theta + 0.5\dot\theta$ (ii) Drive torque = 2 sign $(e + 0.5\dot{e})$; $e = \theta_p - \theta$ (a) Draw the block diagram of the system. (b) Construct a phase trajectory on (e, ė) plane with e(0) = 2 and ė (0) = 0 and comment upon the system stability.

E2.12   (a) Construct phase trajectory for the system shown in fig E2.12a, on the (e, ė) plane and show that with the relay switching on the vertical axis of the phase plane, the system oscillates with increasing frequency and decreasing amplitude. Take [e(0) = 1.4, ė(0) = 0] as the initial condition.

(b) Introduce now a dead-zone of ±0.2 in the relay characteristics. Obtain a phase trajectory for the modified system with [e(0) = 1.4, ė (0) = 0] as the initial state point and comment upon the effect of dead-zone.

(c) The relay with dead-zone is now controlled by the signal (e + (1/3) ė), combining proportional and derivative control as shown in fig E2.12b. Construct a phase trajectory with [e(0) = 1.4, ė(0) = 0] as the initial state point. What is the effect of the derivative control action?

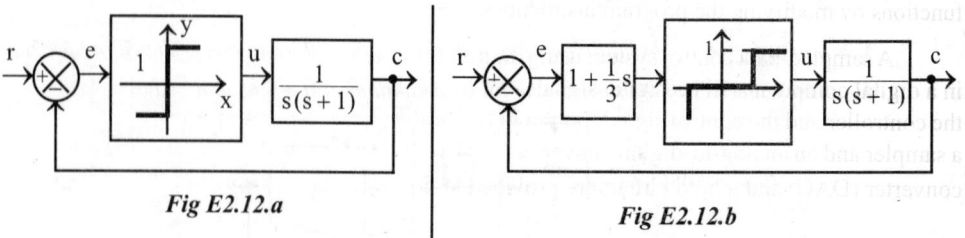

Fig E2.12.a   Fig E2.12.b

Chapter 3

Sampled Data Control Systems

3.1 INTRODUCTION

When the signal or information at any or some points in a system is in the form of discrete pulses, then the system is called discrete data system. In control engineering the discrete data system is popularly known as sampled data system.

The control system becomes a sampled data system in any one of the following situations.

1. *When a digital computer or microprocessor or digital device is employed as a part of the control loop.*

2. *When the control components are used on time sharing basis.*

3. *When the control signals are transmitted by pulse modulation.*

4. *When the output or input of a component in the system is a digital or discrete signal.*

The controllers are provided in control systems to modify the error signal for better control action. If the controllers are constructed using analog elements then they are called analog controllers and their input and output are analog signals, which are continuous functions of time. The analog controllers are complex, costlier and once fabricated it is difficult to alter the controllers.

A digital controller can be employed to implement complex or time shared control functions. [In time shared controller, a single controller will perform more than one function]. The digital controller are simple, versatile, programmable, fast acting and less costlier than analog controllers.

The digital controller can be a special purpose computer (microprocessor based system) or a general purpose computer or it is constructed using non-programmable digital devices. When computer or microprocessor is involved then the controller becomes programmable and its easier to alter the control functions by modifying the program instructions.

A sampled-data control system using digital controller is shown in fig 3.1. The input and output signal in a digital computer will be digital signals, but the error signal (input to the controller) to be modified by the controller and the control signal (output of the controller) to drive the plant are analog in nature. Hence a sampler and an analog-to-digital converter (ADC) are provided at the computer input. A digital to analog converter (DAC) and a hold circuit are provided at the computer output.

The sampler converts the continuous time-error signal into a sequence of pulses and ADC produces a binary code (binary number) for each sample. These codes are the input data to the digital computer which process the binary codes and produces another stream of binary codes as output. The DAC and hold circuit converts the output binary codes to continuous time signal (Analog signal) called control signal. This output control signal is used to drive the plant.

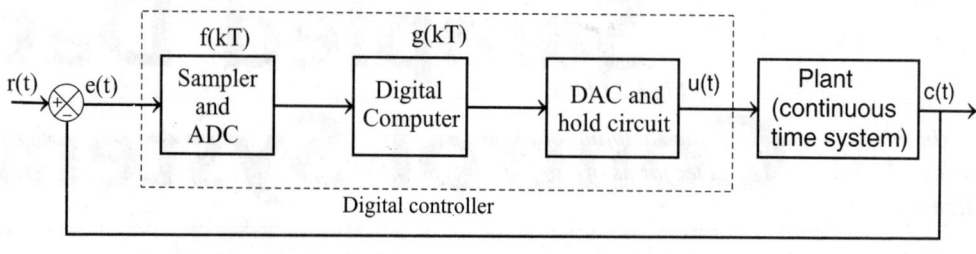

e(t)-Error signal(Analog)   f(kT)-Digital error signal
u(t)-Control signal(Analog)   g(kT)-Digital control signal

Fig 3.1 : Sampled-data control system.

ADVANTAGES OF DIGITAL CONTROLLERS

1. *The digital controllers can perform large and complex computation with any desired degree of accuracy at very high speed. In analog controllers the cost of controllers increases rapidly with the increase in complexity of computation and desired accuracy.*

2. *The digital controllers are easily programmable and so they are more versatile.*

3. *Digital controllers have better resolution.*

ADVANTAGES OF SAMPLED DATA CONTROL SYSTEMS

1. *The sampled data control systems are highly accurate, fast and flexible.*

2. *Use of time sharing concept of digital computer results in economical cost and space.*

3. *Digital transducers used in the system have better resolution.*

4. *The digital components used in the system are less affected by noise, nonlinearities and transmission errors of noisy channel.*

5. *The sampled data system require low power instruments which can be built to have high sensitivity.*

6. *Digital coded signals can be stored, transmitted, retransmitted, detected, analysed or processed as desired.*

7. *The system performance can be modified by compensation techniques.*

3.2   SAMPLING PROCESS

Sampling is the conversion of a continuous-time signal (or analog signal) into a discrete-time signal obtained by taking samples of the continuous time signal (or analog signal) at discrete time instants. Thus if f(t) is the input to the sampler as shown in fig 3.2, the output is f(kT), where T is called the sampling interval or sampling period. The reciprocal of T, i.e., $1/T = F_s$ is called the sampling rate (or samples per second or sampling frequency). This type of sampling is called periodic sampling, since samples are obtained uniformly at intervals of T seconds.

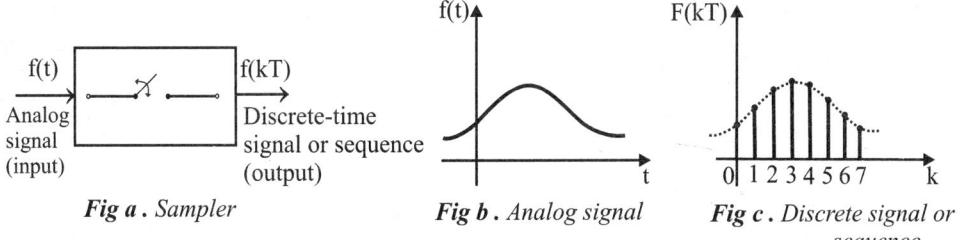

Fig a . *Sampler*   **Fig b .** *Analog signal*   **Fig c .** *Discrete signal or sequence*

Fig 3.2 : *Periodic sampling of an analog signal.*

(In this book only periodic sampling of signals is considered, because periodic sampling is most widely used in practice. The other forms of sampling are multiple-order sampling, multiple-rate sampling and Random sampling).

Multiple-order sampling : *A particular sampling pattern is repeated periodically.*

Multiple-rate sampling : *In this method two simultaneous sampling operations with different time periods are carried out on the signal to produce the sampled output.*

Random sampling : *(In this case the sampling instants are random.)*

The sampling frequency F_s (=1/T) must be selected large enough such that the sampling process will not result in any loss of spectral information (i.e., if the spectrum of the analog signal can be recovered from the spectrum of the discrete - time signal, there is no loss of information). A guideline for choosing the sampling frequency is given by sampling theorem.

SAMPLING THEOREM : *A band limited continuous time signal with highest frequency (bandwidth) f_m hertz, can be uniquely recovered from its samples provided that the sampling rate F_s is greater than or equal to $2f_m$ samples per second.*

From the sampling theorem we can infer that the knowledge of frequency content of a signal is essential while choosing the sampling frequency.

For processing the sampled signals by digital means, it has to be converted to binary codes and this conversion process is called quantization and coding. The process of converting a discrete time continuous valued signal into a discrete time discrete valued signal is called quantization. In quantization the value of each signal sample is represented by a value selected from a finite set of possible values called quantization levels. The difference between the unquantized sample and the quantized output is called the quantization error. The coding is the process of representing each discrete value by an n-bit binary sequence (or code or number). The process of sampling, quantization and coding are performed by sample/hold circuit and ADC.

3.3   ANALYSIS OF SAMPLING PROCESS IN FREQUENCY DOMAIN

The sampling process explained in the previous section is equivalent to multiplying the analog signal, f(t) with a impulse train, $\delta_T(t)$ to produce the sampled signal, $f_s(t)$. Let the impulse train consists of pulses of area, Δ. Hence the impulse sampled signal, $f_s(t)$ can be expressed as,

$$f_s(t) = f(t) \, \Delta\delta_T(t) \qquad\qquad\qquad .....(3.1)$$

Mathematically, the impulse train, $\delta_T(t)$ can be expressed as,

$$\delta_T(t) = \sum_{k=-\infty}^{+\infty} \delta(t-kT) \qquad \qquad .....(3.2)$$

$$f_s(t) = \Delta f(t) \sum_{k=-\infty}^{+\infty} \delta(t-kT) \qquad \qquad .....(3.3)$$

where, T is the sampling period.

A typical analog signal, f(t) [Fig a]; the impulse train, $\delta_T(t)$ [Fig b] and the impulse sampled signal, $f_s(t)$ [Fig c] are shown in figure 3.3.

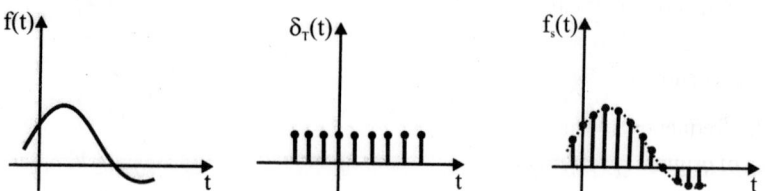

Fig a . Analog signal   **Fig b . Impulse train**   **Fig c . Impulse sampled analog signal**

Fig 3.3 : Impulse sampling of an analog signal.

The frequency content (frequency response) of a signal can be obtained from the fourier transform of the signal [i.e., Fourier transform converts the time domain signal to frequency domain signal]. Hence the frequency response of the impulse sampled signal can be obtained by taking fourier transform of equation (3.3).

The fourier transform of a single-valued function, f(t) is defined as

$$F\{f(t)\} = F(\omega) = \int_{-\infty}^{\infty} f(t)\, e^{-j\omega t}\, dt \qquad \qquad .....(3.4)$$

On taking fourier transform of $f_s(t)$ using the definition of fourier transform we get,

$$F\{f_s(t)\} = F_s(\omega) = \int_{-\infty}^{\infty} f_s(t)\, e^{-j\omega t}\, dt$$

$$= \int_{-\infty}^{\infty} \Delta f(t) \sum_{k=-\infty}^{+\infty} \delta(t-kT)e^{-j\omega t}\, dt \qquad \qquad .....(3.5)$$

Mathematically the equation (3.5) represents, the convolution of two signals, f(t) and $\delta(t-kT)$. The convolution theorem of fourier transform says that, the convolution of two time domain signals is equivalent to the product of their individual fourier transforms. Therefore, fourier transform of $f_s(t)$ can be expressed as a product of fourier transform of f(t) and $\delta(t-kT)$.

$$\therefore F_s(\omega) = \frac{\Delta}{2\pi} F\{f(t)\} . F\left\{ \sum_{k=-\infty}^{+\infty} \delta(t-kT) \right\} \qquad \qquad .....(3.6)$$

Let, $F\{f(t)\} = F(\omega)$ $\qquad \qquad .....(3.7)$

$$F\left\{ \sum_{k=-\infty}^{+\infty} \delta(t-kT) \right\} = \omega_s \sum_{k=-\infty}^{+\infty} \delta(\omega - k\omega_s) \qquad \qquad .....(3.8)$$

where, $\omega_s = 2\pi/T = $ Sampling frequency in rad/sec.

Using equations (3.7) and (3.8), the equation (3.6) can be written as,

$$F_s(\omega) = \frac{\Delta}{2\pi} \times F(\omega) \times \omega_s \sum_{k=-\infty}^{+\infty} \delta(\omega - k\omega_s) = \frac{\Delta}{2\pi} \frac{2\pi}{T} \sum_{k=-\infty}^{+\infty} F(\omega)\delta(\omega - k\omega_s)$$

Since $F(\omega) \delta(\omega - k\omega_s) = F(\omega - k\omega_s)$

$$F_s(\omega) = \frac{\Delta}{T} \sum_{k=-\infty}^{+\infty} F(\omega - k\omega_s) \qquad\qquad .....(3.9)$$

The equation (3.9) gives the frequency spectrum of the impulse sampled signal.

Let $F(\omega)$ be a band-limited signal with a maximum frequency of ω_m. The frequency spectrum of $F(\omega)$ is shown in fig 3.4(a), which is a plot of $|F(\omega)|$ Vs ω. The frequency spectrum of impulse sampled signal, i.e, $|F_s(\omega)|$ Vs ω, is shown in fig 3.4(b), when $\omega_s > 2\,\omega_m$ and in fig 3.4(c), when $\omega_s < 2\,\omega_m$.

In fig 3.4(b) the frequency spectrum of original signal is repeated periodically with period ω_s and there is no overlapping of original spectrum. In fig 3.4(c) the periodic repetition of original spectrum overlaps.

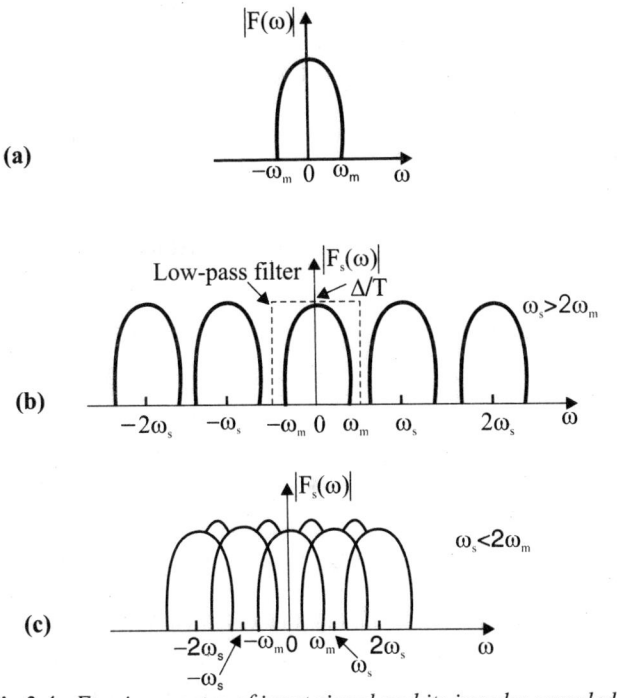

Fig 3.4 : *Fourier spectra of input signal and its impulse sampled version.*

From fig 4.4 it is observed that, as long as $\omega_s \geq \omega_m$, the original spectrum is preserved (since there is no overlapping) in the sampled signal and can be extracted from it by low-pass filtering. This fact was proposed as shanon's sampling theorem, which states that the information contained in a signal is fully preserved in the sampled version as long as the sampling frequency is at least twice the maximum frequency in the signal.

3.4   RECONSTRUCTION OF SAMPLED SIGNALS USING HOLD CIRCUITS

The hold circuits are popularly used in the process of analog-to-digital conversion (ADC) and digital-to-analog conversion (DAC). In ADC process the hold circuit is used to hold the sample until the quantization and coding for the current sample is complete.

In DAC process various types of hold circuits are used to convert the discrete time signal to analog signal. The simplest hold circuit is the zero order hold (ZOH). In zero order hold circuits the signal is reconstructed such that the value of reconstructed signal for a sampling period is same as the value of last received sample. The schematic diagram of sampler and zero order hold (ZOH) is shown in fig 3.5. The signal reconstruction by zero order hold (ZOH) circuit is illustrated in fig 3.6.

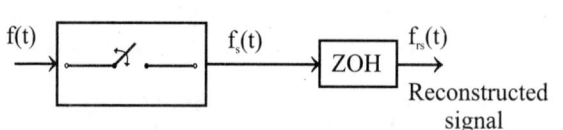

Fig 3.5 : Sampler and ZOH.

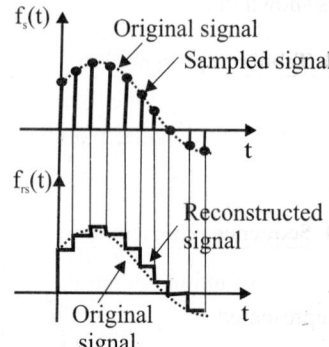

Fig 3.6 : Signal reconstruction by ZOH.

The high frequencies present in the reconstructed signal are easily filtered out by the various elements of the control system, because the control system is basically a low-pass filter.

In a first-order hold, the last two signal samples (current and previous sample) are used to reconstruct the signal for the current sampling period. Similarly higher order hold circuits can be devised. First or higher-order hold circuits offer no particular advantage over the zero-order hold. In sampled data control systems, the zero-order hold when used in conjunction with a high sampling rate provides a satisfactory performance.

An ideal sample/hold circuit introduces no distortion in the conversion process. However, in practical sample/hold circuits the following problems may be encountered.

1. *Errors in the periodicity of sampling process.*
2. *Nonlinear variations in the duration of sampling aperture.*
3. *Droop (changes) in the voltage held during conversion.*

3.5   DISCRETE SEQUENCE (DISCRETE TIME SIGNAL)

A discrete sequence or discrete time signal, $f(k)$, is a function of an independent variable, k, which is an integer. It is important to note that a Discrete time signal is not defined at instants between two successive samples. Also, it is incorrect to think that $f(k)$ is equal to zero if k is not an integer. Simply the signal $f(k)$ is not defined for non-integer values of k. A discrete-time signal is defined for every integer value of k in the range $-\infty < k < \infty$.

Since a digital signal is represented by a set of numbers it is also called a sequence. (i.e., the terms signal and sequence refers the digital or discrete time signal).

METHODS OF REPRESENTING A DISCRETE-TIME SIGNAL OR SEQUENCE

1. Functional representation

$$f(k) = 1 \quad ; \quad k = 1, 3$$
$$4 \quad ; \quad k = 2$$
$$0 \quad ; \quad \text{other } k$$

2. Graphical representation

The graphical representation of a discrete sequence is shown in fig 3.7.

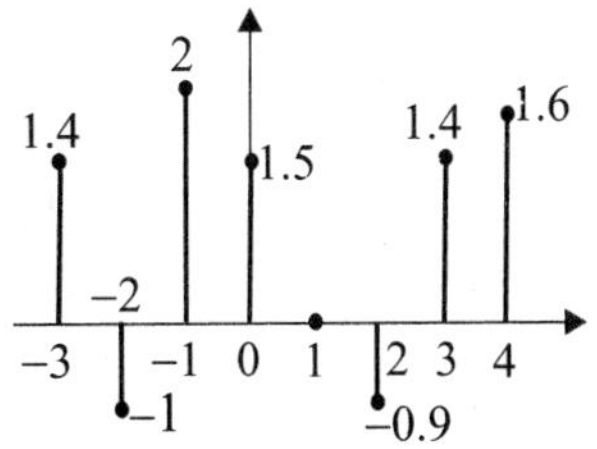

Fig 3.7 : Graphical representation of a discrete-time signal.

3. Tabular representation

k	.....	−2	−1	0	1	2	....
f(k)	.....	0	0	0	1	4	....

4. Sequence representation

An infinite duration signal or sequence with the time origin (k = 0) indicated by the symbol ↑ is represented as

$$f(k) = \{......1, 2, 1, 4, 1, 0, 0.....\}$$
$$\uparrow$$

An infinite sequence f(k), which is zero for k < 0, may be represented as

$$f(k) = \{2, 1, 4, 1, 0, 0.....\} \qquad \text{(or)} \qquad f(k) = \{2, 1, 4, 1....\}$$
$$\uparrow$$

A finite duration sequence with the time origin (k = 0) indicated by the symbol ↑ is represented as

$$f(k) = \{3, -1, -2, 5, 0, 4, ...\}$$
$$\uparrow$$

A finite duration sequence that satisfies the condition f(k) = 0 for k < 0 may be represented as

$$f(k) = \{2, 1, 4, 1\} \qquad \text{(or)} \qquad f(k) = \{2, 1, 4, 1\}$$
$$\uparrow$$

SOME ELEMENTARY DISCRETE-TIME SIGNALS

1. Digital impulse signal or unit sample sequence

$$\delta(k) = \begin{cases} 1 & ; \ k = 0 \\ 0 & ; \ k \neq 0 \end{cases}$$

An impulse delayed by k_0,

$$\delta_1(k) = \delta(k - k_0) = \begin{cases} 1 & ; \ k = k_0 \\ 0 & ; \ k \neq k_0 \end{cases}$$

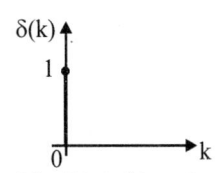

Fig 3.8 : Digital impulse signal

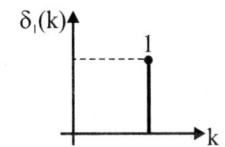

Fig 3.9 : Delayed impulse signal

2. Unit step signal

$$u(k) = \begin{cases} 1 \; ; \; k \geq 0 \\ 0 \; ; \; k < 0 \end{cases}$$

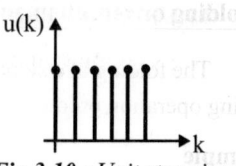

Fig 3.10 : Unit step signal

An unit step signal delayed by k_0

$$u_1(k) = u(k - k_0) = \begin{cases} 1 \; ; \; k \geq k_0 \\ 0 \; ; \; k < k_0 \end{cases}$$

The unit step is related to digital impulse by the summation relation

$$u(k) = \sum_{m=0}^{\infty} \delta(k - m)$$

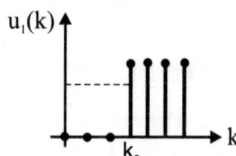

Fig 3.11 : Delayed Unit step signal

3. Ramp signal

$$u_r(k) = \begin{cases} k \; ; \; k \geq 0 \\ 0 \; ; \; k < 0 \end{cases}$$

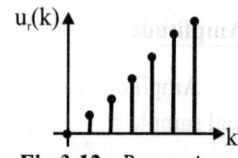

Fig 3.12 : Ramp signal

4. Exponential signal

$$g(k) = \begin{cases} a^k \; ; \; k \geq 0 \\ 0 \; ; \; k < 0 \end{cases}$$

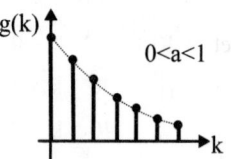

Fig 3.13 : Exponential signal

MATHEMATICAL OPERATIONS ON DISCRETE TIME SIGNALS

1. Shifting in time

A signal f(k) may be shifted in time by replacing the independent variable k by (k − m), where m is an integer. If m is a positive integer, the time shift results in a delay by m units of time. If m is a negative integer, the time shift results in an advance of the signal by |m| units in time. The delay results in shifting each sample of f(k) to right. The advance results in shifting each sample of f(k) to left.

Example

Let, f(k) = 1 k = 0 | Now, $f_1(k)$ = f(k + 2) = 1 k = −2 | and $f_2(k)$ = f(k − 2) = 1 k = 2
         2 k = 1 | 2 k = −1 | 2 k = 3
         3 k = 2 | 3 k = 0 | 3 k = 4

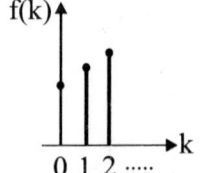

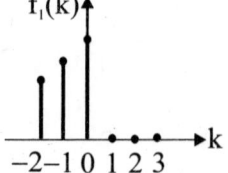

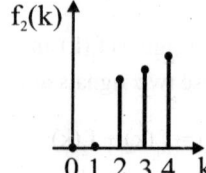

2. Folding or reflection or Transpose

The folding of a signal f(k) is performed by changing the sign of the time base k in the signal f(k). The folding operation produces a signal f(–k) which is mirror image of f(k) with respect to time origin k = 0.

Example

Let $f(k) = k$ $-3 \le k \le 3$; $f_1(k) = f(-k) = -k$ $-3 \le k \le 3$

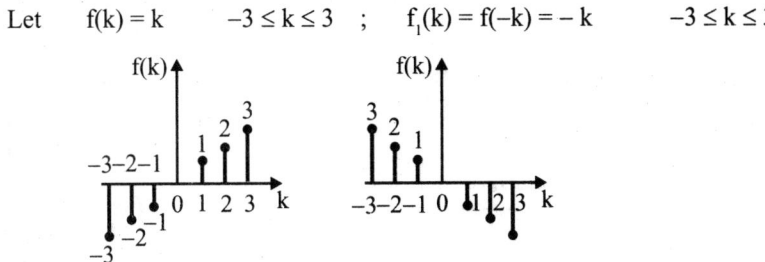

3. Amplitude scaling or scalar multiplication

Amplitude scaling of a signal by a constant A is accomplished by multiplying the value of every signal sample by A.

Let c(k) be amplitude scaled signal of f(k), then c(k) = Af(k)

Let $f(k) = 20$ $k = 0$ and $A = 0.1$; $c(k) = 2.0$ $k = 0$

              36 $k = 1$ 3.6 $k = 1$

              40 $k = 2$ 4.0 $k = 2$

            -15 $k = 3$ -1.5 $k = 3$

4. Time scaling or down sampling

In a signal, f(k), if k is replaced by μk, where μ is an integer, then it is called time scaling or down sampling.

Example : If $f(k) = a^k$; $k \ge 0$, then $f_1(k) = f(2k) = a^k$ for even values of k

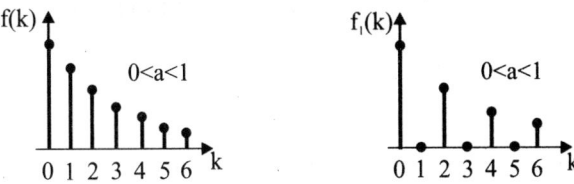

5. Signal (or vector) addition

The sum of two signals $f_1(k)$ and $f_2(k)$ is a signal c(k), whose value at any instant is equal to the sum of the samples of these two signals at that instant.

i.e. $c(k) = f_1(k) + f_2(k)$; $-\infty < k < \infty$.

Example

Let $f_1(k) = \{1, 2, -1, 2\}$ and $f_2(k) = \{-2, 1, 3, 1\}$

$c(k) = f_1(k) + f_2(k) = \{-1, 3, 2, 3\}$

6. Signal (or vector) multiplication

Signal multiplication results in the product of two signals on a sample-by-sample basis. The product of two signals $f_1(k)$ and $f_2(k)$ is a signal $c(k)$, whose value at any instant is equal to the product of the sample of these two signals at that instant. The product is also called modulation.

Example

Let $f_1(k) = \{1, 2, -1, 2\}$ and $f_2(k) = \{-2, 1, 3, 1\}$

$c(k) = f_1(k) \cdot f_2(k) = \{-2, 2, -3, 2\}$

3.6 Z-TRANSFORM

Transform techniques are an important tool in the analysis of signals and linear time invariant systems. The Laplace transforms are popularly used for analysis of continuous time signals and systems. Similarly Z-transform plays an important role in analysis and representation of linear discrete time systems. The Z-transform provides a method for the analysis of discrete time systems in the frequency domain which is generally more efficient than its time domain analysis.

DEFINITION OF Z-TRANSFORM

Let, $f(k)$ = Discrete time signal or sequence

$F(z) = Z\{f(k)\} = Z$-transform of $f(k)$

The Z-transform of a discrete time signal or sequence is defined as the power series

$$F(z) = \sum_{k=-\infty}^{\infty} f(k) z^{-k}$$ .....(3.10)

where, z is a complex variable.

The sequence of equation (3.10) is considered to be two sided and the transform is called two sided Z-transform, since the time index k is defined for both positive and negative values. If the sequence $f(k)$ is one sided sequence, (i.e. $f(k)$ is defined only for positive value of k) then the Z-transform is called one sided Z-transform.

The one sided z-transform of $f(k)$ is defined as,

$$F(z) = \sum_{k=0}^{\infty} f(k) \, z^{-k}$$

REGION OF CONVERGENCE

Since the z-transform is an infinite power series, it exists only for those values of z for which the series converges. The region of convergence, (ROC) of $F(z)$ is the set of all values of z, for which $F(z)$ attains a finite value. The ROC of a finite-duration signal is the entire z-plane, except possibly the point $z = 0$ and $z = \infty$. These points are excluded, because z^k (when $k > 0$) becomes unbounded for $z = \infty$ and z^{-k} (when $k < 0$) becomes unbounded for $z = 0$.

The complex variable z can be expressed in the polar form as,

$$z = r \, e^{j\theta} \qquad \qquad \qquad .....(3.11)$$

where, $r = |z|$ and $\theta = \angle z$

On substituting for z from equation (3.11) in equation (3.10) we get,

$$F(z) = \sum_{k=-\infty}^{\infty} f(k) \, z^{-k} \; = \; \sum_{k=-\infty}^{\infty} f(k) \, (r e^{j\theta})^{-k} \; = \; \sum_{k=-\infty}^{\infty} f(k) \, r^{-k} e^{-j\theta k} \qquad .....(3.12)$$

$$\text{Now, } |F(z)| = \sum_{k=-\infty}^{\infty} \left| f(k) \, z^{-k} \right| \qquad \qquad \qquad .....(3.13)$$

In the ROC of $F(z)$, $|F(z)| < \infty$.

From equation (3.13) we observe that $|F(z)|$ is finite, if the sequence $f(k) \, r^{-k}$ is absolutely summable.

To find the ROC, the equation (3.13) can be expressed as,

$$|F(z)| = \sum_{k=-\infty}^{\infty} \left| f(k) \, r^{-k} \right| = \sum_{k=-\infty}^{-1} \left| f(k) \, r^{-k} \right| + \sum_{k=-\infty}^{\infty} \left| f(k) \, r^{-k} \right|$$

$$= \sum_{k=-\infty}^{\infty} \left| f(-k) \, r^{k} \right| + \sum_{k=-\infty}^{\infty} \left| \frac{f(k)}{r^{k}} \right| \qquad \qquad .....(3.14)$$

If F(z) converges in some region of the complex plane, both summations in equation (3.14) must be finite.

If the first sum of equation (3.14) converges, there must exist values of r small enough for f(−k) r^k to be absolutely summable. Hence the ROC for the first sum consists of all points inside a circle of radius, r_1 as shown in fig 3.14

If the second sum of equation (3.14) converges, there must exist large values of r for which f(k) / r^k is absolutely summable. Hence the ROC for the second sum consists of all points outside a circle of radius, r_2 as shown in fig 3.15,

Therefore, the ROC of F(z) is the region in between two circles of radius r_1 and r_2 as shown in fig 3.16, where $r_2 < r < r_1$.

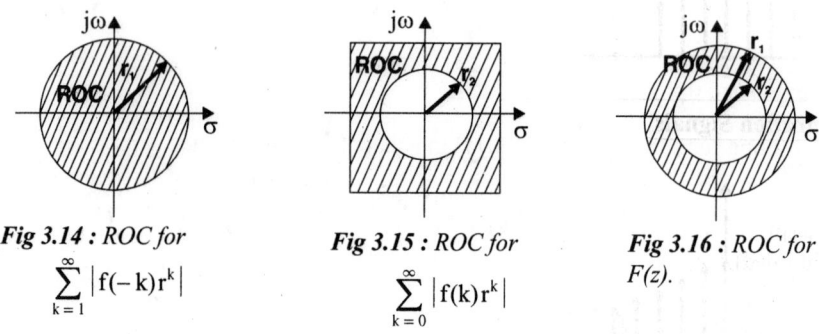

Fig 3.14 : ROC for

$$\sum_{k=1}^{\infty} \left| f(-k) r^k \right|$$

Fig 3.15 : ROC for

$$\sum_{k=0}^{\infty} \left| f(k) r^k \right|$$

Fig 3.16 : ROC for *F(z).*

TABLE-3.1 : Characteristic Families of Signals with Their Corresponding ROC

SIGNAL	ROC
Finite-Duration Signals 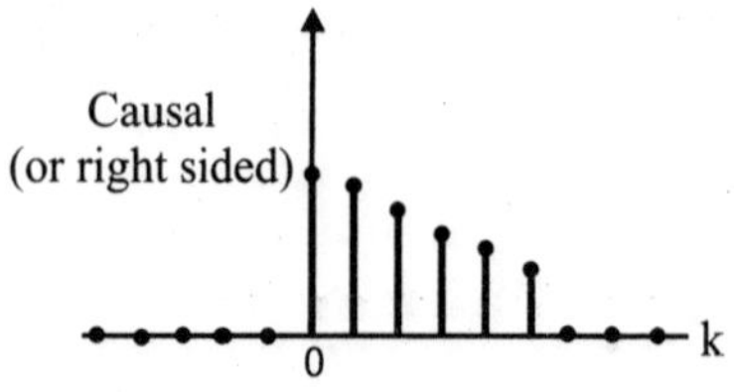 Causal (or right sided)	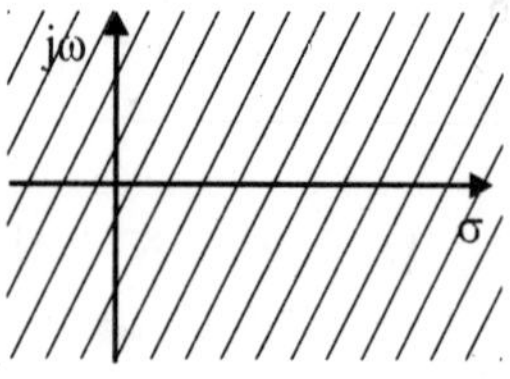 Entire z-plane except z=0

SIGNAL	ROC

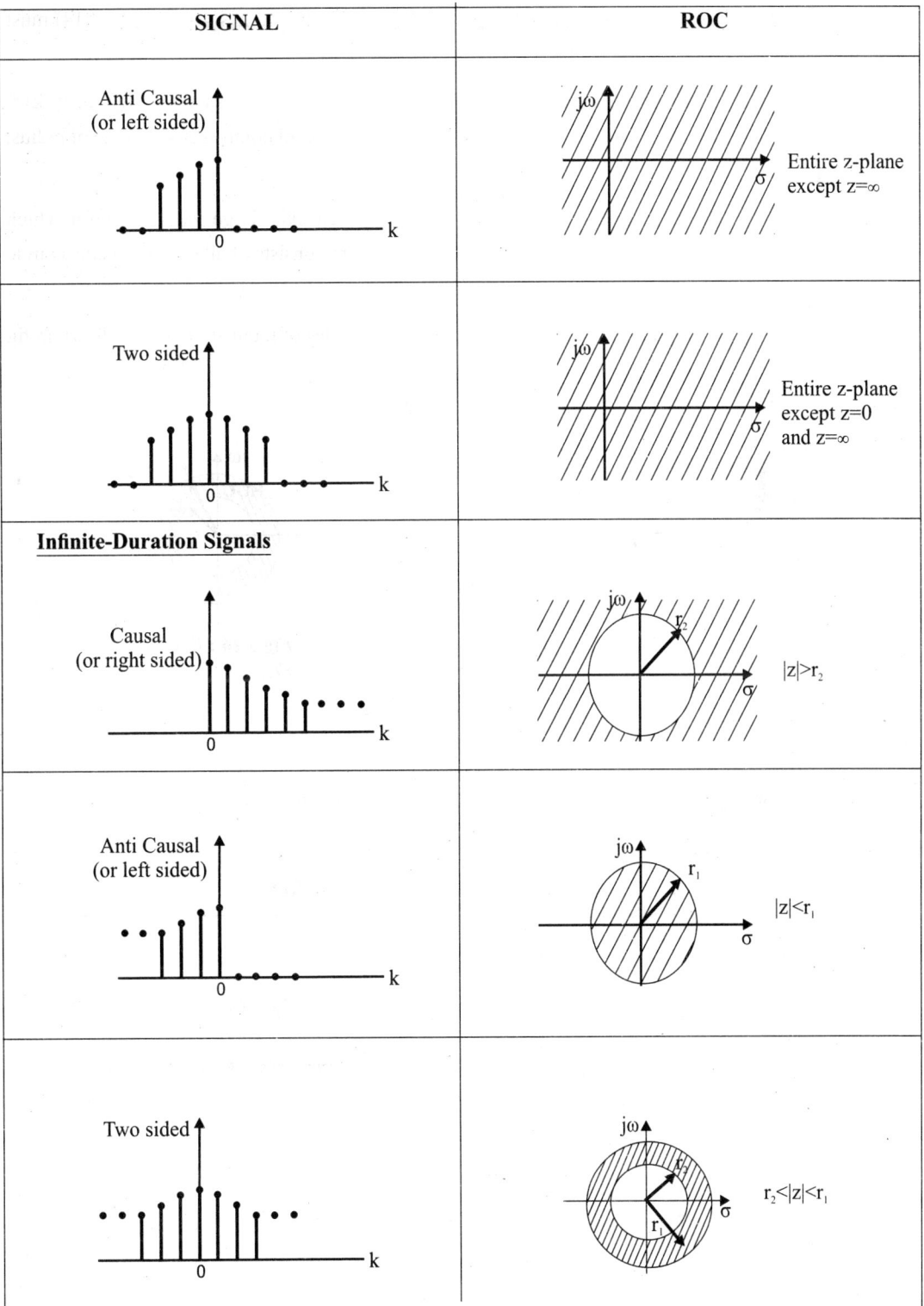

Anti Causal (or left sided)

Entire z-plane except z=∞

Two sided

Entire z-plane except z=0 and z=∞

Infinite-Duration Signals

Causal (or right sided)

$|z| > r_2$

Anti Causal (or left sided)

$|z| < r_1$

Two sided

$r_2 < |z| < r_1$

TABLE-3.2 : Properties of One-Sided Z-Transform

Note : $F(z) = \mathbf{Z}\{f(k)\}$; $F_1(z) = \mathbf{Z}\{f_1(k)\}$; $F_2(z) = \mathbf{Z}\{f_2(k)\}$

Property	Discrete sequence	z-transform
Linearity	$a_1 f_1(k) + a_2 f_2(k)$	$a_1 F_1(z) + a_2 F_2(z)$
Shifting, $m \geq 0$	$f(k+m)$	$z^m F(z) - \displaystyle\sum_{i=0}^{m=-1} f(i) z^{m-1}$
	$f(k-m)$	$z^{-m} F(z)$
Multiplication by k^m (or differentiation in z-domain)	$k^m f(k)$	$\left(-z\dfrac{d}{dz}\right)^m F(z)$
Scaling in z-domain (or multiplication by a^k)	$a^k f(k)$	$F(a^{-1}z)$
Time reversal	$f(-k)$	$F(z^{-1})$
Conjugation	$f^*(k)$	$F^*(z^*)$
Convolution	$\displaystyle\sum_{m=0}^{k} h(k-m)\,r(m)$	$H(z)\,R(z)$
Initial value	$f(0) = \underset{z \to \infty}{Lt}\ F(z)$	
Final value	$f(\infty) = \underset{z \to 1}{Lt}\ (1-z^{-1})\,F(z)$ $= \underset{z \to 1}{Lt}\ (z-1)\,F(z)$ if $F(z)$ is analytic for $|z| > 1$	

TABLE-3.3 : Some Common One Sided Ƶ-transform Pairs

Note : Two sided sequence can be converted to one sided sequence by multiplying by u(k).

f(t) ; t ≥ 0	f(k) or f(kT) ; k ≥ 0	F(z)
	$\delta(k)$	1
	u(k) or 1	$z/(z-1)$
	a^k	$z/(z-a)$
	$k\,a^k$	$\dfrac{az}{(z-a)^2}$
	$k^2\,a^k$	$\dfrac{az(z+a)}{(z-a)^3}$
	$(k+1)\,a^k$	$\dfrac{z^2}{(z-a)^2}$
	$\dfrac{(k+1)(k+2)}{2!}\,a^k$	$\dfrac{z^3}{(z-a)^3}$
	$\dfrac{(k+1)(k+2)(k+3)}{3!}\,a^k$	$\dfrac{z^4}{(z-a)^4}$
	$\dfrac{a^k}{k!}$	$e^{az^{-1}}$
t	kT	$\dfrac{Tz}{(z-1)^2}$
t^2	$(kT)^2$	$\dfrac{T^2 z(z+1)}{(z-1)^3}$
e^{-at}	e^{-akT}	$\dfrac{z}{z-e^{-aT}}$
te^{-aT}	kTe^{-akT}	$\dfrac{zTe^{-aT}}{(z-e^{-aT})^2}$
$\sin \omega t$	$\sin \omega kT$	$\dfrac{z\sin\omega T}{z^2 - 2z\cos\omega T + 1}$
$\cos \omega t$	$\cos \omega kT$	$\dfrac{z(z-\cos\omega T)}{z^2 - 2z\cos\omega T + 1}$

GEOMETRIC SERIES

A geometric series is a series in which consecutive elements differ by a constant ratio. Such a series can be written in the form,

$$f(k) = \sum_{k=M_1}^{M_2} C \qquad\qquad .....(3.15)$$

where, C is a constant and M_1 and M_2 are any two numbers.

If C is a complex number, where $|C| < 1$, then by Taylor's series expansion we can write,

$$\frac{1}{1-C} = 1 + C + C^2 + ... = \sum_{k=0}^{\infty} C^k \qquad\qquad .....(3.16)$$

Applying the result in the reverse direction yields the infinite geometric series sum formula

$$\therefore \sum_{k=0}^{\infty} C^k = \frac{1}{1-C} \qquad\qquad .....(3.17)$$

The equation (3.17) is the infinite geometric series sum formula.

We can also compute the sum of a finite number of elements in a geometric series. Let us consider the following sum,

$$1 + C + C^2 + ... + C^{M-1} = \sum_{k=0}^{m-1} C^k \qquad\qquad .....(3.18)$$

The sum of the finite duration sequence in equation (3.18) can be expressed as the difference between the sum of two infinite duration sequences as shown in equation (3.19).

$$\sum_{k=0}^{m-1} C^k = \sum_{k=0}^{\infty} C^k - \sum_{k=M}^{\infty} C^k \qquad\qquad .....(3.19)$$

Now, $\sum_{K=M}^{\infty} C^k = C^M + C^{M+1} + C^{M+2} + ...$

$$= C^M + C^M C + C^M C^2 + ... = C^M (1 + C + C^2 + C^3 + ...)$$

$$= C^M \left(\sum_{K=0}^{\infty} C^k \right) \qquad\qquad .....(3.20)$$

From equations (3.19) and (3.20) we can write,

$$\sum_{K=0}^{M-1} C^k = \sum_{K=0}^{\infty} C^k - C^M \sum_{K=0}^{\infty} C^k = (1 - C^M) \sum_{K=0}^{\infty} C^k$$

$$= (1 - C^M)\left(\frac{1}{1-C}\right) = \frac{1 - C^M}{1 - C} = \frac{C^M - 1}{C - 1} \quad \text{except } C = 1 \qquad\qquad .....(3.21)$$

$$\text{When } C = 1, \sum_{K=0}^{M-1} C^k = M \qquad\qquad .....(3.22)$$

The equations (3.21) and (3.22) are finite geometric series sum formula.

Note : The infinite geometric series sum formula requires that the magnitude of C be strictly less than unity, but the finite geometric series sum formula is valid for any value of C.

EXAMPLE 3.1

Determine the z-transform and their ROC of the following discrete sequences

(a) $f(k) = \{3, 2, 5, 7\}$ (b) $f(k) = \{2, 4, 5, 7, 3\}$
$$\uparrow$$

SOLUTION

(a)   **Given that, f(k) = {3, 2, 5, 7}**

i.e., $f(0) = 3$; $f(1) = 2$; $f(2) = 5$; $f(3) = 7$

and $f(k) = 0$ for $k < 0$ and for $k > 3$

By the definition of z-transforms

$$z\{f(k)\} = F(z) = \sum_{k=-\infty}^{\infty} f(z)\, z^{-k}$$

The given sequence is a finite duration sequence, hence the limits of summation can be changed as $k = 0$ to $k = 3$.

$$\therefore F(z) = \sum_{k=0}^{3} f(k)\, z^{-k}$$

On expanding the summation we get,

$$F(z) = f(0)z^0 + f(1)z^{-1} + f(2)z^{-2} + f(3)z^{-3}$$

$$= 3 + 2z^{-1} + 5z^{-2} + 7z^{-3}$$

Here, $F(z)$ is bounded (i.e., finite) except when $z = 0$, therefore the ROC is entire z-plane except $z = 0$.

(b)   **Given that, f(k) = {2, 4, 5, 7, 3}**
$$\uparrow$$

i.e, $f(-2) = 2$; $f(-1) = 4$; $f(0) = 5$; $f(1) = 7$; $f(2) = 3$

and $f(k) = 0$ for $k < -2$ and for $k > 2$.

By the definition of z-transform,

$$z\{f(k)\} = F(z) = \sum_{k=-\infty}^{\infty} f(k)z^{-k}$$

The given sequence is a finite duration sequence, hence the limits of summation can be changed as $k = -2$ to $k = 2$.

$$\therefore F(z) = \sum_{k=-2}^{2} f(k)z^{-k}$$

On expanding the summation we get,

$$F(z) = f(-2)\, z^2 + f(-1)\, z^1 + f(0)\, z^0 + f(1)\, z^{-1} + f(2)\, z^{-2}$$

$$= 2z^2 + 4z + 5 + 7z^{-1} + 3z^{-2}$$

Here, $F(z)$ is bounded (i.e., finite) except when $z = 0$ and $z = \infty$, therefore the ROC is entire z-plane except $z = 0$ and $z = \infty$.

EXAMPLE 3.2

Determine the $\mathbb{Z}$-transform of the following discrete sequences.

(a) $f(k) = u(k)$ (b) $f(k) = (1/2)^k\, u(k)$ (c) $f(k) = \alpha^k\, u(-k-1)$

SOLUTION

(a)   Given that, f(k) = u(k)

u(k) is a discrete unit step sequence, which is defined as

$$u(k) = 1 \text{ for } k \geq 0$$
$$= 0 \text{ for } k < 0$$

By the definition of $\mathbb{Z}$-transform,

$$\mathbb{Z}\{f(k)\} = F(z) = \sum_{k=-\infty}^{\infty} f(k)\, z^{-k}$$

$$= \sum_{k=0}^{\infty} u(k) z^{-k} = \sum_{k=0}^{\infty} z^{-k} = \sum_{k=0}^{\infty} (z^{-1})^k$$

Here, F(z) is an infinite geometric series and it converges if $|z| > 1$ (i.e., $|z^{-1}| < 1$). Using infinite geometric series sum formula we get,

$$F(z) = \frac{1}{1-z^{-1}} = \frac{1}{1-1/z} = \frac{z}{z-1}$$

(b)   Given that, f(k) = (1/2)^k u(k)

u(k) is a discrete unit step sequence, which is defined as

$$u(k) = 1 \text{ for } k \geq 0$$
$$= 0 \text{ for } k < 0$$
$$\therefore\ f(k) = (1/2)^k \text{ for } k \geq 0$$
$$= 0 \quad \text{ for } k < 0$$

By the definition of $\mathbb{Z}$-transform,

$$\mathbb{Z}\{f(k)\} = F(z) = \sum_{k=-\infty}^{\infty} f(k)\, z^{-k}$$

$$= \sum_{k=0}^{\infty} (1/2)^k z^{-k} = \sum_{k=0}^{\infty} \left(\frac{1}{2} z^{-1}\right)^k$$

Here, F(z) is an infinite geometric series and it converges if $|z| > 1$ (i.e., $|z^{-1}| < 1$). Using infinite geometric series sum formula we get,

$$F(z) = \frac{1}{1-\frac{1}{2}z^{-1}} = \frac{1}{1-\frac{1}{2}\cdot\frac{1}{z}} = \frac{2z}{2z-1}$$

(c) **Given that, f(k) = α^k u(−k−1)**

u(−k−1) is a discrete unit step sequence, which is defined as

$$u(-k-1) = 0 \text{ for } k \geq 0$$

$$= 1 \text{ for } k \leq -1$$

∴ f(k) = 0 for k ≥ 0

$$= \alpha^k \text{ for } k \leq -1$$

By the definition of Z-transform,

$$\mathbf{Z}\{f(k)\} = F(z) = \sum_{k=-\infty}^{\infty} f(k) z^{-k}$$

$$= \sum_{k=-\infty}^{-1} (\alpha)^k z^{-k} = \sum_{k=1}^{\infty} \alpha^{-k} z^{k}$$

$$= \sum_{k=1}^{\infty} (\alpha^{-1} z)^k = \sum_{k=0}^{\infty} (\alpha^{-1} z)^k - 1$$

Using infinite geometric series sum formula we get,

$$F(z) = \frac{1}{1 - \alpha^{-1} z} - 1 = \frac{1}{1 - \frac{z}{\alpha}} - 1 = \frac{\alpha}{\alpha - z} - 1$$

$$= \frac{\alpha - \alpha + z}{\alpha - z} = \frac{z}{\alpha - z}$$

EXAMPLE 3.3

Find the one sided Z-transform of the following discrete sequences.

(a) f(k) = k $a^{(k-1)}$ (b) f(k) = k^2

SOLUTION

(a) **Given that f(k) = k $a^{(k-1)}$**

The one sided Z-transform of a^k is given by

$$\mathbf{Z}\{a^k\} = \sum_{k=0}^{\infty} a^k z^{-k} = \sum_{k=0}^{\infty} (az^{-1})^k \qquad \dots(3.3.1)$$

Using infinite geometric series sum formula,

$$\mathbf{Z}\{a^k\} = \frac{1}{1 - az^{-1}} = \frac{1}{1 - a/z} = \frac{z}{z - a} \qquad \dots(3.3.2)$$

From equations (3.3.1) and (3.3.2) we get

$$\sum_{k=0}^{\infty} a^k z^{-k} = \frac{z}{z - a}$$

On expanding the summation in the above equation we get,

$$1 + az^{-1} + a^2z^{-2} + a^3z^{-3} + \ldots = \frac{z}{z-a} \qquad \ldots(3.3.3)$$

On differentiating the equation (3.3.3) we get,

$$-az^{-2} - 2a^2z^{-3} - 3a^3z^{-4} \ldots = \frac{(z-a) \times 1 - z \times 1}{(z-a)^2}$$

$$-az^{-2} - 2a^2z^{-3} - 3a^3z^{-4} \ldots = \frac{-a}{(z-a)^2} \qquad \ldots(3.3.4)$$

On multiplying the equation (3.3.4) by $-(z/a)$ we get,

$$z^{-1} + 2az^{-2} + 3a^2z^{-3} + \ldots = \frac{z}{(z-a)^2} \qquad \ldots(3.3.5)$$

The infinite series on the left hand side of the equation (3.3.5) can be expressed as a summation and the equation (3.3.5) is written as shown below.

$$\sum_{k=1}^{\infty} k\, a^{(k-1)} z^{-k} = \frac{z}{(z-a)^2} \qquad \ldots(3.3.6)$$

By definition of $\mathbf{Z}$-transform, the one sided $\mathbf{Z}$-transform of $k\, a^{(k-1)}$ is given by,

$$\mathbf{Z}\{ka^{(k-1)}\} = \sum_{k=0}^{\infty} k\, a^{(k-1)} z^{-k} = \sum_{k=1}^{\infty} k\, a^{(k-1)} z^{-k} \qquad \ldots(3.3.7)$$

(Because, $k\, a^{(k-1)} = 0$ when $k = 0$)

On comparing equations (3.3.6) and (3.3.7) we get,

$$\mathbf{Z}\{k\, a^{(k-1)}\} = \frac{z}{(z-a)^2}$$

(b)   Given that, f(k) = k²

Let us multiply the given discrete sequence by a discrete unit step sequence,

$$\therefore \ f(k) = k^2\, u(k)$$

Note : *Multiplying a one sided sequence by u(k) will not alter its value.*

By the property of $\mathbf{Z}$-transform, we get,

$$\mathbf{Z}\{k^m u(k)\} = \left(-z\frac{d}{dz}\right)^m U(z)$$

where, $U(z) = \mathbf{Z}\{u(k)\} = \dfrac{z}{z-1}$

$$\therefore -z\frac{d}{dz}U(z) = -z\left[\frac{d}{dz}\left(\frac{z}{z-1}\right)\right] = -z\left[\frac{z-1-z}{(z-1)^2}\right] = \frac{z}{(z-1)^2}$$

$$\left(-z\frac{d}{dz}\right)^2 U(z) = -z\frac{d}{dz}\left[-z\frac{d}{dz}U(z)\right]$$

$$= -z \frac{d}{dz}\left(\frac{z}{(z-1)^2}\right) = -z\left(\frac{(z-1)^2 - z \times 2(z-1)}{(z-1)^4}\right)$$

$$= -z\left(\frac{(z-1)(z-1-2z)}{(z-1)^4}\right) = -z\left(\frac{-(z+1)}{(z-1)^3}\right) = \frac{z(z+1)}{(z-1)^3}$$

$$\therefore Z\{f(k)\} = Z\{k^2 u(k)\} = \left(-z\frac{d}{dz}\right)^2 U(z) = \frac{z(z+1)}{(z-1)^3}$$

EXAMPLE 3.4

Find the one sided Z-transform of the discrete sequences generated by mathematically sampling the following continuous time functions

    (a) t^2 (b) $\sin \omega t$ (c) $\cos \omega t$

SOLUTION

(a)   **Given that, f(t) = t²**

The discrete sequence is generated by replacing t by kT, where T is the sampling time period.

    $\therefore$ $f(k) = (kT)^2 = k^2 T^2 = k^2 g(k)$

where, $g(k) = T^2$

By the definition of one sided Z-transform we get,

$$G(z) = Z\{g(k)\} = Z\{T^2\} = \sum_{k=0}^{\infty} T^2 z^{-k} = T^2 \sum_{k=0}^{\infty} (z^{-1})^k = T^2\left(\frac{1}{1-z^{-1}}\right) = \frac{T^2 z}{z-1}$$

By the property of Z-transform we get,

$$Z\{f(k)\} = F(z) = \left(-z\frac{d}{dz}\right)^2 G(z) = -z\frac{d}{dz}\left(-z\frac{d}{dz} G(z)\right)$$

$$= -z\frac{d}{dz}\left(-z\frac{d}{dz}\frac{T^2 z}{z-1}\right) = -z\frac{d}{dz}\left(-z\times \frac{(z-1)T^2 - T^2 z}{(z-1)^2}\right)$$

$$= -z\frac{d}{dz}\left(\frac{zT^2}{(z-1)^2}\right) = -z\times \frac{(z-1)^2 T^2 - zT^2 \times 2(z-1)}{(z-1)^4}$$

$$= -z\times \frac{(z-1)(zT^2 - T^2 - 2zT^2)}{(z-1)^4} = -z\times \frac{-zT^2 - T^2}{(z-1)^3} = \frac{zT^2(z+1)}{(z-1)^3}$$

(b)   **Given that, f(t) = sinωt**

The discrete sequence is generated by replacing t by kT, where T is the sampling time period.

    $\therefore$ $f(k) = \sin(\omega kT)$

By the definition of one sided Z-transform,

$$Z\{f(k)\} = F(z) = \sum_{k=0}^{\infty} f(k) z^{-k} = \sum_{k=0}^{\infty} \sin \omega kT \times z^{-k}$$

We know that, $\sin\theta = (e^{j\theta} - e^{-j\theta})/2j$

$$\therefore F(z) = \sum_{k=0}^{\infty} \frac{e^{j\omega kT} - e^{-j\omega kT}}{2j} z^{-k} = \frac{1}{2j}\sum_{k=0}^{\infty} e^{j\omega kT} z^{-k} - \frac{1}{2j}\sum_{k=0}^{\infty} e^{-j\omega kT} z^{-k}$$

We know that, $\mathbb{Z}\{e^{\pm akT}\} = \sum_{k=0}^{\infty} e^{\pm akT} z^{-k} = \frac{z}{z - e^{\pm aT}}$

$$\therefore F(z) = \frac{1}{2j}\frac{z}{z - e^{j\omega T}} - \frac{1}{2j}\frac{z}{z - e^{-j\omega T}}$$

$$= \frac{z(z - e^{-j\omega T}) - z(z - e^{j\omega T})}{2j(z - e^{j\omega T})(z - e^{-j\omega T})} = \frac{z^2 - ze^{-j\omega T} - z^2 + ze^{j\omega T}}{2j(z^2 - ze^{-j\omega T} - ze^{j\omega T} + e^{j\omega T}.e^{-j\omega T})}$$

$$= \frac{z(e^{j\omega T} - e^{-j\omega T})/2j}{z^2 - z(e^{j\omega T} + e^{-j\omega T}) + 1}$$

We know that, $\sin\theta = (e^{j\theta} - e^{-j\theta})/2j$ and $\cos\theta = (e^{j\theta} + e^{-j\theta})/2$

$$\therefore F(z) = \frac{z\sin\omega T}{z^2 - 2z\cos\omega T + 1}$$

(c)   **Given that, f(t) = cos ωt**

The discrete sequence is generated by replacing t by kT, where T is the sampling time period.

$\therefore f(k) = \cos(\omega kT)$

By the definition of one sided $\mathbb{Z}$-transform,

$$\mathbb{Z}\{f(k)\} = F(z) = \sum_{k=0}^{\infty} f(k)z^{-k} = \sum_{k=0}^{\infty} \cos\omega kT \times z^{-k}$$

We know that, $\cos\theta = (e^{j\theta} + e^{-j\theta})/2$

$$\therefore F(z) = \sum_{k=0}^{\infty} \frac{e^{j\omega kT} + e^{-j\omega kT}}{2} z^{-k} = \frac{1}{2}\sum_{k=0}^{\infty} e^{j\omega kT} z^{-k} + \frac{1}{2}\sum_{k=0}^{\infty} e^{-j\omega kT} z^{-k}$$

We know that, $\mathbb{Z}\{e^{\pm akT}\} = \sum_{k=0}^{\infty} e^{\pm akT} z^{-k} = \frac{z}{z - e^{\pm aT}}$

$$\therefore F(z) = \frac{1}{2}\frac{z}{z - e^{j\omega T}} + \frac{1}{2}\frac{z}{z - e^{-j\omega T}}$$

$$= \frac{z(z - e^{-j\omega T}) + z(z - e^{j\omega T})}{2(z - e^{j\omega T})(z - e^{-j\omega T})} = \frac{z^2 - ze^{-j\omega T} + z^2 - ze^{j\omega T}}{2(z^2 - ze^{-j\omega T} - ze^{j\omega T} + e^{j\omega T}.e^{-j\omega T})}$$

$$= \frac{2z^2 - z(e^{j\omega T} + e^{-j\omega T})}{2[z^2 - z(e^{j\omega T} + e^{-j\omega T}) + 1]} = \frac{z^2 - z(e^{j\omega T} + e^{-j\omega T})/2}{z^2 - z(e^{j\omega T} + e^{-j\omega T}) + 1}$$

We know that, $\cos\theta = (e^{j\theta} + e^{-j\theta})/2$

$$\therefore F(z) = \frac{z(z - \cos\omega T)}{z^2 - 2z\cos\omega T + 1}$$

EXAMPLE 3.5

Find the one sided $\mathcal{Z}$-transform of the discrete sequences generated by mathematically sampling the following continuous time function,

(a) $e^{-at}\cos\omega t$ (b) $e^{-at}\sin\omega t$

SOLUTION

(a)   **Given that, $f(t) = e^{-at}\cos\omega t$**

The discrete sequence is generated by replacing t by kT, where T is the sampling time period.

$$\therefore f(k) = e^{-akt}\cos\omega kT$$

By the definition of one sided $\mathcal{Z}$-transform we get,

$$F(z) = \mathcal{Z}\{f(k)\} = \sum_{k=0}^{\infty} e^{-akT}\cos\omega kT\, z^{-k} = \sum_{k=0}^{\infty} e^{-akT}\left(\frac{e^{j\omega kT} + e^{-j\omega kT}}{2}\right)z^{-k}$$

$$= \frac{1}{2}\sum_{k=0}^{\infty}\left(e^{-akT}e^{j\omega T}z^{-1}\right)^k + \frac{1}{2}\sum_{k=0}^{\infty}\left(e^{-aT}e^{-j\omega T}z^{-1}\right)^k$$

From infinite geometric sum series formula we know that, $\displaystyle\sum_{k=0}^{\infty} C^k = \frac{1}{1-C}$

$$\therefore F(z) = \frac{1}{2}\frac{1}{1 - e^{-aT}e^{j\omega T}z^{-1}} + \frac{1}{2}\frac{1}{1 - e^{-aT}e^{-j\omega T}z^{-1}}$$

$$= \frac{1}{2}\frac{1}{1 - e^{j\omega T}/ze^{aT}} + \frac{1}{2}\frac{1}{1 - e^{-j\omega T}/ze^{aT}}$$

$$= \frac{1}{2}\left[\frac{ze^{aT}}{ze^{aT} - e^{j\omega T}} + \frac{ze^{aT}}{ze^{aT} - e^{-j\omega T}}\right]$$

$$= \frac{1}{2}\left[\frac{ze^{aT}(ze^{aT} - e^{-j\omega T}) + ze^{aT}(ze^{aT} - e^{-j\omega T})}{(ze^{aT} - e^{j\omega T})(ze^{aT} - e^{-j\omega T})}\right]$$

$$= \frac{ze^{aT}}{2}\left[\frac{ze^{aT} - e^{-j\omega T} + ze^{aT} - e^{j\omega T}}{(ze^{aT})^2 - ze^{aT}e^{-j\omega T} - ze^{aT}e^{j\omega T} + e^{j\omega T}e^{-j\omega T}}\right]$$

$$= \frac{ze^{aT}}{2}\left[\frac{2ze^{aT} - (e^{j\omega T} + e^{-j\omega T})}{z^2e^{2aT} - ze^{aT}(e^{j\omega T} + e^{-j\omega T}) + 1}\right]$$

$$= \left[\frac{ze^{aT}(ze^{aT} - \cos\omega T)}{z^2e^{2aT} - 2ze^{aT}\cos\omega T + 1}\right] \qquad \boxed{\because \cos\theta = \frac{e^{j\theta} + e^{-j\theta}}{2}}$$

(b)   **Given that, $f(t) = e^{-at} \sin \omega t$**

The discrete sequence f(k) is generated by replacing t by kT, where T is the sampling time period.

$$\therefore f(k) = e^{-akt} \sin \omega kT$$

By the definition of one sided $\mathbf{Z}$-transform we get,

$$F(z) = \mathbf{Z}\{f(k)\} = \sum_{k=0}^{\alpha} e^{-akT} \sin \omega kT \, z^{-k}$$

$$= \sum_{k=0}^{\infty} e^{-akT} \left(\frac{e^{j\omega kT} - e^{-j\omega kT}}{2j} \right) z^{-k}$$

$$= \frac{1}{2j} \sum_{k=0}^{\infty} \left(e^{-aT} e^{j\omega T} z^{-1} \right)^k - \frac{1}{2j} \sum_{k=0}^{\infty} \left(e^{-aT} e^{-j\omega T} z^{-1} \right)^k$$

From infinite geometric sum series formula we know that, $\displaystyle\sum_{k=0}^{\infty} C^k = \frac{1}{1-C}$

$$\therefore F(z) = \frac{1}{2j} \frac{1}{1 - e^{-aT} e^{j\omega T} z^{-1}} - \frac{1}{2j} \frac{1}{1 - e^{-aT} e^{-j\omega T} z^{-1}}$$

$$= \frac{1}{2j} \frac{1}{1 - e^{j\omega T}/z e^{aT}} - \frac{1}{2j} \frac{1}{1 - e^{-j\omega T}/z e^{aT}}$$

$$= \frac{1}{2j} \frac{z e^{aT}}{z e^{aT} - e^{j\omega T}} - \frac{1}{2j} \frac{z e^{aT}}{z e^{aT} - e^{-j\omega T}}$$

$$= \frac{1}{2j} \left[\frac{z e^{aT} \left(z e^{aT} - e^{-j\omega T} \right) - z e^{aT} \left(z e^{aT} - e^{j\omega T} \right)}{\left(z e^{aT} - e^{j\omega T} \right)\left(z e^{aT} - e^{-j\omega T} \right)} \right]$$

$$= \frac{1}{2j} \left[\frac{\left(z e^{aT} \right)\left(z e^{aT} - e^{-j\omega T} - z e^{aT} + e^{j\omega T} \right)}{\left(z e^{aT} \right)^2 - z e^{aT} e^{-j\omega T} - z e^{aT} e^{j\omega T} + e^{j\omega T} e^{-j\omega T}} \right]$$

$$= \left[\frac{z e^{aT} \left[e^{j\omega T} - e^{-j\omega T} \right]/2j}{z^2 e^{2aT} - z e^{aT} \left(e^{j\omega T} + e^{-j\omega T} \right) + 1} \right]$$

$$= \frac{z e^{aT} \sin \omega T}{z^2 e^{2aT} - 2z e^{aT} \cos \omega T + 1}$$

INVERSE $\mathbf{Z}$-TRANSFORM

The following methods are employed to recover the original discrete sequence from its $\mathbf{Z}$-transform

1. *Direct evaluation by contour integration (or) complex inversion integral.*

2. *Partial fraction expansion.*

3. *Power series expansion.*

The inverse $\mathbf{Z}$-transform by partial fraction expansion method and power series expansion method are presented in this section. The inverse $\mathbf{Z}$-transform by contour integration is beyond the scope of the book.

PARTIAL FRACTION EXPANSION METHOD

Let f(k) = Discrete sequence

and $F(z) = \mathcal{Z}\{f(k)\} = \mathcal{Z}$-Transform of f(k).

The function F(z) can be expressed as a ratio of two polynomials in z as shown below.

$$F(z) = \frac{b_0 z^m + b_1 z^{m-1} + b_2 z^{m-2} + \,.....\, + b_{m-1}z + b_m}{z^n + a_1 z^{n-1} + a_2 z^{n-2} + \,.....\, + a_{n-1}z + a_n} \;;\; \text{where } m \leq n$$

The function F(z) can be expressed as a series of sum terms by partial fraction expansion technique.

$$\therefore\; F(z) = A_0 + \sum_{i=1}^{n} \frac{A_i}{(z + p_i)} \qquad\qquad .....(3.23)$$

where, A_0 is a constant, $A_1, A_2, ... A_n$ are residues and $p_1, p_2, ... p_n$ are poles of F(z).

> *Note : Sometimes it will be convenient to express F(z)/z as a series of sum terms instead of F(z).*

Once the function F(z) is expressed as a series of sum terms, the inverse $\mathcal{Z}$-transform of F(z) is given by the sum of inverse $\mathcal{Z}$-transform of each term in equation (3.23) [The inverse $\mathcal{Z}$-transform of each term of equation (3.23) can be obtained from standard $\mathcal{Z}$-transform pairs].

The coefficients of the polynomials of F(z) are assumed real and so the roots of the polynomial are real and/or complex conjugate pairs (i.e., complex roots will occur only in conjugate pairs). Hence on factorizing the denominator polynomial we get the following cases. (The roots of the denominator polynomial are poles of F(z)).

Case (i)   : When roots (or poles) are real and distinct

Case (ii) : When roots (or poles) have multiplicity

Case (iii) : When roots (or poles) are complex conjugate.

Case (i) When roots (or poles) are real and distinct

In this case F(z) can be expressed as,

$$F(z) = \frac{b_0 z^m + b_1 z^{m-1} + b_2 z^{m-2} + \,.....\, + b_{m-1}z + b_m}{(z + p_1)(z + p_2) .... (z + p_n)}$$

$$= A_0 + \frac{A_1}{(z + p_1)} + \frac{A_2}{(z + p_2)} + \,.......\, + \frac{A_n}{(z + p_n)}$$

where, A_0 is a constant ; $A_1, A_2 ...... A_n$ are residues and $p_1, p_2, ... p_n$ are poles.

The constant A_0 is present when m = n (i.e., when the order of numerator and denominator polynomial are equal). The value of A_0 is obtained by dividing the numerator polynomial by denominator polynomial.

The residue A_i is evaluated by multiplying both sides of H(z) by $(z + p_i)$ and letting $z = -p_i$.

$$\therefore\; A_i = (z + p_i)F(z)\Big|_{z = -p_i}$$

Case (ii) When roots (or poles) have multiplicity

Let one of pole has a multiplicity of q. (i.e., repeats q times). In this case F(z) can be expressed as,

$$F(z) = \frac{b_0 z^m + b_1 z^{m-1} + b_2 z^{m-2} + \dots + b_{m-1} z + b_m}{(z+p_1)(z+p_2)\dots(z+p_x)^q\dots(z+p_n)}$$

$$= A_0 + \frac{A_1}{(z+p_1)} + \frac{A_2}{(z+p_2)} + \dots + \frac{A_{x0}}{(z+p_x)^q} + \frac{A_{x1}}{(z+p_x)^{q-1}} + \dots$$

$$\dots + \frac{A_{x(q-1)}}{(z+p_x)} + \dots + \frac{A_n}{(z+p_n)}$$

where, $A_{x0}, A_{x1}, \dots A_{x(q-1)}$ are residues of repeated root (or pole), $z = -p_x$.

The constant A_0 and residues of distinct real roots are evaluated as explained in case(i).

The residue A_{xr} of repeated root is obtained as shown below.

$$A_{xr} = \frac{1}{r!}\frac{d^r}{dz^r}\left[(z+p_x)^q F(z)\right]\Big|_{z=-p_x} \quad ; \text{ where } r = 0, 1, 2, \dots(q-1)$$

Case (iii) When roots (or poles) are complex conjugate

Let F(z) has one pair of complex conjugate pole. In this case F(z) can be expressed as,

$$F(z) = \frac{b_0 z^m + b_1 z^{m-1} + b_2 z^{m-2} + \dots + b_{m-1} z + b_m}{(z+p_1)(z+p_2)\dots(z^2 + az + b)\dots(z+p_n)}$$

$$= A_0 + \frac{A_1}{(z+p_1)} + \frac{A_2}{(z+p_2)} + \dots + \frac{A_x}{z+\sigma+j\omega} + \frac{A_x^*}{z+\sigma-j\omega} + \dots + \frac{A_n}{(z+p_n)}$$

The constant A_0 and residues of real and non-repeated roots are evaluated as explained in case(i).

The residue A_x is evaluated as that of case(i) and the residue A_x^* is conjugate of A_x.

POWER SERIES EXPANSION METHOD

Let f(k) = Discrete sequence

and $F(z) = Z\{f(k)\} = Z$-transform of f(k).

By the definition of Z-transform we get,

$$F(z) = \sum_{k=-\infty}^{\infty} f(k)\, z^{-k}$$

On expanding the summation we get,

$$F(z) = \dots f(-3)z^{-(-3)} + f(-2)z^{-(-2)} + f(-1)z^{-(-1)} + f(0)z^0 \qquad \dots(3.24)$$
$$+ f(1)z^{-1} + f(2)z^{-2} + f(3)z^{-3} + \dots$$

If the given function, F(z) can be expressed as a power series of z by long division then on comparing the coefficients of z with that of equation (3.24), the samples of f(k) are determined. [i.e., the coefficient of z^i is the ith sample f(i) of the sequence f(k)].

Note : *The different method of evaluation of inverse Z-transform of a function F(z) will result in different type of mathematical expressions. But on evaluating the expressions for each value of k, we may get a same sequence.*

EXAMPLE 3.6

Determine the inverse Z-transform of the following function,

(a) $F(z) = \dfrac{1}{1 - 1.5z^{-1} + 0.5z^{-2}}$

(b) $F(z) = \dfrac{z^2}{z^2 - z + 0.5}$

(c) $F(z) = \dfrac{1 + z^{-1}}{1 - z^{-1} + 0.5z^{-2}}$

(d) $F(z) = \dfrac{1}{(1 + z^{-1})(1 - z^{-1})^2}$

SOLUTION

(a) Given that, $F(z) = \dfrac{1}{1 - 1.5z^{-1} + 0.5z^{-2}}$

$$F(z) = \frac{1}{1 - 1.5z^{-1} + 0.5z^{-2}} = \frac{1}{1 - \dfrac{1.5}{z} + \dfrac{0.5}{z^2}}$$

$$= \frac{z^2}{z^2 - 1.5z + 0.5} = \frac{z^2}{(z - 1)(z - 0.5)}$$

$$\therefore \frac{F(z)}{z} = \frac{z}{(z - 1)(z - 0.5)}$$

By partial fraction expansion, F(z)/z can be expressed as

$$\frac{F(z)}{z} = \frac{A_1}{z - 1} + \frac{A_2}{z - 0.5}$$

$$A_1 = \frac{F(z)}{z}(z - 1)\Big|_{z=1} = \frac{z}{(z - 1)(z - 0.5)}(z - 1)\Big|_{z=1}$$

$$= \frac{z}{(z - 0.5)}\Big|_{z=1} = \frac{1}{1 - 0.5} = 2$$

$$A_2 = \frac{F(z)}{z}(z - 0.5)\Big|_{z=0.5} = \frac{z}{(z - 1)(z - 0.5)}(z - 0.5)\Big|_{z=0.5}$$

$$= \frac{z}{(z - 1)}\Big|_{z=0.5} = \frac{0.5}{0.5 - 1} = -1$$

$$\therefore \frac{F(z)}{z} = \frac{2}{z - 1} - \frac{1}{z - 0.5}$$

$$F(z) = \frac{2z}{z - 1} - \frac{z}{z - 0.5}$$

We know that $\mathcal{Z}\{a^k\} = \dfrac{z}{z - a}$ and $\mathcal{Z}\{u(k)\} = \dfrac{z}{z - 1}$

On taking inverse $\mathcal{Z}$-transform of F(z) we get,

$$f(k) = 2\,u(k) - (0.5)^k \; ; \; k \geq 0$$

(Here we consider only one sided $\mathcal{Z}$-transform)

(b) **Given that, $F(z) = \dfrac{z^2}{z^2 - z + 0.5}$**

$$F(z) = \frac{z^2}{z^2 - z + 0.5} = \frac{z^2}{(z - 0.5 - j0.5)(z - 0.5 + j0.5)}$$

$$\therefore \ \frac{F(z)}{z} = \frac{z}{(z - 0.5 - j0.5)(z - 0.5 + j0.5)}$$

By partial fraction expansion, we can write,

$$\frac{F(z)}{z} = \frac{A}{z - 0.5 - j0.5} + \frac{A^*}{z - 0.5 - j0.5}$$

The roots of the quadratic $z^2 - z + 0.5 = 0$ are $z = \dfrac{1 \pm \sqrt{1 - 4 \times 0.5}}{2}$ $= 0.5 \pm j0.5$

$$A = \frac{F(z)}{z}(z - 0.5 - j0.5)\Big|_{z = 0.5 + j0.5}$$

$$= \frac{z}{(z - 0.5 - j0.5)(z - 0.5 + j0.5)}(z - 0.5 - j0.5)\Big|_{z = 0.5 + j0.5}$$

$$= \frac{0.5 + j0.5}{j1} = 0.5 - j0.5$$

$$\therefore A^* = (0.5 - j0.5)^* = 0.5 + j0.5$$

$$\therefore \ \frac{F(z)}{z} = \frac{0.5 - j0.5}{z - 0.5 - j0.5} + \frac{0.5 + j0.5}{z - 0.5 + j0.5}$$

$$F(z) = \frac{(0.5 - j0.5)z}{z - (0.5 - j0.5)} + \frac{(0.5 + j0.5)z}{z - (0.5 + j0.5)}$$

We know that $\mathbf{Z}\{a^k\} = \dfrac{z}{z - a}$

On taking inverse $\mathbf{Z}$-transform of F(z) we get,

$$f(k) = (0.5 - j0.5)(0.5 + j0.5)^k + (0.5 + j0.5)(0.5 - j0.5)^k$$

$$= -j\left(\frac{0.5}{-j} + 0.5\right)(0.5 + j0.5)^k + j\left(\frac{0.5}{j} + 0.5\right)(0.5 - j0.5)^k$$

$$= -j(0.5 + j0.5)(0.5 + j0.5)^k + j(0.5 - j0.5)(0.5 - j0.5)^k$$

$$= -j(0.5 + j0.5)^{(k+1)} + j(0.5 + j0.5)^{(k+1)}$$

(c) **Given that, $F(z) = \dfrac{1 + z^{-1}}{1 - z^{-1} + 0.5z^{-2}}$**

$$F(z) = \frac{1 + z^{-1}}{1 - z^{-1} + 0.5z^{-2}} = \frac{1 + 1/z}{1 - \dfrac{1}{z} + \dfrac{0.5}{z^2}}$$

The roots of the quadratic $z^2 - z + 0.5 = 0$ are $z = \dfrac{1 \pm \sqrt{1 - 4 \times 0.5}}{2}$ $= 0.5 \pm j0.5$

$$= \frac{\dfrac{z + 1}{z}}{\dfrac{z^2 - z + 0.5}{z^2}} = \frac{z(z + 1)}{(z^2 - z - 0.5)} = \frac{z(z + 1)}{(z - 0.5 - j0.5)(z - 0.5 + j0.5)}$$

By partial fraction expansion, we can write,

$$\frac{F(z)}{z} = \frac{(z+1)}{(z-0.5-j0.5)(z-0.5+j0.5)} = \frac{A}{z-0.5-j0.5} + \frac{A^*}{z-0.5+j0.5}$$

$$A = \frac{F(z)}{z}(z-0.5-j0.5)\Big|_{z=0.5+j0.5}$$

$$= \frac{(z+1)}{(z-0.5-j0.5)(z-0.5+j0.5)}(z-0.5-j0.5)\Big|_{z=0.5+j0.5}$$

$$= \frac{0.5+j0.5+1}{0.5+j0.5-0.5+j0.5} = \frac{1.5+j0.5}{j1} = -j1.5+0.5 = 0.5-j1.5$$

$$A^* = (0.5-j1.5)^* = 0.5+j1.5$$

$$\therefore \frac{F(z)}{z} = \frac{0.5-j1.5}{z-0.5-j0.5} + \frac{0.5+j1.5}{z-0.5+j0.5}$$

$$\therefore F(z) = (0.5-j1.5)\frac{z}{z-(0.5-j0.5)} + (0.5+j1.5)\frac{z}{z-(0.5+j0.5)}$$

We know that $\mathbf{z}\{a^k\} = \dfrac{z}{z-a}$

On taking inverse $\mathbf{z}$-transform of F(z) we get,

$$f(k) = (0.5-j1.5)(0.5+j0.5)^k + (0.5+j1.5)(0.5-j0.5)^k \; ; \; \text{for } k \geq 0$$

(d) Given that, $F(z) = \dfrac{1}{(1+z^{-1})(1-z^{-1})^2}$

$$F(z) = \frac{1}{(1+z^{-1})(1-z^{-1})^2} = \frac{1}{\left(1+\dfrac{1}{z}\right)\left(1-\dfrac{1}{z}\right)^2}$$

$$= \frac{1}{\dfrac{(z+1)}{z}\left(\dfrac{z-1}{z}\right)^2} = \frac{z^3}{(z+1)(z-1)^2}$$

$$\therefore \frac{F(z)}{z} = \frac{z^2}{(z+1)(z-1)^2}$$

By partial fraction expansion, we can write,

$$\frac{F(z)}{z} = \frac{A_1}{z+1} + \frac{A_2}{(z-1)^2} + \frac{A_3}{z-1}$$

$$A_1 = \frac{F(z)}{z}(z+1)\Big|_{z=-1} = \frac{z^2}{(z+1)(z-1)^2}(z+1)\Big|_{z=-1} = \frac{z^2}{(z-1)^2}\Big|_{z=-1} = \frac{(-1)^2}{(-1-1)^2} = 0.25$$

$$A_2 = \frac{F(z)}{z}(z-1)^2\Big|_{z=1} = \frac{z^2}{(z+1)(z-1)^2}(z-1)^2\Big|_{z=1}$$

$$= \frac{z^2}{z+1}\Big|_{z=1} = \frac{1}{1+1} = 0.5$$

$$A_3 = \frac{d}{dz}\left[\frac{F(z)}{z}(z-1)^2\right]\Bigg|_{z=1} = \frac{d}{dz}\left[\frac{z^2}{(z+1)(z-1)^2}(z-1)^2\right]\Bigg|_{z=1}$$

$$= \frac{d}{dz}\left[\frac{z^2}{z+1}\right]\Bigg|_{z=1} = \frac{(z+1)2z-z^2}{(z+1)^2}\Bigg|_{z=1} = \frac{(1+1)\times 2-1}{(1+1)^2} = \frac{3}{4} = 0.75$$

$$\therefore \quad \frac{F(z)}{z} = \frac{0.25}{z+1} + \frac{0.5}{(z-1)^2} + \frac{0.75}{z-1}$$

$$F(z) = 0.25\frac{z}{z+1} + 0.5\frac{z}{(z-1)^2} + 0.75\frac{z}{z-1}$$

$$= 0.25\frac{z}{z-(-1)} + 0.5\frac{z}{(z-1)^2} + 0.75\frac{z}{z-1}$$

We know that $\mathbf{Z}\{a^k\} = \dfrac{z}{z-a}$; $\mathbf{Z}\left\{\dfrac{az}{(z-a)^2}\right\} = ka^k$ and $\mathbf{Z}\{u(k)\} = \dfrac{z}{z-1}$

On taking inverse $\mathbf{Z}$-transform of F(z) we get,

$$f(k) = 0.25(-1)^k + 0.5k(1)^k + 0.75\ u(k)$$

$$f(k) = 0.25(-1)^k + 0.5k + 0.75\ u(k)\ ;\ \text{for } k \geq 0$$

EXAMPLE 3.7

Determine the inverse $\mathbf{Z}$-transform of the following z-domain functions.

(a) $F(z) = \dfrac{3z^2+2z+1}{z^2-3z+2}$

(b) $F(z) = \dfrac{3z^2+2z+1}{z^2+3z+2}$

(c) $F(z) = \dfrac{z-0.4}{z^2+z+2}$

(d) $F(z) = \dfrac{z-4}{(z-1)(z-2)^2}$

SOLUTION

(a)   Given that, $F(z) = \dfrac{3z^2+2z+1}{z^2-3z+2}$

$$F(z) = \frac{3z^2+2z+1}{z^2-3z+2} = 3 + \frac{11z-5}{z^2-3z+2}$$

$$= 3 + \frac{11z-5}{(z-1)(z-2)}$$

$$
\begin{array}{r}
3 \\
z^2-3z+2\ \overline{\smash{\big)}\ 3z^2+2z+1} \\
3z^2-9z+6 \\
\hline
11z-5
\end{array}
$$

By partial fraction expansion we get, $F(z) = 3 + \dfrac{A_1}{z-1} + \dfrac{A_2}{z-2}$

$$A_1 = \frac{11z-5}{(z-1)(z-2)}(z-1)\Bigg|_{z=1} = \frac{11z-5}{(z-2)}\Bigg|_{z=1} = \frac{11-5}{1-2} = -6$$

$$A_2 = \frac{11z-5}{(z-1)(z-2)}(z-2)\Bigg|_{z=2} = \frac{11z-5}{(z-1)}\Bigg|_{z=2} = \frac{11\times 2-5}{2-1} = 17$$

$$\therefore F(z) = 3 - \frac{6}{z-1} + \frac{17}{z-2}$$

$$= 3 - 6\frac{1}{z}\frac{z}{z-1} + 17\frac{1}{z}\frac{z}{z-2}$$

$$= 3 - 6z^{-1}\frac{z}{z-1} + 17z^{-1}\frac{z}{z-2}$$

We know that, $\mathbb{Z}\{\delta(k)\} = 1$; $\mathbb{Z}\{u(k)\} = \frac{z}{z-1}$ and $\mathbb{Z}\{a^k\} = \frac{z}{z-a}$

By time shifting property we get,

$$\mathbb{Z}\{u(k-1)\} = z^{-1}\frac{z}{z-1} \text{ and } \mathbb{Z}\{a^{(k-1)}\} = z^{-1}\frac{z}{z-a}$$

On taking inverse $\mathbb{Z}$-transform of $F(z)$ we get,

$$f(k) = 3\,\delta(k) - 6\,u(k-1) + 17 \times 2^{(k-1)}\,u(k-1) \text{ ; for } k \geq 0$$

Note : The term $2^{(k-1)}$ is multiplied by u(k−1), because this term have samples only after $k \geq 1$.

(b) **Given that,** $F(z) = \frac{3z^2 + 2z + 1}{z^2 + 3z + 2}$

$$F(z) = \frac{3z^2 + 2z + 1}{z^2 + 3z + 2} = 3 - \frac{7z+5}{z^2+3z+2}$$

$$= 3 - \frac{7z+5}{(z+1)(z+2)}$$

$$
\begin{array}{r}
3 \\
z^2 - 3z + 2 \overline{\smash{\big)}\ 3z^2 + 2z + 1} \\
\underline{3z^2 + 9z + 6} \\
-7z - 5
\end{array}
$$

By partial fraction expansion we get, $F(z) = 3 - \frac{A_1}{z+1} - \frac{A_2}{z+2}$

$$A_1 = \frac{7z+5}{(z+1)(z+2)}(z+1)\bigg|_{z=-1} = \frac{7z+5}{(z+2)}\bigg|_{z=-1} = \frac{7\times(-1)+5}{-1+2} = -2$$

$$A_2 = \frac{7z+5}{(z+1)(z+2)}(z+2)\bigg|_{z=-2} = \frac{7z+5}{(z+1)}\bigg|_{z=-2} = \frac{7\times(-2)+5}{-2+1} = 9$$

$$\therefore F(z) = 3 + \frac{2}{z+1} - \frac{9}{z+2}$$

$$= 3 + 2\frac{1}{z}\frac{z}{z-(-1)} - 9\frac{1}{z}\frac{z}{z-(-2)}$$

$$= 3 + 2z^{-1}\frac{z}{z-(-1)} - 9z^{-1}\frac{z}{z-(-2)}$$

We know that, $\mathbb{Z}\{\delta(k)\} = 1$ and $\mathbb{Z}\{a^k\} = \frac{z}{z-a}$

By time shifting property,

$$\mathbb{Z}\{a^{(k-1)}\} = z^{-1}\frac{z}{z-a}$$

On taking inverse $\mathbb{Z}$-transform of $F(z)$ we get,

$$f(k) = 3\,\delta(k) + 2(-1)^{(k-1)}\,u(k-1) + 9(-2)^{(k-1)}\,u(k-1) \text{ ; for } k \geq 0$$

Note: The term $2^{(k-1)}$ is multiplied by u(k−1), because these term have samples only after $k \geq 1$.

(c)   **Given that,** $F(z) = \dfrac{z-0.4}{z^2+z+2}$

$$F(z) = \frac{z-0.4}{z^2+z+2}$$

$$= \frac{z-0.4}{(z+0.5-j\sqrt{7}/2)(0.5+j\sqrt{7}/2)}$$

The roots of the quadratic
$z^2 = z + 2 = 0$ are,
$z = \dfrac{-1 \pm \sqrt{1-4\times 2}}{2}$
$= -0.5 \pm j\sqrt{7}/2$

By partial fraction expansion we get, $F(z) = \dfrac{A}{z+0.5-j\sqrt{7}/2} + \dfrac{A^*}{z+0.5-j\sqrt{7}/2}$

$$A = \frac{z-0.4}{(z+0.5-j\sqrt{7}/2)(z+0.5+j\sqrt{7}/2)}(z+0.5-j\sqrt{7}/2)\bigg|_{z=-0.5+j\sqrt{7}/2}$$

$$= \frac{z-0.4}{(z+0.5+j\sqrt{7}/2)}\bigg|_{z=-0.5+j\sqrt{7}/2} = \frac{-0.5+j\sqrt{7}/2-0.4}{-0.5+j\sqrt{7}/2+0.5+j\sqrt{7}/2}$$

$$= \frac{-0.9+j\sqrt{7}/2}{j\sqrt{7}} = \frac{-0.9}{j\sqrt{7}} + \frac{j\sqrt{7}/2}{j\sqrt{7}} = 0.5 + j\frac{0.9}{\sqrt{7}} = 0.5 + j0.34$$

$$\therefore A^* = (0.5+j0.34)^* = 0.5 - j0.34$$

$$\therefore F(z) = \frac{0.5+j0.34}{z+0.5-j\sqrt{7}/2} + \frac{0.5-j0.34}{z+0.5+j\sqrt{7}/2}$$

$$= (0.5+j0.34)\frac{1}{z}\frac{z}{z+0.5-j\sqrt{7}/2} + (0.5-j0.34)\frac{1}{z}\frac{z}{z+0.5+j\sqrt{7}/2}$$

$$= (0.5+j0.34)z^{-1}\frac{z}{z-(-0.5+j\sqrt{7}/2)} + (0.5-j0.34)z^{-1}\frac{z}{z-(-0.5-j\sqrt{7}/2)}$$

We know that, $\mathbb{Z}\{a^k\} = \dfrac{z}{z-a}$

By time shifting property we get, $\mathbb{Z}\{a^{(k-1)}\} = z^{-1}\dfrac{z}{z-a}$

On taking inverse $\mathbb{Z}$-transform of $F(z)$ we get,

$$f(k) = (0.5+j0.34)(-0.5+j\sqrt{7}/2)^{(k-1)}u(k-1)$$
$$+ (0.5-j0.34)(-0.5-j\sqrt{7}/2)^{(k-1)}u(k-1)\,;\text{ for } k\geq 0$$

Note : *Since the term $a^{(k-1)}$ is valid only for $k\geq 1$, it is multiplied by $u(k-1)$.*

(d)   **Given that,** $F(z) = \dfrac{z-4}{(z-1)(z-2)^2}$

By partial fraction expansion we get,

$$F(z) = \frac{z-4}{(z-1)(z-2)^2} = \frac{A_1}{z-1} + \frac{A_2}{(z-2)^2} + \frac{A_3}{(z-2)}$$

$$A_1 = \frac{z-4}{(z-1)(z-2)^2}(z-1)\Big|_{z=1} = \frac{z-4}{(z-2)^2}\Big|_{z=1} = \frac{1-4}{(1-2)^2} = -3$$

$$A_2 = \frac{z-4}{(z-1)(z-2)^2}(z-2)^2\Big|_{z=2} = \frac{z-4}{z-1}\Big|_{z=2} = \frac{2-4}{2-1} = -2$$

$$A_3 = \frac{d}{dz}\left[\frac{z-4}{(z-1)(z-2)^2}(z-2)^2\right]\Big|_{z=2} = \frac{d}{dz}\left[\frac{z-4}{z-1}\right]\Big|_{z=2}$$

$$= \frac{(z-1)-(z-4)}{(z-1)^2}\Big|_{z=2} = \frac{3}{(z-1)^2}\Big|_{z=2} = \frac{3}{(2-1)^2} = 3$$

$$\therefore F(z) = \frac{-3}{z-1} - \frac{2}{(z-2)^2} + \frac{3}{z-2} = -3\frac{1}{z}\frac{z}{z-1} - \frac{1}{z}\frac{2z}{(z-2)^2} + 3\frac{1}{z}\frac{z}{z-2}$$

$$= -3z^{-1}\frac{z}{z-1} - z^{-1}\frac{2z}{(z-2)^2} + 3z^{-1}\frac{z}{z-2}$$

We know that, $\mathcal{Z}\{u(k)\} = \frac{z}{z-1}$; $\mathcal{Z}\{a^k\} = \frac{z}{z-a}$ and $\mathcal{Z}\{ka^k\} = \frac{az}{(z-a)^2}$

By time shifting property we get,

$$\mathcal{Z}\{u(k-1)\} = z^{-1}\frac{z}{z-1} \; ; \; \mathcal{Z}\{a^{(k-1)}\} = z^{-1}\frac{z}{z-a} \text{ and } \mathcal{Z}\{(k-1)a^{(k-1)}\} = z^{-1}\frac{az}{(z-a)^2}$$

On taking inverse $\mathcal{Z}$-transform of F(z) we get,

$$f(k) = -3\,u(k-1) - (k-1)2^{(k-1)}u(k-1) + 3\times 2^{(k-1)}u(k-1)$$

Note : Since the term $a^{(k-1)}$ is valid only for $k \geq 1$, it is multiplied by u(k-1).

EXAMPLE 3.8

Determine the inverse $\mathcal{Z}$-transform of $F(z) = \dfrac{1}{1 - \frac{3}{2}z^{-1} + \frac{1}{2}z^{-2}}$

when (a) ROC : $|z| > 1.0$ and   (b) ROC : $|z| < 0.5$.

SOLUTION

(a) Since the ROC is the exterior of a circle, we expect f(k) to be causal signal. Hence we can express F(z) as a power series expansion in negative powers of z. On dividing the numerator of F(z) by its denominator we get,

$$F(z) = \frac{1}{1 - \frac{3}{2}z^{-1} + \frac{1}{2}z^{-2}} = 1 + \frac{3}{2}z^{-1} + \frac{7}{4}z^{-2} + \frac{15}{8}z^{-3} + \frac{31}{16}z^{-4} + \dots\dots \qquad \dots\dots(3.8.1)$$

$$1 + \frac{3}{2}z^{-1} + \frac{7}{4}z^{-2} + \frac{15}{8}z^{-3} + \frac{31}{16}z^{-4} + ...$$

$$1 - \frac{3}{2}z^{-1} + \frac{1}{2}z^{-2} \overline{\smash{\big)}\ 1}$$

$$\underline{1 - \frac{3}{2}z^{-1} + \frac{1}{2}z^{-2}}$$

$$\frac{3}{2}z^{-1} - \frac{1}{2}z^{-2}$$

$$\underline{\frac{3}{2}z^{-1} - \frac{9}{4}z^{-2} + \frac{3}{4}z^{-3}}$$

$$\frac{7}{4}z^{-2} - \frac{3}{4}z^{-3}$$

$$\underline{\frac{7}{4}z^{-2} - \frac{21}{8}z^{-3} + \frac{7}{8}z^{-4}}$$

$$\frac{15}{8}z^{-3} + \frac{7}{8}z^{-4}$$

$$\underline{\frac{15}{8}z^{-3} - \frac{45}{16}z^{-4} + \frac{15}{16}z^{-5}}$$

$$\frac{31}{16}z^{-4} - \frac{15}{16}z^{-5}$$

$$\vdots$$

If F(z) is **Z**-transform of f(k) then, by the definition of **Z**-transform we get,

$$F(z) = \mathbf{Z}\{f(k)\} = \sum_{k=-\infty}^{\infty} f(k)\, z^{-k}$$

For a causal signal,

$$F(z) = \sum_{k=0}^{\infty} f(k)\, z^{-k}$$

On expanding the summation we get,

$$F(z) = f(0)\, z^{-1} + f(1)z^{-2} + f(2)z^{-2} + f(3)z^{-3} + f(4)z^{-4} + ... \qquad .....(3.8.2)$$

On comparing the two power series of F(z) [i.e., equation (3.8.1) & (3.8.2)], we get,

$$f(0) = 1 \ ; \ f(1) = \frac{3}{2}; \ f(2) = \frac{7}{4} \ ; \ f(3) = \frac{15}{8} \ ; \ f(4) = \frac{31}{16} \ ; ....$$

$$f(k) = \left\{ 1, \frac{3}{2}, \frac{7}{4}, \frac{15}{8}, \frac{31}{16}, ... \right\}; \text{ for } k \geq 0$$
$$\uparrow$$

(b) Since the ROC is the interior of a circle, we expect f(k) to be anticausal signal. Hence we can express F(z) as a power series expansion in positive powers of z. Therefore, rewrite the denominator polynomial of F(z) in the reverse order and then the numerator, is divided by the denominator as shown below.

$$\frac{1}{2}z^{-2} - \frac{3}{2}z^{-1} + 1 \overline{\left)\begin{array}{l} 2z^2 + 6z^3 + 14z^4 + 30z^5 + 62z^6 + ... \\ 1 \\ \underline{1 - 3z + 2z^2} \\ 3z - 2z^2 \\ \underline{3z - 9z^2 + 6z^3} \\ 7z^2 - 6z^3 \\ \underline{7z^2 - 21z^3 + 14z^4} \\ 15z^3 - 14z^4 \\ \underline{15z^3 - 45z^4 + 30z^5} \\ 31z^4 - 30z^5 \\ \vdots \end{array}\right.}$$

$$\therefore F(z) = \frac{1}{1 - \frac{3}{2}z^{-1} + \frac{1}{2}z^{-2}} = \frac{1}{\frac{1}{2}z^{-2} - \frac{3}{2}z^{-1} + 1}$$

$$= 2z^2 + 6z^3 + 14z^4 + 30z^5 + 62z^6 + ... \qquad .....(3.8.3)$$

If F(z) is $\mathbb{Z}$-transform of f(k) then, by the definition of $\mathbb{Z}$-transform we get,

$$F(z) = \mathbb{Z}\{f(k)\} = \sum_{k=-\infty}^{\infty} f(k) z^{-k}$$

For an anticausal signal, $F(z) = \sum_{k=-\infty}^{0} f(k) z^{-k}$

On expanding the summation we get,

$$F(z) = ...f(-6) z^6 + f(-5) z^5 + f(-4) z^4 + f(-3) z^3 + f(-2) z^2 + f(-1) z + f(0) \qquad .....(3.8.4)$$

On comparing the two power series of F(z) [i.e., equations (3.8.3) & (3.8.4)] , we get,

f(−6) = 62 ; f(−5) = 30 ; f(−4) = 14 ; f(−3) = 6

f(−2) = 2 ; f(−1) = 0 and f(0) = 0

$$\therefore f(k) = \{........, 62, 30, 14, 6, 2, 0, 0\}$$
$$\uparrow$$

3.7   LINEAR DISCRETE-TIME SYSTEMS

A discrete-time system is a device or algorithm that operates on a discrete-time signal called the input or excitation, according to some well-defined rule, to produce another discrete-time signal called the output or the response of the system. We can say that the input signal r(k) is transformed by the system into a signal c(k) and expressed as

$$c(k) = H\,[r(k)] \qquad\qquad \text{.....(3.25)}$$

where, H denotes the transformation

(also called an operator)

A discrete time system is linear if it obeys the principle of superposition and it is time invariant if its input-output relationship do not change with time.

When the input to a discrete time system is unit impulse, $\delta(k)$ then the output is called impulse response of the system and denoted by h(k)

$$h(k) = H\,[\delta(k)] \qquad\qquad \text{.....(3.26)}$$

A linear time invariant discrete time system is characterized by its impulse response h(k) and so the impulse response h(k) is also called weighing sequence.

The input-output description of a discrete-time system consists of mathematical expression or a rule, which explicitly defines the relation between the input and output signals (input-output relationship). It is denoted by

$$r(k) \xrightarrow{\;\;H\;\;} c(k) \qquad\qquad \text{.....(3.27)}$$

The input-output relationship of a linear-time invariant discrete time system, (LDS) can be expressed by N^{th} order constant coefficient difference equation given below.

$$c(k) = -\sum_{m=1}^{N} a_m c\,(k-m) + \sum_{m=0}^{M} b_m r\,(k-m) \qquad\qquad \text{.....(3.28)}$$

The integer N is called the order of the system and $M \le N$.

Here c(k − m) are past outputs, r(k − m) are past inputs, r(k) is present input and a_k and b_k are constant coefficients.

ANALYSIS OF LINEAR DISCRETE TIME SYSTEM (LDS)

There are two methods of analysing the behaviour or response of a LDS system.

Method 1

The input-output relation of the LDS system is governed by the constant coefficient difference equation of the form shown in equation (3.28). Mathematically the direct solution of equation (3.28) can be obtained to analyse the performance of the system.

Method 2

The given input signal is first decomposed or resolved into a sum of elementary signals. Then using the linearity property of the system, the responses of the system to the elementary signals are added to obtain the total response of the system to the given input signal.

Resolution of discrete time signal (or sequence) into impulses

  Let r(k) = Discrete time signal

      δ(k) = Unit impulse signal

  and δ(k – m) = Delayed unit impulse signal

  Consider the product of r(k) and δ(k – m)

$$r(k)\,\delta(k-m) = \begin{cases} r(m)\,\delta(k-m) & ; \text{ at } k = m \\ 0 & ; \text{ for } k \neq m \end{cases} \qquad \text{.....(3.29)}$$

$$\therefore r(k)\,\delta(k-m) = \begin{cases} r(m) & ; \text{ at } k = m\,(\because \delta(k-m) = 1 \text{ at } k = m) \\ 0 & ; \text{ for } k \neq m \end{cases} \qquad \text{.....(3.30)}$$

  The product r(k) δ(k – m) has zero everywhere except at k = m. The value of the signal at k = m is the mth sample of the signal r(k) and it is denoted by r(m).

  Therefore each multiplication of the signal r(k) by an unit impulse at some delay m, in essence picks out the single value r(m) of the signal r(k) at k = m, where the unit impulse is nonzero. Consequently if we repeat this multiplication over all possible delays in the range of, $0 \leq m < \infty$ and sum all the product sequences, the result will be a sequence that is equal to the sequence r(k). Hence r(k) can be expressed as

$$\mathbf{r(k) = \sum_{m=0}^{\infty} r(m)\,\delta(k-m)} \qquad \text{.....(3.31)}$$

> **Note :** *Each product r(k) δ (k – m) is an impulse and the summation of impulses gives r(k). Here r(k) is considered as one sided sequence. If r(k) is two sided sequence then the range of m is $-\infty$ to $+\infty$.*

RESPONSE OF LDS SYSTEM FOR ARBITRARY INPUT - THE CONVOLUTION SUM

  In a LDS system the response c(k) of the system for arbitrary input r(k) is given by convolution of the input r(k) with the impulse response h(k) of the system. It is expressed as

    c(k) = r(k) * h(k)   .....(3.32)

       where, the symbol * represents convolution operation.

Proof

    Let c(k) be the response of the system H for an input r(k). [Let r(k) be a one sided sequence].

    c(k) = H[r(k)]   .....(3.33)

    The signal r(k) can be expressed as a summation of impulses as,

$$r(k) = \sum_{m=0}^{\infty} r(m)\,\delta(k-m) \qquad \text{.....(3.34)}$$

    where, δ(k – m) is the delayed unit impulse signal.

    From equations (3.33) and (3.34) we get,

$$c(k) = H\left[\sum_{m=0}^{\infty} r(m)\,\delta(k-m)\right] \qquad \text{.....(3.35)}$$

The system H is a function of k and not a function of m. Hence by linearity property the equation (3.35) can be written as,

$$c(k) = \sum_{m=0}^{\infty} r(m) H[\delta(k-m)] \qquad \qquad .....(3.36)$$

Let the response of the LDS system to the unit impulse input $\delta(k)$ be denoted by h(k).

$$h(k) = H[\delta(k)] \qquad \qquad .....(3.37)$$

Then by time invariance property the response of the system to the delayed unit impulse input $\delta(k - m)$ is

$$h(k - m) = H[\delta(k - m)] \qquad \qquad .....(3.38)$$

Using equation (3.38), the equation (3.36) can be expressed as

$$c(k) = \sum_{m=0}^{\infty} r(m) h(k-m) \qquad \qquad .....(3.39)$$

The equation of c(k) [equation (3.39)] is called convolution sum. We can say that the input r(k) is convoluted with the impulse response h(k) to yield the output c(k).

$$\therefore c(k) = \sum_{m=0}^{\infty} r(m) h(k-m) = r(k) * h(k) \qquad \qquad .....(3.40)$$

PROPERTIES OF CONVOLUTION

Commutative property   : $r(k) * h(k) = h(k) * r(k)$

Associative property   : $[r(k) * h_1(k)] * h_2(k) = r(k) * [h_1(k) * h_2(k)]$

Distributive property   : $r(k) * [h_1(k) + h_2(k)] = [r(k) * h_1(k)] + [r(k) * h_2(k)]$

3.8   TRANSFER FUNCTION OF LDS SYSTEM (PULSE TRANSFER FUNCTION)

The transfer function of LDS system is given by z-transform of its impulse response. The transfer function of LDS system is also called z-transfer function or pulse transfer function.

Let h(k) = Impulse response of a LDS system

Now, $\mathcal{Z}$-transform of h(k) = $\mathcal{Z}\{h(k)\}$ = H(z)

∴ Transfer function of LDS system = H(z) $\qquad \qquad .....(3.41)$

The input-output relationship of a LDS system is governed by a convolution sum of equation (3.40). By taking $\mathcal{Z}$-transform of this convolution sum it can be shown that, H(z) is given by the ratio C(z)/R(z), where C(z) is the $\mathcal{Z}$-transform of output c(k) of LDS system and R(z) is the $\mathcal{Z}$-transform of input r(k) to the LDS system.

Proof

By the definition of one sided Z-transform,

$$C(z) = \mathcal{Z}\{c(k)\} = \sum_{k=0}^{\infty} c(k) z^{-k} \qquad \qquad .....(3.42)$$

From equation (3.40), we get, $C(k) = \sum_{m=0}^{\infty} r(m) h(k-m)$

On substituting this convolution sum in equation (3.42) we get,

$$C(z) = \sum_{k=0}^{\infty} \left[\sum_{m=0}^{\infty} r(m) h(k-m) \right] z^{-k} \qquad \qquad .....(3.43)$$

The order of summation in equation (3.43) can be interchanged. Therefore equation (3.43) can be written as

$$C(z) = \sum_{m=0}^{\infty} r(m) \sum_{k=0}^{\infty} h(k-m) z^{-k} \qquad .....(3.44)$$

Let, p = (k − m),   When   k = 0, p = −m

                     and when   k = ∞, p = ∞

                     Also,   k = p + m

On replacing (k − m) by p in equation (3.44) we get

$$C(z) = \sum_{m=0}^{\infty} r(m) \sum_{p=-m}^{\infty} h(p) z^{-(p+m)}$$

$$= \sum_{m=0}^{\infty} r(m) \sum_{m=0}^{\infty} h(p) z^{-p} z^{m} \quad (\because h(p) = 0 \; ; \; \text{for } p < 0)$$

$$= \sum_{m=0}^{\infty} r(m) z^{-m} \sum_{p=0}^{\infty} h(p) z^{-p} \qquad .....(3.45)$$

By the definition of one sided Z-transform,

$$\sum_{m=0}^{\infty} r(m) z^{-m} = R(z) \text{ and } \sum_{p=0}^{\infty} h(p) z^{-p} = H(z)$$

Hence equation (3.45) can be written as

$$C(z) = R(z) H(z) \quad \text{(or)} \quad H(z) = \frac{C(z)}{R(z)} \qquad .....(3.46)$$

From equation (3.46) we can conclude that the transfer function of the system is given by the ratio C(z)/R(z).

From the above analysis we can define the transfer function of the LDS system is the ratio of the Z-transform of the output of a system to the Z-transform of the input to the system with zero initial conditions.

     Let r(k) = Input of LDS system

and c(k) = Output of a LDS system

Now, Z{r(k)}=R(z) and Z{c(k)} = C(z)

$$\therefore \text{ Transfer function of LDS system} = \frac{C(z)}{R(z)} \qquad .....(3.47)$$

The input-output relation of LDS system is governed by the constant coefficient difference equation,

$$c(k) = -\sum_{m=1}^{N} a_m c(k-m) + \sum_{m=0}^{M} b_m r(k-m) \qquad .....(3.48)$$

where, N is the order of the system and M ≤ N.

On taking Z-transform of equation (3.48) we get,

[By time shifting property, Z{c(k − m)} = z⁻ᵐ C(z) andZ{r(k − m)} = z⁻ᵐ R(z)]

$$C(z) = -\sum_{m=1}^{N} a_m z^{-m} C(z) + \sum_{m=0}^{M} b_m z^{-m} R(z)$$

$$\therefore C(z) + \sum_{m=1}^{N} a_m z^{-m} C(z) = \sum_{m=0}^{N} b_m z^{-m} R(z) \qquad .....(3.49)$$

On expanding the equation (3.49) with M = N, we get,

$$C(z) + a_1 z^{-1} C(z) + a_2 z^{-2} C(z) + \ldots + a_N z^{-N} C(z)$$

$$= b_0 R(z) + b_1 z^{-1} R(z) + b_2 z^{-2} R(z) + \ldots + b_N z^{-N} R(z)$$

$$C(z)\left[1 + a_1 z^{-1} + a_2 z^{-2} + \ldots + a_N z^{-N}\right] = R(z)\left[b_0 + b_1 z^{-1} + b_2 z^{-2} + \ldots + b_N z^{-N}\right]$$

$$\therefore \frac{C(z)}{R(z)} = \frac{b_0 + b_1 z^{-1} + b_2 z^{-2} + \ldots + b_N z^{-N}}{1 + a_1 z^{-1} + a_2 z^{-2} + \ldots + a_N z^{-N}} \qquad \ldots (3.50)$$

From the above discussions it is evident that the transfer function of the LDS system can be obtained by taking Z-transform of the difference equation governing the system.

EXAMPLE 3.9

The input-output relation of a sampled data system is described by the equation

$$c(k + 2) + 3\ c(k + 1) + 4c(k) = r(k + 1) - r(k).$$

Determine the z-transfer function. Also obtain the weighting sequence of the system.

SOLUTION

Let $R(z) = \mathcal{Z}\{r(k)\}$ and $C(z) = \mathcal{Z}\{c(k)\}$

By time shifting property, when initial conditions are zero, we get,

$$\mathcal{Z}\{c(k + m)\} = z^m\ C(z) \quad \text{and} \quad \mathcal{Z}\{r(k + m)\} = z^m\ R(z)$$

Given that, $c(k + 2) + 3\ c(k + 1) + 4c(k) = r(k + 1) - r(k)$

On taking $\mathcal{Z}$-transform of the above equation we get,

$$z^2\ C(z) + 3z\ C(z) + 4\ C(z) = z\ R(z) - R(z)$$

$$(z^2 + 3z + 4)\ C(z) = (z - 1)\ R(z)$$

$$\therefore \frac{C(z)}{R(z)} = \frac{z - 1}{z^2 + 3z + 4}$$

> The roots of the quadratic $z^2 + 3z + 4 = 0$ are,
>
> $$z = \frac{-3 \pm \sqrt{9 - 4 \times 4}}{2}$$
>
> $$= -\frac{3}{2} \pm j\frac{\sqrt{7}}{2}$$

The transfer function of the system, $H(z) = \dfrac{C(z)}{R(z)} = \dfrac{z - 1}{z^2 + 3z + 4}$

The weighing sequence is the impulse response, $h(k)$ of the system. It is given by inverse $\mathcal{Z}$-transform of $H(z)$.

$$H(z) = \frac{z - 1}{z^2 + 3z + 4} = \frac{z - 1}{\left(z + \dfrac{3}{2} + j\dfrac{\sqrt{7}}{2}\right)\left(z + \dfrac{3}{2} - j\dfrac{\sqrt{7}}{2}\right)}$$

By partial fraction technique $H(z)$ can be expressed as

$$H(z) = \frac{A}{z + \dfrac{3}{2} + j\dfrac{\sqrt{7}}{2}} + \frac{A^*}{z + \dfrac{3}{2} - j\dfrac{\sqrt{7}}{2}}$$

$$A = \frac{z - 1}{\left(z + \dfrac{3}{2} + j\dfrac{\sqrt{7}}{2}\right)\left(z + \dfrac{3}{2} - j\dfrac{\sqrt{7}}{2}\right)}\left(z + \dfrac{3}{2} + j\dfrac{\sqrt{7}}{2}\right)\Bigg|_{z = -\frac{3}{2} - j\frac{\sqrt{7}}{2}} = \frac{z - 1}{z + \dfrac{3}{2} - j\dfrac{\sqrt{7}}{2}}\Bigg|_{z = -\frac{3}{2} - j\frac{\sqrt{7}}{2}}$$

$$A = \frac{-\frac{3}{2} - j\frac{\sqrt{7}}{2} - 1}{-\frac{3}{2} - j\frac{\sqrt{7}}{2} + \frac{3}{2} - j} = \frac{-\frac{5}{2} - j\frac{\sqrt{7}}{2}}{-j\sqrt{7}} = -j\frac{5}{2\sqrt{7}} + \frac{1}{2} = \frac{1}{2} - j\frac{5}{2\sqrt{7}}$$

$$A^* = \left(\frac{1}{2} - j\frac{5}{2\sqrt{7}}\right)^* = \frac{1}{2} + j\frac{5}{2\sqrt{7}}$$

$$H(z) = \frac{\left(\frac{1}{2} - j\frac{5}{2\sqrt{7}}\right)}{z + \frac{3}{2} + j\frac{\sqrt{7}}{2}} + \frac{\frac{1}{2} + j\frac{5}{2\sqrt{7}}}{z + \frac{3}{2} - j\frac{\sqrt{7}}{2}}$$

$$H(z) = \left(\frac{1}{2} - j\frac{5}{2\sqrt{7}}\right)z^{-1}\frac{z}{z - \left(-\frac{3}{2} - j\frac{\sqrt{7}}{2}\right)} + \left(\frac{1}{2} + j\frac{5}{2\sqrt{7}}\right)z^{-1}\frac{z}{z - \left(-\frac{3}{2} + j\frac{\sqrt{7}}{2}\right)}$$

We know that $\mathcal{Z}\{a^k\} = \dfrac{z}{z-a}$

By time shifting property, $\mathcal{Z}\{a^{(k-1)}\} = z^{-1}\dfrac{z}{z-a}$

On taking inverse $\mathcal{Z}$-transform of H(z) we get,

$$h(k) = \left(\frac{1}{2} - j\frac{5}{2\sqrt{7}}\right)\left(-\frac{3}{2} - j\frac{\sqrt{7}}{2}\right)^{(k-1)}u(k-1) + \left(\frac{1}{2} + j\frac{5}{2\sqrt{7}}\right)\left(-\frac{3}{2} + j\frac{\sqrt{7}}{2}\right)^{(k-1)}u(k-1)$$

EXAMPLE 3.10

Solve the difference equation c(k + 2) + 3 c(k + 1) + 2 c(k) = u(k)

Given that c(0) = 1 ; c(1) = −3 ; c(k) = 0 for k < 0

SOLUTION

Let $\mathcal{Z}\{c(k)\} = C(z)$ and $\mathcal{Z}\{u(k)\} = U(z)$

Since u(k) is unit step signal, $U(z) = \dfrac{z}{z-1}$

We know that, if $F(z) = \mathcal{Z}\{f(k)\}$ then

$$\mathcal{Z}\{f(k+m)\} = z^m F(z) - \sum_{i=0}^{m-1} f(i)\, z^{m-1}$$

Given that, c(k + 2) + 3 c(k + 1) + 2 c(k) = u(k)

On taking $\mathcal{Z}$-transform of the above equation we get,

$$\mathcal{Z}\{c(k+2)\} + \mathcal{Z}\{3\, c(k+1)\} + \mathcal{Z}\{2\, c(k)\} = \mathcal{Z}\{u(k)\}$$

$$z^2 C(z) - z^2 c(0) - z\, c(1) + 3\,[z\, C(z) - z\, c(0)] + 2\, C(z) = \frac{z}{z-1}$$

On substituting the initial conditions, $c(0) = 1$ and $c(1) = -3$ we get,

$$z^2 C(z) - z^2 + 3z + 3z\,C(z) - 3z + 2\,C(z) = \frac{z}{z-1}$$

$$(z^2 + 3z + 2)C(z) - z^2 = \frac{z}{z-1}$$

$$(z^2 + 3z + 2)C(z) = \frac{z}{z-1} + z^2$$

$$(z+1)(z+2)C(z) = \frac{z + z^2(z-1)}{(z-1)}$$

$$\therefore C(z) = \frac{z\left[1 + z^2 - z\right]}{(z-1)(z+1)(z+2)}$$

$$\therefore \frac{C(z)}{z} = \frac{z^2 - z + 1}{(z-1)(z+1)(z+2)}$$

By partial fraction expansion technique we can write C(z)/z as,

$$\frac{C(z)}{z} = \frac{A_1}{z-1} + \frac{A_2}{z+1} + \frac{A_3}{z+2}$$

$$A_1 = \frac{z^2 - z + 1}{(z-1)(z+1)(z+2)}\,(z-1)\bigg|_{z=1} = \frac{z^2 - z + 1}{(z+1)(z+2)}\bigg|_{z=1} = \frac{1 - 1 + 1}{(1+1)(1+2)} = \frac{1}{6}$$

$$A_2 = \frac{z^2 - z + 1}{(z-1)(z+1)(z+2)}\,(z-1)\bigg|_{z=-1} = \frac{z^2 - z + 1}{(z-1)(z+2)}\bigg|_{z=-1} = \frac{(-1)-(-1)+1}{(-1-1)(-1+2)} = -\frac{3}{2}$$

$$A_3 = \frac{z^2 - z + 1}{(z-1)(z+1)(z+2)}\,(z+2)\bigg|_{z=-2} = \frac{z^2 - z + 1}{(z-1)(z+1)}\bigg|_{z=-2} = \frac{(-2)^2 - (-2)+1}{(-2-1)(-2+1)} = \frac{7}{3}$$

$$\therefore \frac{C(z)}{z} = \frac{1}{6}\frac{1}{z-1} - \frac{3}{2}\frac{1}{z+1} + \frac{7}{3}\frac{1}{z+2}$$

$$C(z) = \frac{1}{6}\frac{1}{z-1} - \frac{3}{2}\frac{z}{z-(-1)} + \frac{7}{3}\frac{z}{z-(-2)}$$

We know that $\mathbb{Z}\{u(k)\} = \frac{z}{z-1}$ and $\mathbb{Z}\{a^k\} = \frac{z}{z-a}$

On taking inverse $\mathbb{Z}$-transform of C(z) we get,

$$c(k) = \frac{1}{6}u(k) - \frac{3}{2}(-1)^k + \frac{7}{3}(-2)^k \ ; \ k \geq 0$$

The above equation of c(k) is the solution of the given difference equation.

3.9 ANALYSIS OF SAMPLER AND ZERO - ORDER HOLD

Consider a pulse sampler with zero-order hold (ZOH) shown in fig 7.19. Let the output of sampler be a pulse train of pulse width Δ. For each input pulse, the ZOH produces a pulse of duration T, where T is the sampling period.

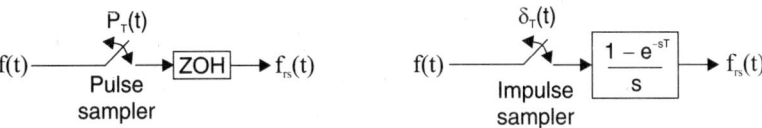

Fig 3.19 : Pulse sampler with ZOH.   *Fig 3.20 : Equivalent representation of Pulse sampler with ZOH.*

It can be proved that the output of pulse sampler with ZOH can be produced by impulse sampled f(t) when passed through a transfer function,

$$G_0(s) = \frac{1-e^{-sT}}{s} \qquad\qquad .....(3.51)$$

Hence the pulse sampler with ZOH can be replaced by an equivalent system consisting of an impulse sampler and a block with transfer function, $(1 - e^{-sT})/s$ as shown in fig 3.20. This equivalent representation offers easier analysis of sampled data control systems.

FREQUENCY RESPONSE CHARACTERISTICS OF ZERO ORDER HOLDING DEVICE

The sinusoidal transfer function of ZOH can be obtained from $G_0(s)$ by replacing s by jω.

$$\therefore G_0(j\omega) = \frac{1-e^{-j\omega T}}{j\omega} \qquad\qquad .....(3.52)$$

We know that, $e^{-\frac{j\omega T}{2}}\, e^{+\frac{j\omega T}{2}} = 1$ .....(3.53)

Hence from equation (3.52) and (3.53) we get,

$$G_0(j\omega) = \frac{e^{-\frac{j\omega T}{2}}\, e^{\frac{j\omega T}{2}} - e^{-j\omega T}}{j\omega} = \frac{e^{-\frac{j\omega T}{2}}\, e^{\frac{j\omega T}{2}} - e^{-\frac{j\omega T}{2}}\, e^{-\frac{j\omega T}{2}}}{j\omega}$$

$$= \left(\frac{e^{\frac{j\omega T}{2}} - e^{-\frac{j\omega T}{2}}}{j\omega}\right) e^{-\frac{j\omega T}{2}} = \frac{2}{\omega}\sin\frac{\omega T}{2}\, e^{-\frac{j\omega T}{2}}$$

$$\boxed{\textit{Note :} \ \sin\theta = \frac{e^{j\theta} - e^{-j\theta}}{2j}}$$

$$= T\, \frac{\sin\frac{\omega T}{2}}{\frac{\omega T}{2}}\, e^{-\frac{j\omega T}{2}} \qquad\qquad .....(3.54)$$

We know that, sampling frequency, $\omega_s = \dfrac{2\pi}{T}$

$$\therefore T = \frac{2\pi}{\omega_s}$$

On substituting $T = 2\pi/\omega_s$ in equation (3.54) we get,

$$G_0(j\omega) = \frac{2\pi}{\omega_s}\, \frac{\sin(\pi\omega/\omega_s)}{(\pi\omega/\omega_s)}\, e^{-\frac{j\pi\omega}{\omega_s}}$$

Magnitude of $G_0(j\omega) = |G_0(j\omega)| = \dfrac{2\pi}{\omega_s} \dfrac{\sin(\pi\omega/\omega_s)}{(\pi\omega/\omega_s)}$ .....(3.55)

Argument (or phase) of $G_0(j\omega) = \angle G_0(j\omega) = \dfrac{-\pi\omega}{\omega_s}$ .....(3.56)

The frequency response characteristics consists of magnitude response and phase response characteristics. The magnitude and phase response of ZOH device are given by equations (3.55) and (3.56) respectively. The fig (3.21) shows the frequency response curve of ZOH device. From the frequency response curve we can conclude that ZOH device has low pass filtering characteristics.

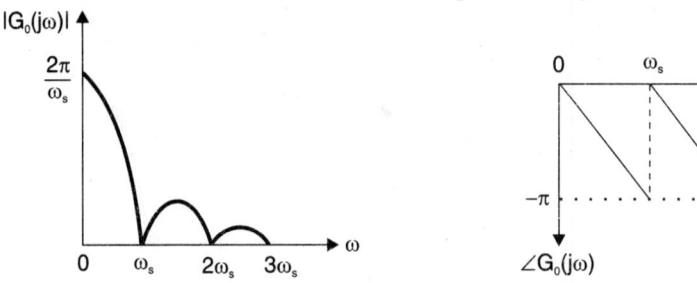

Fig a : Magnitude response of ZOH device.   *Fig b : Phase response of ZOH device.*

Fig 3.21 : Frequency response of ZOH device.

3.10   ANALYSIS OF SYSTEMS WITH IMPULSE SAMPLING

Consider a linear continuous time system fed from an impulse sampler as shown in fig 3.22a. Let H(s) be the transfer function of the system in s-domain. In such a system we are interested in reading the output at sampling instants. This can be achieved by means of a mathematical sampler or read-out sampler.

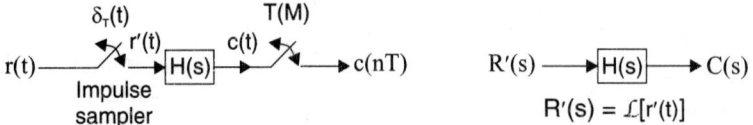

Fig 3.22 : Linear continuous time system with impulse sampled input.

For the system shown in fig 3.22b, it can be shown that the z-domain transfer function H(z) can be directly obtained from s-domain transfer function by taking $\mathcal{Z}$-transform of H(s)

i.e., $H(z) = \mathcal{Z}\{H(s)\}$ .....(3.57)

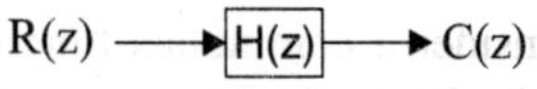

The fig 3.23 shows the $\mathcal{Z}$-transform equivalent of the s-domain system of fig 3.22b.

Fig 3.23 : The $\mathcal{Z}$-transform equivalent of the system shown in fig 3.22b.

The output in z-domain is given by, $C(z) = H(z) R(z)$ .....(3.58)

Procedure to find z-transfer function from s-domain transfer function

1. Determine h(t) from H(s), where $h(t) = \mathcal{L}^{-1}[H(s)]$

2. Determine the discrete sequence h(kT) by replacing t by kT in h(t)

3. Take $\mathcal{Z}$-transform of h(kT), which is z-transfer function of the system (i.e., $H(z) = \mathcal{Z}\{h(kT)\}$).

TABLE-3.4. Laplace and $\bar{Z}$-Transform Pairs

H(s)	H(z)
$\dfrac{1}{s}$	$\dfrac{z}{z-1}$
$\dfrac{1}{s^2}$	$\dfrac{Tz}{(z-1)^2}$
$\dfrac{1}{s^3}$	$\dfrac{T^2 z(z+1)}{2(z-1)^3}$
$\dfrac{1}{s+a}$	$\dfrac{z}{z-e^{-aT}}$
$\dfrac{1}{(s+a)^2}$	$\dfrac{Tze^{-aT}}{(z-e^{-aT})^2}$
$\dfrac{a}{s(s+a)}$	$\dfrac{z(1-e^{-aT})}{(z-1)(z-e^{-aT})}$
$\dfrac{\omega}{s^2+\omega^2}$	$\dfrac{z\sin\omega T}{z^2-2z\cos\omega T+1}$
$\dfrac{s}{s^2+\omega^2}$	$\dfrac{z(z-\cos\omega T)}{z^2-2z\cos\omega T+1}$
$\dfrac{\omega}{(s+a)^2+\omega^2}$	$\dfrac{ze^{-aT}\sin\omega T}{z^2-2ze^{-aT}\cos\omega T+e^{-2aT}}$
$\dfrac{s+a}{(s+a)^2+\omega^2}$	$\dfrac{z^2-ze^{-aT}\cos\omega T}{z^2-2ze^{-aT}\cos\omega T+e^{-2aT}}$

Alternatively, by partial fraction technique if H(s) can be expressed as a summation of first order terms then using standard transform pairs listed in table-3.4, the $\bar{Z}$-transform of H(s) can be directly obtained.

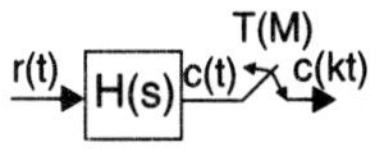

Fig 3.24a

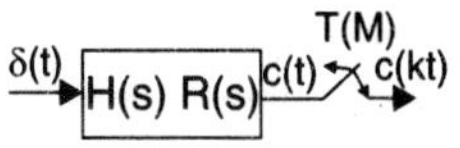

Consider a continuous time system with transfer function H(s) as shown in fig 3.24a. Let the input r(t) be a continuous time input. To read the continuous output at sampling instants, let us imagine a mathematical sampler at the output stage.

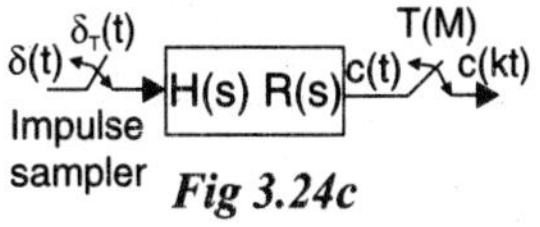

Fig 3.24b

The system shown in fig 3.24a can be equivalently represented by a block of H(s) R(s) with impulse input $\delta(t)$ as shown in fig 3.24b. Now the input and so the output does not change by imagining a fictious impulse sampler through which $\delta(t)$ is applied to H(s) R(s) as shown in fig 3.24c.

Fig 3.24c

Fig 3.24

For such a system we can prove that

$$C(z) = \bar{Z}\{H(s)\,R(s)\} \qquad\qquad .....(3.59)$$

Hence, if C(s) = H(s) R(s) then C(z) = $\bar{Z}\{H(s)\,R(s)\}$ = HR(z)   • .....(3.60)

The function $Z\{H(s)\,R(s)\}$ is also denoted as HR(z).

When the impulse sampled input is applied to two or more s-domain transfer function in cascade as shown in fig 3.25a, then z-transfer function of the system is given by

$$H(z) = Z\{H_1(s)\,H_2(s)\}$$ .....(3.61)

and $C(z) = Z\{H_1(s)\,H_2(s)\}\,R(z)$ .....(3.62)

where, $R(z) = Z\{R(s)\}$ and $R(s) = L[r(t)]$

The function $Z\{H_1(s)\,H_2(s)\}$ is also denoted as $H_1H_2(z)$. The equivalent z-domain system is shown in fig 3.25b.

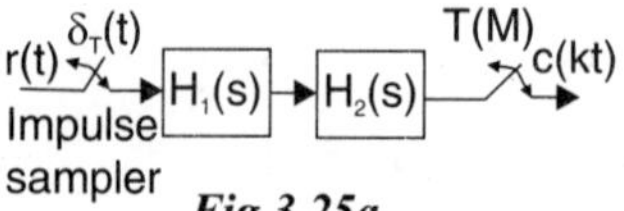

Fig 3.25a 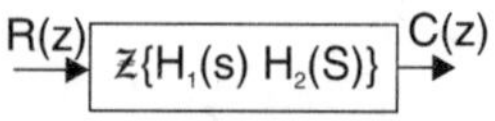   Fig 3.25b

Consider a system in which impulse sampler is introduced at the input of each block as shown in fig 3.26a.

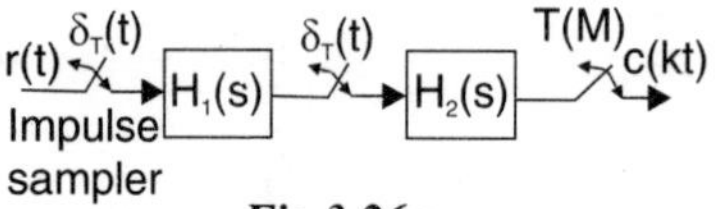

Fig 3.26a 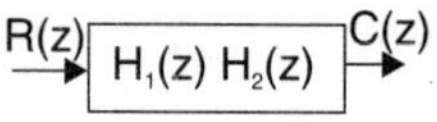   Fig 3.26b

Now the z-transfer function of the system is given by,

$$H(z) = H_1(z)\,H_2(z)$$ .....(3.63)

where, $H_1(z) = Z\{H_1(s)\}$ and $H_2(z) = Z\{H_2(s)\}$

and $C(z) = H_1(z)\,H_2(z)\,R(z)$ .....(3.64)

where, $R(z) = Z\{R(s)\}$ and $R(s) = L[r(t)]$.

The equivalent z-domain system is shown in fig 3.26b.

EXAMPLE 3.11

Determine the z-domain transfer function for the following s-domain transfer functions.

(a) $H(s) = \dfrac{a}{(s+a)^2}$ (b) $H(s) = \dfrac{s}{s^2+\omega^2}$ (c) $H(s) = \dfrac{a}{s^2-a^2}$

(d) $H(s) = \dfrac{s+b}{(s+b)^2+a^2}$ (e) $H(s) = \dfrac{a}{(s+b)^2+a^2}$

SOLUTION

(a)   **Given that, $H(s) = \dfrac{a}{(s+a)^2}$** .

$$h(t) = L^{-1}[H(s)] = L^{-1}\left[\frac{a}{(s+a)^2}\right] = a\,t\,e^{-at}$$

The discrete sequence h(kT) is obtained by letting $t = kT$ in h(t)

$$\therefore\ h(kT) = akT\,e^{-akT}$$

z-transfer function, $H(z) = Z\{h(kT)\}$

$$\therefore H(z) = Z\{h(kT)\} = Z\{akT\,e^{-akT}\} = aT \times Z\{k\,e^{-akT}\}$$

Let $f(k) = e^{-akT}$, $\qquad\qquad \therefore F(z) = Z\{f(k)\}$

By the definition of Z-transform,

$$F(z) = \sum_{k=0}^{\infty} f(k)\,z^{-k} = \sum_{k=0}^{\infty} e^{-akT}z^{-k} = \sum_{k=0}^{\infty}(e^{-aT}z^{-1})^k = \frac{1}{1-e^{-aT}z^{-1}}$$

$$\left(\because \sum_{k=0}^{\infty} C^k = \frac{1}{1-C}\ \text{if}\ |C|<1\right)\binom{\text{Infinite geometric}}{\text{series sum formula}}$$

$$\therefore F(z) = \frac{1}{1-e^{-aT}/z} = \frac{z}{z-e^{-aT}}$$

By the property of Z-transform we get, $Z\{k\,f(k)\} = -z\dfrac{d}{dz}F(z)$

$$\therefore Z\{ke^{-akT}\} = Z\{k\,f(k)\} = -z\frac{d}{dz}\frac{z}{z-e^{-aT}} = -z\times\frac{(z-e^{-aT})-z}{(z-e^{-aT})^2} = \frac{ze^{-aT}}{(z-e^{-aT})^2}$$

$$\therefore H(z) = aT\times Z\{ke^{-akT}\} = aT\times\frac{z\,e^{-aT}}{(z-e^{-aT})^2} = \frac{aT\,z\,e^{-aT}}{(z-e^{-aT})^2}$$

(b) **Given that,** $H(s) = \dfrac{s}{s^2+\omega^2}$

$$h(t) = \mathcal{L}^{-1}[H(s)] = \mathcal{L}^{-1}\left[\frac{s}{s^2+\omega^2}\right] = \cos\omega t$$

The discrete sequence $h(kT)$ is obtained by letting $t = kT$ in $h(t)$

$$\therefore h(kT) = \cos\omega kT$$

z-transfer function, $H(z) = Z\{h(kT)\} = Z\{\cos\omega kT\} = \dfrac{z(z-\cos\omega T)}{z^2 - 2z\cos\omega T + 1}$

[Refer table-3.3 and example 3.4(c)]

(c) **Given that,** $H(s) = \dfrac{a}{s^2-a^2}$

$$h(t) = \mathcal{L}^{-1}[H(s)] = \mathcal{L}^{-1}\left[\frac{a}{s^2-a^2}\right] = \mathcal{L}^{-1}\left[\frac{a}{(s+a)(s-a)}\right]$$

By partial fraction expansion,

$$\frac{a}{(s+a)(s-a)} = \frac{A_1}{s+a} + \frac{A_2}{s-a}$$

$$A_1 = \frac{a}{(s+a)(s-a)}(s-a)\bigg|_{s=-a} = \frac{a}{s-a}\bigg|_{s=-a} = \frac{a}{-a-a} = \frac{-a}{2a} = -\frac{1}{2}$$

$$A_2 = \frac{a}{(s+a)(s-a)}(s-a)\bigg|_{s=a} = \frac{a}{s+a}\bigg|_{s=a} = \frac{a}{a+a} = \frac{a}{2a} = \frac{1}{2}$$

$$\therefore h(t) = \mathcal{L}^{-1}\left[-\frac{1}{2}\frac{1}{(s+a)} + \frac{1}{2}\frac{1}{(s-a)}\right] = -\frac{1}{2}e^{-at} + \frac{1}{2}e^{at}$$

The discrete sequence h(kT) is obtained by letting t = kT in h(t)

$$\therefore h(kT) = -\frac{1}{2}e^{-akT} + \frac{1}{2}e^{akT}$$

By the definition of one sided **Z**-transform,

$$H(z) = \sum_{k=0}^{\infty} h(kT)z^{-k} = \sum_{k=0}^{\infty}\left[-\frac{1}{2}e^{-akT} + \frac{1}{2}e^{akT}\right]z^{-k}$$

$$= -\frac{1}{2}\sum_{k=0}^{\infty} e^{-akT}z^{-k} + \frac{1}{2}\sum_{k=0}^{\infty} e^{akT}z^{-k}$$

$$= -\frac{1}{2}\sum_{k=0}^{\infty} (e^{-aT}z^{-1})^k + \frac{1}{2}\sum_{k=0}^{\infty} (e^{aT}z^{-1})^k$$

From infinite geometric sum series formula we know that,

$$\sum_{k=0}^{\infty} C^k = \frac{1}{1-C} \quad ; \quad \text{when} \mid C \mid < 1$$

$$\therefore H(z) = -\frac{1}{2}\frac{1}{1-e^{-aT}z^{-1}} + \frac{1}{2}\frac{1}{1-e^{aT}z^{-1}} = -\frac{1}{2}\frac{1}{1-e^{-aT}/z} + \frac{1}{2}\frac{1}{1-e^{aT}/z}$$

$$= -\frac{1}{2}\frac{z}{z-e^{-aT}} + \frac{1}{2}\frac{z}{z-e^{aT}} = \frac{z}{2}\left[\frac{-(z-e^{aT})+z-e^{-aT}}{(z-e^{-aT})(z-e^{aT})}\right]$$

$$= \frac{z}{2}\left[\frac{-z+e^{aT}+z-e^{-aT}}{z^2-ze^{aT}-ze^{-aT}+e^{-aT}e^{aT}}\right] = \frac{z}{2}\left[\frac{e^{aT}-e^{-aT}}{z^2-z(e^{aT}+e^{-aT})+1}\right]$$

Since, $\cosh\theta = \frac{e^\theta + e^{-\theta}}{2}$ and $\sinh\theta = \frac{e^\theta - e^{-\theta}}{2}$

$$H(z) = \frac{z}{2}\left(\frac{2\sinh aT}{z^2-2z\cosh aT+1}\right) = \frac{z\sinh aT}{z^2-2z\cosh aT+1}$$

(d) **Given that, H(s) =** $\dfrac{(s+b)}{(s+b)^2+a^2}$

$$h(t) = \mathcal{L}^{-1}[H(s)] = \mathcal{L}^{-1}\left[\frac{(s+b)}{(s+b)^2+a^2}\right] = e^{-bt}\cos at$$

The discrete sequence h(kT) is obtained by letting t = kT in h(t)

$$\therefore h(kT) = e^{-bkT}\cos akT$$

z-transfer function, $H(z) = \mathbf{Z}\{h(kT)\} = \mathbf{Z}\{e^{-bkT}\cos akT\}$

From example 3.5(a) we get

$$H(z) = \frac{ze^{bT}(ze^{bT} - \cos aT)}{z^2 e^{2bT} - 2\,z\,e^{bT}\cos aT + 1}$$

(e) **Given that,** $H(s) = \dfrac{a}{(s+b)^2 + a^2}$

$$h(t) = \mathcal{L}^{-1}[H(s)] = \mathcal{L}^{-1}\left[\frac{a}{(s+b)^2 + a^2}\right] = e^{-bt}\sin at$$

The discrete sequence $h(kT)$ is obtained by letting $t = kT$ in $h(t)$

$$\therefore h(kT) = e^{-bkT}\sin akT$$

z-transform function, $H(z) = \mathcal{Z}\{h(kT)\} = \mathcal{Z}\{e^{-bkT}\sin akT\}$

From example 3.5(b) we get, $H(z) = \dfrac{z\,e^{bT}\sin aT}{z^2 e^{2bT} - 2\,z\,e^{bT}\cos aT + 1}$

3.11 ANALYSIS OF SAMPLED DATA CONTROL SYSTEMS USING $\mathcal{Z}$-TRANSFORM

The analysis of sampled data control systems are performed using the concepts developed in section 3.9 and 3.10. The following points serve as guidelines to determine the output in z-domain and hence the z-transfer function of the sampled data control systems.

1. The pulse sampling is approximated as impulse sampling.

2. The ZOH is replaced by a black with transfer function, $G_0(s) = (1 - e^{-sT})/s$.

3. When the input to a block is impulse sampled signal then the $\mathcal{Z}$-transform of the output of the block can be obtained from the $\mathcal{Z}$-transform of the input and $\mathcal{Z}$-transform of the s-domain transfer function of the block. In determining the output of a block one may come across the following cases.

Case (i) The impulse sampler is located at the input of a block as shown in fig 3.27.

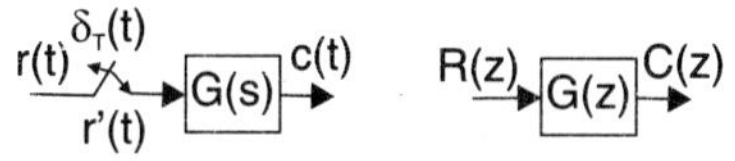

Fig 3.27

In this case, $C(z) = G(z)\,R(z)$ .....(3.65)

Here, $G(z) = \mathcal{Z}\{G(s)\}$; $R(z) = \mathcal{Z}\{R'(s)\}$ and $R'(s) = \mathcal{L}[r'(t)]$

Case (ii) The impulse sampler is located at the input of two s-domain cascaded blocks as shown in fig 3.28.

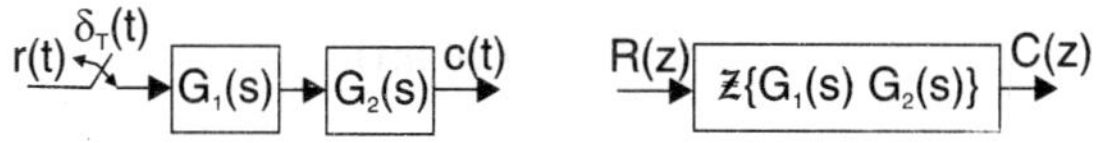

Fig 3.28

In this case, $C(z) = Z\{G_1(s)\,G_2(s)\}\,R(z) = G_1 G_2(z)\,R(z)$ .....(3.66)

Case (iii) The impulse sampler is located at the input of each block as shown in fig 3.29.

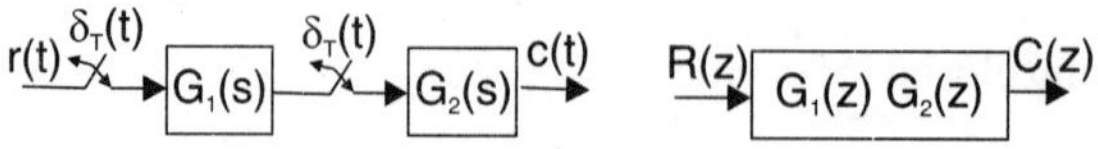

Fig 3.29

In this case, $C(z) = G_1(z)\, G_2(z)\, R(z)$ .....(3.67)

Here, $G_1(z) = Z\{G_1(s)\}$ and $G_2(z) = Z\{G_2(s)\}$

Case(iv) The impulse sampler is located at the input of ZOH in cascade with G(s) as shown in fig 3.30.

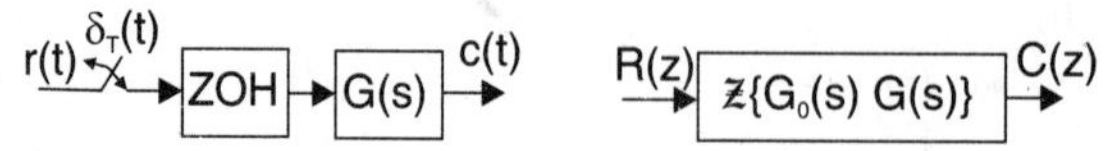

Fig 3.30

In this case, $C(z) = Z\{G_0(s)\,G(s)\}\,R(z) = (1 - z^{-1})\,Z\{G(s)/s\}\,R(z)$ .....(3.68)

The table-3.5 shows some configurations of the closed loop sampled data control systems and their corresponding z-domain outputs.

TABLE-3.5

Closed loop sampled data control system	Output in z-domain
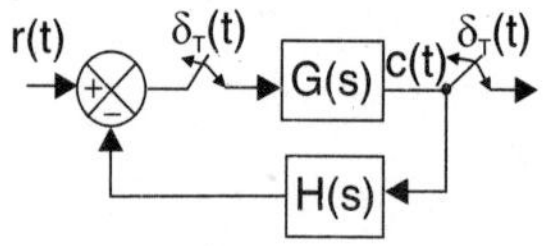	$C(z) = \dfrac{G(z)\,R(z)}{1 + Z\{G(s)\,H(s)\}}$ $= \dfrac{G(z)R(z)}{1 + GH(z)}$
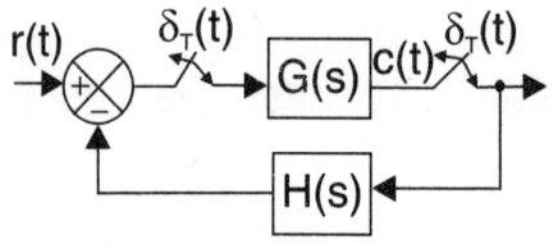	$C(z) = \dfrac{G(z)\,R(z)}{1 + G(z)\,H(z)}$
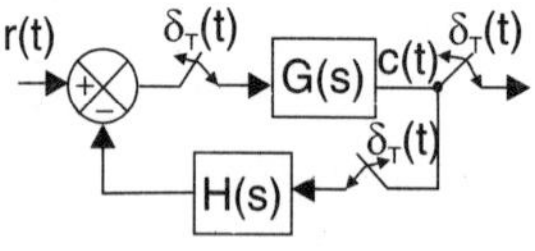	$C(z) = \dfrac{Z\{G(s)\,R(s)\}}{1 + Z\{G(s)\,H(s)\}}$ $= \dfrac{GR(z)}{1 + GH(z)}$
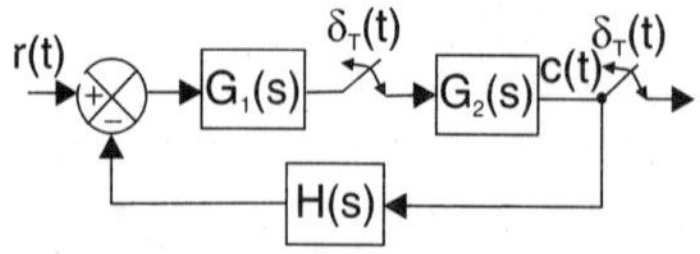	$C(z) = \dfrac{Z\{G_1(s)\,R(s)\}\,G_2(z)}{1 + Z\{G_1(s)G_2(s)\,H(s)\}}$ $= \dfrac{G_1 R(z) G_2(z)}{1 + G_1 G_2 H(z)}$
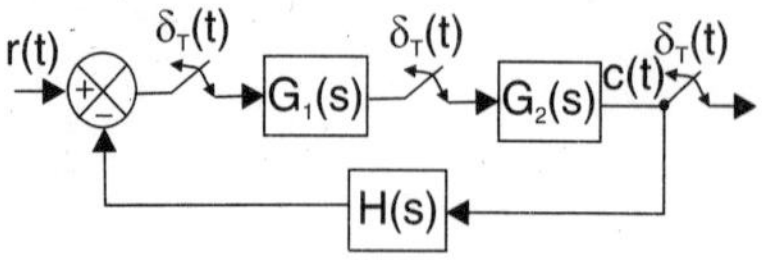	$C(z) = \dfrac{G_1(z)\,G_2(z)R(z)}{1 + G_1(z)Z\{G_2(z)\,H(s)\}}$ $= \dfrac{G_1(z)G_2(z)R(z)}{1 + G_1(z)G_2 H(z)}$

EXAMPLE 3.12

Find C(z)/R(z) for the following closed loop sampled data control systems. Assume all the samplers to be of impulse type.

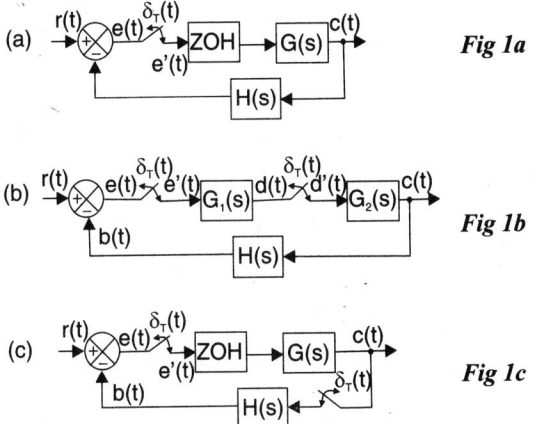

Fig 1a

Fig 1b

Fig 1c

SOLUTION

(a) **The ZOH in the system is replaced by $G_0(s)$ as shown in fig 1, where $G_0(s) = (1 - e^{-sT})/s$.**

Let   e(t) = Error signal

e'(t) = Impulse sampled error signal

b(t) = Feedback signal

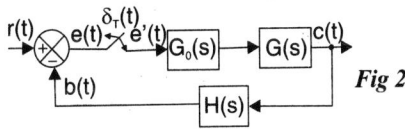

Fig 2

The input to the cascaded blocks of $G_0(s)$ and G(s) is an impulse sampled signal as shown in fig 2a. It's z-domain equivalent is shown in fig 2b.

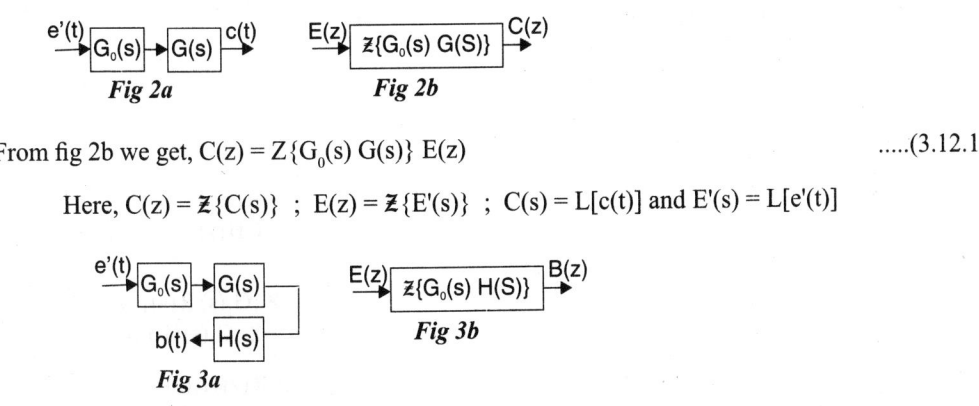

Fig 2a   Fig 2b

From fig 2b we get, $C(z) = Z\{G_0(s)\ G(s)\}\ E(z)$ .....(3.12.1)

Here, $C(z) = Z\{C(s)\}$; $E(z) = Z\{E'(s)\}$; $C(s) = L[c(t)]$ and $E'(s) = L[e'(t)]$

Fig 3a

Fig 3b

The input to the cascaded blocks of $G_0(s)$, G(s) an l H(s) is an impulse sampled signal as shown in fig 3a. It's z-domain equivalent is shown in fig 3b.

From fig 3b we get, $B(z) = Z\{G_0(s)\ G(s)\ H(s)\}\ E(z)$ .....(3.12.2)

Here, $B(z) = Z\{B(s)\}$ and $B(s) = L[b(t)]$

With reference to fig 1.1, at the summing point we get,

$$e(t) = r(t) - b(t) \qquad .....(3.12.3)$$

Since $e'(t) = e(kT)$ is an impulse sampled signal, by superposition principle the equation (3.12.3) can be written as,

$$e(kT) = r(kT) - b(kT) \qquad .....(3.12.4)$$

where, $e(kT)$, $r(kT)$ and $b(kT)$ are impulse sampled signals of $e(t)$, $r(t)$ and $b(t)$ respectively.

On taking $\mathcal{Z}$-transform of equation (3.12.4) we get,

$$E(z) = R(z) - B(z)$$

$$\therefore R(z) = E(z) + B(z) \qquad .....(3.12.5)$$

where, $R(z) = \mathcal{Z}\{R(s)\}$ and $R(s) = L[r(t)]$

On substituting for $B(z)$ from equation (3.12.2) in equation (3.12.5) we get,

$$R(z) = E(z) + \mathcal{Z}\{G_0(s)\, G(s)\, H(s)\}\, E(z)$$

$$= [1 + \mathcal{Z}\{G_0(s)\, G(s)\, H(s)\}]\, E(z) \qquad .....(3.12.6)$$

From equations (3.12.1) and (3.12.6) the z-transfer function or pulse transfer function, $C(z)/R(z)$ can be written as,

$$\frac{C(z)}{R(z)} = \frac{\mathcal{Z}\{G_0(s)\, G(s)\}}{1 + \mathcal{Z}\{G_0(s)\, G(s)H(s)\}} = \frac{G_0G(z)}{1 + G_0H(z)} \qquad .....(3.12.7)$$

Here, $\mathcal{Z}\{G_0(s)\, G(s)\}$ is denoted as $G_0G(z)$ and $\mathcal{Z}\{G_0(s)\, G(s)\, H(s)\}$ is denoted as $G_0GH(z)$.

(b)   The input to the block $G_2(s)$ is an impulse sampled signal as shown in fig 4a. It's z-domain equivalent is shown in fig 4b.

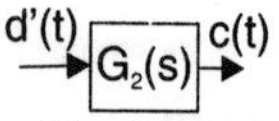

Fig 4a 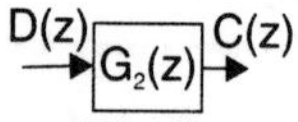   Fig 4b

From fig 4b we get, $C(z) = G_2(z)\, D(z) \qquad .....(3.12.8)$

where, $C(z) = \mathcal{Z}\{C(s)\}$; $G_2(z) = \mathcal{Z}\{G_2(s)\}$; $D(z) = \mathcal{Z}\{D'(s)\}$; $C(s) = L[c(t)]$ and $D'(s) = L[d'(t)]$

The input to the block $G_1(s)$ is an impulse sampled signal as shown in fig 5a. It's z-domain equivalent is shown in fig 5b.

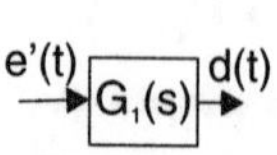

Fig 5a 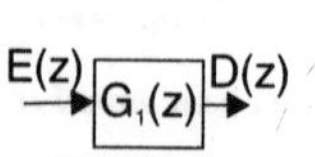   Fig 5b

From fig 5b we get, $D(z) = G_1(z)\, E(z) \qquad .....(3.12.9)$

From equations (3.12.8) and (3.12.9) we get,

$$C(z) = G_2(z)\, G_1(z)\, E(z) \qquad .....(3.12.10)$$

where, $G_1(z) = Z\{G_1(s)\}$; $E(z) = Z\{E'(s)\}$ and $E'(s) = L[e'(t)]$

The input to the cascaded blocks $G_2(s)$ and $H(s)$ is an impulse sampled signal as shown in fig 6a. It's z-domain equivalent is shown in fig 6b.

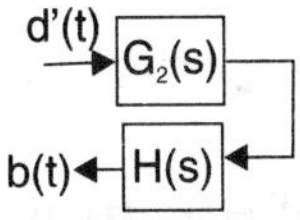

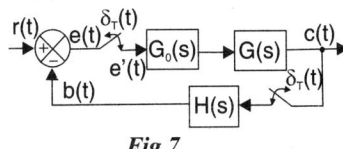

Fig 6b

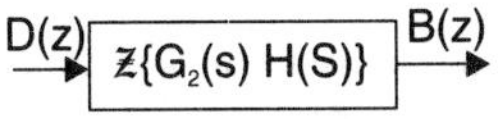

Fig 6a

From fig 6b we get,

$$B(z) = Z\{G_2(s)\ H(s)\}\ D(z) \qquad .....(3.12.11)$$

On substituting for D(z) from equation (3.12.9) in equation (3.12.11) we get,

$$B(z) = \mathbf{Z}\{G_2(s)\ H(s)\}\ G_1(z)\ E(z) \qquad .....(3.12.12)$$

With reference to fig 1b, at the summing point we get,

$$e(t) = r(t) - b(t) \qquad .....(3.12.13)$$

Since $e'(t) = e(kT)$ is an impulse sampled signal, by superposition principle the equation (3.12.13) can be written as,

$$e(kT) = r(kT) - b(kT) \qquad .....(3.12.14)$$

where, e(kT), r(kT) and b(kT) are impulse sampled signals of e(t), r(t) and b(t) respectively.

On taking $\mathbf{Z}$-transform of equation (3.12.14) we get,

$$E(z) = R(z) - B(z)$$

$$\therefore R(z) = E(z) + B(z) \qquad .....(3.12.15)$$

On substituting for B(z) from equation (3.12.12) in equation (3.12.15) we get,

$$R(z) = E(z) + \mathbf{Z}\{G_2(s)\ H(s)\}\ G_1(z)\ E(z)$$

$$= [1 + \mathbf{Z}\{G_2(s)\ H(s)\}\ G_1(z)]\ E(z) \qquad .....(3.12.16)$$

From equations (3.12.10) and (3.12.16) the z-transfer function or pulse transfer function C(z)/R(z) can be written as,

$$\frac{C(z)}{R(z)} = \frac{G_1(z)\ G_2(z)}{1 + \mathbf{Z}\{G_2(s)\ H(s)\ G_1(z)\}} = \frac{G_1(z)\ G_2(z)}{1 + G_2H(z)G_1(z)} \qquad .....(3.12.17)$$

Here $\mathbf{Z}\{G_2(s)\ H(s)\}$ is denoted as $G_2H(z)$

(c)   **The ZOH in the system is replaced by $G_0(s)$ as shown in fig 7, where $G_0(s) = (1 - e^{-sT})/s$.**

Fig 7

The input to the cascaded blocks of $G_0(s)$ and $G(s)$ is an impulse sampled signal as shown in fig 8a. It's z-domain equivalent is shown in fig 8b.

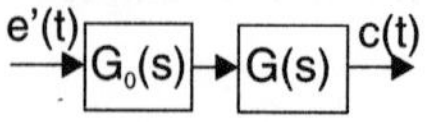

Fig 8a 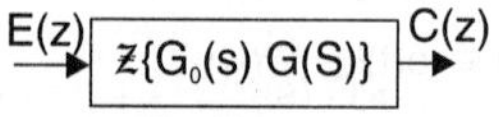   Fig 8b

From 8b, we get, $C(z) = Z\{G_0(s)\ G(s)\}\ E(z)$ .....(3.12.18)

where, $C(z) = Z\{C(s)\}$; $E(z) = Z\{E'(s)\}$; $C(s) = L[c(t)]$ and $E'(s) = L[e'(t)]$.

The input to the block $H(s)$ is an impulse sampled signal as shown in fig 9a. It's z-domain equivalent is shown in fig 9b.

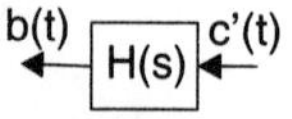

Fig 9a 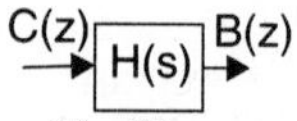   Fig 9b

From fig 9b, we get, $B(z) = H(z)\ C(z)$ .....(3.12.19)

With reference to fig 7, at the summing point we get,

$$e(t) = r(t) - b(t)$$ .....(3.12.20)

Since $e'(t) = e(kT)$ is an impulse sampled signal, by principle of superposition the equation (3.12.20) can be written as,

$$e(kT) = r(kT) - b(kT)$$ .....(3.12.21)

where, $e(kT)$, $r(kT)$ and $b(kT)$ are impulse sampled signals of $e(t)$, $r(t)$ and $b(t)$ respectively.

On taking Z-transform of equation (3.12.21) we get,

$$E(z) = R(z) - B(z)$$ .....(3.12.22)

On substituting for $B(z)$ from equation (3.12.19) in equation (3.12.22) we get,

$$E(z) = R(z) - H(z)\ C(z)$$ .....(3.12.23)

On substituting for $E(z)$ from equation (3.12.23) in equation (3.12.18) we get,

$$C(z) = Z\{G_0(s)\ G(s)\}\ [R(z) - H(z)\ C(z)]$$

$$C(z) = Z\ Z\{G_0(s)\ G(s)\}\ R(z) - Z\{G_0(s)\ G(s)\}\ H(z)\ C(z)$$

$$C(z) + Z\{G_0(s)\ G(s)\}\ H(z)\ C(z) = Z\{G_0(s)\ G(s)\}\ R(z)$$

$$C(z)\ [1 + Z\{G_0(s)\ G(s)\}\ H(z)\ ] = Z\{G_0(s)\ G(s)\}\ R(z)$$

$$\therefore \frac{C(z)}{R(z)} = \frac{Z\{G_0(s)\ G(s)\}}{1 + Z\{G_0(s)\ G(s)H(z)\}} = \frac{G_0G(z)}{1 + G_0G(z)H(z)}$$ .....(3.12.24)

The equation (3.12.24) is the z-transfer function of the system.

Here $Z\{G_0(s)\ G(s)\}$ is denoted as $G_0G(z)$.

EXAMPLE 3.13

Find the output C(z) in z-domain for the closed loop sampled data control system shown in fig 3.13.1.

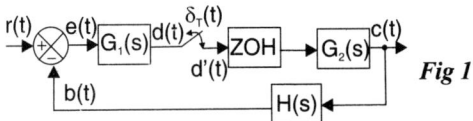

Fig 1

SOLUTION

The ZOH in fig 1 is replaced by a block with transfer function $G_0(s)$ as shown in fig 2 where $G_0(s) = (1 - e^{-sT})/s$.

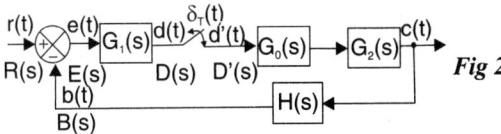

Fig 2

Here, d'(t) = Impulse sampled signal of d(t).

The input to the cascaded blocks of $G_0(s)$ and $G_2(s)$ is an impulse sampled signal as shown in fig 2a. It's z-domain equivalent is shown in fig 2b.

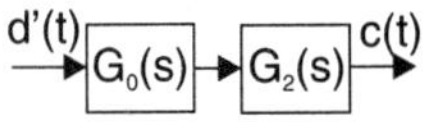

$$\xrightarrow{d'(t)} \boxed{G_0(s)} \rightarrow \boxed{G_2(s)} \xrightarrow{c(t)} \qquad \xrightarrow{D(z)} \boxed{Z\{G_0(s)\,G_2(S)\}} \xrightarrow{C(z)}$$

Fig 2a 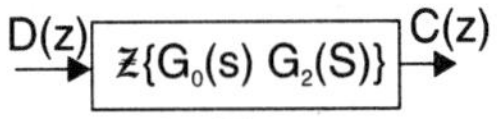   *Fig 2b*

From fig 2b we get, $C(z) = Z\{G_0(s)\,G_2(s)\}\,D(z)$ .....(3.13.1)

where, $C(z) = Z\{C(s)\}$; $D(z) = Z\{D'(s)\}$; $C(s) = \mathcal{L}[c(t)]$ and $D'(s) = \mathcal{L}[d'(t)]$.

With reference to fig 1 the following s-domain equations can be obtained.

$\qquad E(s) = R(s) - B(s)$ .....(3.13.2)

$\qquad D(s) = E(s)\,G_1(s)$ .....(3.13.3)

$\qquad B(s) = G_0(s)\,G_2(s)\,H(s)\,D'(s)$ .....(3.13.4)

On substituting for E(s) from equation (3.13.2) in equation (3.13.3) we get,

$\qquad D(s) = [R(s) - B(s)]\,G_1(s) = G_1(s)\,R(s) - G_1(s)\,B(s)$ .....(3.13.5)

On substituting for B(s) from equation (3.13.4) in equation (3.13.5) we get,

$\qquad D(s) = G_1(s)\,R(s) - G_1(s)\,G_0(s)\,G_2(s)\,H(s)\,D'(s)$ .....(3.13.6)

On taking Z-transform of equation (3.13.6) we get,

$\qquad D(z) = Z\{G_1(s)\,R(s)\} - Z\{G_1(s)\,G_0(s)\,G_2(s)\,H(s)\}\,D(z)$

$\qquad D(z) + Z\{G_0(s)\,G_1(s)\,G_2(s)\,H(s)\}\,D(z) = Z\{G_1(s)\,R(s)\}$

$\qquad D(z)\,[1 + Z\{G_0(s)\,G_1(s)\,G_2(s)\,H(s)\}] = Z\{G_1(s)\,R(s)\}$

$$\therefore D(z) = \frac{Z\{G_1(s)R(s)\}}{1 + Z\{G_0(s)G_1(s)G_2(s)H(s)\}} \qquad .....(3.13.7)$$

> **Note :** *The term $G_0(s)\, G_1(s)\, G_2(s)\, H(s)\, D'(s)$ represents the output of a block with transfer function $G_0(s)\, G_1(s)\, G_2(s)\, H(s)$ when the input is $D'(s)$.*

On substituting for D(z) from equation (3.13.7) in equation (3.13.1) we get,

Output in z-domain, $C(z) = \dfrac{Z\{G_0(s)G_2(s)\}\, Z\{G_1(z)G_1R(z)\}}{1 + Z\{G_0(s)G_1(s)G_2(s)H(s)\}} = \dfrac{G_0G_2(z)G_1R(z)}{1 + G_0G_1G_2H(z)}$

where, $Z\{G_0(s)\, G_2(s)\}$ is represented as $G_0G_2(z)$

$Z\{G_1(s)\, R(s)\}$ is represented as $G_1R(z)$ and

$Z\{G_0(s)\, G_1(s)\, G_2(s)\, H(s)\}$ is represented as $G_0G_1G_2H(z)$

EXAMPLE 3.14

For the sampled data control system shown in fig 1, find the response to unit step input, where $G(s) = 1/(s+1)$.

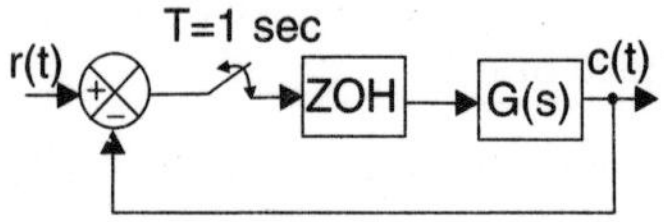

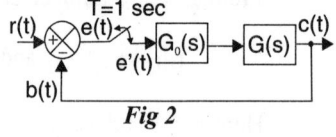

Fig 1

SOLUTION

The ZOH in the system is replaced by $G_0(s)$ as shown in fig 1, where $G_0(s) = (1 - e^{-sT})/s$.

The input to the cascaded blocks of $G_0(s)$ and $G(s)$ is an impulse sampled signal as shown in fig 2a. It's z-domain equivalent is shown in fig 2b.

Fig 2

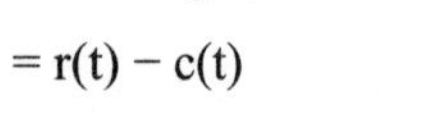

Fig 2a

Fig 2b

From fig 2b we get, $C(z) = Z\{G_0(s)\, G(s)\}\, E(z)$ ......(3.14.1)

With reference to fig 1, at the summing point we get,

$e(t) = r(t) - c(t)$ ......(3.14.2)

Since $e'(t) = e(kT)$ is an impulse sampled signal, the equation (3.14.2) can be written as,

$e(kT) = r(kT) - c(kT)$ ......(3.14.3)

where, $e(kT)$, $r(kT)$ and $c(kT)$ are impulse sampled signals of $e(t)$, $r(t)$ and $c(t)$ respectively.

On taking Z-transform of equation (3.14.3) we get,

$E(z) = R(z) - C(z)$ ......(3.14.4)

On substituting for E(z) from equation (3.14.4) in equation (3.14.1) we get,

$C(z) = Z\{G_0(s)\, G(s)\}\, [R(z) - C(z)]$

$$C(z) = \mathcal{Z}\{G_0(s)\,G(s)\}\,R(z) - \mathcal{Z}\{G_0(s)\,G(s)\}\,C(z)$$

$$C(z) + \mathcal{Z}\,Z\{G_0(s)\,G(s)\}\,C(z) = \mathcal{Z}\{G_0(s)\,G(s)\}\,R(z)$$

$$C(z)\,[1 + \mathcal{Z}\{G_0(s)\,G(s)\}] = \mathcal{Z}\{G_0(s)\,G(s)\}\,R(z)$$

$$\therefore C(z) = \frac{\mathcal{Z}\{G_0(s)G(s)R(z)\}}{1 + \mathcal{Z}\{G_0(s)G(s)\}} \qquad\qquad .....(3.14.5)$$

We know that, $\mathcal{Z}\{G_0(s)G(s)\} = (1 - z^{-1})\,\mathcal{Z}\left\{\dfrac{G(s)}{s}\right\}$

Here, $G(s) = \dfrac{1}{s+1}$ and $\dfrac{G(s)}{s} = \dfrac{1}{s(s+1)}$

By partial fraction expansion,

$$\frac{G(s)}{s} = \frac{1}{s(s+1)} = \frac{A}{s} + \frac{B}{s+1}$$

$$A = \frac{1}{s(s+1)}\,s\bigg|_{s=0} = \frac{1}{s+1}\bigg|_{s=0} = 1$$

$$B = \frac{1}{(s+1)}\,(s+1)\bigg|_{s=-1} = \frac{1}{s}\bigg|_{s=-1} = -1$$

$$\mathcal{Z}\left\{\frac{G(s)}{s}\right\} = \mathcal{Z}\left\{\frac{1}{s} - \frac{1}{s+1}\right\} = \mathcal{Z}\left\{\frac{1}{s}\right\} - \mathcal{Z}\left\{\frac{1}{s+1}\right\}$$

From standard Laplace and $\mathcal{Z}$-transform pairs we get,

$$\mathcal{Z}\left\{\frac{1}{s}\right\} = \frac{z}{z-1} \quad \text{and} \quad \mathcal{Z}\left\{\frac{1}{s+a}\right\} = \frac{z}{z - e^{-aT}}$$

Here, $a = 1$ and $T = 1$

$$\therefore \mathcal{Z}\left\{\frac{G(s)}{s}\right\} = \frac{z}{z-1} - \frac{z}{z - e^{-T}}$$

Now, $\mathcal{Z}\{G_0(s)G(s)\} = (1 - z^{-1})\mathcal{Z}\left\{\dfrac{G(s)}{s}\right\} = (1 - z^{-1})\left(\dfrac{z}{z-1} - \dfrac{z}{z - e^{-1}}\right)$

$$= \left(1 - \frac{1}{z}\right)\left(\frac{z(z - e^{-1}) - z(z-1)}{(z-1)(z - e^{-1})}\right)$$

$$= \left(\frac{z-1}{z}\right)\left(\frac{z(z - e^{-1} - z + 1)}{(z-1)(z - e^{-1})}\right) = \frac{1 - e^{-1}}{z - e^{-T}} = \frac{0.632}{z - 0.368} \qquad .....(3.14.6)$$

Given that input is unit step

$$\therefore R(z) = U(z) = \frac{z}{z-1} \qquad\qquad .....(3.14.7)$$

From equations (3.14.5), (3.14.6) and (3.14.7) we get,

$$C(z) = \frac{\mathcal{Z}\{G_0(s)G(s)R(z)\}}{1 + \mathcal{Z}\{G_0(s)G(s)\}} = \frac{\left(\dfrac{0.632}{z - 0.368}\right)\dfrac{z}{z-1}}{1 + \left(\dfrac{0.632}{z - 0.368}\right)}$$

$$= \frac{\dfrac{0.632z}{(z-1)(z-0.368)}}{\dfrac{(z-0.368)+0.632}{(z-0.368)}} = \frac{0.632z}{(z-1)(z-0.368+0.632)} = \frac{0.632z}{(z-1)(z+0.264)}$$

$$\therefore \frac{C(z)}{z} = \frac{0.632}{(z-1)(z+0.264)}$$

By partial fraction expansion,

$$\frac{C(z)}{z} = \frac{A}{z-1} + \frac{B}{z+0.264}$$

$$A = \frac{0.632}{(z-1)(z+0.264)}(z-1)\bigg|_{z=1} = \frac{0.632}{z+0.264}\bigg|_{z=1} = \frac{0.632}{1+0.264} = 0.5$$

$$B = \frac{0.632}{(z-1)(z+0.264)}(z+0.264)\bigg|_{z=-0.264} = \frac{0.632}{z-1}\bigg|_{z=-0.264} = \frac{0.632}{-0.264-1} = -0.5$$

$$\therefore \frac{C(z)}{z} = \frac{0.5}{z-1} - \frac{0.5}{z+0.264}$$

$$C(z) = 0.5\frac{z}{z-1} - 0.5\frac{z}{z-(-0.264)} \qquad\qquad .....(3.14.8)$$

We know that,

$$\mathcal{Z}\{1\} = \frac{z}{z-1} \text{ and } \mathcal{Z}\{a^k\} = \frac{z}{z-a}$$

On taking inverse $\mathcal{Z}$-transform of equation (3.14.8) we get,

$$c(k) = 0.5 - 0.5\,(-0.264)^k = 0.5\,[1-(-0.264)^k] \qquad\qquad .....(3.14.9)$$

The equation (3.14.9) is the response of given system for unit step input.

3.12 THE z AND s - DOMAIN RELATIONSHIP

Let r(kT) be a discrete sequence which has been obtained by sampling r(t) at a sampling rate of 1/T. On taking $\mathcal{Z}$-transform of r(kT) we get,

$$\mathcal{Z}\{r(kT)\} = R(z) = \sum_{k=0}^{\infty} r(kT)\,z^{-k} \qquad\qquad .....(3.69)$$

Let, r(t) = Impulse sampled signal of r(t) at the sampling rate of 1/T.

and R'(s) = $\mathcal{L}[r'(t)]$ = Laplace transform of r'(t).

Now, $$r'(t) = \sum_{k=0}^{\infty} r(kT)\,\delta(t-kT) \qquad\qquad .....(3.70)$$

On taking laplace transform of equation (3.70) we get,

$$R'(s) = \sum_{k=0}^{\infty} r(kT)\,e^{-ksT} \qquad\qquad .....(3.71)$$

Let us choose a transformation such that,

$$z = e^{sT} \qquad\qquad .....(3.72)$$

$$\therefore \; ln\, z = sT \quad \text{(or)} \quad s = \frac{1}{T} ln\, z \qquad\qquad .....(3.73)$$

On substituting for s from equation (3.73) in equation (3.71) we get,

$$R'(s) = \sum_{k=0}^{\infty} r(kT)\, e^{(-kT\frac{1}{T} ln z)}$$

$$= \sum_{k=0}^{\infty} r(kT)\, e^{(ln\, z^{-k})} = \sum_{k=0}^{\infty} r(kT)\, z^{-k} = R(z) \qquad\qquad .....(3.74)$$

From equation (3.74) it is obvious that z-transform of a discrete sequence can be obtained from the Laplace transform of its impulse sampled version, by choosing a transformation, s = (1/T) *ln* z (or z = e^{sT}).

The transformation, s = (1/T) *ln* z, maps the s-plane into the z-plane. It can be shown that every section of jω-axis of length, Nω, maps into the unit circle in the anticlockwise direction where N is an integer and ω_s is the sampling frequency and it can be shown that every strip in the left half s-plane of width ω_s, maps into the interior of the unit circle as shown in fig 3.31.

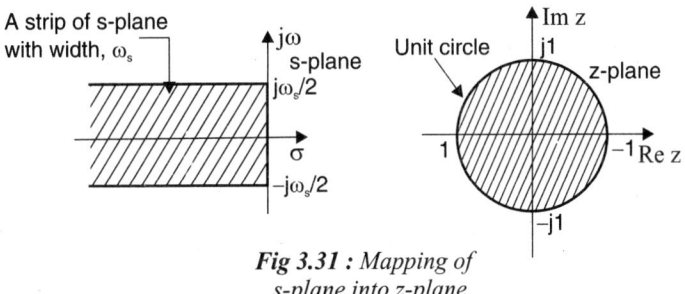

Fig 3.31 : *Mapping of s-plane into z-plane*

The above mapping helps in extending the s-plane stability criterion to z-plane. For stability of a system in s-plane the poles of s-domain transfer function should lie on the left half of s-plane. In this transformation the left half of s-plane maps into interior of unit circle. Hence for the stability of the system in z-domain, the poles of the z-transfer function should lie inside the unit circle.

3.13 STABILITY ANALYSIS OF SAMPLED DATA CONTROL SYSTEMS

The sampled data control system is stable if all the poles of the z-transfer function of the system lies inside the unit circle in z-plane.

The poles of the transfer function are given by the roots of the characteristic equation. Hence the system stability can be determined from the roots of the characteristic equation.

The z-transfer function of the sampled data control system can be expressed as a ratio of two polynomials in z as shown below.

$$\text{z-transfer function, } H(z) = \frac{C(z)}{R(z)} = A_0 \frac{P(z)}{Q(z)} \qquad\qquad .....(3.75)$$

where, A_0 = constant

P(z) = Numerator polynomial

Q(z) = Denominator polynomial

The characteristic equation is the denominator polynomial of H(z). [i.e., characteristic equation is given by Q(z) = 0].

Consider the system shown in fig 3.32. For this system, the z-transfer function is given by,

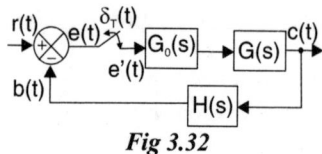

Fig 3.32

$$H(z) = \frac{C(z)}{R(z)} = \frac{\mathbf{Z}\{G_0(s)G(s)\}}{1 + \mathbf{Z}\{G_0(s)G(s)H(s)\}} \qquad \text{.....(3.76)}$$

and the characteristic equation is,

$$1 + \mathbf{Z}\{G_0(s)\ G(s)\ H(s)\} = 0 \qquad \text{.....(3.77)}$$

The following methods are available for the stability analysis of sampled data control systems using the characteristic equation

1. *Jury's stability test*

2. *Bilinear transformation*

3. *Root locus technique*

The Jury's stability test and Bilinear transformation are presented in this book.

JURY'S STABILITY TEST

The Jury's stability test is used to determine whether the roots of the characteristic polynomial lie within a unit circle or not. The Jury's test consists of two parts. One simple test for necessary condition for stability and another test for sufficient condition for stability.

Let F(z) be the nth order characteristic polynomial of a sampled data control system.

$$F(z) = a_n z^n + a_{n-1}z^{n-1} + a_{n-2}z^{n-2} + \ldots\ldots + a_2 z^2 + az + a_0 = 0 \qquad \text{.....(3.78)}$$

where, $a_n > 0$ and $a_0, a_1, a_2, \ldots\ldots a_n$ are constant coefficients.

The necessary conditions to be satisfied for the stability of the system with characteristic polynomial, F(z) are

$$\textbf{F(1)} > \textbf{0 and } (-1)^n \textbf{ F(-1)} > \textbf{0} \qquad \text{.....(3.79)}$$

The sufficient condition for stability can be established through any one of the following two methods.

Method-1 for testing sufficiency

In this method prepare a table as shown below using the coefficients of the characteristic polynomial $F(z)$. The table consists of $(2n - 3)$ rows, where n is the order of the characteristic equation.

Row	z^0	z^1	z^2	......	z^{n-k}	.....	z^{n-2}	z^{n-1}	z^n
1	a_0	a_1	a_2	......	a_{n-k}	.....	a_{n-2}	a_{n-1}	a_n
2	a_n	a_{n-1}	a_{n-2}	......	a_k	.....	a_2	a_1	a_0
3	b_0	b_1	b_2	......	......	.....	b_{n-2}	b_{n-1}	
4	b_{n-1}	b_{n-2}	b_{n-3}	......	......	.....	b_1	b_0	
5	c_0	c_1	c_2	......	......	.....	c_{n-2}		
6	c_{n-2}	c_{n-3}	c_{n-4}	......	......	.....	c_0		
.	.	.	.						
.	.	.	.						
.	.	.	.						
$2n-5$	S_0	S_1	S_2	S_3					
$2n-4$	S_3	S_2	S_1	S_0					
$2n-3$	r_0	r_1	r_2						

In the above table the elements of row-1 are formed using the coefficients of characteristic polynomial and the row-2 is formed by arranging the elements of row-1 in the reverse order.

The kth element of row-3 is given by, $b_k = \begin{vmatrix} a_0 & a_{n-k} \\ a_n & a_k \end{vmatrix}$

$$\therefore b_0 = \begin{vmatrix} a_0 & a_n \\ a_n & a_0 \end{vmatrix} \quad ; \quad b_1 = \begin{vmatrix} a_0 & a_{n-1} \\ a_n & a_1 \end{vmatrix} \quad ; \quad b_2 = \begin{vmatrix} a_0 & a_{n-2} \\ a_n & a_2 \end{vmatrix} \text{ and so on}$$

The row-4 is formed by arranging the elements of row-3 in reverse order.

The kth element of row-5 is given by, $c_k = \begin{vmatrix} b_0 & b_{n-1-k} \\ b_{n-1} & b_k \end{vmatrix}$

$$\therefore c_0 = \begin{vmatrix} b_0 & b_{n-1} \\ b_{n-1} & b_0 \end{vmatrix} \quad ; \quad c_1 = \begin{vmatrix} b_0 & b_{n-2} \\ b_{n-1} & b_1 \end{vmatrix} \quad ; \quad c_2 = \begin{vmatrix} b_0 & b_{n-3} \\ b_{n-1} & b_2 \end{vmatrix} \text{ and so on}$$

The row-6 is formed by arranging the elements of row-5 in reverse order.

In general we can say that the elements of a row with odd row number are calculated using the elements of the two rows just above the concerned row. The row next to the odd numbered row is formed by arranging the elements of the odd numbered row in reverse order. For calculating the elements of a row the logic explained for b_k and c_k are followed.

The first column elements of the table are used to check the following $(n-1)$ conditions. These $(n-1)$ conditions are the sufficient conditions for stability of the system.

$$\left.\begin{array}{l} |a_0| < |a_n| \\ |b_0| > |b_{n-1}| \\ |c_0| > |c_{n-2}| \\ \quad\vdots \\ |r_0| > |r_2| \end{array}\right\} (n-1)\ \textbf{conditions}$$

.....(3.80)

If the necessary and sufficient conditions are satisfied then all the poles of the system lies inside the unit circle in z-plane and so the system is stable.

If even one of the condition is not satisfied then the system is unstable.

Method-2 for testing sufficiency

The coefficients of the characteristic polynomial of equation (3.78) are renamed as shown below.

$$F(z) = a_0 z^n + a_1 z^{n-1} + a_2 z^{n-2} + ... + a_{n-1} z + a_n = 0 \ ; \ a_0 > 0$$

.....(3.81)

Using the coefficients of equation (3.81), construct two square matrices X and Y of order $(n-1)$ as shown below.

$$X = \begin{bmatrix} a_0 & a_1 & a_2 & \cdots & a_{n-2} \\ 0 & a_0 & a_1 & \cdots & a_{n-3} \\ 0 & 0 & a_0 & \cdots & a_{n-4} \\ \vdots & \vdots & \vdots & & \vdots \\ 0 & 0 & 0 & & a_0 \end{bmatrix} \ ; \ Y = \begin{bmatrix} a_2 & a_3 & \cdots & a_{n-2} & a_{n-1} & a_n \\ a_3 & a_4 & \cdots & a_{n-1} & a_n & 0 \\ a_4 & a_5 & \cdots & a_n & 0 & 0 \\ \vdots & \vdots & & \vdots & \vdots & \vdots \\ a_n & 0 & \cdots & 0 & 0 & 0 \end{bmatrix}$$

By taking the sum and difference of the two matrices X and Y, construct two more matrices H_1 and H_2.

where $H_1 = X + Y$ .....(3.82)

and $H_2 = X - Y$ .....(3.83)

In this method the sufficient condition for stability is that the matrices H_1 and H_2 should be positive innerwise.

If the necessary and sufficient condition for stability are satisfied then the system is stable.

> **Note :** *The necessary conditions for stability are same in both the methods.*

A square matrix A is said to be positive innerwise when all the determinants starting with the centre element and proceeding outwards upto the entire matrix are positive. For example consider the matrix A shown below.

$$A = \begin{bmatrix} a_{11} & a_{12} & a_{13} & a_{14} & a_{15} \\ a_{21} & a_{22} & a_{23} & a_{24} & a_{25} \\ a_{31} & a_{32} & a_{33} & a_{34} & a_{35} \\ a_{41} & a_{42} & a_{43} & a_{44} & a_{45} \\ a_{51} & a_{52} & a_{53} & a_{54} & a_{55} \end{bmatrix}$$

The possible determinants of A starting with centre element are shown in dotted lines and these determinants are also shown below.

$$a_{33} \; ; \; \begin{vmatrix} a_{22} & a_{23} & a_{24} \\ a_{32} & a_{33} & a_{34} \\ a_{42} & a_{43} & a_{44} \end{vmatrix} \; ; \; \begin{bmatrix} a_{11} & a_{12} & a_{13} & a_{14} & a_{15} \\ a_{21} & a_{22} & a_{23} & a_{24} & a_{25} \\ a_{31} & a_{32} & a_{33} & a_{34} & a_{35} \\ a_{41} & a_{42} & a_{43} & a_{44} & a_{45} \\ a_{51} & a_{52} & a_{53} & a_{54} & a_{55} \end{bmatrix}$$

STABILITY ANALYSIS USING BILINEAR TRANSFORMATION

The bilinear transformation maps the interior of unit circle in the z-plane into the left half of the r-plane. In this transformation,

$$r = \frac{z-1}{z+1} \quad \text{(or)} \quad z = \frac{1+r}{1-r} \qquad \qquad .....(3.84)$$

The transformation is performed by substituting, in the characteristic equation of the system.

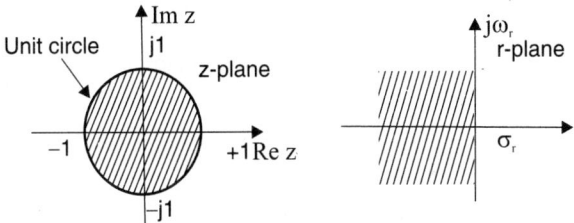

Fig 3.33 : Mapping of unit circle in z-plane
into left half of r-plane

Consider the characteristic equation of the system in z-plane shown below.

$$a_n z^n + a_{n-1} z^{n-1} + a_{n-2} z^{n-2} + ... + az + a_0 = 0 \; ; \; a_n > 0 \qquad .....(3.85)$$

Using bilinear transformation, the equation (3.85) can be written as,

$$\left(\text{i.e., by substituting } z = \frac{1+r}{1-r} \right)$$

$$a_n \left(\frac{1+r}{1-r} \right)^n + a_{n-1} \left(\frac{1+r}{1-r} \right)^{n-1} + a_{n-2} \left(\frac{1+r}{1-r} \right)^{n-2} + ... + a \left(\frac{1+r}{1-r} \right) + a_0 = 0 \qquad .....(3.86)$$

The equation (3.86) can be simplified and organised into the form

$$b_n r^n + b_{n-1} r^{n-1} + ... + b_1 r + b_0 = 0 \qquad .....(7.87)$$

Now the equation (3.87) is the new characteristic equation and for the stability of the system the roots of new characteristic equation should lie on the left half of r-plane. The routh stability criterion can be applied to the new characteristic equation to determine if all its roots lie in the left half of the r-plane.

In this method the necessary condition for stability is that the coefficients of equation (3.87) should be positive. The sufficient condition for stability is that there should not be any sign change in the elements of first column of routh array.

The system is stable if both the necessary and sufficient conditions for stability are satisfied.

EXAMPLE 3.15

Check for stability of the sampled data control systems represented by the following characteristic equation.

(a) $5z^2 - 2z + 2 = 0$

(b) $z^3 - 0.2 z^2 - 0.25 z + 0.05 = 0$

(c) $z^4 - 1.7 z^3 + 1.04 z^2 - 0.268 z + 0.024 = 0$

SOLUTION

(a)   **Given that, $F(z) = a_2 z^2 + a_1 z + a_0 = 5z^2 - 2z + 2$**

Check for necessary condition

$$F(z) = 5z^2 - 2z + 2$$

$$F(1) = 5(1)^2 - 2(1) + 2 = 5$$

$$(-1)^n F(-1) = (-1)^2 [5(-1)^2 - 2(-1)+2] = 9 \quad \text{(Here, n = 2)}$$

Since $F(1) > 0$ and $(-1)^n F(-1) > 0$; the necessary conditions for stability are satisfied.

Check for sufficient condition

The sufficient condition for stability can be checked by constructing a table consisting of $(2n - 3)$ rows as shown below.

Here, $n = 2$; $\therefore$ $(2n - 3) = 1$ and so the table consists of only one row.

Row	z^0	z^1	z^2
1	a_0	a_1	a_2

Here, $a_0 = 2$; $a_1 = -2$ and $a_2 = 5$

Row	z^0	z^1	z^2
1	2	-2	5

The necessary condition to be satisfied is $|a_0| < |a_2|$

Here, $|a_0| = 2$ and $|a_2| = 5$ and so the condition $|a_0| < |a_2|$ is satisfied.

CONCLUSION

The necessary and sufficient conditions for stability are satisfied. Hence the system is stable.

(b)   **METHOD-1 : Stability analysis by Jury's stability test**

Given that $F(z) = a_3 z^3 + a_2 z^2 + a_1 z + a_0 = z^3 - 0.2 z^2 - 0.25 z + 0.05 = 0$

Check for necessary conditon

$$F(z) = z^3 - 0.2 z^2 - 0.25 z + 0.05$$

$$F(1) = (1)^3 - 0.2 (1)^2 - 0.25 (1) + 0.05 = 1 - 0.2 - 0.25 + 0.05 = 0.6$$

$$(-1)^n F(-1) = (-1)^3 [(-1)^3 - 0.2(-1)^2 - 0.25(-1) + 0.05] \qquad (\text{Here } n = 3)$$

$$= (-1)[-1 - 0.2 + 0.25 + 0.05] = 0.9$$

Since $F(1) > 0$ and $(-1)^n F(-1) > 0$; the necessary conditions for stability are satisfied.

Check for sufficient condition

The sufficient condition for stability can be checked by constructing a table consisting of $(2n - 3)$ rows as shown below. Here, $n = 3$, $\therefore (2n - 3) = 3$ and so the table consists of three rows.

Row	z^0	z^1	z^2	z^3
1	a_0	a_1	a_2	a_3
2	a_3	a_2	a_1	a_0
3	b_0	b_1	b_2	

Here, $a_0 = 0.05$; $a_1 = -0.25$; $a_2 = -0.2$ and $a_3 = 1$

$$b_0 = \begin{vmatrix} a_0 & a_3 \\ a_3 & a_0 \end{vmatrix} = \begin{vmatrix} 0.05 & 1 \\ 1 & 0.05 \end{vmatrix} = 0.05^2 - 1 = -0.9975$$

$$b_1 = \begin{vmatrix} a_0 & a_2 \\ a_3 & a_1 \end{vmatrix} = \begin{vmatrix} 0.05 & -0.2 \\ 1 & -0.25 \end{vmatrix} = 0.05 \times (-0.25) - 1 \times (-0.2) = 0.1875$$

$$b_2 = \begin{vmatrix} a_0 & a_1 \\ a_3 & a_2 \end{vmatrix} = \begin{vmatrix} 0.05 & -0.25 \\ 1 & -0.2 \end{vmatrix} = 0.05 \times (-0.2) - 1 \times (-0.25) = 0.24$$

Row	z^0	z^1	z^2	z^3
1	0.05	−0.25	−0.2	1
2	1	−0.2	−0.25	0.05
3	−0.9975	0.1875	0.24	

The necessary condition to be satisfied are, $|a_0| < |a_3|$ and $|b_0| > |b_2|$

$|a_0| < |a_3| \Rightarrow |0.05| < |1|$ - satisfied

$|b_0| > |b_2| \Rightarrow |-0.9975| > |0.25|$ - satisfied

CONCLUSION

The necessary and sufficient conditions for stability are satisfied. Hence the system is stable.

METHOD-2 : Stability analysis by bilinear transformation

Given that, $F(z) = z^3 - 0.2 z^2 - 0.25 z + 0.05 = 0$

Let us choose the transformation, $z = \dfrac{1+r}{1-r}$

$$\therefore F(r) = \left(\frac{1+r}{1-r}\right)^3 - 0.2\left(\frac{1+r}{1-r}\right)^2 - 0.25\left(\frac{1+r}{1-r}\right) + 0.05 = 0$$

On multiplying throughout by $(1-r)^3$ we get,

$$(1-r)^3 - 0.2\,(1+r)^2\,(1-r) - 0.25(1+r)(1-r)^2 + 0.05(1-r)^3 = 0$$

$$(1+r)(1+r^2+2r) - 0.2(1+r)(1-r^2) - 0.25(1-r)(1-r^2) + 0.05(1-r)(1+r^2-2r) = 0$$

$$(1+r)(1+r^2+2r-0.2+0.2r^2) + (1-r)(-0.25+0.25r^2+0.05+0.05r^2-0.1r) = 0$$

$$(1+r)(1.2r^2+2r+0.8) + (1-r)(0.3r^2-0.1r-0.2) = 0$$

$$(1.2r^2+2r+0.8+1.2r^3+2r^2+0.8r) + (0.3r^2-0.1r-0.2-0.3r^3+0.1r^2+0.2r) = 0$$

$$(1.2r^3+3.2r^2+2.8r+0.8) + (-0.3r^3+0.4r^2+0.1r-0.2) = 0$$

$$0.9r^3 + 3.6r^2 + 2.9r + 0.6 = 0$$

The above equation is the new characteristic equation of the system. The coefficients of the new characteristic equation are positive. Hence the necessary condition for stability is satisfied.

The sufficient condition for stability can be determined by constructing routh array as shown below.

		Column-1				
r^3	:	0.9	2.9	......	Row-1	$r^1 : \dfrac{3.6 \times 2.9 - 0.9 \times 0.6}{3.6}$
r^2	:	3.6	0.6	......	Row-2	$r^1 : 2.75$
r^1	:	2.75		......	Row-3	$r^0 : \dfrac{2.75 \times 0.6 - 0 \times 3.6}{2.75}$
r^0	:	0.6		......	Row-4	$r^0 : 0.6$

It is observed that there is no sign change in the elements of first column of routh array. Hence the sufficient condition for stability is satisfied.

CONCLUSION

The necessary and sufficient condition for stability are satisfied. Hence the system is stable.

(c)   **Given that, $F(z) = a_4\,z^4 + a_3\,z^3 + a_2\,z^2 + a_1\,z^1 + a_0$**

$$= z^4 - 1.7z^3 + 1.04\,z^2 - 0.268\,z + 0.024$$

Check for necessary condition

$$F(z) = z^4 - 1.7z^3 + 1.04\,z^2 - 0.268\,z + 0.024$$

$$F(1) = 1 - 1.7 + 1.04 - 0.268 + 0.024 = 0.096$$

$$(-1)^n\,F(-1) = (-1)^4\,[(-1)^4 - 1.7\,(-1)^3 + 1.04\,(-1)^2 - 0.268\,(-1) + 0.024]$$

$$= 1 + 1.7 + 1.04 + 0.268 + 0.024 = 4.032 \quad \text{(Here } n = 4\text{)}$$

Here the condition $F(1) > 0$ and $(-1)^n\,F(-1) > 0$ are satisfied. Therefore the necessary conditions for stability are satisfied.

Check for sufficient condition

The sufficient condition for stability can be checked by constructing a table consisting of $(2n - 3)$ rows as shown below. Here, $n = 4$, $\therefore (2n - 3) = 5$ and so the table consists of five rows.

Row	z^0	z^1	z^2	z^3	z^4
1	a_0	a_1	a_2	a_3	a_4
2	a_4	a_3	a_2	a_1	a_0
3	b_0	b_1	b_2	b_3	
4	b_3	b_2	b_1	b_0	
5	c_0	c_1	c_2		

Here, $a_0 = 0.024$; $a_1 = -0.268$; $a_2 = 1.04$; $a_3 = -1.7$ and $a_4 = 1$

$$b_0 = \begin{vmatrix} a_0 & a_4 \\ a_4 & a_0 \end{vmatrix} = \begin{vmatrix} 0.024 & 1 \\ 1 & 0.024 \end{vmatrix} = 0.024^2 - 1 = -0.9994$$

$$b_1 = \begin{vmatrix} a_0 & a_3 \\ a_4 & a_1 \end{vmatrix} = \begin{vmatrix} 0.024 & -1.7 \\ 1 & -0.268 \end{vmatrix} = 0.024 \times (-0.268) - 1 \times (-1.7) = 1.6936$$

$$b_2 = \begin{vmatrix} a_0 & a_2 \\ a_4 & a_2 \end{vmatrix} = \begin{vmatrix} 0.024 & 1.04 \\ 1 & 1.04 \end{vmatrix} = 0.024 \times 1.04 - 1 \times 1.04 = -1.015$$

$$b_3 = \begin{vmatrix} a_0 & a_1 \\ a_4 & a_3 \end{vmatrix} = \begin{vmatrix} 0.024 & -0.268 \\ 1 & -1.7 \end{vmatrix} = 0.024 \times (-1.7) - 1 \times (-0.268) = 0.2272$$

$$c_0 = \begin{vmatrix} b_0 & b_3 \\ b_3 & b_0 \end{vmatrix} = \begin{vmatrix} -0.9994 & 0.2272 \\ 0.2272 & -0.9994 \end{vmatrix} = 0.9994^2 - 0.2272^2 = 0.9472$$

$$c_1 = \begin{vmatrix} b_0 & b_2 \\ b_3 & b_1 \end{vmatrix} = \begin{vmatrix} -0.9994 & -1.015 \\ 0.2272 & 1.6936 \end{vmatrix} = -0.9994 \times 1.6936 - 0.2272 \times (-1.015) = -1.462$$

$$c_2 = \begin{vmatrix} b_0 & b_1 \\ b_3 & b_2 \end{vmatrix} = \begin{vmatrix} -0.9994 & 1.6936 \\ 0.2272 & -1.015 \end{vmatrix} = -0.9994 \times (-1.015) - 0.2272 \times 1.6936 = 0.6296$$

Row	z^0	z^1	z^2	z^3	z^4
1	0.024	−0.268	1.04	−1.7	1.0
2	1.0	−1.7	1.04	−0.268	0.024
3	−0.9994	1.6936	−1.015	0.2272	
4	0.2272	−1.015	1.6936	−0.9994	
5	0.9472	−1.462	0.6296		

The (n -1) conditions to be satisfied are,

$$|a_0| < |a_4|$$

$$|b_0| > |b_3|$$

$$|c_0| > |c_2|$$

$|a_0| < |a_4| \implies |0.024| < |1|$ - satisfied

$|b_0| > |b_3| \implies |-0.9994| > |0.2272|$ - satisfied

$|c_0| > |c_2| \implies |0.9472| > |0.6296|$ - satisfied

CONCLUSION

The necessary and sufficient conditions for stability are satisfied. Hence the system is stable.

3.14 SHORT-ANSWER QUESTIONS

Q3.1   *What is sampled data control system?*

When the signal or information at any or some points in a system is in the form of discrete pulses, then the system is called discrete data system or sampled data system.

Q3.2   *When the control system is called sampled data system?*

The control system becomes a sampled data system in any one of the following situations.

1. When a digital computer or microprocessor or digital device is employed as a part of the control loop.

2. When the control components are used on time sharing basis.

3. When the control signals are transmitted by pulse modulation.

4. When the output or input of a component in the system is digital or discrete signal.

Q3.3   *Draw the block diagram of a sampled data control system.*

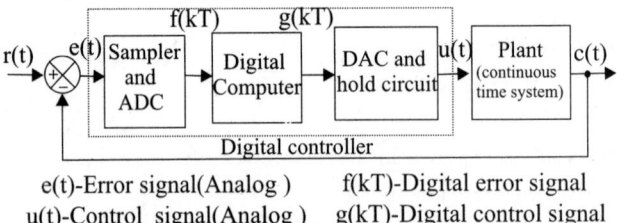

e(t)-Error signal(Analog)   f(kT)-Digital error signal
u(t)-Control signal(Analog)   g(kT)-Digital control signal
Fig Q3.3 : Sampled-data control system

Q3.4   *Distinguish between discrete time systems and continuous time systems.*

The discrete time systems are devices or algorithm that can process (or operate on) discrete-time signals, whereas the continuous time systems are devices that can process (or operator on) analog signals.The input and output signals of discrete-time systems are digital or discrete, but the input and output signals of continuous time systems are analog or continuous time signals.

Q3.5   *Write the advantages and disadvantages of sampled data control systems.*

Advantages of sampled data control system

1. Systems are highly accurate, fast and flexible.
2. Use of time sharing concept of digital computer results in economical cost and space.
3. Digital transducers used in the system have better resolution.
4. The digital components are less affected by noise, non-linearities and transmission errors of noisy channel.

Disadvantages of sampled data control system

1. Conversion of analog signals to discrete-time signals and reconstruction introduce noise and errors in the signal.
2. Additional filters have to be introduced in the system if the components of the system does not have adequate filtering characteristics.

Q3.6   *What is a digital controller?*

A digital controller is a device introduced in the control system to modify the error signal for better control action. The digital controller can be a special purpose computer (microprocessor based system) or a general purpose computer or it is constructed using non-programmable digital devices.

Q3.7   *Compare the analog and digital controller.*

Analog controller	Digital controller
1. Complex	1. Simple
2. Costlier than digital controller	2. Less costlier than analog controller.
3. Slow acting	3. Fast acting
4. Non-programmable	4. Programmable
5. Separate controller should be employed for each control signal	5. A single controller can be used to control more than one signal on time shared basis.

Q3.8   *What are the advantages of digital controllers.*

1. The digital controllers can perform large and complex computation with any desired degree of accuracy at very high speed.
2. The digital controllers are easily programmable and so they are more versatile.
3. Digital controllers have better resolution.

Q3.9   *Explain the terms sampling and sampler.*

Sampling is a process in which the continuous-time signal (or analog signal) is converted into a discrete-time signal by taking samples of the continuous time signal at discrete time instants. Sampler is a device which performs the process of sampling.

Q3.10   *What is periodic sampling?*

The periodic sampling is a sampling process in which the discrete time signal or sequence is obtained by taking samples of continuous time signal periodically or uniformly at intervals of T seconds. Here T is called sampling period and $1/T = F_s$ is called sampling frequency.

Q3.11   *State (shanon's) sampling theorem.*

Sampling theorem states that a bandlimited continuous-time signal with highest frequency f_m, hertz can be uniquely recovered from its samples provided that the sampling rate F_s is greater than or equal to $2f_m$ samples per second.

Q3.12   *What is meant by quantization?*

The process of converting a discrete-time continuous valued signal into a discrete-time discrete valued signal is called quantization. In quantization the value of each signal sample is represented by a value selected from a finite set of possible values called quantization levels.

Q3.13   *What is coding?*

The coding is the process of representing each discrete value by a n-bit binary sequence (or code or number).

Q3.14   *What are hold circuits?*

Hold circuits are devices used to convert discrete time signals to continuous time signals.

Q3.15   *What is zero-order hold?*

The zero-order hold is a hold circuit in which the signal is reconstructed such that the value of reconstructed signal for a sampling period is same as the value of last received sample. The signal reconstruction by zero order hold is shown in fig Q3.15.

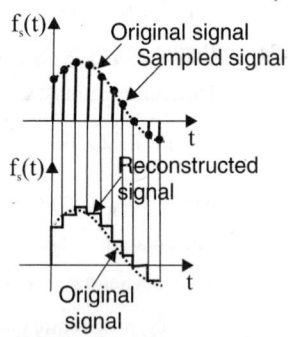

Q3.16   *What is first-order hold?*

The first-order hold is a hold circuit in which the last two signal samples (current and previous sample) are used to reconstruct the signal for the current sampling period. The reconstructed signal will be a straight line in a sampling period, whose slope is determined by the current sample and previous sample.

Fig Q3.15 : Signal reconstruction by ZOH

Q3.17   *Define Acquisition time.*

In analog-to-digital conversion process, the Acquisition time is defined as the total time required for obtaining a signal sample and the time for quantizing and coding. It is also called conversion time.

Q3.18   *Define Aperture time*

The duration of sampling the signal is called aperture time.

Q3.19   *Define settling time*

In digital-to-analog conversion process the settling time is defined as the time required for the output of the D/A converter to reach and remain within a given fraction of the final value, after application of input code word.

Q3.20   *What is "Hold mode droop"?*

The changes in signal voltage level in the hold circuits during hold mode (or hold period) is called hold mode droop.

Q3.21 **What are the problems encountered in a practical hold circuit?**

The problems encountered in practical hold circuit are

1. Errors in the periodicity of sampling process.

2. Nonlinear variations in the duration of sampling aperture.

3. Droop (changes) in the voltage held during conversion.

Q3.22 **How the high frequency noise signals in the reconstructed signal are eliminated?**

The high frequency noise signals (or unwanted signals) introduced by hold circuits in the reconstructed signal are easily filtered out by the various elements of the control system, because the control system is basically a low-pass filter.

Q3.23 **What is discrete sequence?**

A discrete sequence or discrete time signal, f(k) is a function of independent variable, k, which is an integer. A two sided discrete-time signal f(k) is defined for every integer value of k in the range $-\infty < k < \infty$.

A one sided causal discrete-time signal f(k) is defined for every integer value of k in the range of $0 \le k < \infty$.

A one sided anticausal discrete-time signal f(k) is defined for every integer value of k in the range $-\infty < k \le 0$.

Q3.24 **Define one sided and two-sided Z-transform.**

The Z-transform (two-sided Z-transform) of a discrete sequence, f(k) is defined as the power series,

$$F(z) = \mathcal{Z}\{f(k)\} = \sum_{k=0}^{\infty} f(k)z^{-k}$$

where, z is a complex variable.

The notation $\mathcal{Z}\{f(k)\}$ is used to denote Z-transform of $\hat{f}(k)$.

The one sided Z-transform of f(k) is defined as the power series,

$$F(z) = \mathcal{Z}\{f(k)\} = \sum_{k=0}^{\infty} f(k)z^{-k}$$

where, z is a complex variable.

Q3.25 **What is region of convergence (ROC)?**

The Z-transform of a discrete sequence is an infinite power series, hence the Z-transform exists only for those values of z for which the series converges. If F(z) is Z-transform of f(k) then the region of convergence (ROC) of F(z) is the set of all values of z, for which F(z) attains a finite value.

Q3.26 **Write the infinite and finite geometric series sum formula.**

The infinite geometric series sum formula is

$$\sum_{k=0}^{\infty} C^k = \frac{1}{1-C} \quad ; \text{ where } |C| < 1$$

The finite geometric series sum formula is

$$\sum_{k=0}^{M-1} C^k = \frac{C^M - 1}{C - 1} \quad ; \text{ when } C \ne 1$$

$$= M \quad ; \text{ when } C = 1$$

Q3.27   *State the final value theorem with regard to Z-transform.*

If f(k) is causal & stable signal and F(z) exists with z = 1 included in the ROC then the final value theorem is given by

$$f(\infty) = \underset{z \to 1}{Lt}\ (1 - z^{-1})F(z) \qquad ; \quad \text{where } F(z) = Z\{f(k)\}$$

The final value theorem can be applied only if F(z) is analytic for $|z| > 1$.

Q3.28   *State the initial value theorem with regard to Z-transform.*

If f(k) is a causal signal and F(z) exists then the initial value of the signal is given by

$$f(0) = \underset{z \to \infty}{Lt}\ F(z) \qquad ; \quad \text{where } F(z) = Z\{f(k)\}$$

Q3.29   *Define Z-transform of unit step signal.*

The unit step signal, u(k) = 1 for k ≥ 0

The Z-transform at u(k) $= Z\{u(k)\} = \sum\limits_{k=0}^{\infty} z^{-k} = \dfrac{z}{z-1}$

Q3.30   *Find Z-transform of a^k.*

By the definition of z-transform,

$$Z\{a^k\} = \sum_{k=0}^{\infty} a^{-k} z^{-k} = \sum_{k=0}^{\infty} (az^{-1})^k = \frac{1}{1 - az^{-1}} = \frac{1}{1 - a/z} = \frac{z}{z-a}$$

Q3.31   *Find Z-transform of e^{-akT}.*

By the definition of Z-transform,

$$Z\{a^{-akT}\} = \sum_{k=0}^{\infty} a^{-akT} z^{-k} = \sum_{k=0}^{\infty} (e^{-aT} z^{-1})^k = \frac{1}{1 - e^{-aT} z^{-1}} = \frac{1}{1 - e^{-aT}/z} = \frac{z}{z - e^{-aT}}$$

Q3.32   *What are the different methods available for inverse Z-transform?*

The inverse Z-transform of a function, F(z) can be obtained by any one of the following methods.

1.   Direct evaluation by contour integration (or) complex inversion integral.

2.   Partial fraction expansion.

3.   Power series expansion.

Q3.33   *What is linear (time-invariant) discrete time system (LDS)?*

A discrete-time system is a device or algorithm that operates on a discrete-time signal called the input or excitation, according to some well-defined rule, to produce another discrete-time signal called the output or the response of the system.

A discrete time system is linear if it obeys the principle of superposition and it is time invariant if its input-output relationship do not change with time.

Q3.34   *What is weighing sequence?*

The impulse response of a linear discrete-time system is called weighing sequence. The impulse response is the output of the system when the input is unit impulse.

Q3.35   **Explain how a discrete-time signal can be expressed as a summation of impulses.**

Multiplication of a discrete signal, r(k) by an unit impulse at some delay m, picks out the value r(m) of the signal r(k) at k = m, where the unit impulse is non-zero. If we repeat this multiplication over all possible delays in the range of $0 \le m < \infty$ and sum all the product sequences, the result will be a sequence that is equal to the sequence r(k). Hence r(k) can be expressed as

$$r(k) = \sum_{m=0}^{\infty} r(m)\, \delta(k-m)$$

Q3.36   **How the output of a linear discrete-time system (LDS) is related to impulse response?**

The output or response c(k) of a linear discrete time system (LDS) is given by convolution of the input r(k) with the impulse response h(k) of the system. It is expressed as,

$$c(k) = r(k) * h(k)$$

where, the symbol * represents convolution operation.

Q3.37   **What is discrete convolution ?**

The convolution of two discrete-time signals (or sequences) is called discrete convolution. The discrete convolution of sequences $f_1(k)$ and $f_2(k)$ is defined as

$$c(k) = \sum_{m=0}^{\infty} f_1(m)\, f_2(k-m)$$

Q3.38   **Write any two properties of discrete convolution.**

The discrete convolution obeys the commutative property and Associative property.

Commutative property : $r(k) * h(k) = h(k) * r(k)$

Associative property : $[r(k) * h_1(k)] * h_2(k) = r(k) * [h_1(k) * h_2(k)]$

Q3.39   **What is pulse transfer function?**

The transfer function of linear discrete time system is called pulse transfer function or z-transfer function.

It is given by the Z-transform of the impulse response of the system. It is also defined as the ratio of Z-transform of output to Z-transform of input of the linear discrete time system.

Pulse transfer function = H(z) = C(z)/R(z)

where, $H(z) = \mathcal{Z}\{h(k)\}$ = Z-transform of impulse response, h(k)

              C(z) = Z - transform of output of LDS

              R(z) = Z-transform of input to LDS.

Q3.40   **What is the equivalent representation of pulse sampler with ZOH?**

The pulse sampler with ZOH shown in fig Q7.40a can be replaced by an equivalent system consisting of an impulse sampler and a block with transfer function, $G_0(s) = (1 - e^{-sT})/s$ as shown in fig Q3.40b.

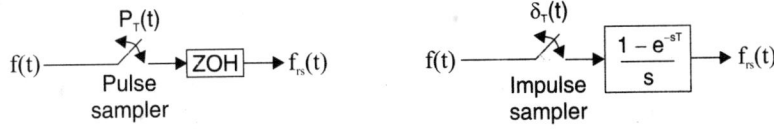

Fig Q3.40 a : Pulse sampler with ZOH.   Fig Q3.40 b : Equivalent representation
                                                of Pulse sampler with ZOH.

Q3.41 *Sketch the frequency response curve of ZOH device.*

The frequency response characteristic curve of ZOH device is shown in fig Q3.41.

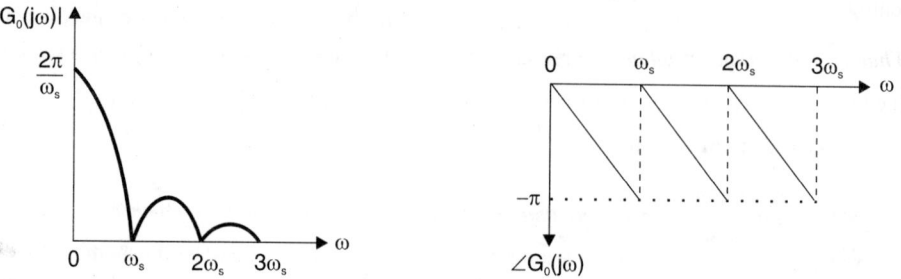

Fig Q3.41a : *Magnitude response of ZOH device.*   **Fig Q3.41b** : *Phase response of ZOH device.*

Fig Q3.41 : *Frequency response of ZOH device.*

Q3.42 *When the z-transfer function of the system can be directly obtained from s-domain transfer function?*

When the input to the system is an impulse sampled signal, the z-transfer function can be directly obtained by taking Z-transform of the s-domain transfer function.

Q3.43 *Give the steps involved in determining the pulse transfer function of G(z) from G(s).*

The following are the steps involved in determing the pulse transfer function G(z) from G(s).

1. Determine g(t) from G(s), where $g(t) = L^{-1} [G(s)]$
2. Determine the discrete sequence g(kT) by replacing t by kT in g(kT).
3. Take Z-transform of g(kT), which is the required z-transfer function, G(z).

Q3.44 *How the s-plane is mapped into z-plane? or What is the relation between s and z domain?*

The transformation, $s = (1/T)$ *ln* z maps the s-plane into the z-plane. Every section of jw axis of length Nw$_s$ of s-plane maps into the unit circle in the anticlockwise direction, when N is integer and ws is sampling frequency. Every strip in s-plane of width w$_s$, maps into the interior of the unit circle. The mapping of s-plane into z-plane is shown in fig Q3.44.

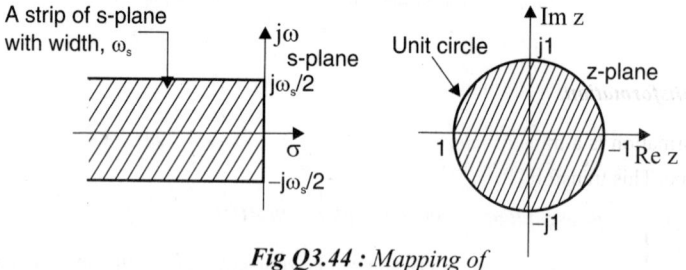

Fig Q3.44 : *Mapping of s-plane into z-plane*

Q3.45 *When a sampled data control system is stable? or what is the stability criterion for sampled data control system?*

The stability criterion for sampled data control system states that the system is stable if all the poles of the z-transfer function of the system lies inside the unit circle in z-plane.

Q3.46 *What is characteristic equation of sampled data system?*

The characteristic equation is the denominator polynomial of the z-transfer function of a sampled data control system.

Q3.47 *What are the methods available for the stability analysis of sampled data control system?*

The following three methods are available for stability analysis of sampled data conrtrol systems.

1. Jury's stability test

2. Bilinear transformation

3. Root locus technique.

Q3.48 *What are the necessary conditions to be satisfied for the stability of sampled data control system?*

Let F(z) be the characteristic equation of the system.

Now, the necessary conditions to be satisfied for the stability of the system are,

$F(1) > 0$ and $(-1)^n F(-1) > 0$.

Q3.49 *How many rows are formed in Jury's table and what are the sufficient conditions to be checked from this table for stability.*

The Jury's table consists of $(2n - 3)$ rows, where n is the order of the system (and order of the characteristic polynomial).

From the table $(n - 1)$ conditions are checked for ascertaining sufficiency. They are

$$|a_0| < |a_n|$$

$$|b_0| > |b_{n-1}|$$

$$|c_0| > |c_{n-2}|$$

$$\vdots$$

$$|r_0| > |r_2|$$

Q3.50 *What is bilinear transformation.*

The bilinear transformation is a transformation used to map the interior of unit circle in the z-plane into the left half of r-plane. This transformation is achieved by choosing,

$$z = \frac{1+r}{1-r}$$

3.15   EXERCISES

I. FILL IN THE BLANKS

1.   When the signal at any point in a system is in the form of discrete pulses the system is called ....

2.   A digital controller can be employed for time shared ....

3.   .... ia the conversion of a continuous time signal into a discrete time signal.

4.   The process of converting a discrete time continuous valued signal into a discrete time discrete valued signal is called ....

5.   The .... is the process of representing each discrete value by an n-bit binary sequence.

6.   The .... states that the information is preserved if the sampling frequency is at least twice the maximum frequency in the signal.

7.   In .... circuits the signal is reconstructed such that the value of reconstructed signal for a sampling period is same as the value of last recieved sample.

8.   The .... of F(z) is the set of all values of z, for which F(z) attains a finite value.

9.   A .... is a device or algorithm that operates on a discrete-time signal.

10.   A discrete time system is linear if it obeys the ....

11.   The impulse response of a discrete-time system is called ....

12.   The transfer function of LDS system is called ....

13.   The transfer function of ZOH is ....

14.   The transfer function .... maps the s-plane into z-plane.

15.   The .... of s-plane can be mapped into the ...... of the unit circle in z-plane.

16.   The sampled data control system is .... if all the poles lies ....the unit circle.

17.   The .... is the denominator polynomial of z-transfer function.

18.   The necessary conditions to be satisfied for the stability of sampled data system are .... and ....

19.   The .... maps the interior of unit circle in the z-plane into the left half of the r-plane.

20.   The bilinear transformation is performed by substituting .... in the characteristic equation.

ANSWERS

1. discrete data system
2. control funcions
3. sampling
4. quantization
5. coding
6. shanon's sampling theorem
7. zero order hold

8. region of convergence
9. discrete time system
10. principle of superposition
11. weighting sequence
12. pulse transfer function
13. $(1-e^{-sT})/s$
14. $s=(1/T)\ln z$

15. left half ; interior
16. stable ; inside
17. characteristic equation
18. $F(1)>0$; $(-1)^n F(-1)>0$
19. bilinear transformation
20. $z=\dfrac{1+r}{1-r}$

II. State whether the following statements are TRUE or FALSE

1. The input and output signals of sampled data control system are analog.

2. An analog controller can be employed for time shared control functions.

3. In periodic sampling, the samples are obtained uniformely at intervals of t seconds.

4. The sampling frequency should be less than the maximum value of frequency in a signal.

5. Discrete time systems can be process only digital signals.

6. The ZOH device has high pass filtering characteristics.

7. The high frequencies present in the reconstructed signal are filtered by various elements of control system.

8. The z-transform provides a method for the analysis of discrete time systems in time domain.

9. The z-transform exists only for those values of z for which F(z) is finite.

10. The linear discrete time system cannot process digital signals.

11. A discrete time system is time invariant if its input-output relationship do not change with time.

12. When the input to a discrete time system is unit impulse, the output is called impulse response.

13. The transfer function of LDS system is given by inverse z transform of its impulse response.

14. The transfer function of LDS system is called pulse transfer function.

15. In analysis of sampled data control systems the pulse sampler is replaced by impulse sampler.

16. The output of pulse sampler with ZOH can be produced by passing the impulse sampled signal through a transfer function,$(1-e^{-st})/s$.

17. When the input is an impulse sampled signal, the z-transfer function can be directly obtained from s-domain transfer function.

18. The jw axis in s-plane maps into the unit circle of z-plane in the clockwise direction.

19. The left half of s-plane maps into the interior of the unit circle in z-plane.

20. The system is unstable if all the poles of z-transfer function lies inside the unit circle in z-plane.

ANSWERS

1. True	5. True	9. True	13. False	17. True
2. False	6. False	10. False	14. True	18. False
3. True	7. True	11. True	15. True	19. True
4. False	8. False	12. True	16. True	20. False

III. UNSOLVED PROBLEM

E3.1   *Determine Ƶ-transform of following discrete time sequences.*

   (a) $\dfrac{(k+1)(k+2)}{2!}$ (b) $\dfrac{(k+1)(k+2)(k+3)}{3!}$ (c) $a^k/k!$ (d) $(kT)^2$

E3.2   *Determine the z-domain transfer function of the following s-domain transfer function.*

   (a) $\dfrac{k}{s(s+4)}$ (b) $\dfrac{(a-b)}{(s+a)(s+b)}$ (c) $\dfrac{b}{s(s+b)}$ (d) $\dfrac{s(2s+3)}{(s+1)^2(s+2)}$

E3.3   *Find the inverse Ƶ-transform of the following function.*

   (a) $\dfrac{5z}{(z-1)(z-2)}$ (b) $\dfrac{5z}{(z-1)^2(z-2)}$ (c) $\dfrac{z(z^2-1)}{(z^2+1)^2}$ (d) $\dfrac{z(z+1)}{(z-1)^3}$

E3.4   *The input output of a sampled data control system is described by the difference equation.*

   $$3c(k+2)+4\,c(k+1)+c(k)=r(k+2)+2r\,(k+1)-3r(k)$$

   where c(0) = 1;   c(1) = −2

   Determine the z-transfer function of the system. Also obtain the weighting sequence of the system.

E3.5   *Determine the z-transfer function of two cascaded systems each described by the difference equation*

   $$c(k)=0.5\,c(k-1)+r(k)$$

E3.6   *Determine the pulse transfer function of a sampled data system shown in fig E3.6.*

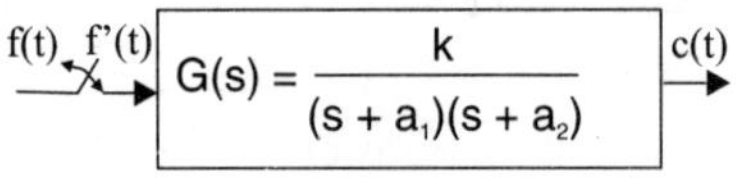

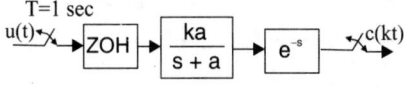

   Fig E3.6

E3.7   *Determine the response c(kT) for the system shown in fig E3.7.*

   T=1 sec
   u(t) → ZOH → $\dfrac{ka}{s+a}$ → e^{-s} → c(kt)

   Fig E3.7

E3.8   *Determine the stability of sampled data control systems described by the following characteristic equation.*

   a. $z^3+3z^2-2.75z+0.75=0$

   b. $z^3+4z^2+4z+1=0$

   c. $z^4-1.4z^3+0.4z^2+0.08z+0.002=0.$

Chapter 4

State Space Analysis

4.1 INTRODUCTION

The state variable approach is a powerful tool/technique for the analysis and design of control systems. The analysis and design of the following systems can be carried using state space method.

1. *Linear system*
2. *Non-linear system*
3. *Time invariant system*
4. *Time varying system*
5. *Multiple input and multiple output system.*

The state space analysis is a modern approach and also easier for analysis using digital computers. The conventional (or old) methods of analysis employs the transfer function of the system. The drawbacks in the transfer function model and analysis are,

1. *Transfer function is defined under zero initial conditions.*
2. *Transfer function is applicable to linear time invariant systems.*
3. *Transfer function analysis is restricted to single input and single output systems.*
4. *Does not provide information regarding the internal state of the system.*

The state variable analysis can be applied for any type of systems. The analysis can be carried with initial conditions and can be carried on multiple input and multiple output systems. In this method of analysis, it is not necessary that the state variables represent physical quantities of the system, but variables that do not represent physical quantities and those that are neither measurable nor observable may be chosen as state variables.

4.2 STATE SPACE FORMULATION

The **state** of a dynamic system is a minimal set of variables (known as state variables) such that the knowledge of these variables at $t = t_0$ together with the knowledge of the inputs for $t \geq t_0$, completely determines the behaviour of the system for $t > t_0$. (or) A set of variables which describes the system at any time instant are called **state variables**.

In the state variable formulation of a system, in general, a system consists of m-inputs, p-outputs and n-state variables. The state space representation of the system may be visualized as shown in fig 4.1.

Let,   State variables   = $x_1(t), x_2(t), x_3(t), \ldots\ldots\ldots\ldots x_n(t)$

        Input variables   = $u_1(t), u_2(t), u_3(t), \ldots\ldots\ldots\ldots u_m(t)$

        Output variables   = $y_1(t), y_2(t), y_3(t), \ldots\ldots\ldots\ldots y_p(t)$

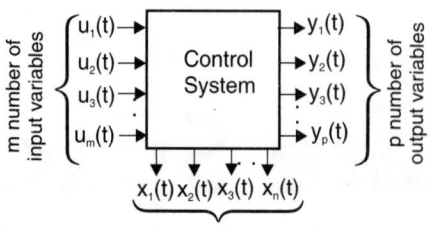

n number of state variables

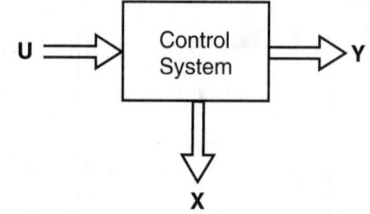

Fig 4.1 : *State space representation of system.*

The different variables may be represented by the vectors (column matrix) as shown below.

$$\text{Input vector } U(t) = \begin{bmatrix} u_1(t) \\ u_2(t) \\ \vdots \\ u_m(t) \end{bmatrix} ; \quad \text{Output vector } Y(t) = \begin{bmatrix} y_1(t) \\ y_2(t) \\ \vdots \\ y_p(t) \end{bmatrix} ; \quad \text{State Variable vector } X(t) = \begin{bmatrix} x_1(t) \\ x_2(t) \\ \vdots \\ x_n(t) \end{bmatrix}$$

STATE EQUATIONS

The state variable representation can be arranged in the form of n number of first order differential equations as shown below.

$$\frac{dx_1}{dt} = \dot{x}_1 = f_1(x_1, x_2, \dots \dots x_n \; ; \; u_1, u_2 \dots \dots u_m)$$

$$\frac{dx_2}{dt} = \dot{x}_2 = f_2(x_1, x_2, \dots \dots x_n \; ; \; u_1, u_2 \dots \dots u_m)$$

$$\vdots$$

$$\frac{dx_n}{dt} = \dot{x}_n = f_n(x_1, x_2, \dots \dots x_n \; ; \; u_1, u_2 \dots \dots u_m) \qquad \qquad \dots(4.1)$$

The n numbers of differential equations may be written in vector notation as,

$$\dot{X}(t) = f(X(t), U(t)) \qquad \qquad \dots(4.2)$$

The set of all possible values which the input vector $U(t)$ can have (assume) at time t forms the input space of the system. Similarly, the set of all possible values which the output vector $Y(t)$ can assume at time t forms the output space of the system and the set of all possible values which the state vector $X(t)$ can assume at time t forms the state space of the system.

4.3   STATE MODEL OF LINEAR SYSTEM

The state model of a system consists of the state equation and output equation. The state equation of a system is a function of state variables and inputs as defined by equation (4.2). For linear time invariant systems the first derivatives of state variables can be expressed as a linear combination of state variables and inputs.

$$\dot{x}_1 = a_{11}x_1 + a_{12}x_2 + \dots \dots + a_{1n}x_n + b_{11}u_1 + b_{12}u_2 + \dots \dots + b_{1m}u_m$$

$$\dot{x}_2 = a_{21}x_1 + a_{22}x_2 + \dots \dots + a_{2n}x_n + b_{21}u_1 + b_{22}u_2 + \dots \dots + b_{2m}u_m \qquad \qquad \dots(4.3)$$

$$\vdots$$

$$\dot{x}_n = a_{n1}x_1 + a_{n2}x_2 + \dots \dots + a_{nn}x_n + b_{n1}u_1 + b_{n2}u_2 + \dots \dots + b_{nm}u_m$$

where, the coefficients a_{ij} and b_{ij} are constants.

In the matrix form the above equations can be expressed as,

$$
\begin{bmatrix} \dot{x}_1 \\ \dot{x}_2 \\ \dot{x}_3 \\ \vdots \\ \dot{x}_n \end{bmatrix} = \begin{bmatrix} a_{11} & a_{12} & \cdots & a_{1n} \\ a_{21} & a_{22} & \cdots & a_{2n} \\ a_{31} & a_{32} & \cdots & a_{3n} \\ \vdots & \vdots & \vdots & \vdots \\ a_{n1} & a_{n2} & \cdots & a_{nn} \end{bmatrix} \begin{bmatrix} x_1 \\ x_2 \\ x_3 \\ \vdots \\ x_n \end{bmatrix} + \begin{bmatrix} b_{11} & b_{12} & \cdots & b_{1m} \\ b_{21} & b_{22} & \cdots & b_{2m} \\ b_{31} & b_{32} & \cdots & b_{3m} \\ \vdots & \vdots & \vdots & \vdots \\ b_{n1} & b_{n2} & \cdots & b_{nm} \end{bmatrix} \begin{bmatrix} u_1 \\ u_2 \\ u_3 \\ \vdots \\ u_m \end{bmatrix}
$$

.....(4.4)

The matrix equation (4.4) can also be written as, $\dot{X}(t) = A\,X(t) + B\,U(t)$ .....(4.5)

$$
\begin{aligned}
\text{where, } X(t) &= \text{State vector of order } (n \times 1) \\
U(t) &= \text{Input vector of order } (m \times 1) \\
A &= \text{System matrix of order } (n \times n) \\
B &= \text{Input matrix of order } (n \times m).
\end{aligned}
$$

> *Note : For convenience the input, output and state variables are denoted as u_1, u_2, ... , y_1, y_2, ... and x_1, x_2, ... ; but actually they are functions of time, t.*

The equation $\dot{X}(t) = A\,X(t) + B\,U(t)$ is called the state equation of Linear Time Invariant (LTI) system. The output at any time are functions of state variables and inputs.

$\therefore$ Output vector, $Y(t) = f(X(t), U(t))$ .....(4.6)

Hence the output variables can be expressed as a linear combination of state variables and inputs.

$$
\begin{aligned}
y_1 &= c_{11}x_1 + c_{12}x_2 + \ldots\ldots + c_{1n}x_n + d_{11}u_1 + d_{12}u_2 + \ldots\ldots + d_{1m}u_m \\
y_2 &= c_{21}x_1 + c_{22}x_2 + \ldots\ldots + c_{2n}x_n + d_{21}u_1 + d_{22}u_2 + \ldots\ldots + d_{2m}u_m \\
&\vdots \\
y_p &= c_{p1}x_1 + c_{p2}x_2 + \ldots\ldots + c_{pn}x_n + d_{p1}u_1 + d_{p2}u_2 + \ldots\ldots + d_{pm}u_m
\end{aligned}
$$

.....(4.7)

where the coefficients c_{ij} and d_{ij} are constants.

In the matrix form the above equations can be expressed as,

$$
\begin{bmatrix} y_1 \\ y_2 \\ y_3 \\ \vdots \\ y_p \end{bmatrix} = \begin{bmatrix} c_{11} & c_{12} & \cdots & c_{1n} \\ c_{21} & c_{22} & \cdots & c_{2n} \\ c_{31} & c_{32} & \cdots & c_{3n} \\ \vdots & \vdots & \vdots & \vdots \\ c_{p1} & c_{p2} & \cdots & c_{pn} \end{bmatrix} \begin{bmatrix} x_1 \\ x_2 \\ x_3 \\ \vdots \\ x_n \end{bmatrix} + \begin{bmatrix} d_{11} & d_{12} & \cdots & d_{1m} \\ d_{21} & d_{22} & \cdots & d_{2m} \\ d_{31} & d_{32} & \cdots & d_{3m} \\ \vdots & \vdots & \vdots & \vdots \\ d_{p1} & d_{p2} & \cdots & d_{pm} \end{bmatrix} \begin{bmatrix} u_1 \\ u_2 \\ u_3 \\ \vdots \\ u_m \end{bmatrix}
$$

.....(4.8)

The matrix equation (4.8) can also be written as, $Y(t) = C\,X(t) + D\,U(t)$ .....(4.9)

$$
\begin{aligned}
\text{where, } X(t) &= \text{State vector of order } (n \times 1) \\
U(t) &= \text{Input vector of order } (m \times 1) \\
Y(t) &= \text{Output vector of order } (p \times 1) \\
C &= \text{Output matrix of order } (p \times n) \\
D &= \text{Transmission matrix of order } (p \times m)
\end{aligned}
$$

The equation $Y(t) = C\,X(t) + D\,U(t)$ is called the output equation of Linear Time Invariant (LTI) system.

The state model of a system consists of state equation and output equation. (or) The state equation and output equation together called as state model of the system. Hence the state model of a linear time invariant system (LTI) system is given by the following equations.

$$\dot{X}(t) = A\,X(t) + B\,U(t) \quad \text{................. State equation.}$$

$$Y(t) = C\,X(t) + D\,U(t) \quad \text{................. Output equation.}$$

4.4   STATE DIAGRAM

The pictorial representation of the state model of the system is called *State diagram.* The State diagram of the system can be either in Block Diagram form or in Signal flow graph form.

The state diagram describes the relationships among the state variables and provides physical interpretations of the state variables. The time domain state diagram may be obtained directly from the differential equation governing the system and this diagram can be used for simulation of the system in analog computers.

The s-domain state diagram can be obtained from the transfer function of the system. The state diagram provides a direct relation between time domain and s-domain. [i.e., the time domain equations can be directly obtained from the s-domain state diagram].

The state diagram (Block diagram and signal flow graph) of a state model is constructued using three basic elements *Scalar, Adder* and *Integrator*.

Scalar :   The scalar is used to multiply a signal by a constant. The input signal x(t) is multiplied by the scalar **a** to give the output, **a x(t)**.

Adder : The adder is used to add two or more signals. The output of the adder is the sum of incoming signals.

Integrator : The integrator is used to integrate the signal. They are used to integrate the derivatives of state variables to get the state variables. The initial conditions of the state variable can be added by using an adder after integrator.

The time domain and s-domain elements of block diagram are shown in table-4.1. The time domain and s-domain elements of signal flow graph are shown in table-4.2.

TABLE-4.1 : Elements of Block Diagram

Element	Time domain	s-domain
Scalar	$x(t)$ → $\boxed{a}$ → $ax(t)$	$X(s)$ → $\boxed{a}$ → $aX(s)$
Adder	$x_1(t)$ → ⊗ → $x_1(t)+x_2(t)$; $x_2(t)$ ↑	$X_1(s)$ → ⊗ → $X_1(s)+X_2(s)$; $X_2(s)$ ↑
Integrator	$\dot{x}(t)$ → $\boxed{\int}$ → ⊗ → $\int_0^t x(t)dt + x(0)$; $x(0)$ ↓	$X(s)$ → $\boxed{1/s}$ → ⊗ → $\dfrac{X(s)}{s}+\dfrac{X(0)}{s}$; $X(0)/s$ ↓

Table-4.2 : Elements of Signal Flow Graph

Element	Time domain	s-domain
Scalar	$x(t) \xrightarrow{\ a\ } a\,x(t)$	$X(s) \xrightarrow{\ a\ } a\,X(s)$
Adder	$x_1(t)$ and $x_2(t)$ with gains 1, 1 giving $x_1(t) + x_2(t)$	$X_1(s)$ and $X_2(s)$ with gains 1, 1 giving $X_1(s) + X_2(s)$
Integrator	$\dot{x}(t) \xrightarrow{\int dt} x(t)+x(0)$ with branch $x(0)$ gain 1	$X(s) \xrightarrow{\frac{1}{s}} \dfrac{X(s)}{s} + \dfrac{X(0)}{s}$ with branch $\dfrac{X(0)}{s}$ gain 1

The state model of linear time invariant System is given by the equations.

$$\dot{X}(t) = A\,X(t) + B\,U(t) \quad \dots\dots\dots\dots \text{State equation.}$$

$$Y(t) = C\,X(t) + D\,U(t) \quad \dots\dots\dots\dots \text{Output equation.}$$

The time domain block diagram representation of the state model is shown in fig 4.2. and the time domain signal flow graph representation of the system is shown in fig 4.3.

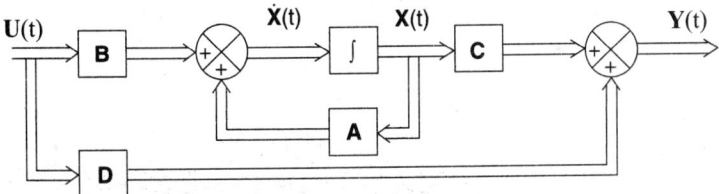

Fig 4.2 : Block diagram of state model.

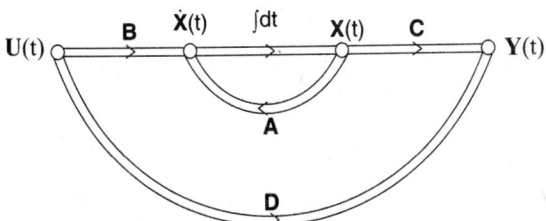

Fig 4.3 : Signal flow graph of state model.

CONSTRUCTION OF TIME DOMAIN STATE DIAGRAM

In state space modelling, n-numbers of first order differential equations are formed for an nth order system. In order to integrate n-numbers of first derivatives, the state diagram requires n-numbers of integrators. Therefore the first step in constructing the state diagram is to draw n-numbers of integrators. Mark the input to the integrators as first derivatives of state variables and so the output of the integrators are state variables. [If initial conditions are given, then they can be added at the output of integrators using adders].

In each state equation, the first derivative of state variable is expressed as a function of state variables and inputs. Therefore from the knowledge of a state equation, the state variables and inputs are multiplied by appropriate scalars and then added to get the first derivative of a state variable. Now, the first derivative of the state variable is given as input to the corresponding integrator. Similarly the input of all other integrators are obtained by considering the state equations one by one.

Each output equation is a function of state variables and inputs. Therefore from the knowledge of an output equation, the state variables and inputs are multiplied by appropriate scalars and then added to get an output. Similar procedure is followed to generate all other outputs.

4.5 STATE - SPACE REPRESENTATION USING PHYSICAL VARIABLES

In state-space modelling of systems, the choice of state variables is arbitrary. One of the possible choice of state variables is the physical variables. The physical variables of electrical systems are current or voltage in the R, L and C elements. The physical variables of mechanical systems are displacement, velocity and acceleration. The advantages of choosing the physical variables (or quantities) of the system as state variables are the following,

1. *The state variables can be utilized for the purpose of feedback.*
2. *The implementation of design with state variable feedback becomes straight forward.*
3. *The solution of state equation gives time variation of variables which have direct relevance to the physical system.*

The drawback in choosing the physical quantities as state variables is that the solution of state equation may become a difficult task.

In state space modelling using physical variables, the state equations are obtained from the differential equations governing the system. The differential equations governing a system are obtained from a basic model of the system which is developed using the fundamental elements of the system.

ELECTRICAL SYSTEM

The basic model of an electrical system can be obtained by using the fundamental elements Resistor, Capacitor and Inductor. Using these elements the electrical network or equivalent circuit of the system is drawn. Then the differential equations governing the electrical systems can be formed by writing Kirchoff's Current Law equations by choosing various nodes in the network or Kirchoff's Voltage Law by choosing various closed path in the network. The current-voltage relation of the basic elements R, L and C are given in table 4.3.

TABLE-4.3

Element	Voltage across the element	Current through the element
i(t), R, + v(t) −	$v(t) = Ri(t)$	$i(t) = \dfrac{v(t)}{R}$
i(t), L, + v(t) −	$v(t) = L\dfrac{d}{dt}i(t)$	$i(t) = \dfrac{1}{L}\displaystyle\int v(t)\,dt$
i(t), C, + v(t) −	$v(t) = \dfrac{1}{C}\displaystyle\int i(t)\,dt$	$i(t) = C\dfrac{dv(t)}{dt}$

    A minimal number of state variables are chosen for obtaining the state model of the system. The best choice of state variables in electrical system are currents and voltages in energy storage elements. The energy storage elements are inductance and capacitance. The physical variables in the differential equations are replaced by state variables and the equations are rearranged as first order differential equations. These set of first order equations constitutes the state equation of the system.

    The inputs to the system are exciting voltage sources or current sources. The outputs in electrical system are usually voltages or currents in energy dissipating elements. The resistance is energy dissipating element in electrical network. In general the output variables can be any voltage or current in the network.

MECHANICAL TRANSLATIONAL SYSTEM

    The basic model of mechanical translational system can be obtained by using three basic elements mass, spring and dash-pot. When a force is applied to a mechanical translational system, it is opposed by opposing forces due to mass, friction and elasticity of the system. The forces acting on a body are governed by Newton's second law of motion.

    The differential equations governing the system are obtained by writing force balance equations at various nodes in the system. A node is a meeting point of elements. The table-4.4 shows the force balance equations of idealized elements.

List of symbols used in mechanical translational system are

    y = Displacement, m

    v = dy/dt = Velocity, m/sec

    a = dv/dt = d²y/dt² = Acceleration, m/sec²

    f = Applied force, N (Newton)

    f_m = Opposing force offered by mass of the body, N

    f_k = Opposing force offered by the elasticity of the body (spring), N

    f_b = Opposing force offered by the friction of the body (dash-pot), N

    M = Mass, Kg

    K = Stiffness of spring, N/m

    B = Viscous friction coefficient, N/(m/sec).

Guidelines to form the state model of mechanical translational systems

1. *For each node in the system one differential equation can be framed by equating the sum of applied forces to the sum of opposing forces. Generally, the nodes are mass elements of the system, but in some cases the nodes may be without mass element.*

2. *Assign a displacement to each node and draw a free body diagram for each node. The free body diagram is obtained by drawing each mass of node separately and then marking all the forces acting on it.*

3. *In the free body diagram, the opposing forces due to mass, spring and dash-pot are always act in a direction opposite to applied force. The displacement, velocity and acceleration will be in the direction of applied force or in the direction opposite to that of opposing force.*

4. *For each free body diagram write one differential equation by equating the sum of applied forces to the sum of opposing forces.*

5. *Choose a minimum number of state variables. The choice of state variables are displacement, velocity or acceleration.*

Table 4.4 : Force Balance Equations of Idealized Elements

Elements	Force balance equations
$\vdash\!\!\rightarrow y$ $f \rightarrow$ M Reference	$f = f_m = M\dfrac{d^2y}{dt^2}$
$\vdash\!\!\rightarrow y$ $f \rightarrow$ B Reference	$f = f_b = B\dfrac{dy}{dt}$
$\vdash\!\!\rightarrow y_1 \quad \vdash\!\!\rightarrow y_2$ $f \rightarrow$ B	$f = f_b = B\dfrac{d}{dt}(y_1 - y_2)$
$\vdash\!\!\rightarrow$ y $f \rightarrow$ K Reference	$f = f_k = Ky$
$\vdash\!\!\rightarrow y_1 \quad \vdash\!\!\rightarrow y_2$ $f \rightarrow$ K	$f = f_k = K(y_1 - y_2)$

6. *The physical variables in differential equations are replaced by state variables and the equations are rearranged as first order differential equations. These set of first order equations constitute the state equation of the system.*

7. *The inputs are the applied forces and the outputs are the displacement, velocity or acceleration of the desired nodes.*

MECHANICAL ROTATIONAL SYSTEM

The basic model of mechanical rotational system can be obtained by using three basic elements moment of inertia of mass, rotational dash-pot and rotational spring. When a torque is applied to a mechanical rotational system, it is opposed by opposing torques due to moment of inertia, friction and elasticity of the system. The torque acting on a body are governed by Newton's second law of motion.

The differential equations governing the system are obtained by writing torque balance equations at various nodes in the system. A node is a meeting point of elements. The table-4.5 shows the torque balance equations of the idealized elements.

List of symbols used in mechanical rotational system

$$
\begin{aligned}
\theta &= \text{Angular displacement, rad} \\
d\theta/dt &= \text{Angular velocity, rad/sec} \\
d^2\theta/dt^2 &= \text{Angular acceleration, rad/sec}^2 \\
T &= \text{Applied torque, N-m} \\
J &= \text{Moment of inertia, Kg-m}^2/\text{rad} \\
B &= \text{Rotational frictional coefficient, N-m/(rad/sec)} \\
K &= \text{Stiffness of the spring, N-m/rad.}
\end{aligned}
$$

Guidelines to form the state model of mechanical rotational systems

1. *For each node in the system one differential equation can be framed by equating the sum of applied torques to the sum of opposing torques. Generally the nodes are mass elements but in some cases the nodes may be without mass element.*

2. *Assign an angular displacement to each node and draw a free body diagram for each node. The free body diagram is obtained by drawing each node separately and then drawing all the torques acting on it.*

3. *In the free body diagram, the opposing torques due to moment of inertia, spring and dash-pot are always act in a direction opposite to applied force. The angular displacement, velocity and acceleration will be in the direction of applied torque or in the direction opposite to that of opposing torque.*

Table-4.5 : Torque Balance Equations of Idealized Elements

Elements	Torque balance equations
	$T = T_j = J \dfrac{d^2\theta}{dt^2}$
	$T = T_b = B \dfrac{d\theta}{dt}$
	$T = T_b = B \dfrac{d}{dt}(\theta_1 - \theta_2)$

Table-4.5 : Continued....

Elements	Torque balance equations
![spring element with T, θ, K]	$T_k = K\theta$
![spring element with T, θ₁, K, θ₂]	$T = T_k = K(\theta_1 - \theta_2)$

4.   *For each free body diagram write one differential equation by equating the sum of applied torques to the sum of opposing torques.*

5.   *Choose a minimum number of state variables. The choice of state variables are angular displacement, velocity or acceleration.*

6.   *The physical variables in differential equations are replaced by state variables and the equations are rearranged as first order differential equations. These set of first order equations constitute the state equation of the system.*

7.   *The inputs are the applied torques and the outputs are the angular displacement, velocity or acceleration of the desired nodes.*

EXAMPLE 4.1

Obtain the state model of the electrical network shown in fig 1 by choosing minimal number of state variables.

SOLUTION

Let us choose the current through the inductances i_1, i_2 and voltage across the capacitor v_c as state variables. The assumed directions of currents and polarity of the voltage are shown in fig 2.

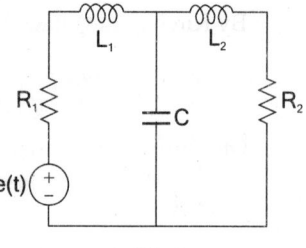

Fig 1.

> **Note :** *The best choice of state variables in electrical network are currents and voltages in energy storage elements.*

Let the three state variables x_1, x_2 and x_3 be related to physical quantities as shown below.

$x_1 = i_1 = $ Current through L_1

$x_2 = i_2 = $ Current through L_2

$x_3 = v_c = $ Voltage across capacitor.

At node A, by Kirchoff's Current Law (refer fig 3.),

$$i_1 + i_2 + C\frac{dv_c}{dt} = 0 \qquad .....(4.1.1)$$

On substituting the state variables for physical variables in equation (7.1.1) we get,

(i.e., $i_1 = x_1$, $i_2 = x_2$ and $\frac{dv_c}{dt} = \dot{x}_3$)

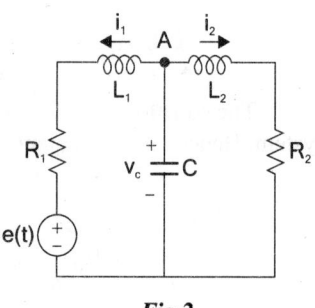

Fig 2.

$$x_1 + x_2 + C\dot{x}_3 = 0$$

$$C\dot{x}_3 = -x_1 - x_2$$

$$\dot{x}_3 = -\frac{1}{C}x_1 - \frac{1}{C}x_2 \qquad .....(4.1.2)$$

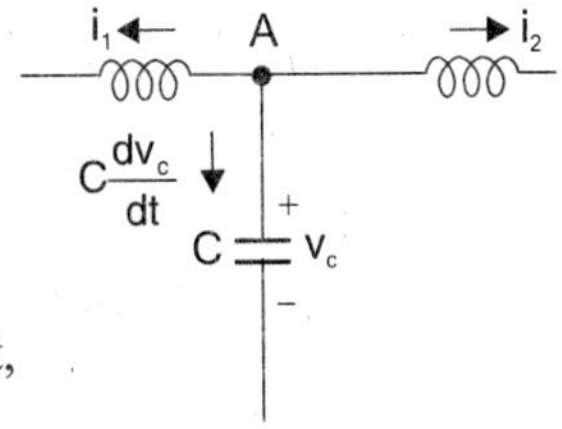

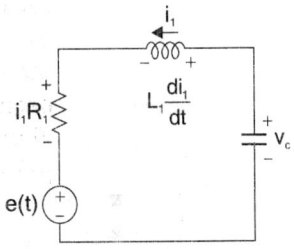

Fig 3.

By Kirchoff's Voltage Law in the closed path shown in fig 4 we get,

$$e(t) + i_1R_1 + L_1\frac{di_1}{dt} = v_c \qquad .....(4.1.3)$$

On substituting the state variables for physical variables in equation (4.1.3) we get,

(i.e., $i_1 = x_1$, $\dfrac{di_1}{dt} = \dot{x}_1$ and $v_c = x_3$)

$$e(t) + x_1R_1 + L_1\dot{x}_1 = x_3$$

Also, let u(t) = e(t) = input to the system

$$\therefore u + x_1R_1 + L_1\dot{x}_1 = x_3$$

$$L_1\dot{x}_1 = x_3 - x_1R_1 - u$$

$$\dot{x}_1 = -\frac{R_1}{L_1}x_1 + \frac{1}{L_1}x_3 - \frac{1}{L_1}u \qquad .....(4.1.4)$$

By Kirchoff's Voltage Law in the closed path shown in fig 5 we get,

$$v_c = L_2\frac{di_2}{dt} + i_2R_2 \qquad ......(4.1.5)$$

On substituting the state variables for physical variables in equation (4.1.5) we get,

(i.e., $i_2 = x_2$, $\dfrac{di_2}{dt} = \dot{x}_2$ and $v_c = x_3$)

$$x_3 = L_2\dot{x}_2 + x_2R_2$$

$$\therefore L_2\dot{x}_2 = -x_2R_2 + x_3$$

$$\dot{x}_2 = -\frac{R_2}{L_2}x_2 + \frac{1}{L_2}x_3 \qquad .....(4.1.6)$$

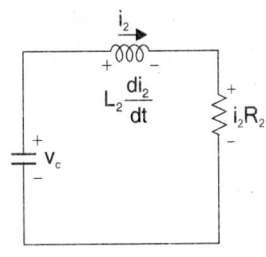

Fig 5.

The equations (4.1.2), (4.1.4) and (4.1.6) are the state equations of the system. Hence the state equations of the system are,

$$\dot{x}_1 = -\frac{R_1}{L_1}x_1 + \frac{1}{L_1}x_3 - \frac{1}{L_1}u$$

$$\dot{x}_2 = -\frac{R_2}{L_2}x_2 + \frac{1}{L_2}x_3$$

$$\dot{x}_3 = -\frac{1}{C}x_1 - \frac{1}{C}x_2$$

Fig 4. appears beside the middle section.

On arranging the state equations in the matrix form we get,

$$\begin{bmatrix} \dot{x}_1 \\ \dot{x}_2 \\ \dot{x}_3 \end{bmatrix} = \begin{bmatrix} -\dfrac{R_1}{L_1} & 0 & \dfrac{1}{L_1} \\ 0 & -\dfrac{R_2}{L_2} & \dfrac{1}{L_2} \\ -\dfrac{1}{C} & -\dfrac{1}{C} & 0 \end{bmatrix} \begin{bmatrix} x_1 \\ x_2 \\ x_3 \end{bmatrix} + \begin{bmatrix} -\dfrac{1}{L_1} \\ 0 \\ 0 \end{bmatrix} [u] \quad \text{State equation}$$

.....(4.1.7)

Let us choose the voltage across the resistances as output variables and the output variables are denoted by y_1 and y_2.

$$\therefore \quad y_1 = i_1 R_1 \qquad\qquad\qquad .....(4.1.8)$$

$$\text{and} \quad y_2 = i_2 R_2 \qquad\qquad\qquad .....(4.1.9)$$

On substituting the state variables in equations (4.1.8) and (4.1.9) we get,

(i.e., $i_1 = x_1$ and $i_2 = x_2$)

$$y_1 = x_1 R_1 \qquad ; \qquad y_2 = x_2 R_2$$

On arranging the above equations in the matrix form we get

$$\begin{bmatrix} y_1 \\ y_2 \end{bmatrix} = \begin{bmatrix} R_1 & 0 \\ 0 & R_2 \end{bmatrix} \begin{bmatrix} x_1 \\ x_2 \end{bmatrix} \qquad \text{Output equation} \qquad .....(4.1.10)$$

The state equation [equation (4.1.7)] and output equation [equation (4.1.10)] together constitute the state model of the system.

EXAMPLE 4.2

Obtain the state model of the electrical network shown in fig 1. by choosing $v_1(t)$ and $v_2(t)$ as state variables.

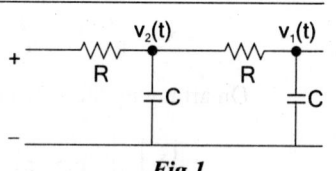

Fig 1.

SOLUTION

Connect a voltage source at the input as shown in fig 2.

Convert the voltage source to current source as shown in fig 3.

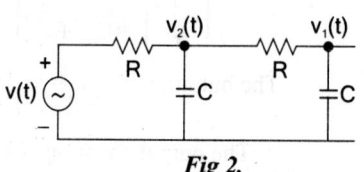

Fig 2.

At node 1, by Kirchoff's Current Law we can write (refer fig 4)

$$\frac{v_1 - v_2}{R} + C \frac{dv_1}{dt} = 0 \qquad .....(4.2.1)$$

At node 2, by Kirchoff's Current Law, we can write (Refer fig 5)

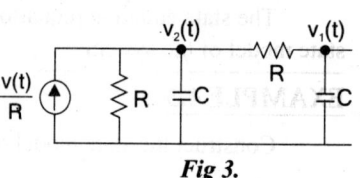

Fig 3.

$$\frac{v_2 - v_1}{R} + \frac{v_2}{R} + C \frac{dv_2}{dt} = \frac{v(t)}{R} \qquad .....(4.2.2)$$

Let the state variables be x_1 and x_2 and they are related to physical variables as shown below.

$$v_1 = x_1 \text{ and } v_2 = x_2$$

Also Let $v(t) = u = \text{input}.$

On substituting the state variables in equations (4.2.1) and (4.2.2) we get,

$$\frac{x_1 - x_2}{R} + C\frac{dx_1}{dt} = 0 \qquad .....(4.2.3)$$

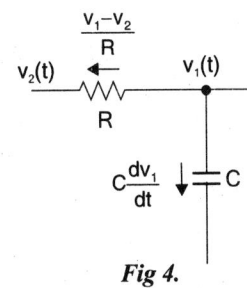

$$\frac{x_2 - x_1}{R} + \frac{x_2}{R} + C\frac{dx_2}{dt} = \frac{u}{R} \qquad .....(4.2.4)$$

From equation (4.2.3) we get, $\frac{x_1}{R} - \frac{x_2}{R} + C\dot{x}_1 = 0$

$$\therefore C\dot{x}_1 = -\frac{x_1}{R} + \frac{x_2}{R}$$

Fig 4.

$$\dot{x}_1 = -\frac{1}{RC}x_1 + \frac{1}{RC}x_2 \qquad ......(4.2.5)$$

From equation (4.2.4) we get, $\frac{x_2}{R} - \frac{x_1}{R} + \frac{x_2}{R} + C\dot{x}_2 = \frac{u}{R}$

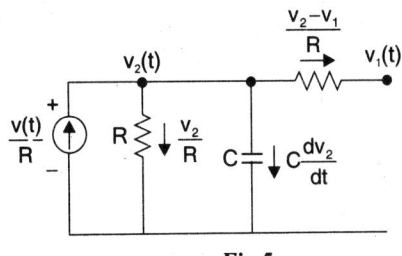

$$\therefore C\dot{x}_2 = \frac{x_1}{R} - \frac{x_2}{R} - \frac{x_2}{R} + \frac{u}{R}$$

$$\dot{x}_2 = \frac{1}{RC}x_1 - \frac{2}{RC}x_2 + \frac{1}{RC}u \qquad ......(4.2.6)$$

The equation (4.2.5) and (4.2.6) are state equations of the system. Hence the state equations of the system are

Fig 5.

$$\dot{x}_1 = -\frac{1}{RC}x_1 + \frac{1}{RC}x_2$$

$$\dot{x}_2 = \frac{1}{RC}x_1 - \frac{2}{RC}x_2 + \frac{1}{RC}u$$

On arranging the state equations in the matrix form,

$$\begin{bmatrix} \dot{x}_1 \\ \dot{x}_2 \end{bmatrix} = \begin{bmatrix} -\frac{1}{RC} & \frac{1}{RC} \\ \frac{1}{RC} & \frac{-2}{RC} \end{bmatrix} \begin{bmatrix} x_1 \\ x_2 \end{bmatrix} + \begin{bmatrix} 0 \\ \frac{1}{RC} \end{bmatrix} [u] \qquad .....(4.2.7)$$

The output, $y = v_1(t) = x_1$

$$\therefore \text{ The output equation is } y = \begin{bmatrix} 1 & 0 \end{bmatrix} \begin{bmatrix} x_1 \\ x_2 \end{bmatrix} \qquad .....(4.2.8)$$

The state equation [equation (4.2.7)] and output equation [equation (4.2.8)] together constitute the state model of the system.

EXAMPLE 4.3

Construct the state model of mechanical system shown in fig 1.

SOLUTION

Free body diagram of M_1 is shown in fig 2.

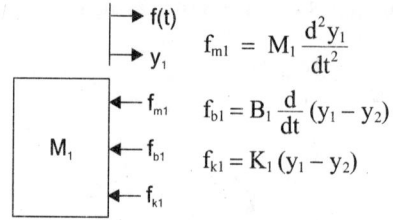

$$f_{m1} = M_1 \frac{d^2 y_1}{dt^2}$$

$$f_{b1} = B_1 \frac{d}{dt}(y_1 - y_2)$$

$$f_{k1} = K_1 (y_1 - y_2)$$

Fig 2.

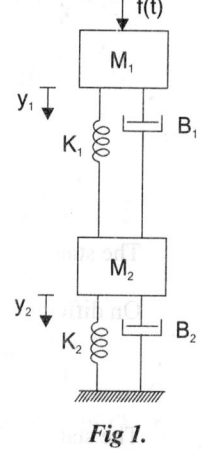

Fig 1.

By Newton's second law, the force balance equation at node M_1 is,

$$f(t) = f_{m1} + f_{b1} + f_{k1}$$

$$f(t) = M_1 \frac{d^2 y_1}{dt^2} + B_1 \frac{d}{dt}(y_1 - y_2) + K_1(y_1 - y_2)$$

$$f(t) = M_1 \frac{d^2 y_1}{dt^2} + B_1 \frac{dy_1}{dt} - B_1 \frac{dy_2}{dt} + K_1 y_1 - K_1 y_2 \qquad .....(4.3.1)$$

Free body diagram of M_2 is shown in fig 3

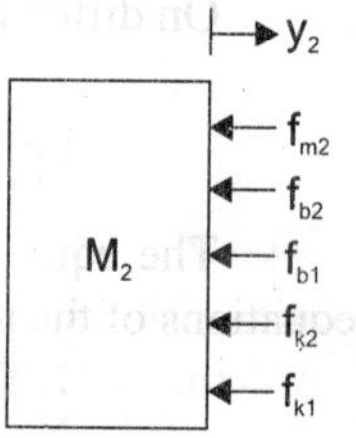

$$f_{m2} = M_2 \frac{d^2 y_2}{dt^2} \quad ; \quad f_{b2} = B_2 \frac{dy_2}{dt}$$

$$f_{b1} = B_1 \frac{d}{dt}(y_2 - y_1); \quad f_{k2} = K_2 y_2$$

$$f_{k1} = K_1 (y_2 - y_1)$$

Fig 3.

By Newton's second law, the force balance equation at node M_2 is,

$$f_{m2} + f_{b2} + f_{b1} + f_{k2} + f_{k1} = 0$$

$$\therefore M_2 \frac{d^2 y_2}{dt^2} + B_2 \frac{dy_2}{dt} + B_1 \frac{d}{dt}(y_2 - y_1) + K_2 y_2 + K_1(y_2 - y_1) = 0$$

$$M_2 \frac{d^2 y_2}{dt^2} + B_2 \frac{dy_2}{dt} + B_1 \frac{dy_2}{dt} - B_1 \frac{dy_1}{dt} + K_2 y_2 + K_1 y_2 - K_1 y_1 = 0 \qquad .....(4.3.2)$$

Let us choose four state variables x_1, x_2, x_3 and x_4. Also, let the input $f(t) = u$. The state variables are related to physical variables as follows

$$x_1 = y_1 \ ; \ x_2 = y_2 \ ; \ x_3 = \frac{dy_1}{dt} \ ; \qquad x_4 = \frac{dy_2}{dt} \ ; \ \dot{x}_3 = \frac{d^2 y_1}{dt^2} \ ; \ \dot{x}_4 = \frac{d^2 y_2}{dt^2}$$

On substituting $y_1 = x_1 \ ; \ y_2 = x_2 \ ; \ \dfrac{dy_1}{dt} = x_3 \ ; \ \dfrac{dy_2}{dt} = x_4 \ ; \ \dfrac{d^2 y_1}{dt^2} = \dot{x}_3$ and $f(t) = u$ in equation (4.3.1) we get,

$$u = M_1 \dot{x}_3 + B_1 x_3 - B_1 x_4 + K_1 x_1 - K_1 x_2$$

$$M_1 \dot{x}_3 = - B_1 x_3 + B_1 x_4 - K_1 x_1 + K_1 x_2 + u$$

$$\therefore \dot{x}_3 = - \frac{K_1}{M_1} x_1 + \frac{K_1}{M_1} x_2 - \frac{B_1}{M_1} x_3 + \frac{B_1}{M_1} x_4 + \frac{1}{M_1} u \qquad .....(4.3.3)$$

On substituting $y_1 = x_1 \ ; \ y_2 = x_2 \ ; \ \dfrac{dy_1}{dt} = x_3 \ ; \ \dfrac{dy_2}{dt} = x_4$ and $\dfrac{d^2 y_2}{dt^2} = \dot{x}_4$ in equation (4.3.2) we get,

$$M_2 \dot{x}_4 + B_2 x_4 + B_1 x_4 - B_1 x_3 + K_1 x_2 + K_2 x_2 - K_1 x_1 = 0$$

$$\therefore M_2 \dot{x}_4 = - B_2 x_4 - B_1 x_4 + B_1 x_3 - K_1 x_2 - K_2 x_2 + K_1 x_1$$

$$= - (B_2 + B_1) x_4 + B_1 x_3 - (K_2 + K_1) x_2 + K_1 x_1$$

$$\therefore \dot{x}_4 = \frac{K_1}{M_2} x_1 - \frac{(K_1 + K_2)}{M_2} x_2 + \frac{B_1}{M_2} x_3 - \frac{(B_1 + B_2)}{M_2} x_4 \qquad .....(4.3.4)$$

The state variable $x_1 = y_1$.

On differentiating $x_1 = y_1$ with respect to t we get, $\dfrac{dx_1}{dt} = \dfrac{dy_1}{dt}$

Let $\dfrac{dx_1}{dt} = \dot{x}_1$ and $\dfrac{dy_1}{dt} = x_3$; $\therefore \dot{x}_1 = x_3$ $\qquad ....(4.3.5)$

The state variable, $x_2 = y_2$.

On differentiating $x_2 = y_2$ with respect to t we get, $\dfrac{dx_2}{dt} = \dfrac{dy_2}{dt}$

Let $\dfrac{dx_2}{dt} = \dot{x}_2$ and $\dfrac{dy_2}{dt} = x_4$; $\therefore \dot{x}_2 = x_4$ $\qquad ....(4.3.6)$

The equations (4.3.3) to (4.3.6) are state equations of the mechanical system. Hence the state equations of the mechanical system are,

$$\dot{x}_1 = x_3$$

$$\dot{x}_2 = x_4$$

$$\dot{x}_3 = - \frac{K_1}{M_1} x_1 + \frac{K_2}{M_2} x_2 - \frac{B_1}{M_1} x_3 + \frac{B_1}{M_1} x_4 + \frac{1}{M_1} u$$

$$\dot{x}_4 = \frac{K_1}{M_2} x_1 - \frac{(K_1 + K_2)}{M_2} x_2 + \frac{B_1}{M_2} x_3 - \frac{(B_1 + B_2)}{M_2} x_4$$

On arranging the state equations in the matrix form, we get,[u]

$$\begin{bmatrix} \dot{x}_1 \\ \dot{x}_2 \\ \dot{x}_3 \\ \dot{x}_4 \end{bmatrix} = \begin{bmatrix} 0 & 0 & 1 & 0 \\ 0 & 0 & 0 & 1 \\ -\dfrac{K_1}{M_1} & \dfrac{K_1}{M_1} & -\dfrac{B_1}{M_1} & \dfrac{B_1}{M_1} \\ \dfrac{K_1}{M_2} & -\dfrac{(K_1+K_2)}{M_2} & \dfrac{B_1}{M_2} & -\dfrac{(B_1+B_2)}{M_2} \end{bmatrix} \begin{bmatrix} x_1 \\ x_2 \\ x_3 \\ x_4 \end{bmatrix} + \begin{bmatrix} 0 \\ 0 \\ \dfrac{1}{M_1} \\ 0 \end{bmatrix} [u] \qquad .....(4.3.7)$$

Let the displacements y_1 and y_2 be the outputs of the system.

$\qquad \therefore y_1 = x_1$ and $\quad y_2 = x_2$.

The output equation in matrix form is given by,

$$\begin{bmatrix} y_1 \\ y_2 \end{bmatrix} = \begin{bmatrix} 1 & 0 & 0 & 0 \\ 0 & 1 & 0 & 0 \end{bmatrix} \begin{bmatrix} x_1 \\ x_2 \\ x_3 \\ x_4 \end{bmatrix} \qquad .....(4.3.8)$$

The state equation [equation (4.3.7)] and the output equation [equation (4.3.8)] together called state model of the system.

EXAMPLE 4.4

Obtain the state model of the mechanical system shown in fig 1 by choosing a minimum of three state variables.

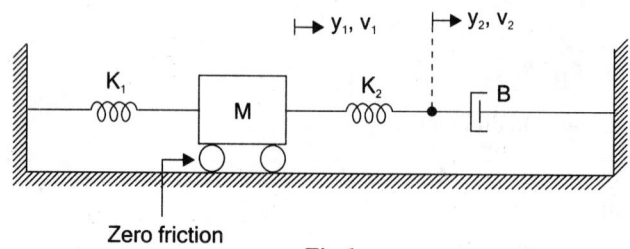

Zero friction

Fig 1.

SOLUTION

Let the three state variables be x_1, x_2 and x_3 and they are related to physical variables as shown below.

$$x_1 = y_1 \; ; \; x_2 = y_2 \; ; \; x_3 = \frac{dy_1}{dt} = v_1$$

Free body diagram of mass M is shown in fig 2.

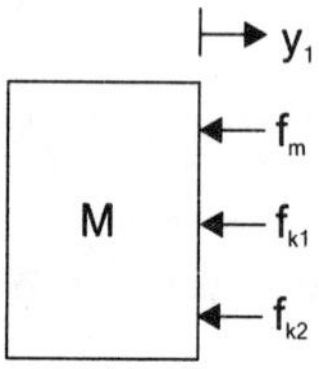

$$f_m = M \frac{d^2 y_1}{dt^2} \; ; \; f_{k1} = K_1 y_1 \; ; \; f_{k2} = K_2(y_1 - y_2)$$

By Newton's second law, the force balance equation at node M is,

$$f_m + f_{k1} + f_{k2} = 0$$

$$M \frac{d^2 y_1}{dt^2} + K_1 y_1 + K_2(y_1 - y_2) = 0$$

$$M \frac{d^2 y_1}{dt^2} + K_1 y_1 + K_2 y_1 - K_2 y_2 = 0 \qquad \text{.....(4.4.1)}$$

Put $\dfrac{d^2 y_1}{dt^2} = \dot{x}_3 \; ; \; y_1 = x_1 \, , \; y_2 = x_2$ in equation (4.4.1)

$$M \dot{x}_3 + K_1 x_1 + K_2 x_1 - K_2 x_2 = 0$$

$$M \dot{x}_3 + (K_1 + K_2) x_1 - K_2 x_2 = 0$$

$$\dot{x}_3 = -\frac{K_1 + K_2}{M} x_1 + \frac{K_2}{M} x_2 \qquad \text{.....(4.4.2)}$$

The free body diagram of node 2 (meeting point of K_2 and B) is shown in fig 3.

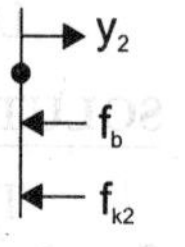

$$f_b = B \frac{dy_2}{dt} \; ; \; f_{k2} = K_2(y_2 - y_1)$$

Writing force balance equation at the meeting point of K_2 and B we get,

$$f_b + f_{k2} = 0$$

$$B \frac{dy_2}{dt} + K_2(y_2 - y_1) = 0$$

Fig 2.

Fig 3.

$$\therefore \frac{dy_2}{dt} = \frac{K_2}{B} y_1 - \frac{K_2}{B} y_2$$

Put $\dfrac{dy_2}{dt} = \dot{x}_2$, $y_1 = x_1$ and $y_2 = x_2$

$$\therefore \dot{x}_2 = \frac{K_2}{B} x_1 - \frac{K_2}{B} x_2 \qquad\qquad .....(4.4.3)$$

The state variable, $x_1 = y_1$. On differentiating this expression with respect to t we get.

$$\frac{dx_1}{dt} = \frac{dy_1}{dt}$$

Let $\dfrac{dx_1}{dt} = \dot{x}_1$ and $\dfrac{dy_1}{dt} = x_3$; $\therefore \dot{x}_1 = x_3$ .....(4.4.4)

The state equations are given by equations (4.4.4), (4.4.3) and (4.4.2)

$$\dot{x}_1 = x_3$$

$$\dot{x}_2 = \frac{K_2}{B} x_1 - \frac{K_2}{B} x_2$$

$$\dot{x}_3 = -\frac{K_1 + K_2}{M} x_1 + \frac{K_2}{M} x_2$$

On arranging the state equations in the matrix form,

$$\begin{bmatrix} \dot{x}_1 \\ \dot{x}_2 \\ \dot{x}_3 \end{bmatrix} = \begin{bmatrix} 0 & 0 & 1 \\ \dfrac{K_2}{B} & -\dfrac{K_2}{B} & 0 \\ -\dfrac{K_1 + K_2}{M} & 0 & \dfrac{K_2}{M} \end{bmatrix} \begin{bmatrix} x_1 \\ x_2 \\ x_3 \end{bmatrix} \qquad .....(4.4.5)$$

If the desired outputs are y_1 and y_2, then $y_1 = x_1$ and $y_2 = x_2$

The output equation in the matrix form is given by

$$\begin{bmatrix} y_1 \\ y_2 \end{bmatrix} = \begin{bmatrix} 1 & 0 \\ 0 & 1 \end{bmatrix} \begin{bmatrix} x_1 \\ x_2 \end{bmatrix} \qquad .....(4.4.6)$$

The state equation [equation(4.4.5)] and the output equation [equation(4.4.6)] together constitute the state model of the system.

EXAMPLE 4.5

Determine the state model of armature controlled dc motor.

SOLUTION

The speed of DC motor is directly proportional to armature voltage and inversely proportional to flux. In armature controlled DC motor the desired speed is obtained by varying the armature voltage. This speed control system is an electro-mechanical control system. The electrical system consists of the armature and the field circuit but for analysis purpose, only the armature circuit is considered because the field is excited by a constant voltage. The mechanical system consist of the rotating part of the motor and load connected to the shaft of the motor. The armature controlled DC motor speed control system is shown in fig 1.

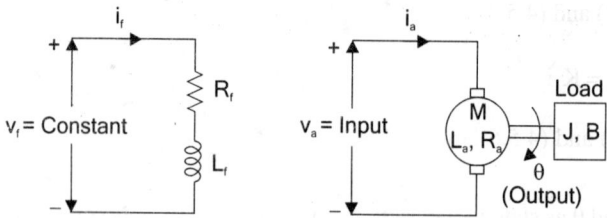

Fig 1 : Armature controlled DC motor.

Let R_a = Armature resistance, Ω

L_a = Armature inductance, H

i_a = Armature current, A

V_a = Armature voltage, V

e_b = Back emf, V

K_t = Torque constant, N-m/A

T = Torque developed by motor, N-m

θ = Angular displacement of shaft, rad

ω = $d\theta\backslash dt$ = Angular velocity of the shaft, rad/sec

J = Moment of inertia of motor and load, Kg-m²/rad

B = Frictional coefficient of motor and load, N-m/(rad/sec)

K_b = Back emf constant, V/(rad/sec).

The equivalent circuit of armature is shown in fig 2

By Kirchoff's Voltage Law, we can write

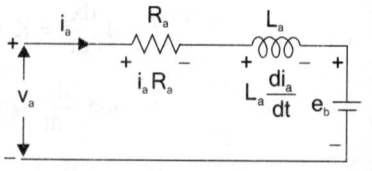

$$i_a R_a + L_a \frac{di_a}{dt} + e_b = v_a \qquad .....(4.5.1)$$

Torque of DC motor is proportional to the product of flux and current. Since flux is constant in this system, the torque is proportional to i_a alone.

Fig 2 : Equivalent circuit of armature.

$$T \propto i_a$$

$$\therefore \text{Torque, } T = K_t i_a \qquad .....(4.5.2)$$

The mechanical system of the motor is shown in fig 3. The differential equation governing the mechanical system of motor is given by,

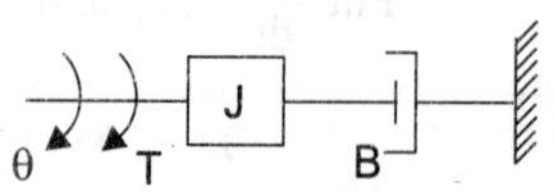

$$J \frac{d^2\theta}{dt^2} + B \frac{d\theta}{dt} = T$$

$$.....(4.5.3)$$

Fig 3.

The back emf of DC machine is proportional to speed (angular velocity) of shaft

$$\therefore e_b \propto \frac{d\theta}{dt} \quad ; \quad \text{Back emf, } e_b = K_b \frac{d\theta}{dt} \qquad .....(4.5.4)$$

From equation (4.5.1) and (4.5.4) we get,

$$i_a R_a + L_a \frac{di_a}{dt} + K_b \frac{d\theta}{dt} = v_a \qquad .....(4.5.5)$$

From equation (4.5.2) and (4.5.3) we get,

$$J \frac{d^2\theta}{dt^2} + B \frac{d\theta}{dt} = K_t i_a \qquad\qquad .....(4.5.6)$$

The equations (4.5.5) and (4.5.6) are the differential equations governing the armature controlled dc motor.

Let us choose i_a, ω and θ as state variables to model the armature controlled dc motor. The physical variables i_a, ω and θ are related to the general notation of state variables x_1, x_2 and x_3 as shown below

$$x_1 = i_a \; ; \; x_2 = \omega = d\theta/dt \text{ and } x_3 = \theta$$

The input to the motor is the armature voltage, v_a and let $v_a = u$, where u is the general notation for input variable.

On substituting the state variables for the physical variables in equation (4.5.5) we get,

$$x_1 R_a + L_a \frac{dx_1}{dt} + K_b x_2 = u$$

Let $\dfrac{dx_1}{dt} = \dot{x}_1$, $\therefore x_1 R_a + L_a \dot{x}_1 + K_b x_2 = u$

$$\dot{x}_1 = -\frac{R_a}{L_a} x_1 - \frac{K_b}{L_a} x_2 + \frac{1}{L_a} u \qquad\qquad .....(4.5.7)$$

On substituting the state variables for physical variables in equation (4.5.6) we get,

$$J \frac{d^2 x_3}{dt^2} + B \frac{dx_3}{dt} = K_t x_1$$

Let $\dfrac{d^2 x_3}{dt^2} = \dot{x}_2$ and $\dfrac{dx_3}{dt} = x_2$, $\qquad \therefore J\dot{x}_2 + B x_2 = K_1 x_1$

$$\dot{x}_2 = \frac{K_1}{J} x_1 - \frac{B}{J} x_2 \qquad\qquad .....(4.5.8)$$

The state variable $x_3 = \theta$. On differentiating $x_3 = \theta$ with respect to t we get,

$$\frac{dx_3}{dt} = \frac{d\theta}{dt}$$

Put $\dfrac{dx_3}{dt} = \dot{x}_3$ and $\dfrac{d\theta}{dt} = x_2$

$$\therefore \dot{x}_3 = x_2 \qquad\qquad .....(4.5.9)$$

The equations (4.5.7), (4.5.8) and (4.5.9) are the state equations of the system.

$$\dot{x}_1 = -\frac{R_a}{L_a} x_1 - \frac{K_b}{L_a} x_2 + \frac{1}{L_a} u$$

$$\dot{x}_2 = \frac{K_t}{J} x_1 - \frac{B}{J} x_2$$

$$\dot{x}_3 = x_2$$

On arranging the state equations in the matrix form,

$$
\begin{bmatrix} \dot{x}_1 \\ \dot{x}_2 \\ \dot{x}_3 \end{bmatrix} = \begin{bmatrix} -\dfrac{R_a}{L_a} & -\dfrac{K_b}{L_a} & 0 \\ \dfrac{K_1}{J} & -\dfrac{B}{J} & 0 \\ 0 & 1 & 0 \end{bmatrix} \begin{bmatrix} x_1 \\ x_2 \\ x_3 \end{bmatrix} + \begin{bmatrix} \dfrac{1}{L_a} \\ 0 \\ 0 \end{bmatrix} [u] \qquad \qquad .....(4.5.10)
$$

Let the desired outputs be i_a, ω and θ. Let us equate the desired output quantities to standard notation y_1, y_2 and y_3 as shown below.

$$ y_1 = i_a \quad ; \quad y_2 = \omega = d\theta/dt \quad ; \quad y_3 = \theta $$

On relating the outputs to state variables we get,

$$ y_1 = x_1 \quad ; \quad y_2 = x_2 \quad ; \quad y_3 = x_3 $$

$\therefore$ The output equation in the matrix form is

$$
\begin{bmatrix} y_1 \\ y_2 \\ y_3 \end{bmatrix} = \begin{bmatrix} 1 & 0 & 0 \\ 0 & 1 & 0 \\ 0 & 0 & 1 \end{bmatrix} \begin{bmatrix} x_1 \\ x_2 \\ x_3 \end{bmatrix} \qquad \qquad .....(4.5.11)
$$

The state equation [equation (4.5.10)] and the output equation [equation (4.5.11)] together constitute the state model of the armature controlled dc motor.

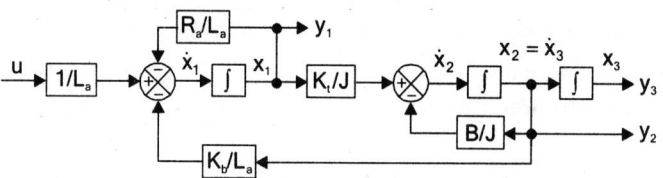

Fig 4 : *Block diagram representation of the state model of armature controlled dc motor.*

EXAMPLE 4.6

Determine the state model of field controlled dc motor.

SOLUTION

The speed of a DC motor is directly proportional to armature voltage and inversely proportional to flux. In field controlled DC motor the armature voltage is kept constant and the speed is varied by varying the flux of the machine. Since flux is directly proportional to field current, the flux is varied by varying field current. The speed control system is an electromechanical control system. The electrical system consists of armature and field circuit but for analysis purpose, only field circuit is considered because the armature is excited by a constant voltage. The mechanical system consists of the rotating part of the motor and the load connected to the shaft of the motor. The field controlled DC motor speed control system is shown in fig 1.

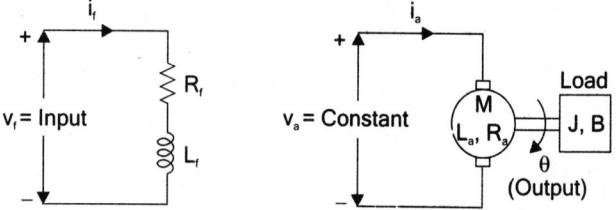

Fig 1 : *Field controlled DC motor.*

Let R_f =   Field resistance, Ω

L_f =   Field inductance, H

i_f =   Field current, A

v_f =   Field voltage, V

θ =   Angular displacement of the motor shaft, rad

ω = $d\theta/dt$ = Angular velocity of the motor shaft, rad/sec

T   =   Torque developed by motor, N-m

K_{tf} =   Torque constant, N-m/A

J   =   Moment of inertia of rotor and load, Kg-m²/rad

B   =   Frictional coefficient of rotor and load, N-m/(rad/sec).

The equivalent circuit of field is shown in fig 2

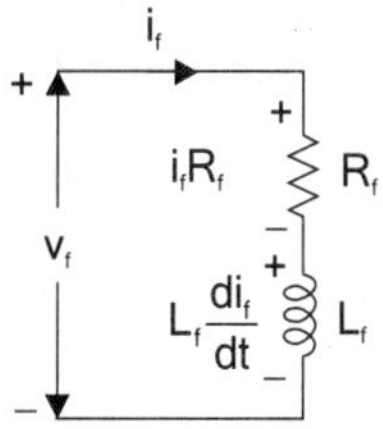

By Kirchoff's voltage law, we can write

Fig 2 : Equivalent circuit of field.

$$R_f i_f + L_f \frac{di_f}{dt} = v_f \qquad\qquad .....(4.6.1)$$

The torque of DC motor is proportional to product of flux and armature current. Since armature current is constant in this system, the torque is proportional to flux alone, but flux is proportional to field current.

$$\therefore T \propto i_f \; ; \; \text{Torque, } T = K_{tf} \, i_f \qquad .....(4.6.2)$$

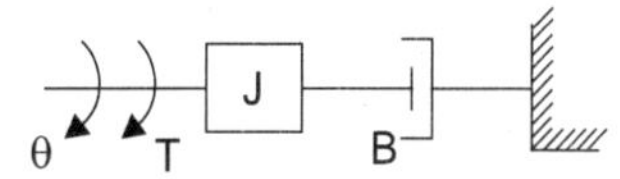

The mechanical system of the motor is shown in fig 3. The differential equation governing the mechanical system of the motor is given by,

Fig 3.

$$J\frac{d^2\theta}{dt^2} + B\frac{d\theta}{dt} = T \qquad ......(4.6.3)$$

From equation (4.6.2) and (4.6.3) we get,

$$J\frac{d^2\theta}{dt^2} + B\frac{d\theta}{dt} = K_{tf} i_f \qquad\qquad .....(4.6.4)$$

The equations (4.6.1) and (4.6.4) are the differential equations governing the field controlled dc motor.

Let us choose i_f, ω and θ as state variable to model the field controlled dc motor. The physical variables i_f, ω and θ are related to the general notation of state variables x_1, x_2 and x_3 as shown below.

$$x_1 = i_f \; ; \quad x_2 = \omega = d\theta/dt \; ; \quad x_3 = \theta$$

The input to the system is the field voltage v_f. Let $v_f = u$, where u is the general notation for input.

On substituting the state variables and input variable for the physical variables in equation (4.6.1) we get,

$$R_f x_1 + L_f \frac{dx_1}{dt} = u$$

Let $\dfrac{dx_1}{dt} = \dot{x}_1$, $\therefore R_f x_1 + L_1 \dot{x}_1 = u$ .....(4.6.5)

$$\dot{x}_1 = -\frac{R_f}{L_f} x_1 + \frac{1}{L_f} u$$

On substituting the state variables for the physical variables in equation (4.6.4) we get,

$$J \frac{d^2 x_3}{dt^2} + B \frac{dx_3}{dt} = K_{tf} x_1$$

Let $\dfrac{d^2 x_3}{dt^2} = \dot{x}_2$ and $\dfrac{dx_3}{dt} = x_2$, $\therefore J \dot{x}_2 + B x_2 = K_{tf} x_1$

$$\dot{x}_2 = \frac{K_{tf}}{J} x_1 - \frac{B}{J} x_2$$ .....(4.6.6)

The state variable $x_3 = \theta$. On differentiating $x_3 = \theta$ with respect to t we get,

$$\frac{dx_3}{dt} = \frac{d\theta}{dt}$$

Put $\dfrac{dx_3}{dt} = \dot{x}_3$ and $\dfrac{d\theta}{dt} = x_2$ $\therefore \dot{x}_3 = x_2$ .....(4.6.7)

The equations (4.6.5), (4.6.6) and (4.6.7) are the state equations of the system.

$$\dot{x}_1 = -\frac{R_f}{L_f} x_1 + \frac{1}{L_f} u$$

$$\dot{x}_2 = \frac{K_{tf}}{J} x_1 - \frac{B}{J} x_2$$

$$\dot{x}_3 = x_2$$

On arranging the state equations in the matrix form,

$$\begin{bmatrix} \dot{x}_1 \\ \dot{x}_2 \\ \dot{x}_3 \end{bmatrix} = \begin{bmatrix} -\dfrac{R_f}{L_f} & 0 & 0 \\ \dfrac{K_{tf}}{J} & -\dfrac{B}{J} & 0 \\ 0 & 1 & 0 \end{bmatrix} \begin{bmatrix} x_1 \\ x_2 \\ x_3 \end{bmatrix} + \begin{bmatrix} \dfrac{1}{L_f} \\ 0 \\ 0 \end{bmatrix} [u]$$ .....(4.6.8)

Let the desired output be ω and θ. Let us equate the desired output quantities to standard notation y_1 and y_2 as shown below.

$$y_1 = \omega \quad ; \quad y_2 = \theta$$

On relating the outputs to state variable we get,

$$y_1 = x_2 \quad ; \quad y_2 = x_3$$

The output equation in the matrix form is

$$\begin{bmatrix} y_1 \\ y_2 \end{bmatrix} = \begin{bmatrix} 0 & 1 & 0 \\ 0 & 0 & 1 \end{bmatrix} \begin{bmatrix} x_1 \\ x_2 \\ x_1 \end{bmatrix}$$ .....(4.6.9)

The state equation [equation (4.6.8)] and the output equation [equation (4.6.9)] together constitute the state model of the system.

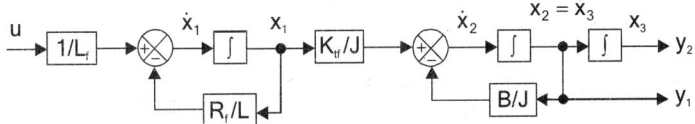

Fig 4 : Block diagram representation of the state model field controlled dc motor.

4.6 STATE SPACE REPRESENTATION USING PHASE VARIABLES

The phase variables are defined as those particular state variables which are obtained from one of the system variables and its derivatives. Usually the variable used in the system output and the remaining state variables are the derivatives of the output. The state model using phase variables can be easily determined if the system model is already known in the differential equation or transfer function form. There are three methods of modelling a system using phase variables and they are explained in the following sections.

Method 1

Consider the following n^{th} order linear differential equation relating the output y(t) to the input u(t) of a system,

$$\overset{n.}{y} + a_1 \overset{(n-1).}{y} + a_2 \overset{(n-2).}{y} + \ldots\ldots + a_{n-2}\overset{..}{y} + a_{n-1}\overset{.}{y} + a_n y = b\,u \qquad .....(4.10)$$

By choosing the output **y** and their derivatives as state variables, we get,

$$x_1 = y$$

$$x_2 = \overset{.}{y}$$

$$x_3 = \overset{..}{y}$$

$$\vdots$$

$$x_n = \overset{(n-1).}{y} \quad ; \quad \therefore \overset{.}{x}_n = \overset{n.}{y}$$

On substituting the state variables in the differential equation governing the system [equation (4.10)], we get,

$$\overset{.}{x}_n + a_1 x_n + a_2 x_{n-1} + \ldots\ldots + a_{n-2}x_3 + a_{n-1}x_2 + a_n x_1 = b\,u$$

$$\therefore \overset{.}{x}_n = -a_n x_1 - a_{n-1}x_2 - a_{n-2}x_3 - \ldots\ldots\ldots - a_2 x_{n-1} - a_1 x_n + b\,u$$

The state equations of the system are

$$\overset{.}{x}_1 = x_2$$

$$\overset{.}{x}_2 = x_3$$

$$\vdots$$

$$\overset{.}{x}_{n-1} = x_n$$

$$\overset{.}{x}_n = -a_n x_1 - a_{n-2}x_3 - \ldots\ldots - a_2 x_{n-1} - a_1 x_n + b\,u$$

On arranging the above equations in the matrix form we get,

$$
\begin{bmatrix} \dot{x}_1 \\ \dot{x}_2 \\ \dot{x}_3 \\ \vdots \\ \dot{x}_{n-1} \\ \dot{x}_n \end{bmatrix} = \begin{bmatrix} 0 & 1 & 0 & 0 & \cdots & 0 \\ 0 & 0 & 1 & 0 & \cdots & 0 \\ 0 & 0 & 0 & 1 & \cdots & 0 \\ & & & & & \\ 0 & 0 & 0 & 0 & \cdots & 1 \\ -a_n & -a_{n-1} & -a_{n-2} & -a_{n-3} & \cdots & -a_1 \end{bmatrix} \begin{bmatrix} x_1 \\ x_2 \\ x_3 \\ \vdots \\ x_{n-1} \\ x_n \end{bmatrix} + \begin{bmatrix} 0 \\ 0 \\ 0 \\ \vdots \\ x_{n-1} \\ x_n \end{bmatrix} + \begin{bmatrix} 0 \\ 0 \\ 0 \\ \vdots \\ 0 \\ b \end{bmatrix} [u] \qquad .....(4.11)
$$

or $\dot{X} = A X + B U$

Here the matrix **A** (system matrix) has a very special form. It has all 1's in the upper off-diagonal, its last row is comprised of the negative of the coefficients of the original differential equation and all other elements are zero. This form of matrix **A** is known as *Bush form* **(or)** *Companion form.*

Also note that **B** matrix has the speciality that all its elements except the last element are zero. The output being $y = x_1$, the output equation is given by,

$$
y = \begin{bmatrix} 1 & 0 & 0 & \cdots\cdots & 0 \end{bmatrix} \begin{bmatrix} x_1 \\ x_2 \\ x_3 \\ \vdots \\ x_n \end{bmatrix} \qquad .....(4.12)
$$

(or) **Y = CX**

The advantage in using phase variables for state space modelling is that the system state model can be written directly by inspection from the differential equation governing the system.

Method 2

Consider the following n^{th} order differential equation governing the output $y(t)$ to the input $u(t)$ of a system.

$$
\overset{n.}{y} + a_1 \overset{(n-1).}{y} + \cdots\cdots\cdots + a_{n-1}\dot{y} + a_n y = b_0 \overset{m.}{u} + b_1 \overset{(m-1).}{u} + \cdots\cdots + b_{m-1}\dot{u} + b_m u \qquad .....(4.13)
$$

let $n = m = 3$

$$
\therefore \overset{...}{y} + a_1 \overset{..}{y} + a_2 \dot{y} + a_3 y = b_0 \overset{...}{u} + b_1 \overset{..}{u} + b_2 \dot{u} + b_3 u \qquad .....(4.14)
$$

On taking Laplace transform of equation (4.14) with zero initial conditions we get,

$$
s^3 Y(s) + a_1 s^2 Y(s) + a_2 s Y(s) + a_3 Y(s) = b_0 s^3 U(s) + b_1 s^2 U(s) + b_2 s U(s) + b_3 U(s)
$$

$$
(s^3 + a_1 s^2 + a_2 s + a_3) \, Y(s) = (b_0 s^3 + b_1 s^2 + b_2 s + b_3) U(s)
$$

$$
\therefore \frac{Y(s)}{U(s)} = \frac{b_0 s^3 + b_1 s^2 + b_2 s + b_3}{s^3 + a_1 s^2 + a_2 s + a_3}
$$

$$= \frac{s^3\left(b_0 + \dfrac{b_1}{s} + \dfrac{b_2}{s^2} + \dfrac{b_3}{s^3}\right)}{s^3\left(1 + \dfrac{a_1}{s} + \dfrac{a_2}{s^2} + \dfrac{a_3}{s^3}\right)} = \frac{b_0 + \dfrac{b_1}{s} + \dfrac{b_2}{s^2} + \dfrac{b_3}{s^3}}{1 - \left(-\dfrac{a_1}{s} - \dfrac{a_2}{s^2} - \dfrac{a_3}{s^3}\right)} \qquad .....(4.15)$$

From the Mason's gain formula, the transfer function of the system is given by,

$$T(s) = \frac{1}{\Delta} \sum_K P_K \Delta_K \qquad\qquad .....(4.16)$$

where, P_K = path gain of Kth forward path.

$\Delta = 1-$ (sum of loop gain of all individual loops)

+ (sum of gain products of all possible combinations
of two non-touching loops) $ - \cdots\cdots$

$\Delta_K = \Delta$ for that part of the graph which is not
touching K^{th} forward path.

The transfer function of a system with four forward paths and with three feedback loops (touching each other) is given by,

$$T(s) = \frac{P_1 + P_2 + P_3 + P_4}{1 - (P_{11} + P_{12} + P_{13})} \qquad\qquad .....(4.17)$$

On comparing equation (4.15) and (4.17) we get,

$$P_1 = b_0 \quad ; \quad P_2 = \frac{b_1}{s} \quad ; \quad P_3 = \frac{b_2}{s^2} \quad \text{and} \quad P_4 = \frac{b_3}{s^3}$$

$$P_{11} = -\frac{a_1}{s} ; \quad P_{12} = -\frac{a_2}{s^2} \quad \text{and} \quad P_{13} = \frac{a_3}{s^3}$$

Hence for the system represented by the transfer function as that of equation (4.15), a signal flow graph can be constructed as shown in the fig 4.4. The signal flow is constructed such that all $\Delta_K = 1$ and all loops are touching loops.

Let us assign state variables at the output of each integrator in the signal flow graph. Hence at the input of each integrator, the first derivative of the state variable will be available. The state equations are formed by summing all the incoming signals to the nodes, whose values correspond to first derivative of state variables.

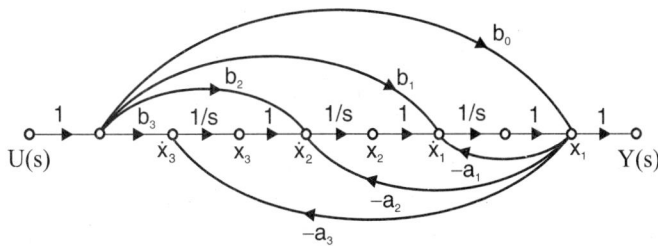

Fig 4.4 : *Signal flow graph of the system represented by the equation 4.15.*

By summing up the incoming signals to node we get, *(Refer fig 4.4a)*

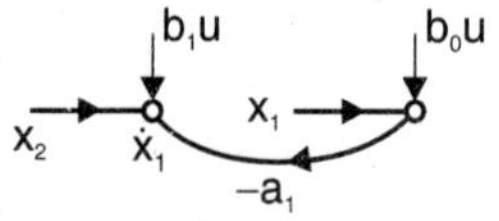

$$\dot{x}_1 = -a_1(x_1 + b_0 u) + x_2 + b_1 u$$

$$\therefore \dot{x}_1 = -a_1 x_1 + x_2 + (b_1 - a_1 b_0) u \qquad .....(4.18)$$

Fig 4.4a.

By summing up the incoming signals to node we get, *(Refer fig 4.4b)*

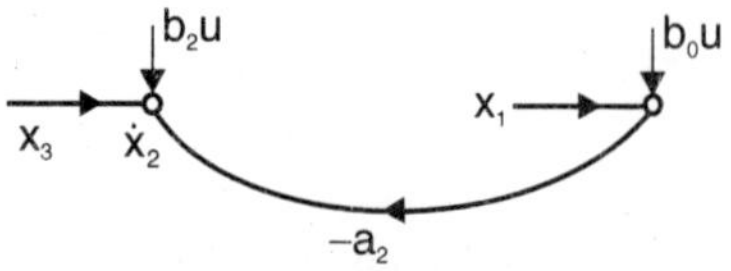

$$\dot{x}_2 = -a_2(x_1 + b_0 u) + x_3 + b_2 u$$

$$\therefore \dot{x}_2 = -a_2 x_1 + x_3 + (b_2 - a_2 b_0) u \qquad .....(4.19)$$

Fig 4.4b.

By summing up the incoming signals to node we get,*(Refer fig 4.4c)*

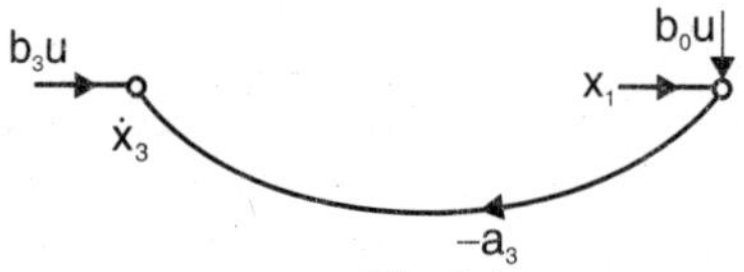

$$\dot{x}_3 = -a_3(x_1 + b_0 u) + b_3 u$$

$$\therefore \dot{x}_3 = -a_3 x_1 + (b_3 - a_3 b_0) u \qquad .....(4.20)$$

Fig 4.4c.

The output equation is given by the sum of incoming signals to output node.

$$\therefore y = x_1 + b_0 u \qquad .....(4.21)$$

On arranging the state equations and the output equations in the matrix form, we get,

$$\begin{bmatrix} \dot{x}_1 \\ \dot{x}_2 \\ \dot{x}_3 \end{bmatrix} = \begin{bmatrix} -a_1 & 1 & 0 \\ -a_2 & 0 & 1 \\ -a_3 & 0 & 0 \end{bmatrix} \begin{bmatrix} x_1 \\ x_2 \\ x_3 \end{bmatrix} + \begin{bmatrix} b_1 - a_1 b_0 \\ b_2 - a_2 b_0 \\ b_3 - a_3 b_0 \end{bmatrix} [u] \qquad .....(4.22)$$

$$y = \begin{bmatrix} 1 & 0 & 0 \end{bmatrix} \begin{bmatrix} x_1 \\ x_2 \\ x_3 \end{bmatrix} + [b_0] u \qquad .....(4.23)$$

The above results can be generalized for an nth order differential equation, and the general state model for m = n is given below.

$$\begin{bmatrix} \dot{x}_1 \\ \dot{x}_2 \\ \vdots \\ \dot{x}_{n-1} \\ \dot{x}_n \end{bmatrix} = \begin{bmatrix} -a_1 & 1 & 0 & ..... & 0 \\ -a_2 & 0 & 1 & ..... & 0 \\ \vdots & \vdots & \vdots & ..... & \vdots \\ -a_{n-1} & 0 & 0 & ..... & 1 \\ -a_n & 0 & 0 & ..... & 0 \end{bmatrix} \begin{bmatrix} x_1 \\ x_2 \\ \vdots \\ x_{n-1} \\ x_n \end{bmatrix} + \begin{bmatrix} b_1 - a_1 b_0 \\ b_2 - a_2 b_0 \\ \vdots \\ b_{n-1} - a_{n-1} b_0 \\ b_n - a_n b_0 \end{bmatrix} [u] \qquad .....(4.24)$$

$$y = \begin{bmatrix} 1 & 0 & 0 & \cdots & 0 \end{bmatrix} \begin{bmatrix} x_1 \\ x_2 \\ x_3 \\ \vdots \\ x_n \end{bmatrix} + [b_0] u \qquad .....(4.25)$$

Method 3

Consider the following n^{th} order differential equation governing the output y(t) to the input u(t) of a system.

$$\overset{n.}{y} + a_1 \overset{(n-1).}{y} + \dots\dots + a_{n-1}\dot{y} + a_n y = b_0 \overset{m.}{u} + b_1 \overset{(m-1).}{u} + \dots\dots + b_{m-1}\dot{u} + b_m u \qquad .....(4.26)$$

let $n = m = 3$

$$\therefore \dddot{y} + a_1 \ddot{y} + a_2 \dot{y} = b_0 \dddot{u} + b_1 \ddot{u} + b_2 \dot{u} + b_3 u \qquad .....(4.27)$$

On taking Laplace transform of equation (4.27) with zero initial conditions, we get

$$s^3 Y(s) + a_1 s^2 Y(s) + a_2 s Y(s) + a_3 Y(s) = b_0 s^3 U(s) + b_1 s^2 U(s) + b_2 s U(s) + b_3 U(s)$$

$$(s^3 + a_1 s^2 + a_2 s + a_3)Y(s) = (b_0 s^3 + b_1 s + b_2 s + b_3)U(s)$$

$$\therefore \frac{Y(s)}{U(s)} = \frac{b_0 s^3 + b_1 s^2 + b_2 s + b_3}{s^3 + a_1 s^2 + a_2 s + a_3}$$

Let $\dfrac{Y(s)}{U(s)} = \dfrac{X_1(s)}{U(s)} \cdot \dfrac{Y(s)}{X_1(s)}$

where, $\dfrac{X_1(s)}{U(s)} = \dfrac{1}{s^3 + a_1 s^2 + a_2 s + a_3}$ .....(4.28)

and $\dfrac{Y(s)}{X_1(s)} = b_0 s^3 + b_1 s^2 + b_2 s + b_3$ .....(4.29)

On cross multiplying the equation (4.28) we get,

$$X_1(s)[s^3 + a_1 s^2 + a_2 s + a_3] = U(s)$$

$$s^3 X_1(s) + a_1 s^2 X_1(s) + a_2 s X_1(s) + a_3 X_1(s) = U(s) \qquad .....(4.30)$$

On taking inverse laplace transform of equation (4.30) we get,

$$\dddot{x}_1 + a_1 \ddot{x}_1 + a_2 \dot{x}_1 + a_3 x_1 = u \qquad .....(4.31)$$

Let the state variable be, x_1, x_2 and x_3

where, $x_2 = \dot{x}_1$

and $x_3 = \ddot{x}_1 = \dot{x}_2$; $\therefore \dot{x}_3 = \dddot{x}_1$

On substituting the state variables in equation (4.31) we get,

$$\dot{x}_3 + a_1 x_3 + a_2 x_2 + a_3 x_1 = u$$

$$\therefore \dot{x}_3 = -a_3 x_1 - a_2 x_2 - a_1 x_3 + u$$

The state equations are,

$$\dot{x}_1 = x_2$$

$$\dot{x}_2 = x_3$$

$$\dot{x}_3 = -a_3 x_1 - a_2 x_2 - a_1 x_3 + u$$

On cross multiplying the equation (4.29) we get,

$$Y(s) = b_0 s^3 X_1(s) + b_1 s^2 X_1(s) + b_2 s X_1(s) + b_3 X_1(s)$$

On taking inverse Laplace transform of equation (4.32), we get,

$$y = b_0 \dddot{x}_1 + b_1 \ddot{x}_1 + b_2 \dot{x}_1 + b_3 x_1 \qquad .....(4.33)$$

On substituting the state variables in equation (4.33) we get,

$$y = b_0 \dot{x}_3 + b_1 x_3 + b_2 x_2 + b_3 x_1 \qquad .....(4.34)$$

Put $\dot{x}_3 = -a_3 x_1 - a_2 x_2 - a_1 x_3 + u$ in equation (4.34)

$$\therefore y = b_0(-a_3 x_1 - a_2 x_2 - a_1 x_3 + u) + b_1 x_3 + b_2 x_2 + b_3 x_1$$

$$y = (b_3 - a_3 b_0)x_1 + (b_2 - a_2 b_0)x_2 + (b_1 - a_1 b_0)x_3 + b_0 u \qquad .....(4.35)$$

The equation (4.35) is the output equation.

On arranging the state equations and output equations in the matrix form, we get,

$$\begin{bmatrix} \dot{x}_1 \\ \dot{x}_2 \\ \dot{x}_3 \end{bmatrix} = \begin{bmatrix} 0 & 1 & 0 \\ 0 & 0 & 1 \\ -a_3 & -a_2 & -a_1 \end{bmatrix} \begin{bmatrix} x_1 \\ x_2 \\ x_3 \end{bmatrix} + \begin{bmatrix} 0 \\ 0 \\ 1 \end{bmatrix} [u] \qquad .....(4.36)$$

$$y = [(b_3 - a_3 b_0) \ (b_2 - a_2 b_0) \ (b_1 - a_1 b_0)] \begin{bmatrix} x_1 \\ x_2 \\ x_3 \end{bmatrix} + [b_0] u \qquad .....(4.37)$$

The above results can be generalized for an nth order differential equation and the general state model for $m = n$ is given below.

$$\begin{bmatrix} \dot{x}_1 \\ \dot{x}_2 \\ \vdots \\ \dot{x}_{n-1} \\ \dot{x}_n \end{bmatrix} = \begin{bmatrix} 0 & 1 & 0 & ..... & 0 \\ 0 & 0 & 1 & ..... & 0 \\ \vdots & \vdots & \vdots & ..... & \vdots \\ 0 & 0 & 0 & ..... & 1 \\ -a_n & -a_{n-1} & -a_{n-2} & ..... & -a_1 \end{bmatrix} \begin{bmatrix} x_1 \\ x_2 \\ \vdots \\ x_{n-1} \\ x_n \end{bmatrix} + \begin{bmatrix} 0 \\ 0 \\ \vdots \\ 0 \\ 1 \end{bmatrix} [u] \qquad .....(4.38)$$

$$y = [(b_n - a_n b_0)(b_{n-1} - a_{n-1} b_0)......(b_2 - a_2 b_0)(b_1 - a_1 b_0)] \begin{bmatrix} x_1 \\ x_2 \\ \vdots \\ x_{n-1} \\ x_n \end{bmatrix} + b_0 u \qquad .....(4.39)$$

ADVANTAGES OF PHASE VARIABLES

The state space model can be directly formed by inspection from the differential equations governing the system. The phase variables provide a link between the transfer function design approach and time-domain design approach.

DISADVANTAGE OF PHASE VARIABLES

The phase variables are not physical variables of the system and therefore are not available for measurement and control purposes.

EXAMPLE 4.7

Construct a state model for a system characterized by the differential equation,

$$\frac{d^3y}{dt^3} + 6\frac{d^2y}{dt^2} + 11\frac{dy}{dt} + 6y + u = 0.$$

Give the block diagram representation of the state model.

SOLUTION

Let us choose y and their derivatives as state variables. The system is governed by third order differential equation and so the number of state variables are three.

The state variables x_1, x_2 and x_3 are related to phase variables as follows

$$x_1 = y$$

$$x_2 = \frac{dy}{dt} = \dot{x}_1$$

$$x_3 = \frac{d^2y}{dt^2} = \dot{x}_2$$

Put $y = x_1$, $\frac{dy}{dt} = x_2$ and $\frac{d^2y}{dt^2} = x_3$ and $\frac{d^3y}{dt^3} = \dot{x}_3$ in the given equation,

$$\therefore \dot{x}_3 + 6x_3 + 11x_2 + 6x_1 + u = 0$$

or $\dot{x}_3 = -6x_1 - 11x_2 - 6x_3 - u$

The state equations are

$$\dot{x}_1 = x_2$$
$$\dot{x}_2 = x_3$$
$$\dot{x}_3 = -6x_1 - 11x_2 - 6x_3 - u$$

On arranging the state equations in the matrix form we get,

$$\begin{bmatrix} \dot{x}_1 \\ \dot{x}_2 \\ \dot{x}_3 \end{bmatrix} = \begin{bmatrix} 0 & 1 & 0 \\ 0 & 0 & 1 \\ -6 & -11 & -6 \end{bmatrix} \begin{bmatrix} x_1 \\ x_2 \\ x_3 \end{bmatrix} + \begin{bmatrix} 0 \\ 0 \\ -1 \end{bmatrix} [u]$$

Here, y = output
But, $y = x_1$

$$\therefore \text{The output equation is, } y = \begin{bmatrix} 1 & 0 & 0 \end{bmatrix} \begin{bmatrix} x_1 \\ x_2 \\ x_3 \end{bmatrix}$$

The state equation and output equation, constitutes the state model of the system.

The block diagram form of the state diagram of the system is shown in fig 1.

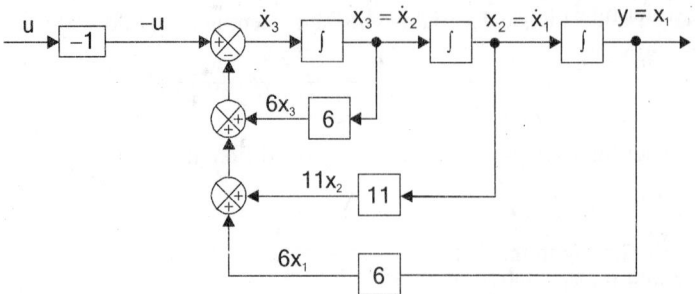

Fig 1 : Block diagram form of state diagram.

EXAMPLE 4.8

The state diagram of a system is shown in fig 1. Assign state variables and obtain the state model of the system.

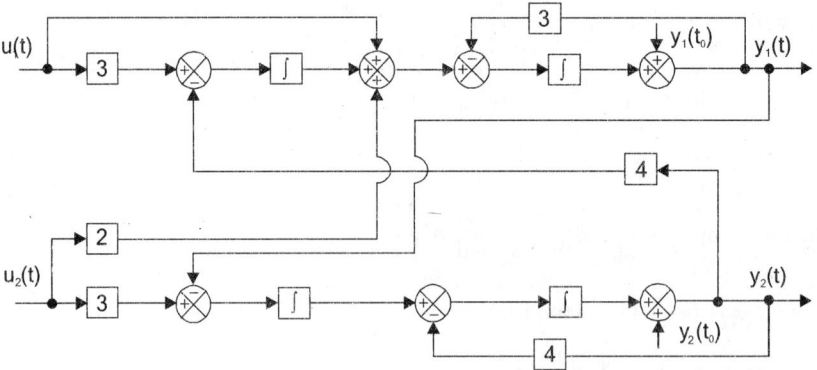

SOLUTION   Fig 1.

Since there are 4-integrators in the state diagram we can assign, 4 state variables. The state variables can be assigned at the output of the integrators as shown in fig 2. Hence at the input of the integrator, the first derivative of the state variable will be available. The state equations are formed by summing all the incoming signals to the input of the integrator and equating to the corresponding first derivative of the state variable.

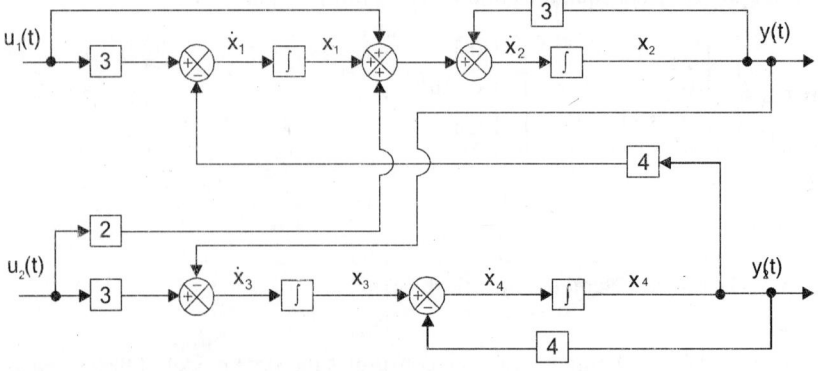

Fig 2.

On adding the signals coming to the 1st
integrator we get, (refer fig 3)

$$\dot{x}_1 = -4x_4 + 3u_1$$

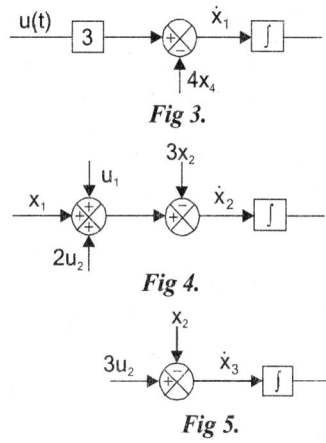

Fig 3.

On adding the signals coming to the 2nd
integrator we get, (refer fig 4)

$$\dot{x}_2 = x_1 - 3x_2 + u_1 + 2u_2$$

Fig 4.

On adding the signals coming to the 3rd
integrator we get, (refer fig 5)

$$\dot{x}_3 = -x_2 + 3u_2$$

Fig 5.

On adding the signals coming to the 4th
integrator we get, (refer fig 6)

$$\dot{x}_4 = x_3 - 4x_4$$

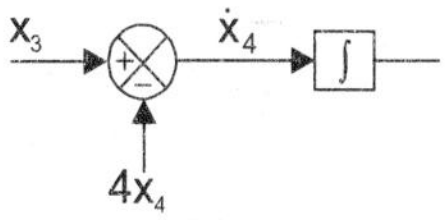

The state equations are

$$\dot{x}_1 = -4x_4 + 3u_1$$
$$\dot{x}_2 = x_1 - 3x_2 + u_1 + 2u_2$$
$$\dot{x}_3 = -x_2 + 3u_2$$
$$\dot{x}_4 = x_3 - 4x_4$$

Fig 6.

The output equations are, $y_1 = x_2$ and $y_2 = x_4$.

The state equations and output equations are arranged in the matrix form as shown below. The state equations and output equations together constitute the state model of the system.

$$\begin{bmatrix} \dot{x}_1 \\ \dot{x}_2 \\ \dot{x}_3 \\ \dot{x}_4 \end{bmatrix} = \begin{bmatrix} 0 & 0 & 0 & -4 \\ 1 & -3 & 0 & 0 \\ 0 & -1 & 0 & 0 \\ 0 & 0 & 1 & -4 \end{bmatrix} \begin{bmatrix} x_1 \\ x_2 \\ x_3 \\ x_4 \end{bmatrix} + \begin{bmatrix} 3 & 0 \\ 1 & 2 \\ 0 & 3 \\ 0 & 0 \end{bmatrix} \begin{bmatrix} u_1 \\ u_2 \end{bmatrix}$$

$$\begin{bmatrix} y_1 \\ y_2 \end{bmatrix} = \begin{bmatrix} 0 & 1 & 0 & 0 \\ 0 & 0 & 0 & 1 \end{bmatrix} \begin{bmatrix} x_1 \\ x_2 \\ x_3 \\ x_4 \end{bmatrix}$$

EXAMPLE 4.9

The state diagram of a linear system is given below. Assign state variables and obtain the state model.

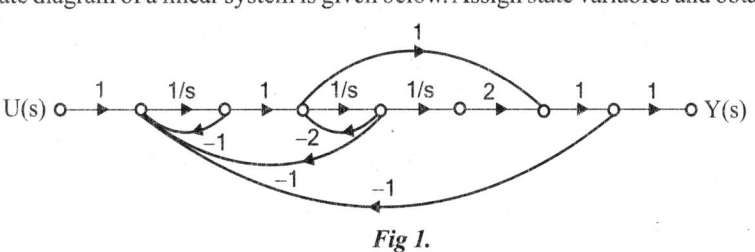

Fig 1.

SOLUTION

Since there are three integrators (1/s) we can assign three state variables. The state variables are assigned at the output of the integrator as shown in fig 2. At the input of the integrator we have the first derivative of the state variable. The state equations are formed by summing all the signals at the input of integrator and equating to the corresponding first derivative of state variable.

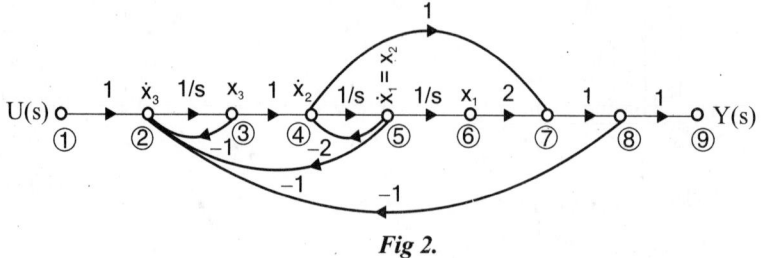

Fig 2.

On adding the signals coming to node-5, we get, (refer fig 3).

$$\dot{x}_1 = x_2$$

On adding the signals coming to node-4, we get, (refer (fig 4)

$$\dot{x}_2 = -2x_2 + x_3$$

On adding the signals coming to node-2, we get, (refer fig 5)

$$\dot{x}_3 = -(\dot{x}_2 + 2x_1) - x_2 - x_3 + u = -\dot{x}_2 - 2x_1 - x_2 - x_3 + u$$

Put $\dot{x}_2 = -2x_2 + x_3$

$$\therefore \ \dot{x}_3 = +2x_2 - x_3 - 2x_1 - x_2 - x_3 + u = -2x_1 + x_2 - 2x_3 + u$$

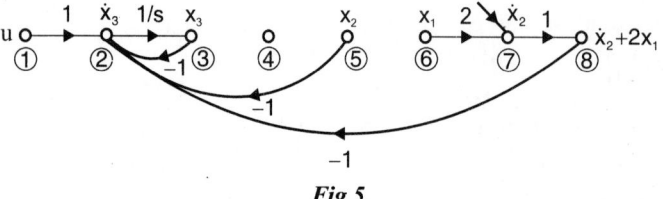

Fig 5.

The state equations are

$$\dot{x}_1 = x_2$$
$$\dot{x}_2 = -2x_2 + x_3$$
$$\dot{x}_3 = -2x_1 + x_2 - 2x_3 + u$$

The output equation is obtained by adding the signals coming to output node (refer fig 6)

$$y = 2x_1 + \dot{x}_2$$

Put $\dot{x}_2 = -2x_2 + x_3$

$$y = 2x_1 + (-2x_2 + x_3)$$
$$y = 2x_1 - 2x_2 + x_3$$

The state equations and the output equation are arranged in matrix form as shown below.

$$\begin{bmatrix} \dot{x}_1 \\ \dot{x}_2 \\ \dot{x}_3 \end{bmatrix} = \begin{bmatrix} 0 & 1 & 0 \\ 0 & -2 & 1 \\ -2 & 1 & -2 \end{bmatrix} \begin{bmatrix} x_1 \\ x_2 \\ x_3 \end{bmatrix} + \begin{bmatrix} 0 \\ 0 \\ 1 \end{bmatrix} [u]$$

$$y = \begin{bmatrix} 2 & -2 & 1 \end{bmatrix} \begin{bmatrix} x_1 \\ x_2 \\ x_3 \end{bmatrix}$$

EXAMPLE 4.10

Obtain the state model of the system whose transfer function is given as,

$$\frac{Y(s)}{U(s)} = \frac{10}{s^3 + 4s^2 + 2s + 1}$$

SOLUTION

Method 1

Given that, $\dfrac{Y(s)}{U(s)} = \dfrac{10}{s^3 + 4s^2 + 2s + 1}$ .....(4.10.1)

On cross multiplying the equation (4.10.1) we get,

$$Y(s)[s^3 + 4s^2 + 2s + 1] = 10\ U(s)$$

$$s^3 Y(s) + 4s^2 Y(s) + 2s\ Y(s) + Y(s) = 10\ U(s) \qquad \qquad .....(4.10.2)$$

On taking inverse Laplace transform of equation (4.10.2) we get,

$$\dddot{y} + 4\ddot{y} + 2\dot{y} + y = 10u \qquad \qquad .....(4.10.3)$$

Let us define state variables as follows,

$$x_1 = y \quad ; \quad x_2 = \dot{y} \quad ; \quad x_3 = \ddot{y}$$

Put $\dddot{y} = \dot{x}_3$; $\ddot{y} = x_3$; $\dot{y} = x_2$ and $y = x_1$ in the equation (4.10.3)

$$\therefore \dot{x}_3 + 4x_3 + 2x_2 + x_1 = 10u$$
$$\text{or } \dot{x}_3 = -x_1 - 2x_2 - 4x_3 + 10u$$

The state equations are

$$\dot{x}_1 = x_2 \quad ; \quad \dot{x}_2 = x_3 \quad ; \quad \dot{x}_3 = -x_1 - 2x_2 - 4x_3 + 10u$$

The output equation is $y = x_1$

The state model in the matrix form is,

$$\begin{bmatrix} \dot{x}_1 \\ \dot{x}_2 \\ \dot{x}_3 \end{bmatrix} = \begin{bmatrix} 0 & 1 & 0 \\ 0 & 0 & 1 \\ -1 & -2 & -4 \end{bmatrix} \begin{bmatrix} x_1 \\ x_2 \\ x_3 \end{bmatrix} + \begin{bmatrix} 0 \\ 0 \\ 10 \end{bmatrix} [u] \quad ; \quad y = \begin{bmatrix} 1 & 0 & 0 \end{bmatrix} \begin{bmatrix} x_1 \\ x_2 \\ x_3 \end{bmatrix}$$

Method 2

$$\frac{Y(s)}{U(s)} = \frac{10}{s^3 + 4s^2 + 2s + 1} = \frac{10}{s^3\left(1 + \dfrac{4}{s} + \dfrac{2}{s^2} + \dfrac{1}{s^3}\right)}$$

$$= \frac{10/s^3}{1 - \left(-\dfrac{4}{s} - \dfrac{2}{s^2} - \dfrac{1}{s^3}\right)}$$

The signal flow graph for the above transfer function can be constructed as shown in fig 1 with a single forward path consisting of three integrators and with path gain $10/s^3$. The graph will have three individual loops with loop gains $-4/s$, $-2/s^2$, and $-1/s^3$.

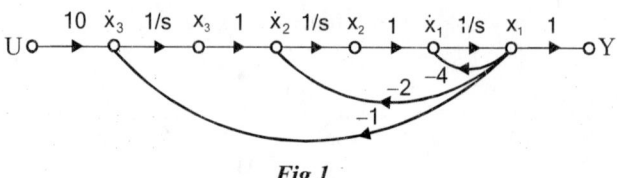

Fig 1.

Assign state variables at the output of the integrator $(1/s)$. The state equations are obtained by summing the incoming signals to the input of the integrators and equating them to the corresponding first derivative of the state variable. [Refer fig 2 to fig 4]

The state equations are

$$\dot{x}_1 = -4x_1 + x_2$$

$$\dot{x}_2 = -2x_1 + x_3$$

$$\dot{x}_3 = -x_1 + 10u$$

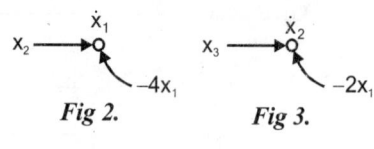

Fig 2.   Fig 3.

Fig 4.

The output equation is, $y = x_1$

The state model in the matrix form is,

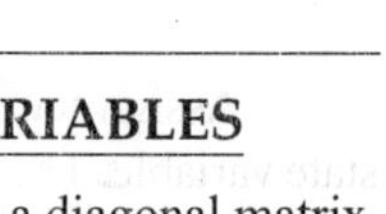

$$\begin{bmatrix} \dot{x}_1 \\ \dot{x}_2 \\ \dot{x}_3 \end{bmatrix} = \begin{bmatrix} -4 & 1 & 0 \\ -2 & 0 & 1 \\ -1 & 0 & 0 \end{bmatrix} \begin{bmatrix} x_1 \\ x_2 \\ x_3 \end{bmatrix} + \begin{bmatrix} 0 \\ 0 \\ 10 \end{bmatrix} [u] \; ; \; y = \begin{bmatrix} 1 & 0 & 0 \end{bmatrix} \begin{bmatrix} x_1 \\ x_2 \\ x_3 \end{bmatrix}$$

4.7   STATE SPACE REPRESENTATION USING CANONICAL VARIABLES

In canonical form (or normal form) of state model, the system matrix A will be a diagonal matrix. The elements on the diagonal are the poles of the transfer function of the system.

By partial fraction expansion, the transfer function $Y(s)/U(s)$ of the nth order system can be expressed as shown in equation (4.40).

$$\frac{Y(s)}{U(s)} = b_0 + \frac{C_1}{s + \lambda_1} + \frac{C_2}{s + \lambda_2} + \dots\dots + \frac{C_n}{s + \lambda_n} \qquad\qquad \dots(4.40)$$

where $C_1, C_2, C_3, \dots\dots C_n$ are residues and $\lambda_1, \lambda_2, \dots\dots \lambda_n$ are roots of denominator polynomial (or poles of the system).

The equation (4.40) can be rearranged as shown below.

$$\frac{Y(s)}{U(s)} = b_0 + \frac{C_1}{s\left(1 + \frac{\lambda_1}{s}\right)} + \frac{C_2}{s\left(1 + \frac{\lambda_2}{s}\right)} + \ldots\ldots + \frac{C_n}{s\left(1 + \frac{\lambda_n}{s}\right)}$$

$$= b_0 + \frac{C_1/s}{1 + \lambda_1/s} + \frac{C_2/s}{1 + \lambda_2/s} + \ldots\ldots + \frac{C_n/s}{1 + \lambda_n/s}$$

$$\therefore \; Y(s) = b_0 U(s) + \left[\frac{1/s}{1 + (1/s) \times \lambda_1} \times C_1\right] U(s) + \left[\frac{1/s}{1 + (1/s) \times \lambda_2} \times C_2\right] U(s)$$

$$+ \ldots\ldots + \left[\frac{1/s}{1 + (1/s) \times \lambda_n} \times C_n\right] U(s) \qquad \ldots\ldots(4.41)$$

The equation (4.41) can be represented by a block diagram as shown in fig 4.5

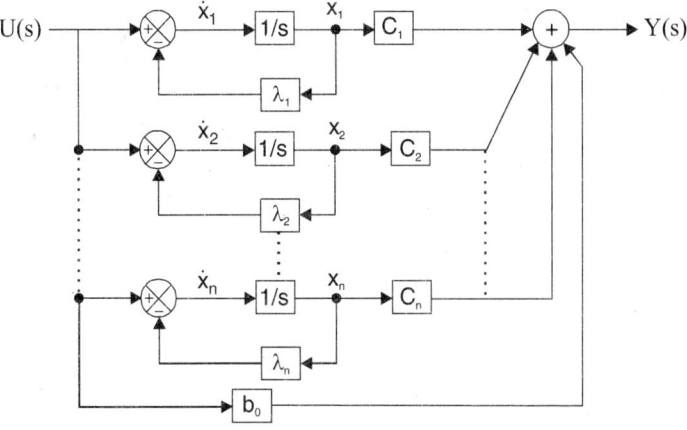

Fig 4.5 : *Block diagram of canonical state model.*

Assign state variables at the output of integrator. The input of the integrator will be first derivative of state variable. The state equations are formed by adding the incoming signals to the integrator and equating to first derivative of state variable. The state equations are,

$$\dot{x}_1 = -\lambda_1 x_1 + u$$
$$\dot{x}_2 = -\lambda_2 x_2 + u$$
$$\vdots$$
$$\dot{x}_n = -\lambda_n x_n + u$$

The output equation is, $y = C_1 x_1 + C_2 x_2 + \ldots\ldots C_n x_n + b_0 u$

The canonical form of state model in the matrix form is given below

$$\begin{bmatrix} \dot{x}_1 \\ \dot{x}_2 \\ \dot{x}_3 \\ \vdots \\ \dot{x}_n \end{bmatrix} = \begin{bmatrix} -\lambda_1 & 0 & 0 & \cdots & 0 \\ 0 & -\lambda_2 & 0 & \cdots & 0 \\ 0 & 0 & -\lambda_3 & \cdots & 0 \\ \vdots & \vdots & \vdots & \cdots & \vdots \\ 0 & 0 & 0 & \cdots & -\lambda_n \end{bmatrix} \begin{bmatrix} x_1 \\ x_2 \\ x_3 \\ \vdots \\ x_n \end{bmatrix} + \begin{bmatrix} 1 \\ 1 \\ 1 \\ \vdots \\ 1 \end{bmatrix} [u]$$

$$y = [C_1 C_2 C_3 \ldots \ldots C_n] \begin{bmatrix} x_1 \\ x_2 \\ x_3 \\ \vdots \\ x_n \end{bmatrix} + [b_0] [u] \qquad \qquad \ldots\ldots(4.42)$$

The advantage of canonical form is that the state equations are independent of each other. The disadvantage is that the canonical variables are not physical variables and so they are not available for measurement and control.

When a pole of the transfer function has multiplicity, the canonical state model will be in a special form called Jordan canonical form. In this form the system matrix **A** will have a Jordan block of size q × q, correspond to a pole of value λ_i with multiplicity q. In the Jordan block the diagonal element will be the poles and the element just above the diagonal is one.

Consider a system with poles $\lambda_1, \lambda_1, \lambda_1, \lambda_4, \lambda_5, \ldots\ldots \lambda_n$, where λ_1 has multiplicity of three. The input matrix (**B**) and system matrix for this case will be as shown in equation (7.42a). The system matrix is also denoted as **J**.

Jordan block of size 3×3

$$B = \begin{bmatrix} 0 \\ 0 \\ 1 \\ 1 \\ \vdots \\ 1 \end{bmatrix} \quad ; \quad A = J = \begin{bmatrix} -\lambda_1 & 1 & 0 & 0 & \cdots & 0 \\ 0 & -\lambda_1 & 1 & 0 & \cdots & 0 \\ 0 & 0 & -\lambda_1 & 0 & \cdots & 0 \\ 0 & 0 & 0 & -\lambda_4 & \cdots & 0 \\ \vdots & \vdots & \vdots & \vdots & & \vdots \\ 0 & 0 & 0 & 0 & \cdots & -\lambda_n \end{bmatrix} \qquad \ldots\ldots(4.42a)$$

The transfer function of the system for this case is given by equation (4.42a) and the block diagram is shown in fig 4.5a.

$$\frac{Y(s)}{U(s)} = b_0 + \frac{C_1}{(s+\lambda_1)^3} + \frac{C_2}{(s+\lambda_1)^2} + \frac{C_3}{s+\lambda_1} + \frac{C_4}{s+\lambda_4} + \ldots\ldots + \frac{C_n}{s+\lambda_n} \qquad \ldots\ldots(4.42b)$$

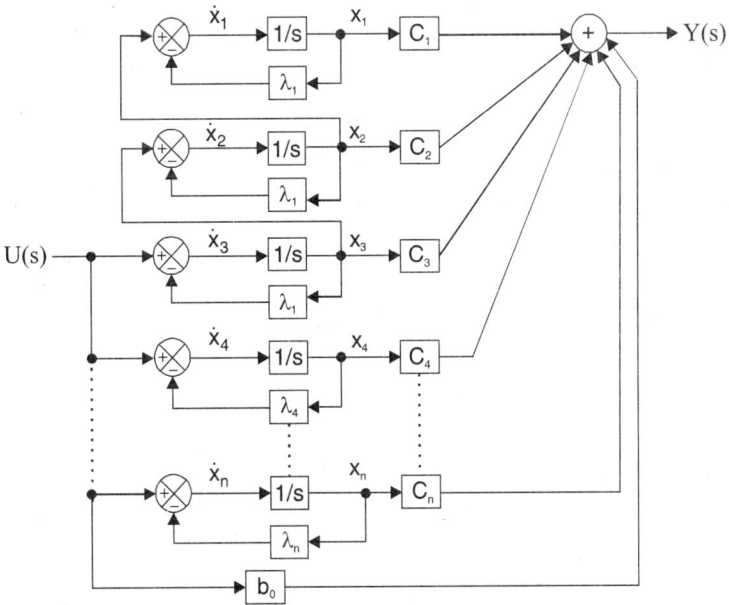

Fig 4.5a : Block diagram of Jordan canonical state model.

EXAMPLE 4.11

A feedback system has a closed-loop transfer function

$$\frac{Y(s)}{U(s)} = \frac{10(s+4)}{s(s+1)(s+3)}$$

Construct three different state models for this system and give block diagram representation for each state model.

SOLUTION

Model 1

$$\frac{Y(s)}{U(s)} = \frac{10(s+4)}{s(s+1)(s+3)} = \frac{10s+40}{s(s^2+4s+3)} = \frac{10s+40}{s^3+4s^2+3s} = \frac{10s+40}{s^3\left(1+\dfrac{4}{s}+\dfrac{3}{s^2}\right)} = \frac{\dfrac{10}{s^2}+\dfrac{40}{s^3}}{1-\left(-\dfrac{4}{s}-\dfrac{3}{s^2}\right)}$$

A signal flow graph for the above transfer function can be constructed as shown fig 1, with two forward paths and two individual loops. The forward path gains are $10/s^2$ and $40/s^3$. The loop gains are $-4/s$ and $-3/s^2$.

Assign state variables at the output of integrator as shown in fig 1 and so the input of integrator is first derivative of state variable. The state equations are obtained by summing all the incoming signals to the integrator and equating to the corresponding first derivative of the state variable. [Refer fig 2 to 4].

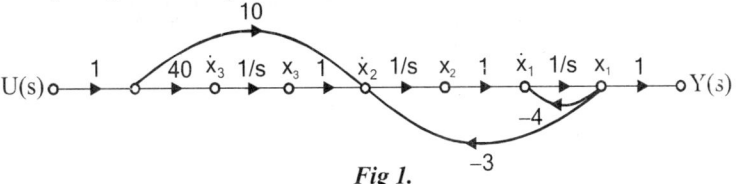

Fig 1.

The state equations are

$$\dot{x}_1 = -4x_1 + x_2$$
$$\dot{x}_2 = -3x_1 + x_3 + 10u$$
$$\dot{x}_3 = 40u$$

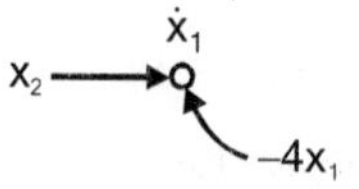

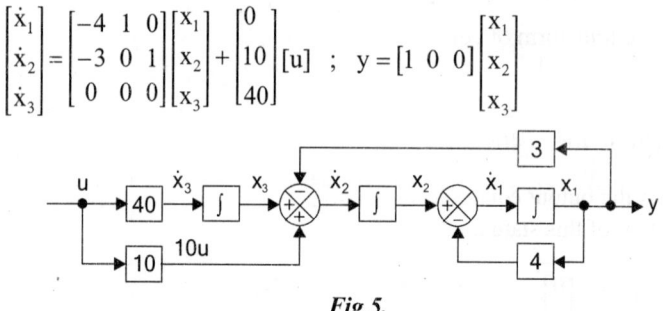

Fig 2. 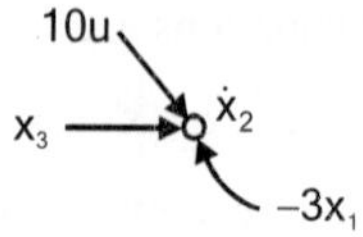   *Fig 3.* 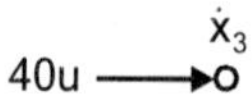   *Fig 4.*

The output equation is, $y = x_1$

The state model is obtained by arranging the state equations and the output equation in the matrix form as shown below. The block diagram representation of this state model is shown in fig 5.

$$\begin{bmatrix} \dot{x}_1 \\ \dot{x}_2 \\ \dot{x}_3 \end{bmatrix} = \begin{bmatrix} -4 & 1 & 0 \\ -3 & 0 & 1 \\ 0 & 0 & 0 \end{bmatrix} \begin{bmatrix} x_1 \\ x_2 \\ x_3 \end{bmatrix} + \begin{bmatrix} 0 \\ 10 \\ 40 \end{bmatrix} [u] \quad ; \quad y = \begin{bmatrix} 1 & 0 & 0 \end{bmatrix} \begin{bmatrix} x_1 \\ x_2 \\ x_3 \end{bmatrix}$$

Fig 5.

Model 2

Given that, $\dfrac{Y(s)}{U(s)} = \dfrac{10(s+4)}{s(s+1)(s+3)}$

Let, $\dfrac{Y(s)}{U(s)} = \dfrac{X_1(s)}{U(s)} \cdot \dfrac{Y(s)}{X_1(s)} = \dfrac{10(s+4)}{s(s+1)(s+3)}$

Let, $\dfrac{X_1(s)}{U(s)} = \dfrac{1}{s(s+1)(s+3)}$ and $\dfrac{Y(s)}{X_1(s)} = 10(s+4)$

$$\dfrac{X_1(s)}{U(s)} = \dfrac{1}{s(s+1)(s+3)} = \dfrac{1}{s(s^2+4s+3)} \quad ; \quad \therefore \dfrac{X_1(s)}{U(s)} = \dfrac{1}{s^3+4s^2+3s} \qquad \text{......(4.11.1)}$$

On cross multiplying the equation (4.11.1) we get,

$$X_1(s)[s^3 + 4s^2 + 3s] = U(s)$$

$$\therefore \ s^3 X_1(s) + 4s^2 X_1(s) + 3s\, X_1(s) = U(s) \qquad \text{......(4.11.2)}$$

On taking inverse Laplace transform of equation (4.11.2) we get,

$$\dddot{x}_1 + 4\ddot{x}_1 + 3\dot{x}_1 = u \qquad \text{......(4.11.3)}$$

Let the state variables be x_1, x_2 and x_3 ; where $x_2 = \dot{x}_1$ and $x_3 = \ddot{x}_1$.

Put $\dot{x}_1 = x_2$, $\ddot{x}_1 = x_3$ and $\dddot{x}_1 = \dot{x}_3$ in equation (4.11.3),

$$\therefore \ \dot{x}_3 + 4x_3 + 3x_2 = u \ \text{ (or) } \ \dot{x}_3 = -3x_2 - 4x_3 + u$$

The state equations are

$$\dot{x}_1 = x_2 \quad ; \quad \dot{x}_2 = x_3 \quad ; \quad \dot{x}_3 = -3x_2 - 4x_3 + u$$

Consider the second part of transfer function,

$$\frac{Y(s)}{X_1(s)} = 10(s+4) = 10s + 40 \qquad\qquad .....(4.11.4)$$

On cross multiplying equation (4.11.4) we get,

$$Y(s) = 10s\, X_1(s) + 40\, X_1(s) \qquad\qquad .....(4.11.5)$$

On taking inverse Laplace transform of equation (4.11.5) we get,

$$y = 10\dot{x}_1 + 40x_1$$

Put $\dot{x}_1 = x_2$, $\therefore y = 10x_2 + 40x_1 = 40x_1 + 10x_2$

Here, $y = 40x_1 + 10x_2$ is the output equation. The state model in the matrix form is shown below. The block diagram representation of this state model is shown in fig 6.

$$\begin{bmatrix} \dot{x}_1 \\ \dot{x}_2 \\ \dot{x}_3 \end{bmatrix} = \begin{bmatrix} 0 & 1 & 0 \\ 0 & 0 & 1 \\ 0 & -3 & -4 \end{bmatrix} \begin{bmatrix} x_1 \\ x_2 \\ x_3 \end{bmatrix} + \begin{bmatrix} 0 \\ 0 \\ 1 \end{bmatrix}[u] \quad ; \quad y = \begin{bmatrix} 40 & 10 & 0 \end{bmatrix} \begin{bmatrix} x_1 \\ x_2 \\ x_3 \end{bmatrix}$$

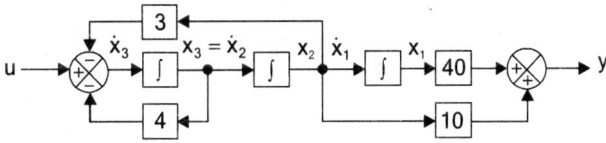

Fig 6.

Model 3

$$\frac{Y(s)}{U(s)} = \frac{10(s+4)}{s(s+1)(s+3)}$$

By partial fraction expansion Y(s)/U(s) can be expressed as,

$$\frac{Y(s)}{U(s)} = \frac{10(s+4)}{s(s+1)(s+3)} = \frac{A}{s} + \frac{B}{s+1} + \frac{C}{s+3}$$

$$A = \frac{10(s+4)}{(s+1)(s+3)}\bigg|_{s=0} = \frac{10 \times 4}{1 \times 3} = \frac{40}{3}$$

$$B = \frac{10(s+4)}{s(s+3)}\bigg|_{s=-1} = \frac{10(-1+4)}{1(-1+3)} = \frac{10 \times 3}{-1 \times 2} = -15$$

$$C = \frac{10(s+4)}{s(s+1)}\bigg|_{s=-3} = \frac{10(-3+4)}{-3(-3+1)} = \frac{10 \times 1}{-3 \times (-2)} = \frac{5}{3}$$

$$\frac{Y(s)}{U(s)} = \frac{40/3}{s} - \frac{15}{s+1} + \frac{5/3}{s+3} \qquad\qquad .....(4.11.6)$$

The equation (4.11.6) can be rearranged as shown below

$$\frac{Y(s)}{U(s)} = \frac{40/3}{s} - \frac{15}{s(1+1/s)} + \frac{5/3}{s(1+3/s)}$$

$$\therefore \ Y(s) = \left[\frac{1}{s} \times \frac{40}{3}\right]U(s) - \left[\frac{1/s}{1+\frac{1}{s}\times 1} \times 15\right]U(s) + \left[\frac{1/s}{1+\frac{1}{s}\times 3} \times \frac{5}{3}\right]U(s) \qquad .....(4.11.7)$$

The block diagram of the equation (4.11.7) is shown in fig 7.

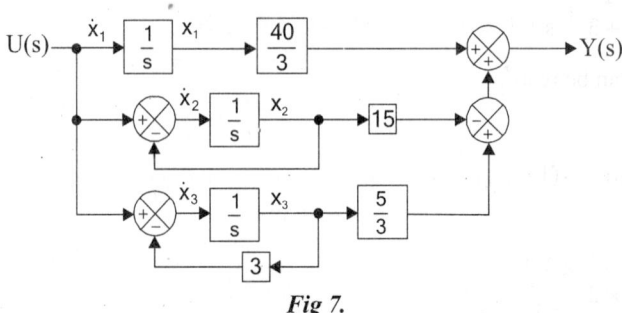

Fig 7.

Assign state variables at the output of the integrator as shown in fig 7. At the input of the integrator, the first derivative of the state variables will be available. The state equations are obtained by adding incoming signals to the integrator and equating to the corresponding first derivative of the state variable.

The state equations are

$$\dot{x}_1 = u$$
$$\dot{x}_2 = -x_2 + u$$
$$\dot{x}_3 = -3x_3 + u$$

The output equation is $y = \frac{40}{3}x_1 - 15x_2 + \frac{5}{3}x_3$

The state model in the matrix form is shown below. The fig 7 is the block diagram representation of this state model.

$$\begin{bmatrix} \dot{x}_1 \\ \dot{x}_2 \\ \dot{x}_3 \end{bmatrix} = \begin{bmatrix} 0 & 0 & 0 \\ 0 & -1 & 0 \\ 0 & 0 & -3 \end{bmatrix}\begin{bmatrix} x_1 \\ x_2 \\ x_3 \end{bmatrix} + \begin{bmatrix} 1 \\ 1 \\ 1 \end{bmatrix}[u] \ ; \ \ y = \begin{bmatrix} \frac{40}{3} & -15 & \frac{5}{3} \end{bmatrix}\begin{bmatrix} x_1 \\ x_2 \\ x_3 \end{bmatrix}$$

EXAMPLE 4.12

Determine the canonical state model of the system, whose transfer function is

T(s) = 2(s+5)/[(s+2)(s+3)(s+4)].

SOLUTION

By partial fraction expansion,

$$\frac{Y(s)}{U(s)} = \frac{2(s+5)}{(s+2)(s+3)(s+4)} = \frac{A}{s+2} + \frac{B}{s+3} + \frac{C}{s+4}$$

$$A = \frac{2(s+5)}{(s+3)(s+4)}\bigg|_{s=-2} = \frac{2(-2+5)}{(-2+3)(-2+4)} = \frac{2 \times 3}{1 \times 2} = 3$$

$$B = \frac{2(s+5)}{(s+2)(s+4)}\bigg|_{s=-3} = \frac{2(-3+5)}{(-3+2)(-3+4)} = \frac{2 \times 2}{-1 \times 1} = -4$$

$$C = \frac{2(s+5)}{(s+2)(s+4)}\bigg|_{s=-4} = \frac{2(-4+5)}{(-4+2)(-4+3)} = \frac{2 \times 1}{-2 \times (-1)} = 1$$

$$\therefore \frac{Y(s)}{U(s)} = \frac{3}{s+2} - \frac{4}{s+3} + \frac{1}{s+4} \qquad\qquad .....(4.12.1)$$

The equation (4.12.1) can be rearranged as shown below

$$\frac{Y(s)}{U(s)} = \frac{3}{s(1+2/s)} - \frac{4}{s(1+3/s)} + \frac{1}{s(1+4/s)}$$

$$\therefore Y(s) = \left[\frac{\frac{1}{s}}{1+\frac{1}{s}\times 2}\times 3\right] U(s) - \left[\frac{\frac{1}{s}}{1+\frac{1}{s}\times 3}\times 4\right] U(s) + \left[\frac{\frac{1}{s}}{1+\frac{1}{s}\times 4}\right] U(s) \qquad .....(4.12.2)$$

The equation (4.12.2) can be represented by the block diagram in fig 1.

Assign state variables at the output of the integrators as shown in fig 1. At the input of the integrators we have first derivative of the state variables. The state equations are formed by adding all the incoming signals to the integrator and equating to the corresponding first derivative of state variable.

The state equations are

$$\dot{x}_1 = -2x_1 + u \quad ; \quad \dot{x}_2 = -3x_2 + u \quad ; \quad \dot{x}_3 = -4x_3 + u$$

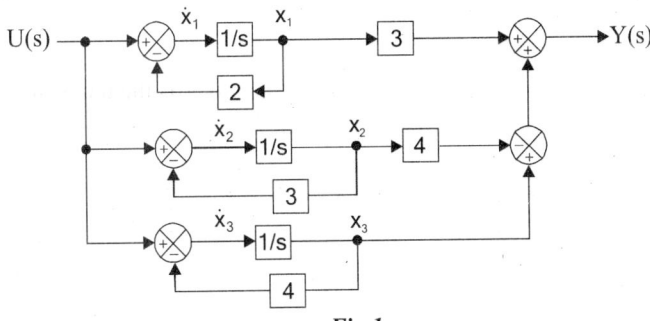

Fig 1.

The output equation is, $y = 3x_1 - 4x_2 + x_3$

The state model in matrix form is given below

$$\begin{bmatrix} \dot{x}_1 \\ \dot{x}_2 \\ \dot{x}_3 \end{bmatrix} = \begin{bmatrix} -2 & 0 & 0 \\ 0 & -3 & 0 \\ 0 & 0 & -4 \end{bmatrix} \begin{bmatrix} x_1 \\ x_2 \\ x_3 \end{bmatrix} + \begin{bmatrix} 1 \\ 1 \\ 1 \end{bmatrix} [u] \quad ; \quad y = \begin{bmatrix} 3 & -4 & 1 \end{bmatrix} \begin{bmatrix} x_1 \\ x_2 \\ x_3 \end{bmatrix}$$

4.8   SOLUTION OF STATE EQUATIONS

SOLUTION OF HOMOGENEOUS STATE EQUATIONS (Solution of state equations without input or excitation)

Consider a first order differential equation, with initial condition, $x(0) = x_0$.

$$\frac{dx}{dt} = ax \quad ; \quad x(0) = x_0 \qquad \qquad .....(4.43)$$

On rearranging equation (4.43) we get, $\frac{dx}{x} = a \, dt$ .....(4.44)

On integrating equation (4.44) we get,

$$\log x = at + C$$

$$\therefore x = e^{(at + C)} = e^{at}.e^C \qquad \qquad .....(4.45)$$

When $t = 0$, from equation (4.45) we get, $x = x(0) = e^C$

Given that, $x(0) = x_0$; $\therefore e^C = x_0$

On substituting the initial condition in equation (4.45), we get the solution of first order differential equation as

$$x = e^{at} x_0 \qquad \qquad .....(4.46)$$

We know that, $\quad e^x = \left[1 + x + \frac{1}{2!}x^2 + ... + \frac{1}{n}x^n + ... \right]$ .....(4.47)

From equations (4.46) and (4.47) we get,

$$x = e^{at}x_0 = \left(1 + at + \frac{1}{2!}a^2t^2 + \frac{1}{3!}a^3t^3 + ... + \frac{1}{i!}a^it^i + ... \right)x_0 \qquad .....(4.48)$$

Consider the state equations without input vector, (i.e., homogeneous state equation)

$$\dot{X}(t) = A\,X(t) \quad ; \quad X(0) = X_0 \qquad \qquad .....(4.49)$$

where, $X(0)$ is the initial condition vector.

By analogy of the solution of first order differential equation [equation (4.48)], the solution of the matrix or vector equation can be assumed as shown in equation (4.50)

$$X(t) = A_0 + A_1t + A_2t^2 + A_3t^3 + ...+A_it^i +......... \qquad .....(4.50)$$

where, $A_0, A_1, A_2, ... A_i, ...$ are matrices and the elements of the matrices are constants.

On differentiating the equation (4.50) we get,

$$\dot{X}(t) = A_1 + 2A_2t + 3A_3t^2 + ... + iA_it^{i-1} + ... \qquad .....(4.51)$$

On multiplying equation (4.50) by A, we get,

$$A\,X(t) = A[A_0 + A_1t + A_2t^2 + A_3t^3 + ... + A_it^i + ...] \qquad .....(4.52)$$

From equation (4.49), we know that $\dot{X}(t) = A\,X(t)$. Therefore we can equate the coefficients of equal powers of t in equations (4.51) and (4.52) as shown below.

On equating constants we get,

$$A_1 = A \, A_0$$

On equating coefficients of t we get,

$$2A_2 = A A_1$$

$$\therefore A_2 = \frac{1}{2} \, A \, A_1$$

Put $A_1 = A \, A_0$

$$\therefore A_2 = \frac{1}{2} \, A \, A A_0$$

$$A_2 = \frac{1}{2} \, A^2 \, A_0$$

On equating coefficients of t^2 we get,

$$3A_3 = A A_2$$

$$\therefore A_3 = \frac{1}{3} A A_2$$

Put $A_2 = \frac{1}{2} A^2 \, A_0$

$$\therefore A_3 = \frac{1}{3} \, A \times \frac{1}{2} A^2 A_0$$

$$A_3 = \frac{1}{3!} \, A^3 A_0$$

Similarly, on equating coefficient of t^i we get,

$$A_i = \frac{1}{i!} \, A^i A_0$$

In the above analysis, the matrices A_1, A_2, A_3, etc., are expressed in terms of A and A_0. Hence replace the matrices A_1, A_2, A_3 ...... A_i in the assumed solution of X(t) [i.e., equation (4.50)] by the equivalent terms obtained above.

$$\therefore X(t) = A_0 + AA_0t + \frac{1}{2!} \, A^2 A_0 t^2 + \frac{1}{3!} \, A^3 A_0 t^3 + \dots\dots + \frac{1}{i!} \, A^i A_0 t^i + \dots\dots$$

$$= \left[I + At + \frac{1}{2!} \, A^2 t^2 + \frac{1}{3!} \, A^3 t^3 + \dots\dots + \frac{1}{i!} \, A^i t^i + \dots\dots \right] A_0 \qquad\qquad .....(4.53)$$

where **I** is the unit matrix.

It is given that, when t = 0, $X(t) = X(0) = X_0$ .....(4.54)

From equation (4.53) when t = 0, we get

$$X(t)\big|_{t=0} = X(0) = A_0 \qquad\qquad .....(4.55)$$

From equations (4.54) and (4.55) we get,

$$A_0 = X_0 \qquad\qquad .....(4.56)$$

On substituting for A_0 from equation (4.56) in equation (4.53) we get,

$$X(t) = \left[I + At + \frac{1}{2!} \, A^2 t^2 + \frac{1}{3!} \, A^3 t^3 + \dots\dots + \frac{1}{i!} \, A^i t^i + \dots\dots \right] X_0 \qquad .....(4.57)$$

Each of the term inside the brackets is an n × n matrix. Because of the similarity of the entity inside the bracket with a scalar exponential of e^{at}, we call it a matrix exponential, which may be written as,

$$e^{At} = I + At + \frac{1}{2!} \, A^2 t^2 + \frac{1}{3!} \, A^3 t^3 + \dots\dots + \frac{1}{i!} \, A^i t^i + \dots\dots \qquad .....(4.58)$$

Hence the solution of the state equation is

$$X(t) = e^{At} \, X_0 \qquad\qquad .....(4.59)$$

The matrix e^{At} is called state transition matrix and denoted by $\phi(t)$. From the solution of the state equations it is observed that the initial state X_0 at t = 0, is driven to state X(t) at time t by state transition matrix.

SOLUTION OF NON-HOMOGENEOUS STATE EQUATIONS (Solution of state equations with input or excitation)

The state equation of nth order system is given by

$$\dot{X}(t) = A\,X(t) + B\,U(t) \ ; \ X(0) = X_0 \qquad\qquad .....(4.60)$$

where X_0 is initial condition vector.

The state equation of equation (4.60) can be rearranged as shown below,

$$\dot{X}(t) - A\,X(t) = B\,U(t) \qquad\qquad .....(4.61)$$

Premultiply both sides of equation (4.61) by e^{-At}

$$e^{-At}\left[\dot{X}(t) - A\,X(t)\right] = e^{-At}\,B\,U(t)$$

$$e^{-At}\dot{X}(t) + e^{-At}(-A)X(t) = e^{-At}\,B\,U(t) \qquad\qquad .....(4.62)$$

Consider the differential of $e^{-At}\,X(t)$

$$\frac{d}{dt}\left(e^{-At}\,X(t)\right) = e^{-At}\,\dot{X}(t) + e^{-At}(-A)X(t) \qquad\qquad .....(4.63)$$

On comparing equations (4.62) and (4.63) we can write,

$$\frac{d}{dt}\left(e^{-At}\,X(t)\right) = e^{-At}\,B\,U(t)$$

$$d\left(e^{-At}\,X(t)\right) = e^{-At}\,B\,U(t)\,dt \qquad\qquad .....(4.64)$$

On integrating the equation (4.64) between limits 0 to t we get,

$$e^{-At}\,X(t) = X_0 + \int_0^t e^{-A\tau}\,B\,U(\tau)\,d\tau \qquad\qquad .....(4.65)$$

where X_0 = Initial condition vector = Integral constant

τ = Dummy variable substituted for t.

Premultiply both sides of equation (4.65) by e^{At},

$$e^{At}e^{-At}\,X(t) = e^{At}X_0 + e^{At}\int_0^t e^{-A\tau}\,B\,U(\tau)\,dt$$

$$X(t) = e^{At}X_0 + e^{At}\int_0^t e^{-A\tau}\,B\,U(\tau)\,dt \qquad\qquad .....(4.66)$$

The term e^{At} independent of the integral variable τ, and so e^{At} can be brought inside the integral function.

$$\therefore X(t) = e^{At}X_0 + \int_0^t e^{At}\,e^{-A\tau}\,B\,U(\tau)\,dt$$

$$X(t) = e^{At}X_0 + \int_0^t e^{A(t-\tau)}\,B\,U(\tau)\,d\tau \qquad\qquad .....(4.67)$$

The equation (4.67) is the solution of state equation, when the initial conditions are known at $t = 0$. If initial conditions are known at $t = t_0$ then the solution of state equation is given by equation (4.68).

$$X(t) = e^{A(t-t_0)} X(t_0) + \int_{t_0}^{t} e^{A(t-\tau)} B U(\tau) d\tau \qquad \qquad .....(4.68)$$

The state transition matrix e^{At} is denoted by the symbol $\phi(t)$, i.e., $\phi(t) = e^{At}$

Hence, $e^{A(t-t_0)}$ can be expressed as, $e^{A(t-t_0)} = \phi(t - t_0)$ .....(4.69)

and, $e^{A(t-\tau)}$ can be expressed as, $e^{A(t-\tau)} = \phi(t-\tau)$ .....(4.70)

The equations (4.67) and (4.68) can also be expressed as

$$X(t) = \phi(t) X(0) + \int_{0}^{t} \phi(t-\tau) B U(\tau) d\tau \text{ if the initial conditions are known at } t = 0 \qquad .....(4.71)$$

$$X(t) = \phi(t-t_0) X(t_0) + \int_{t_0}^{t} \phi(t-\tau) B U(\tau) d\tau \text{ if the initial conditions are known at } t = t_0 \qquad .....(4.72)$$

PROPERTIES OF STATE TRANSITION MATRIX

1. $\phi(0) = e^{A \times 0} = I$ (unit matrix)

2. $\phi(t) = e^{At} = (e^{-At})^{-1} = [\phi(-t)]^{-1}$ or $\phi^{-1}(t) = \phi(t)$

3. $\phi(t_1 + t_2) = e^{A(t_1 + t_2)} = (e^{At_1})(e^{At_2}) = \phi(t_1)\phi(t_2)$

COMPUTATION OF STATE TRANSITION MATRIX

The state transition matrix (e^{At}) can be computed by any one of the following two methods.

Method 1 : Computation of e^{At} using matrix exponential.

Method 2 : Computation of e^{At} using Laplace transform.

Method 3 : Computation of e^{At} by canonical transformation.

Method 4 : Computation of e^{At} using Sylvester's interpolation formula (or computation based on Cayley-Hamilton theorem.

COMPUTATION OF STATE TRANSITION MATRIX USING MATRIX EXPONENTIAL

In this method, the e^{At} is computed using the matrix exponential of equation (4.58), which is also given below.

$$e^{At} = I + At + \frac{1}{2!} A^2 t^2 + \frac{1}{3!} A^3 t^3 + ... + \frac{1}{i!} A^i t^i + ...$$

where, $e^{At} =$ State transition matrix of order n × n

$A =$ System matrix of order n × n

$I =$ Unit matrix of order n × n.

The disadvantage in this method is that each term of e^{At} will be an infinite series and the convergence of the infinite series are obtained by trial and error.

COMPUTATION OF STATE TRANSITION MATRIX BY LAPLACE TRANSFORM METHOD

Consider the state equation without input vector, $\dot{X}(t) = A\,X(t)$ .....(4.73)

On taking Laplace transform of equation (4.73) we get,

$s\,X(s) - X(0) = A\,X(s)$

$s\,X(s) - A\,X(s) = X(0)$

$s\,I\,X(s) - A\,X(s) = X(0),$ where, I is a unit matrix.

$(sI - A)\,X(s) = X(0)$

Premultiply both sides by $(sI - A)^{-1}$

$X(s) = (sI - A)^{-1}\,X(0)$

On taking inverse Laplace transform we get,

$X(t) = \mathcal{L}^{-1}\,[(sI - A)^{-1}\,X(0)]$

$X(t) = \mathcal{L}^{-1}\,[(s\,I - A)^{-1}]\,X(0)$ .....(4.74)

On comparing equation (4.74) with the solution of state equation, $X(t) = e^{At}\,X(0)$ we get,

$e^{At} = \mathcal{L}^{-1}\,[(sI - A)^{-1}]$ or $\mathcal{L}[e^{At}] = (sI - A)^{-1}$ .....(4.75)

We know that, $e^{At} = \phi(t),$

$\therefore\ \mathcal{L}[e^{At}] = \mathcal{L}[\phi(t)] = \phi\,(s)$ .....(4.76)

where, $\phi(s) = (sI - A)^{-1}$ and it is called resolvant matrix.

From the system matrix, A the resolvant matrix, $\phi(s)$ can be computed. By taking inverse Laplace transform of resolvant matrix, the state transition matrix is computed, from which the solution of state equation is obtained.

The solution of state equation is given by,

$X(t) = e^{At}\,X(0)$

$\therefore\ X(t) = \mathcal{L}^{-1}\,[\phi(s)]\,X(0)$ .....(4.77)

where, $\phi(s) = (sI - A)^{-1}$

Consider the state equation with forcing function (input or excitation),

$\dot{X} = AX + BU$ .....(4.78)

On taking Laplace transform of equation (4.78) we get,

$s\,X(s) - X(0) = A\,X(s) + B\,U(s)$

$sI\,X(s) - A\,X(s) = X(0) + B\,U(s),$ where I is the unit matrix.

$(sI - A)\,X(s) = X(0) + B\,U(s)$ .....(4.79)

Premultiply the equation (4.79) by $(sI - A)^{-1}$

$\therefore\ X(s) = (sI - A)^{-1}\,X(0) + (sI - A)^{-1}\,B\,U(s)$

$= \phi(s)\,X(0) + \phi(s)\,B\,U(s)$ .....(4.80)

On taking inverse Laplace transform of equation (4.80) we get,

$$X(t) = \phi(t) \, X(0) + \mathcal{L}^{-1} \, [\phi(s) \, \mathbf{B} \, U(s)] \qquad\qquad .....(4.81)$$

The equation (4.81) is the solution of state equation with forcing function.

EXAMPLE 4.13

Consider the matrix A. Compute e^{At} by two methods.

$$\mathbf{A} = \begin{bmatrix} 0 & 1 \\ -2 & -3 \end{bmatrix}$$

SOLUTION

Method 1

$$e^{At} = \left[I + At + \frac{1}{2!} A^2 t^2 + \frac{1}{3!} A^3 t^3 + ... \right]$$

$$\mathbf{A} = \begin{bmatrix} 0 & 1 \\ -2 & -3 \end{bmatrix}$$

$$\mathbf{A}^2 = \mathbf{AA} = \begin{bmatrix} 0 & 1 \\ -2 & -3 \end{bmatrix}\begin{bmatrix} 0 & 1 \\ -2 & -3 \end{bmatrix} = \begin{bmatrix} -2 & -3 \\ 6 & 7 \end{bmatrix}$$

$$\mathbf{A}^3 = \mathbf{A}^2\mathbf{A} = \begin{bmatrix} -2 & -3 \\ 6 & 7 \end{bmatrix}\begin{bmatrix} 0 & 1 \\ -2 & -3 \end{bmatrix} = \begin{bmatrix} 6 & 7 \\ -14 & -15 \end{bmatrix}$$

$$\mathbf{A}^4 = \mathbf{A}^3\mathbf{A} = \begin{bmatrix} 6 & 7 \\ -14 & -15 \end{bmatrix}\begin{bmatrix} 0 & 1 \\ -2 & -3 \end{bmatrix} = \begin{bmatrix} -14 & -15 \\ 30 & 31 \end{bmatrix}$$

$$e^{At} = I + At + \frac{1}{2!} A^2 t^2 + \frac{1}{3!} A^3 t^3 + \frac{1}{4!} A^4 t^4 + ...$$

$$= \begin{bmatrix} 1 & 0 \\ 0 & 1 \end{bmatrix} + \begin{bmatrix} 0 & 1 \\ -2 & -3 \end{bmatrix} t + \frac{1}{2}\begin{bmatrix} -2 & -3 \\ 6 & 7 \end{bmatrix} t^2 + \frac{1}{6}\begin{bmatrix} 6 & 7 \\ -14 & -15 \end{bmatrix} t^3 + \frac{1}{24}\begin{bmatrix} -14 & -15 \\ 30 & 31 \end{bmatrix} t^4 + ...$$

$$= \begin{bmatrix} 1 & 0 \\ 0 & 1 \end{bmatrix} + \begin{bmatrix} 0 & t \\ -2t & -3t \end{bmatrix} + \begin{bmatrix} -t^2 & -\frac{3}{2}t^2 \\ 3t^2 & \frac{7}{2}t^2 \end{bmatrix} + \begin{bmatrix} t^3 & \frac{7}{6}t^3 \\ -\frac{7}{3}t^3 & -\frac{5}{2}t^3 \end{bmatrix} + \begin{bmatrix} -\frac{7}{12}t^4 & -\frac{5}{8}t^4 \\ \frac{5}{4}t^4 & \frac{31}{24}t^4 \end{bmatrix} + ...$$

$$= \begin{bmatrix} 1 - t^2 + t^3 - \frac{7}{12}t^4 + ... & t - \frac{3}{2}t^2 + \frac{7}{6}t^3 - \frac{5}{8}t^4 + ... \\ -2t + 3t^2 - \frac{7}{3}t^3 + \frac{5}{4}t^4 + ... & 1 - 3t + \frac{7}{2}t^2 - \frac{5}{2}t^3 + \frac{31}{24}t^4 + ... \end{bmatrix}$$

The each term in the matrix is an expansion of e^{at}. The convergence of series is obtained by trial and error. Consider the expansion of e^{-t} and e^{-2t}.

$$e^{-t} = 1 - t + \frac{1}{2!}t^2 - \frac{1}{3!}t^3 + \frac{1}{4!}t^4 \ldots\ldots = 1 - t + \frac{1}{2}t^2 - \frac{1}{6}t^3 + \frac{1}{24}t^4 \ldots\ldots$$

$$e^{-2t} = 1 - 2t + \frac{1}{2!}2^2t^2 - \frac{1}{3!}2^3t^3 + \frac{1}{4!}2^4t^4 \ldots\ldots = 1 - 2t + 2t^2 - \frac{4}{3}t^3 + \frac{2}{3}t^4 \ldots\ldots$$

$$2e^{-t} - e^{-2t} = 2 - 2t + t^2 - \frac{1}{3}t^3 + \frac{1}{12}t^4 \ldots\ldots - 1 + 2t - 2t^2 + \frac{4}{3}t^3 - \frac{2}{3}t^4 \ldots\ldots$$

$$= 1 - t^2 + t^3 - \frac{7}{12}t^4 + \ldots\ldots$$

$$e^{-t} - e^{-2t} = 1 - t + \frac{1}{2}t^2 - \frac{1}{6}t^3 + \frac{1}{24}t^4 \ldots\ldots - 1 + 2t - 2t^2 + \frac{4}{3}t^3 - \frac{2}{3}t^4 \ldots\ldots$$

$$= t - \frac{3}{2}t^2 + \frac{7}{6}t^3 - \frac{5}{8}t^4 + \ldots\ldots$$

$$-2e^{-t} + 2e^{-2t} = -2 + 2t - t^2 + \frac{1}{3}t^3 - \frac{1}{12}t^4 \ldots\ldots + 2 - 4t + 4t^2 - \frac{8}{3}t^3 + \frac{4}{3}t^4 \ldots\ldots$$

$$= -2t + 3t^2 - \frac{7}{3}t^3 + \frac{5}{4}t^4 + \ldots\ldots$$

$$-e^{-t} + 2e^{-2t} = -1 + t - \frac{1}{2}t^2 + \frac{1}{6}t^3 - \frac{1}{24}t^4 \ldots\ldots + 2 - 4t + 4t^2 - \frac{8}{3}t^3 + \frac{4}{3}t^4 \ldots\ldots$$

$$= 1 - 3t + \frac{7}{2}t^2 - \frac{5}{2}t^3 + \frac{31}{24}t^4 + \ldots\ldots$$

$$\therefore e^{At} = \begin{bmatrix} 2e^{-t} - e^{-2t} & e^{-t} - e^{-2t} \\ -2e^{-t} + 2e^{-2t} & -e^{-t} + 2e^{-2t} \end{bmatrix}$$

Method 2

$$A = \begin{bmatrix} 0 & 1 \\ -2 & -3 \end{bmatrix}$$

$$e^{At} = \phi(t) = \mathcal{L}^{-1}[(sI - A)^{-1}]$$

$$sI - A = s\begin{bmatrix} 1 & 0 \\ 0 & 1 \end{bmatrix} - \begin{bmatrix} 0 & 1 \\ -2 & -3 \end{bmatrix} = \begin{bmatrix} s & 0 \\ 0 & s \end{bmatrix} - \begin{bmatrix} 0 & 1 \\ -2 & -3 \end{bmatrix} = \begin{bmatrix} s & -1 \\ 2 & s+3 \end{bmatrix}$$

Let, $\Delta = |sI - A| = $ determinant of $(sI - A)$

$$\therefore \Delta = |sI - A| = \begin{vmatrix} s & -1 \\ 2 & s+3 \end{vmatrix} = s(s+3) + 2 = s^2 + 3s + 2 = (s+2)(s+1)$$

$$\phi(s) = [sI - A]^{-1} = \frac{[\text{Cofactor of } (sI - A)]^T}{\text{determinant of } (sI - A)} = \frac{[\text{Cofactor of } (sI - A)]^T}{\Delta}$$

$$\therefore \phi(s) = \frac{1}{(s+1)(s+2)} \begin{bmatrix} s+3 & 1 \\ -2 & s \end{bmatrix}$$

$$\phi(s) = \begin{bmatrix} \dfrac{s+3}{(s+1)(s+2)} & \dfrac{1}{(s+1)(s+2)} \\[3mm] \dfrac{-2}{(s+1)(s+2)} & \dfrac{s}{(s+1)(s+2)} \end{bmatrix}$$

By partial fraction expansion, $\phi(s)$ can be written as,

$$\phi(s) = \begin{bmatrix} \dfrac{A_1}{s+1} + \dfrac{B_1}{s+2} & \dfrac{A_2}{s+1} + \dfrac{B_2}{s+2} \\[3mm] \dfrac{A_3}{s+1} + \dfrac{B_3}{s+2} & \dfrac{A_4}{s+1} + \dfrac{B_4}{s+2} \end{bmatrix}$$

$$\frac{s+3}{(s+1)(s+2)} = \frac{A_1}{s+1} + \frac{B_1}{s+2} \qquad\qquad \frac{1}{(s+1)(s+2)} = \frac{A_2}{s+1} + \frac{B_2}{s+2}$$

$$A_1 = \frac{s+3}{s+2}\bigg|_{s=-1} = 2 \qquad\qquad A_2 = \frac{1}{s+2}\bigg|_{s=-1} = 1$$

$$B_1 = \frac{s+3}{s+1}\bigg|_{s=-2} = -1 \qquad\qquad B_2 = \frac{1}{s+1}\bigg|_{s=-2} = -1$$

$$\frac{-2}{(s+1)(s+2)} = \frac{A_3}{s+1} + \frac{B_3}{s+2} \qquad\qquad \frac{s}{(s+1)(s+2)} = \frac{A_4}{s+1} + \frac{B_4}{s+2}$$

$$A_3 = \frac{-2}{s+2}\bigg|_{s=-1} = -2 \qquad\qquad A_4 = \frac{s}{s+2}\bigg|_{s=-1} = -1$$

$$B_3 = \frac{-2}{s+1}\bigg|_{s=-2} = 2 \qquad\qquad B_4 = \frac{s}{s+1}\bigg|_{s=-2} = 2$$

$$\therefore \phi(s) = \begin{bmatrix} \dfrac{2}{s+1} - \dfrac{1}{s+2} & \dfrac{1}{s+1} - \dfrac{1}{s+2} \\[3mm] \dfrac{-2}{s+1} + \dfrac{2}{s+2} & \dfrac{-1}{s+1} + \dfrac{2}{s+2} \end{bmatrix}$$

On taking inverse Laplace transform of $\phi(s)$ we get $\phi(t)$, where $\phi(t) = e^{At}$

$$\therefore e^{At} = \phi(t) = \begin{bmatrix} 2e^{-t} - e^{-2t} & e^{-t} - e^{-2t} \\[2mm] -2e^{-t} + 2e^{-2t} & -e^{-t} + 2e^{-2t} \end{bmatrix}$$

It is observed that the results of both the methods are same.

EXAMPLE 4.14

Given that, $A_1 = \begin{bmatrix} \sigma & 0 \\ 0 & \sigma \end{bmatrix}$; $A_2 = \begin{bmatrix} 0 & \omega \\ -\omega & 0 \end{bmatrix}$; $A = \begin{bmatrix} \sigma & \omega \\ -\omega & \sigma \end{bmatrix}$ compute e^{At}.

SOLUTION

Here $A = A_1 + A_2$

$$\therefore e^{At} = e^{(A_1 + A_2)t} = e^{A_1 t}\, e^{A_2 t}$$

$$sI - A_1 = s\begin{bmatrix} 1 & 0 \\ 0 & 1 \end{bmatrix} - \begin{bmatrix} \sigma & 0 \\ 0 & \sigma \end{bmatrix} = \begin{bmatrix} s & 0 \\ 0 & s \end{bmatrix} - \begin{bmatrix} \sigma & 0 \\ 0 & \sigma \end{bmatrix} = \begin{bmatrix} s-\sigma & 0 \\ 0 & s-\sigma \end{bmatrix}$$

$$\Delta_1 = |sI - A_1| = \begin{vmatrix} s-\sigma & 0 \\ 0 & s-\sigma \end{vmatrix} = (s-\sigma)^2$$

$$[sI - A_1]^{-1} = \frac{1}{\Delta_1}\begin{bmatrix} s-\sigma & 0 \\ 0 & s-\sigma \end{bmatrix} = \frac{1}{(s-\sigma)^2}\begin{bmatrix} s-\sigma & 0 \\ 0 & s-\sigma \end{bmatrix} = \begin{bmatrix} \dfrac{1}{s-\sigma} & 0 \\ 0 & \dfrac{1}{s-\sigma} \end{bmatrix}$$

$$e^{A_1 t} = \mathcal{L}^{-1}[(sI - A_1)^{-1}] = \begin{bmatrix} e^{-\sigma t} & 0 \\ 0 & e^{-\sigma t} \end{bmatrix}$$

$$sI - A_2 = s\begin{bmatrix} 1 & 0 \\ 0 & 1 \end{bmatrix} - \begin{bmatrix} 0 & \omega \\ -\omega & 0 \end{bmatrix} = \begin{bmatrix} s & 0 \\ 0 & s \end{bmatrix} - \begin{bmatrix} 0 & \omega \\ -\omega & 0 \end{bmatrix} = \begin{bmatrix} s & -\omega \\ \omega & s \end{bmatrix}$$

$$\Delta_2 = |sI - A_2| = \begin{vmatrix} s & -\omega \\ \omega & s \end{vmatrix} = s^2 + \omega^2$$

$$[sI - A_2]^{-1} = \frac{1}{\Delta_2}\begin{bmatrix} s & \omega \\ -\omega & s \end{bmatrix} = \frac{1}{s^2 + \omega^2}\begin{bmatrix} s & \omega \\ -\omega & s \end{bmatrix} = \begin{bmatrix} \dfrac{s}{s^2+\omega^2} & \dfrac{\omega}{s^2+\omega^2} \\ \dfrac{-\omega}{s^2+\omega^2} & \dfrac{s}{s^2+\omega^2} \end{bmatrix}$$

$$e^{A_2 t} = \mathcal{L}^{-1}[(sI - A_2)^{-1}] = \begin{bmatrix} \cos\omega t & \sin\omega t \\ -\sin\omega t & \cos\omega t \end{bmatrix}$$

$$e^{At} = e^{A_1 t}e^{A_2 t} = \begin{bmatrix} e^{-\sigma t} & 0 \\ 0 & e^{-\sigma t} \end{bmatrix}\begin{bmatrix} \cos\omega t & \sin\omega t \\ -\sin\omega t & \cos\omega t \end{bmatrix}$$

$$\therefore e^{At} = \begin{bmatrix} e^{-\sigma t}\cos\omega t & e^{-\sigma t}\sin\omega t \\ -e^{-\sigma t}\sin\omega t & e^{-\sigma t}\cos\omega t \end{bmatrix}$$

EXAMPLE 4.15

For a system represented by state equation $\dot{X}(t) = A\,X(t)$

The response is $X(t) = \begin{bmatrix} e^{-2t} \\ -2e^{-2t} \end{bmatrix}$ when $X(0) = \begin{bmatrix} 1 \\ -2 \end{bmatrix}$

and $\quad X(t) = \begin{bmatrix} e^{-t} \\ -e^{-t} \end{bmatrix}$ when $X(0) = \begin{bmatrix} 1 \\ -1 \end{bmatrix}$

Determine the system matrix **A** and the state transition matrix.

SOLUTION

The solution of state equation is, $X(t) = e^{At} \, X(0)$ .....(4.15.1)

Premultiply the equation (4.15.1) by e^{-At}

$$e^{-At} \, X(t) = e^{-At} \, e^{At} \, X(0)$$

$$\therefore e^{-At} \, X(t) = X(0) \qquad .....(4.15.2)$$

One of the response is $X(t) = \begin{bmatrix} e^{-2t} \\ -2e^{-2t} \end{bmatrix}$ and $X(0) = \begin{bmatrix} 1 \\ -2 \end{bmatrix}$

On substituting the response in equation (4.15.2) we get,

$$e^{-At} \begin{bmatrix} e^{-2t} \\ -2e^{-2t} \end{bmatrix} = \begin{bmatrix} 1 \\ -2 \end{bmatrix} \qquad .....(4.15.3)$$

Let $e^{-At} = \begin{bmatrix} e_{11} & e_{12} \\ e_{21} & e_{22} \end{bmatrix}$ .....(4.15.4)

From equation (4.15.3) and (4.15.4) we can write

$$\begin{bmatrix} e_{11} & e_{12} \\ e_{21} & e_{22} \end{bmatrix} \begin{bmatrix} e^{-2t} \\ -2e^{-2t} \end{bmatrix} = \begin{bmatrix} 1 \\ -2 \end{bmatrix} \qquad ......(4.15.5)$$

On multiplying the equation (4.15.5) we get the following two equations,

$$e_{11}e^{-2t} - 2e_{12}e^{-2t} = 1 \qquad .....(4.15.6)$$

$$e_{21}e^{-2t} - 2e_{22}e^{-2t} = -2 \qquad .....(4.15.7)$$

The second solution of state equation is $X(t) = \begin{bmatrix} e^{-t} \\ -e^{-t} \end{bmatrix}$ and $X(0) = \begin{bmatrix} 1 \\ -1 \end{bmatrix}$

On substituting this solution in equation (4.15.2) we get,

$$e^{-At} \begin{bmatrix} e^{-t} \\ -e^{-t} \end{bmatrix} = \begin{bmatrix} 1 \\ -1 \end{bmatrix} \qquad .....(4.15.8)$$

From equation (4.15.4) and (4.15.8) we can write,

$$\begin{bmatrix} e_{11} & e_{12} \\ e_{21} & e_{22} \end{bmatrix} \begin{bmatrix} e^{-t} \\ -e^{-t} \end{bmatrix} = \begin{bmatrix} 1 \\ -1 \end{bmatrix} \qquad .....(4.15.9)$$

On multiplying the equation (4.15.9) we get the following two equations,

$$e_{11} e^{-t} - e_{12} e^{-t} = 1 \qquad .....(4.15.10)$$

$$e_{21} e^{-t} - e_{22} e^{-t} = -1 \qquad .....(4.15.11)$$

Equation (4.5.10) $\times e^{-t} \implies e_{11}e^{-2t} - e_{12}e^{-2t} = e^{-t}$

Equation (4.15.6) $\times 1 \implies e_{11}e^{-2t} - 2e_{12}e^{-2t} = 1$

                   $(-)$ $(+)$ $(-)$

On subtracting $e_{12}e^{-2t} = e^{-t} - 1$ .....(4.15.12)

From equation (4.15.12) we get

$$e_{12} = \frac{e^{-t} - 1}{e^{-2t}} = \frac{e^{-t}}{e^{-2t}} - \frac{1}{e^{-2t}} = e^t - e^{2t} \qquad \qquad .....(4.15.13)$$

From equation (4.15.6), $e_{11} = \dfrac{1 + 2e_{12}e^{-2t}}{e^{-2t}}$

Put $e_{12} = e^t - e^{2t}$, $\therefore e_{11} = \dfrac{1 + 2(e^t - e^{2t})e^{-2t}}{e^{-2t}} = \dfrac{1 + 2e^{-t} - 2}{e^{-2t}} = \dfrac{2e^{-t} - 1}{e^{-2t}}$

$$= \frac{2e^{-t}}{e^{-2t}} - \frac{1}{e^{-2t}} = 2e^t - e^{2t}$$

Equation (4.15.11) × e^{-t} $\Rightarrow$ $e_{21}e^{-2t} - e_{22}e^{-2t} = -e^{-t}$

Equation (4.15.7) × 1 $\Rightarrow$ $e_{21}e^{-2t} - 2e_{22}e^{-2t} = -2$

$$\begin{array}{ccc} (-) & (+) & (+) \\ \hline \end{array}$$

On subtracting $\qquad\qquad\qquad\qquad\qquad e_{22}e^{-2t} = 2 - e^{-t}$ $\qquad\qquad\qquad\qquad$ .....(4.15.14)

From equation (4.15.14) we get,

$$e_{22} = \frac{-e^{-t} + 2}{e^{-2t}} = \frac{-e^{-t}}{e^{-2t}} + \frac{2}{e^{-2t}} = -e^t + 2e^{2t}$$

From equation (4.15.11), $e_{21} = \dfrac{-1 + e_{22}e^{-t}}{e^{-t}}$

Put $e_{22} = -e^t + 2e^{2t}$, $\therefore e_{21} = \dfrac{-1 + (-e^t + 2e^{2t})e^{-t}}{e^{-t}}$

$$e_{21} = \frac{-1 - 1 + 2e^{2t}}{e^{-t}} = \frac{-2}{e^{-t}} + \frac{2e^t}{e^{-t}} = -2e^t + 2e^{2t}$$

$$\therefore e^{-At} = \begin{bmatrix} e_{11} & e_{12} \\ e_{21} & e_{22} \end{bmatrix} = \begin{bmatrix} 2e^t - e^{2t} & e^t - e^{2t} \\ -2e^t + 2e^{2t} & -e^t + 2e^{2t} \end{bmatrix}$$

$$\therefore e^{At} = \begin{bmatrix} 2e^t - e^{-2t} & e^{-t} - e^{2t} \\ -2e^{-t} + 2e^{-2t} & -e^{-t} + 2e^{-2t} \end{bmatrix}$$

e^{At} is the state transition matrix.

We know that, $\mathcal{L}[e^{At}] = \phi(s)$

where $\phi(s) = (sI - A)^{-1}$; $\therefore \phi(s)^{-1} = (sI - A)$ or $A = sI - \phi(s)^{-1}$

$$\phi(s) = \mathcal{L}[e^{At}] = \begin{bmatrix} \dfrac{2}{s+1} - \dfrac{1}{s+2} & \dfrac{1}{s+1} - \dfrac{1}{s+2} \\ \dfrac{-2}{s+1} + \dfrac{2}{s+2} & \dfrac{-1}{s+1} + \dfrac{2}{s+2} \end{bmatrix}$$

$$= \begin{bmatrix} \dfrac{2(s+2)-(s+1)}{(s+1)(s+2)} & \dfrac{(s+2)-(s+1)}{(s+1)(s+2)} \\[3mm] \dfrac{-2(s+2)+2(s+1)}{(s+1)(s+2)} & \dfrac{-(s+2)+2(s+1)}{(s+1)(s+2)} \end{bmatrix} = \begin{bmatrix} \dfrac{s+3}{(s+1)(s+2)} & \dfrac{1}{(s+1)(s+2)} \\[3mm] \dfrac{-2}{(s+1)(s+2)} & \dfrac{s}{(s+1)(s+2)} \end{bmatrix}$$

Determinant of $\phi(s) = \dfrac{s(s+3)+2}{(s+1)^2(s+2)^2} = \dfrac{s^2+3s+2}{(s+1)(s+2)^2}$

$$= \dfrac{(s+1)(s+2)}{(s+1)^2(s+2)^2} = \dfrac{1}{(s+1)(s+2)}$$

$$\phi(s)^{-1} = \dfrac{1}{(s+1)(s+2)} \begin{bmatrix} \dfrac{s}{(s+1)(s+2)} & \dfrac{-1}{(s+1)(s+2)} \\[3mm] \dfrac{2}{(s+1)(s+2)} & \dfrac{s+3}{(s+1)(s+2)} \end{bmatrix} = \begin{bmatrix} s & -1 \\ 2 & s+3 \end{bmatrix}$$

$$\mathbf{A} = s\mathbf{I} - \phi(s)^{-1} = s\begin{bmatrix} 1 & 0 \\ 0 & 1 \end{bmatrix} - \begin{bmatrix} s & -1 \\ 2 & s+3 \end{bmatrix} = \begin{bmatrix} s & 0 \\ 0 & s \end{bmatrix} - \begin{bmatrix} s & -1 \\ 2 & s+3 \end{bmatrix} = \begin{bmatrix} 0 & 1 \\ -2 & -3 \end{bmatrix}$$

RESULT

$$\mathbf{A} = \begin{bmatrix} 0 & 1 \\ -2 & -3 \end{bmatrix} ; \ e^{\mathbf{A}t} = \begin{bmatrix} 2e^{-t} - e^{-2t} & e^{-t} - e^{-2t} \\ -2e^{-t} + 2e^{-2t} & -e^{-t} - 2e^{-2t} \end{bmatrix}$$

EXAMPLE 4.16

The state equation and initial condition vector of an linear time-invariant system are given below. Determine the solution of state equation.

$$\begin{bmatrix} \dot{x}_1 \\ \dot{x}_2 \end{bmatrix} = \begin{bmatrix} 1 & 0 \\ 1 & 1 \end{bmatrix} \begin{bmatrix} x_1 \\ x_2 \end{bmatrix} ; \ X_0 = \begin{bmatrix} 1 \\ 0 \end{bmatrix}$$

SOLUTION

Here $\mathbf{A} = \begin{bmatrix} 1 & 0 \\ 1 & 1 \end{bmatrix} ; \ s\mathbf{I} - \mathbf{A} = s\begin{bmatrix} 1 & 0 \\ 0 & 1 \end{bmatrix} - \begin{bmatrix} 1 & 0 \\ 1 & 1 \end{bmatrix} = \begin{bmatrix} s & 0 \\ 0 & s \end{bmatrix} - \begin{bmatrix} 1 & 0 \\ 1 & 1 \end{bmatrix} = \begin{bmatrix} s-1 & 0 \\ -1 & s-1 \end{bmatrix}$

$$|s\mathbf{I} - \mathbf{A}| = \begin{vmatrix} s-1 & 0 \\ -1 & s-1 \end{vmatrix} = (s-1)^2 - 0 = (s-1)^2$$

$$(s\mathbf{I} - \mathbf{A})^{-1} = \dfrac{1}{(s-1)^2} \begin{bmatrix} s-1 & 0 \\ 1 & s-1 \end{bmatrix} = \begin{bmatrix} \dfrac{1}{s-1} & 0 \\[3mm] \dfrac{1}{(s-1)^2} & \dfrac{1}{s-1} \end{bmatrix}$$

$$e^{\mathbf{A}t} = \phi(t) = \mathcal{L}^{-1}[\phi(s)] = \mathcal{L}^{-1}[(s\mathbf{I} - \mathbf{A})^{-1}] \begin{bmatrix} e^t & 0 \\ te^t & e^t \end{bmatrix}$$

The solution of the state equation is, $X(t) = e^{\mathbf{A}t} X_0 = \begin{bmatrix} e^t & 0 \\ te^t & e^t \end{bmatrix} \begin{bmatrix} 1 \\ 0 \end{bmatrix} = \begin{bmatrix} e^t \\ te^t \end{bmatrix}$

4.9   STATE SPACE REPRESENTATION OF DISCRETE TIME SYSTEMS

The state variable analysis techniques of continuous time systems can be extended to the discrete-time system. The discrete form of state space representation is quite analogous to the continuous form.

In the state variable formulation of a discrete time system, in general, a system consists of m-inputs, p-outputs and n-state variables. The state space representation of discrete-time system may be visualized as shown in fig 4.6.

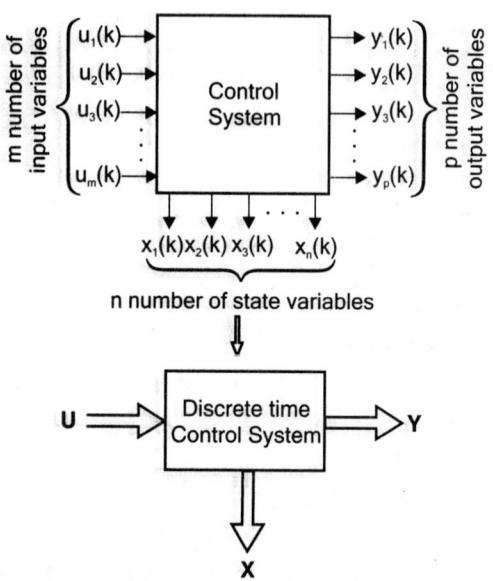

Fig 4.6 : *State space representation of discrete time system.*

Let,   State variables = $x_1(k), x_2(k), x_3(k), \ldots\ldots x_n(k)$
        Input variables = $u_1(k), u_2(k), u_3(k), \ldots\ldots u_m(k)$
        Output variables = $y_1(k), y_2(k), y_3(k), \ldots\ldots y_p(k)$

The different variables may be represented by the vectors (column matrix) as shown below.

Input vector $U(k) = \begin{bmatrix} u_1(k) \\ u_2(k) \\ \vdots \\ u_m(k) \end{bmatrix}$; Output vector $Y(k) = \begin{bmatrix} y_1(k) \\ y_2(k) \\ \vdots \\ y_p(k) \end{bmatrix}$; State variable vector $X(k) = \begin{bmatrix} x_1(k) \\ x_2(k) \\ \vdots \\ x_n(k) \end{bmatrix}$

Note : *The simplified notation x(k), y(k) and u(k) are used to denote x(kT), y(kT) and u(kT) respectively. Also for convenience the variables are denoted $x_1, x_2, x_3, \ldots ; y_1, y_2, y_3, \ldots$ and $u_1, u_2, u_3 \ldots.$*

The state equation of a discrete time system is a set of n-numbers of first order difference equations.

$$x_1(k+1) = f_1(x_1, x_2, \ldots\ldots x_n ; u_1, u_2, \ldots\ldots u_m)$$
$$x_2(k+1) = f_2(x_1, x_2, \ldots\ldots x_n ; u_1, u_2, \ldots\ldots u_m)$$
$$\vdots$$
$$x_n(k+1) = f_n(x_1, x_2, \ldots\ldots x_n ; u_1, u_2, \ldots\ldots u_m)$$

For linear time invariant discrete time systems the above difference equations can be expressed as a linear combination of state variables and inputs.

$$x_1(k+1) = a_{11}x_1 + a_{12}x_2 + \dots + a_{1n}x_n + b_{11}u_1 + b_{12}u_2 + \dots + b_{1m}u_m$$
$$x_2(k+1) = a_{21}x_1 + a_{22}x_2 + \dots + a_{2n}x_n + b_{21}u_1 + b_{22}u_2 + \dots + b_{2m}u_m$$
$$\vdots$$
$$x_n(k+1) = a_{n1}x_1 + a_{n2}x_2 + \dots + a_{nn}x_n + b_{n1}u_1 + b_{n2}u_2 + \dots + b_{nm}u_m$$

where, the coefficients a_{ij} and b_{ij} are constants.

In the matrix form the above equations can be expressed as,

$$\begin{bmatrix} x_1(k+1) \\ x_2(k+1) \\ \vdots \\ x_n(k+1) \end{bmatrix} = \begin{bmatrix} a_{11} & a_{12} & \cdots & a_{1n} \\ a_{21} & a_{22} & \cdots & a_{2n} \\ \vdots & \vdots & & \vdots \\ a_{n1} & a_{n2} & \cdots & a_{nn} \end{bmatrix} \begin{bmatrix} x_1 \\ x_2 \\ \vdots \\ x_n \end{bmatrix} + \begin{bmatrix} b_{11} & b_{12} & \cdots & b_{1m} \\ b_{21} & b_{22} & \cdots & b_{2m} \\ \vdots & \vdots & & \vdots \\ b_{n1} & b_{n2} & \cdots & b_{nm} \end{bmatrix} \begin{bmatrix} u_1 \\ u_2 \\ \vdots \\ u_m \end{bmatrix} \qquad \dots(4.82)$$

The matrix equation (4.82) can be written in the vector notation as

$$X(k+1) = AX(k) + BU(k) \qquad \dots(4.83)$$

where, $X(k)$ = State vector of order $(n \times 1)$

$\quad\quad\ U(k)$ = Input vector of order $(m \times 1)$

$\quad\quad\ \mathbf{A}$ = System matrix of order $(n \times n)$

$\quad\quad\ \mathbf{B}$ = Input matrix of order $(n \times m)$

The equation (4.83) is the state equation of (linear time invariant) discrete time system.

The output at any discrete time instant, k are functions of state variables and inputs. Hence the output variables of linear time invariant system can be expressed as a linear combination of state variables and inputs.

$$y_1 = c_{11}x_1 + c_{12}x_2 + \dots + c_{1n}x_n + d_{11}u_1 + d_{12}u_2 + \dots + d_{1m}u_m$$
$$y_2 = c_{21}x_1 + c_{22}x_2 + \dots + c_{2n}x_n + d_{21}u_1 + d_{22}u_2 + \dots + d_{2m}u_m$$
$$\vdots$$
$$y_p = c_{p1}x_1 + c_{p2}x_2 + \dots + c_{pn}x_n + d_{p1}u_1 + d_{p2}u_2 + \dots + d_{pm}u_m$$

where, the coefficients c_{ij} and d_{ij} are constants.

In the matrix form the above equations can be expressed as,

$$\begin{bmatrix} y_1 \\ y_2 \\ \vdots \\ y_p \end{bmatrix} = \begin{bmatrix} c_{11} & c_{12} & \cdots & c_{1n} \\ c_{21} & c_{22} & \cdots & c_{2n} \\ \vdots & \vdots & & \vdots \\ c_{p1} & c_{p2} & \cdots & c_{pn} \end{bmatrix} \begin{bmatrix} x_1 \\ x_2 \\ \vdots \\ x_n \end{bmatrix} + \begin{bmatrix} d_{11} & d_{12} & \cdots & d_{1m} \\ d_{21} & d_{22} & \cdots & d_{2m} \\ \vdots & \vdots & & \vdots \\ d_{p1} & d_{p2} & \cdots & d_{pm} \end{bmatrix} \begin{bmatrix} u_1 \\ u_2 \\ \vdots \\ u_m \end{bmatrix} \qquad \dots(4.84)$$

The matrix equation (4.84) can be written in the vector notation as,

$$Y(k) = CX(k) + DU(k) \qquad \dots(4.85)$$

where, $X(k)$ = State vector of order $(n \times 1)$

$\quad\quad\ U(k)$ = Input vector of order $(m \times 1)$

$\quad\quad\ Y(k)$ = Output vector of order $(p \times 1)$

$\quad\quad\ \mathbf{C}$ = Output matrix of order $(p \times n)$

$\quad\quad\ \mathbf{D}$ = Transmission matrix of order $(p \times m)$

The equation (4.85) is the output equation of (linear time invariant) discrete time system.

The state equation and output equation are together called as state model of the system. Hence the state model of discrete time system is given by the following equations.

$$X(k+1) = A\,X(k) + B\,U(k) \quad .... \text{ State equation}$$

$$Y(k) = C\,X(k) + D\,U(k) \quad .... \text{ Output equation}$$

STATE DIAGRAM OF DISCRETE TIME SYSTEM

The state diagram of discrete time system can be either in block diagram form or signal flow graph form. The three fundamental elements, scalar, adder and unit delay elements are used to construct the state diagram. These basic elements are shown in table-4.6.

Scalar   :   The scalar is used to multiply a signal by a constant.

Adder   :   The adder is used to add two or more signals.

Unit delay :   The unit delay element will delay the signal passing through it by one sample time.

TABLE-4.6 : Basic Elements of State Diagram of Discrete Time System

Element	Block diagram representation	Signal flow graph representation
Scalar	$x(k)$ →[a]→ $ax(k)$	$x(k)$ o——a——o $a\,x(k)$
Adder	$x_1(k)$ →(+)→ $x_1(k)+x_2(k)$, $x_2(k)$	$x_1(k)$ o, $x_2(k)$ o ——1, 1——o $x_1(k) + x_2(k)$
Unit delay	$x(k+1)$ →[z^{-1}]→ $x(k)$	$x(k+1)$ o——z^{-1}——o $x(k)$

The block diagram representation of the state model of discrete time system is shown in fig 4.7. and the signal flow graph representation is shown in fig 4.8.

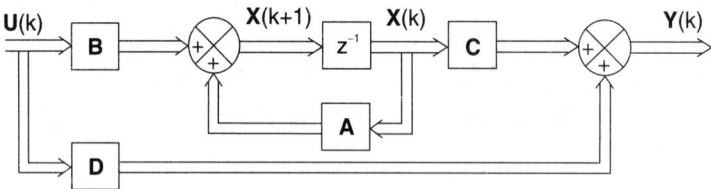

Fig 4.7 : Block diagram representation of discrete time system.

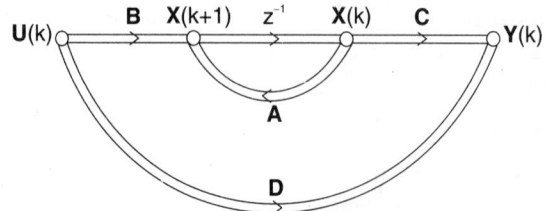

Fig 4.8 : Signal flow graph representation of discrete time system.

PHASE VARIABLE FORM OF STATE MODEL

The discrete time system is governed by n^{th} order difference equation. The general form of nth order difference equation is,

$$y(k) = -\sum_{j=1}^{n} a_j\, y(k-j) + \sum_{j=0}^{m} b_j\, u(k-j) \hspace{2cm} .....(4.86)$$

where, a_j and b_j are constants for time invariant system.

On expanding the summation of equation (4.86) and rearranging we get,

$$y(k) + a_1\, y(k-1) + a_2\, y(k-2) +.........+a_n\, y(k-n) = b_0\, u(k) + b_1\, u(k-1) + b_2\, u(k-2) +......+ b_m u(k-m)$$
$$.....(4.87)$$

On taking $\mathcal{Z}$-transform of equation (4.87) with zero initial conditions we get,

$$y(z) + a_1\, z^{-1} y(z) + a_2\, z^{-2} y(z) +.........+ a_n\, z^{-n} y(z) = b_0\, U(z) + b_1\, z^{-1} U(z) + b_2\, z^{-2} U(z) +......+ b_m z^{-m} U(z)$$

$$Y(z)\,[\,1 + a_1\, z^{-1} + a_2\, z^{-2} +.......+ a_n\, z^{-n}\,] = [b_0 + b_1\, z^{-1} + b_2\, z^{-2} +........+ b_m\, z^{-m}]\, U(z)$$

When $m = n$,

$$\frac{Y(z)}{U(z)} = \frac{b_0 + b_1 z^{-1} + b_2 z^{-2} + ... + b_m z^{-m}}{1 + a_1 z^{-1} + a_2 z^{-2} + ... + a_n z^{-n}} \hspace{2cm} .....(4.88)$$

The equation (4.88) is the transfer function of the discrete time system. The equation (4.88) can be expressed as shown below.

$$\frac{Y(z)}{U(z)} = \frac{b_0 + b_1 z^{-1} + b_2 z^{-2} + ... + b_m z^{-m}}{1 - (-a_1 z^{-1} - a_2 z^{-2} - ..... - a_n z^{-n})} \hspace{2cm} .....(4.89)$$

On comparing the equation (4.89) with Mason's gain formula, a signal flow graph can be constructed. Each numerator term of equation (4.89) represent a forward path gain and each denominator term of equation (4.89) represent a individual loop gain. The signal flow graph will not have any non-touching loops. The signal flow graph when n=3 is shown in fig 4.9.

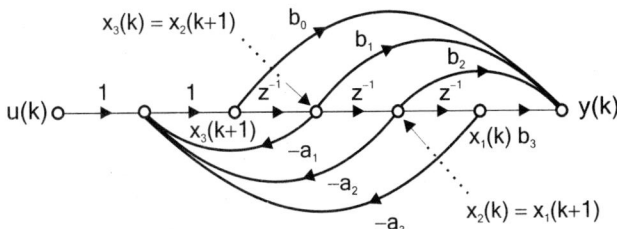

Fig 4.9 : *Signal flow graph representation of discrete time system.*

Let us assign state variables at the output of each unit delay element. Hence the signal at the input of unit delay element will be a signal advanced by one sampling time. The state equations are obtained by summing up the signals at the input of each unit delay element.

In fig 4.9, at node x_2 and x_3 we get,

$$x_1(k+1) = x_2 \hspace{5cm} .....(4.90)$$

$$x_2(k+1) = x_3 \hspace{5cm} .....(4.91)$$

By summing up the incoming signals to node $x_3(k+1)$ we get, (refer fig 4.9a)

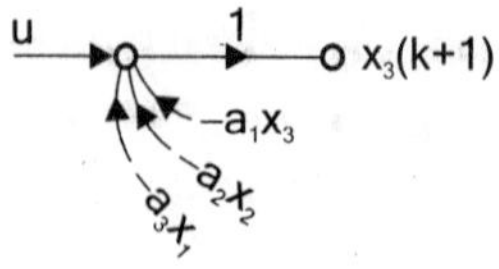

$$x_3(k+1) = -a_3x_1 - a_2x_2 - a_1x_3 + u \qquad .....(4.92)$$

Fig 4.9a.

By summing up the incoming signals to node y we get, (refer fig 4.9b).

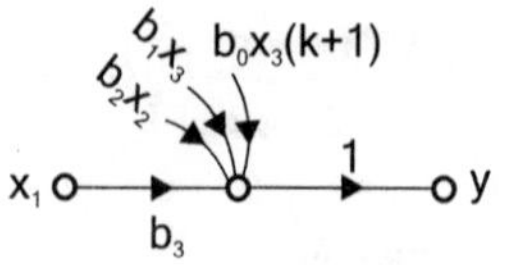

$$y = b_3x_1 + b_2x_2 + b_1x_3 + b_0x_3(k+1) \qquad .....(4.93)$$

Fig 4.9b.

Substitute for $x_3(k+1)$ from equation (4.92) in equation (4.93).

$$y = b_3 x_1 + b_2x_2 + b_1x_3 + b_0 (-a_3 x_1 - a_2x_2 - a_1x_3 + u)$$

$$\therefore y = (b_3 - b_0a_3) x_1 + (b_2 - b_0a_2) x_2 + (b_1 - b_0a_1) x_3 + b_0u \qquad .....(4.94)$$

The equations (4.90), (4.91) and (4.92) are state equations and equation (4.94) is the output equation. On arranging the state equations and output equations in the matrix form we get,

$$\begin{bmatrix} x_1(k+1) \\ x_2(k+1) \\ x_3(k+1) \end{bmatrix} = \begin{bmatrix} 0 & 1 & 0 \\ 0 & 0 & 1 \\ -a_3 & -a_2 & -a_1 \end{bmatrix} \begin{bmatrix} x_1 \\ x_2 \\ x_3 \end{bmatrix} + \begin{bmatrix} 0 \\ 0 \\ 1 \end{bmatrix} u \qquad .....(4.95)$$

$$y = [(b_3 - b_0a_3)(b_2 - b_0a_2)(b_1 - b_0a_1)] \begin{bmatrix} x_1 \\ x_2 \\ x_3 \end{bmatrix} + b_0u \qquad .....(4.96)$$

The equations (4.95) and (4.96) can be extended to n^{th} order system as shown below.

$$\begin{bmatrix} x_1(k+1) \\ x_2(k+1) \\ x_3(k+1) \\ \vdots \\ x_n(k+1) \end{bmatrix} = \begin{bmatrix} 0 & 1 & 0 & \cdots & 0 \\ 0 & 0 & 1 & \cdots & 0 \\ 0 & 0 & 0 & \cdots & 0 \\ \vdots & \vdots & \vdots & \vdots & \vdots \\ -a_n & -a_{n-1} & -a_{n-2} & \cdots & -a_1 \end{bmatrix} \begin{bmatrix} x_1 \\ x_2 \\ x_3 \\ \vdots \\ x_n \end{bmatrix} + \begin{bmatrix} 0 \\ 0 \\ 0 \\ \vdots \\ 1 \end{bmatrix} u \qquad .....(4.97)$$

$$y = [(b_n - b_0a_n)(b_{n-1} - b_0a_{n-1}) .... (b_1 - b_0a_1)] \begin{bmatrix} x_1 \\ x_2 \\ \vdots \\ x_n \end{bmatrix} + b_0u \qquad .....(4.98)$$

The equations (4.97) and (4.98) is the phase variable form of state model of discrete time n^{th} order system.

CANONICAL FORM OF STATE MODEL

The transfer function of equation (4.88) can be expressed as a summation using partial fraction technique as shown below.

$$\frac{Y(z)}{U(z)} = b_0 + \frac{C_1}{z + \lambda_1} + \frac{C_2}{z + \lambda_2} + ... + \frac{C_n}{z + \lambda_n}$$

where C_1, C_2 ,...... C_n are residues and λ_1 , λ_2 , ... λ_n are poles of the system.

$$\frac{Y(z)}{U(z)} = b_0 + \frac{C_1}{z\left(1 + \dfrac{\lambda_1}{z}\right)} + \frac{C_2}{z\left(1 + \dfrac{\lambda_2}{z}\right)} + ... + \frac{C_n}{z\left(1 + \dfrac{\lambda_n}{z}\right)}$$

$$\frac{Y(z)}{U(z)} = b_0 + \frac{z^{-1}C_1}{1 + z^{-1}\lambda_1} + \frac{z^{-1}C_2}{1 + z^{-1}\lambda_2} + ... + \frac{z^{-1}C_n}{1 + z^{-1}\lambda_n} \qquad .....(4.99)$$

The equation (4.99) can be used to construct the block diagram of state model as shown in fig (4.10).

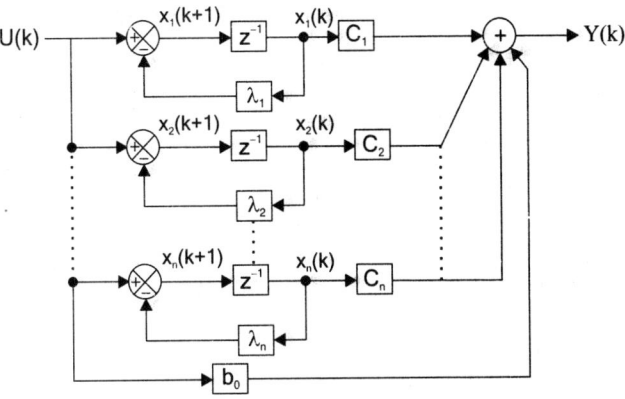

Fig 4.10 : Block diagram of canonical state model of discrete time system.

Let us assign the state variables at the output of unit delay elements. Hence the signal at the input of unit delay element will be a signal advanced by one sampling time. The state equations are obtained by summing up the signals at the input of each unit delay element.

The state equations are

$$x_1(k+1) = -\lambda_1 x_1 + u$$
$$x_2(k+1) = -\lambda_2 x_2 + u$$
$$x_3(k+1) = -\lambda_3 x_3 + u$$
$$\vdots \qquad \vdots$$
$$x_n(k+1) = -\lambda_n x_n + u$$

The output equation is

$$y = c_1 x_1 + c_2 x_2 + c_3 x_3 + ... + c_n x_n + b_0 u$$

The state equations and output equation can be expressed in the matrix form as shown below. The equations (4.100) and (4.101) is the canonical form of state model of discrete time n^{th} order system.

$$\begin{bmatrix} x_1(k+1) \\ x_2(k+1) \\ x_3(k+1) \\ \vdots \\ x_n(k+1) \end{bmatrix} = \begin{bmatrix} -\lambda_1 & 0 & 0 & \cdots & 0 \\ 0 & -\lambda_2 & 0 & \cdots & 0 \\ 0 & 0 & -\lambda_3 & \cdots & 0 \\ \vdots & \vdots & \vdots & & \vdots \\ -a_n & -a_{n-1} & -a_{n-2} & \cdots & -\lambda_n \end{bmatrix} \begin{bmatrix} x_1 \\ x_2 \\ x_3 \\ \vdots \\ x_n \end{bmatrix} + \begin{bmatrix} 1 \\ 1 \\ 1 \\ \vdots \\ 1 \end{bmatrix} u \qquad .....(4.100)$$

$$y = \begin{bmatrix} C_1 & C_2 & C_3 & \cdots & C_n \end{bmatrix} \begin{bmatrix} x_1 \\ x_2 \\ x_3 \\ \vdots \\ x_n \end{bmatrix} + b_0 u \qquad\qquad .....(4.101)$$

When a pole of the transfer function has multiplicity then the state model will be in a special form called Jordan canonical form. In this form the system matrix **A** will have a Jordan block of size q×q, correspond to a pole of value λ_i with multiplicity q. In the Jordan block the diagonal element will be the poles and the element just above the diagonal is one. Consider a system with poles $\lambda_1, \lambda_2, \lambda_3, \lambda_4, \ldots \lambda_n$, where λ_1 has multiplicity of three. The input matrix and system matrix for this case will be as shown in equation (4.100a). The system matrix is also denoted as **J**. The transfer function of the system for this case is given by equation (4.101) and the block diagram is shown in fig (4.10a).

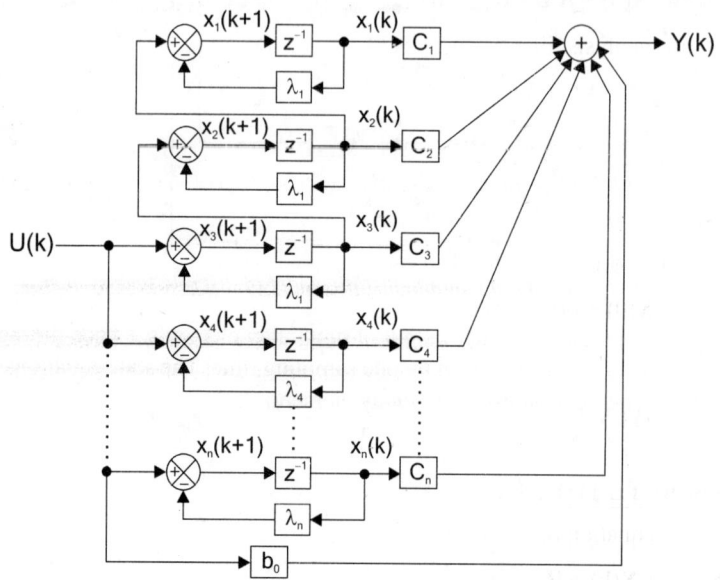

Fig 4.10a : *Block diagram of Jordan canonical state model of discrete time system.*

Jordan block of size 3×3

$$B = \begin{bmatrix} 0 \\ 0 \\ 1 \\ 1 \\ \vdots \\ 1 \end{bmatrix} \; ; \quad A = J = \begin{bmatrix} -\lambda_1 & 1 & 0 & 0 & \cdots & 0 \\ 0 & -\lambda_1 & 1 & 0 & \cdots & 0 \\ 0 & 0 & -\lambda_1 & 0 & \cdots & 0 \\ 0 & 0 & 0 & -\lambda_4 & \cdots & 0 \\ \vdots & \vdots & \vdots & \vdots & & \vdots \\ 0 & 0 & 0 & 0 & \cdots & -\lambda_n \end{bmatrix} \qquad .....(4.100a)$$

$$\frac{Y(z)}{U(z)} = b_0 + \frac{C_1}{(z+\lambda_1)^3} + \frac{C_2}{(z+\lambda_1)^2} + \frac{C_3}{z+\lambda_1} + \frac{C_4}{z+\lambda_4} + \ldots + \frac{C_n}{z+\lambda_n} \qquad .....(4.101)$$

SOLUTION OF DISCRETE TIME STATE EQUATION

The state equation of discrete time system is given by

$$X(k+1) = A X(k) + B U(k) \qquad\qquad .....(4.102)$$

When k = 0, the equation (4.102) can be written as

$$X(1) = A X(0) + B U(0) \qquad\qquad .....(4.103)$$

When k = 1, the equation (4.102) can be written as

$$X(2) = A X(1) + B U(1) \qquad\qquad .....(4.104)$$

On substituting for $X(1)$ from equation (4.103) in equation (4.104) we get,

$$X(2) = A [A X(0) + B U(0)] + BU(1)$$
$$\therefore X(2) = A^2 X(0) + A B U(0) + BU(1) \qquad\qquad .....(4.105)$$

On continuing this analysis till $X(k)$ we get,

$$X(k) = A^k X(0) + A^{(k-1)} B U(0) + A^{(k-2)} BU(1) + ...$$
$$......+ ABU(k-2) + BU(k-1) \qquad\qquad .....(4.106)$$

The matrix equation (4.106) is the solution of discrete time state equation. The matrix A^k is called the state transition matrix of discrete time system and it is also denoted by $\phi(k)$.

On substituting, $A^k = \phi(k)$ and $\phi(0) = I$ in equation (4.106) we get

$$X(k) = \phi(k) X(0) + \phi(k-1) B U(0) + \phi(k-2) BU(1) + ...$$
$$......+ \phi(1) BU(k-2) + \phi(0) BU(k-1)$$
$$\therefore X(k) = \phi(k) X(0) + \sum_{j=0}^{k-1} \phi(k-1-j) BU(j) \qquad\qquad .....(4.107)$$

SOLUTION OF DISCRETE TIME STATE EQUATION USING Z-TRANSFORM

Consider the state equation of the discrete time system

$$X(k+1) = A X(k) + B U(k) \qquad\qquad .....(4.108)$$

On taking $\mathbb{Z}$-transform of equation (4.108) we get

$$z X(z) - z X(0) = A X(z) + B U(z)$$
$$z X(z) - A X(z) = z X(0) + B U(z)$$
$$(zI - A) X(z) = z X(0) + B U(z) \qquad\qquad .....(4.109)$$

On premultiplying the equation (4.109) by $(zI-A)^{-1}$ we get

$$X(z) = (zI - A)^{-1} z X(0) + (zI - A)^{-1} B U(z) \qquad\qquad .....(4.110)$$

On taking inverse $\mathbb{Z}$-transform of equation (4.110) we get $X(k)$

$$\therefore X(k) = \mathbb{Z}^{-1}\{X(z)\} = \mathbb{Z}^{-1}\{(zI - A)^{-1} zX(0) + (zI - A)^{-1} BU(z)\}$$
$$X(k) = \mathbb{Z}^{-1}\{(zI - A)^{-1}z\}X(0) + \mathbb{Z}^{-1}\{(zI - A)^{-1} BU(z)\} \qquad\qquad .....(4.111)$$

The equation (4.111) is the solution of discrete time state equation.

PROPERTIES OF STATE TRANSITION MATRIX OF DISCRETE TIME SYSTEM

1. $\phi(0) = \mathbf{I}$

2. $\phi^{-1}(k) = \phi(-k)$

3. $\phi(k, k_0) = \phi(k-k_0) = \mathbf{A}^{(k-k_0)}$; where, $k > k_0$

COMPUTATION OF STATE TRANSITION MATRIX

The state transition matrix $\mathbf{A}^k$ can be computed by any one of the following methods.

Method 1 : Computation of $\mathbf{A}^k$ using Z-transform

Method 2 : Computation of $\mathbf{A}^k$ by canonical transformation

Method 3 : Computation of $\mathbf{A}^k$ by Cayley - Hamilton theorem

The computation of $\mathbf{A}^k$ using Z-transform have been dealt in this section.

On comparing equations (4.111) and (4.107) [or (4.106)] we can write,

State transition matrix, $\mathbf{A}^k = \mathbf{Z}^{-1}\{(z\mathbf{I} - \mathbf{A})^{-1}z\}$ ....(4.112)

The equation (4.112) can be used to compute the state transition matrix, $\mathbf{A}^k$.

EXAMPLE 4.17

A discrete-time system has the transfer function

$$\frac{Y(z)}{U(z)} = \frac{4z^3 - 12z^2 + 13z - 7}{(z-1)^2(z-2)}$$

Determine the state model of the system in (a) Phase variable form and (b) Jordan canonical form.

SOLUTION

a. Phase variable form of state model

Given that, $\dfrac{Y(z)}{U(z)} = \dfrac{4z^3 - 12z^2 + 13z - 7}{(z-1)^2(z-2)}$

$$= \frac{z^3(4 - 12z^{-1} + 13z^{-2} - 7z^{-3})}{(z^2 - 2z + 1)(z - 2)} = \frac{z^3(4 - 12z^{-1} + 13z^{-2} - 7z^{-3})}{z^3 - 2z^2 - 2z^2 + 4z + z - 2}$$

$$= \frac{z^3(4 - 12z^{-1} + 13z^{-2} - 7z^{-3})}{z^3(1 - 4z^{-1} + 5z^{-2} - 2z^{-3})} = \frac{4 - 12z^{-1} + 13z^{-2} - 7z^{-3}}{1 - (4z^{-1} - 5z^{-2} + 2z^{-3})} \qquad .....(4.17.1)$$

The equation (4.17.1) can be used to construct the signal flow graph of the discrete time system shown in fig 4.17.1.

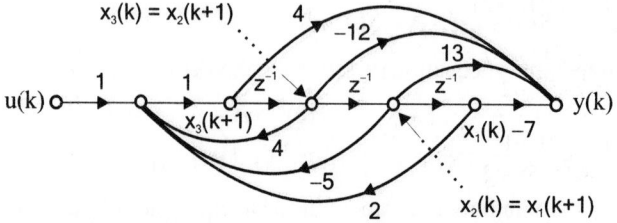

Fig 4.17.1 : Signal flow graph for transfer function of equation (4.17.1).

From node x_2 and x_3 we get,

$$x_1(k+1) = x_2 \qquad\qquad\qquad\qquad .....(4.17.2)$$

$$x_2(k+1) = x_3 \qquad\qquad\qquad\qquad .....(4.17.3)$$

By summing up the incoming signals to node $x_3(k+1)$ we get, (refer fig 4.17.2)

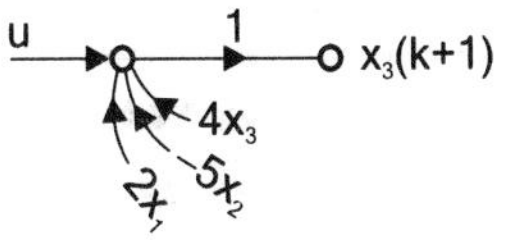

$$x_3(k+1) = 2x_1 - 5x_2 + 4x_3 + u \qquad .....(4.17.4)$$

Fig 4.17.2

The equations (4.17.2) to (4.17.4) are the state equations. The output equation is obtained by summing up the incoming signals to output node (Refer fig 4.17.3)

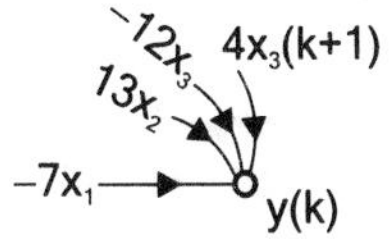

$$y = -7x_1 + 13x_2 - 12x_3 + 4x_3(k+1) \qquad .....(4.17.5)$$

Fig 4.17.3

On substituting for $x_3(k+1)$ from equation (4.17.4) in equation (4.17.5) we get,

$$y = -7x_1 + 13x_2 - 12x_3 + 4\,[2x_1 - 5x_2 + 4x_3 + u]$$

$$y = (-7+8)\,x_1 + (13-20)\,x_2 + (-12+16)\,x_3 + 4u$$

$$y = x_1 - 7x_2 + 4x_3 + 4u \qquad\qquad .....(4.17.6)$$

The state equations and output equation can be arranged in the matrix form as shown below.

$$\begin{bmatrix} x_1(k+1) \\ x_2(k+1) \\ x_3(k+1) \end{bmatrix} = \begin{bmatrix} 0 & 1 & 0 \\ 0 & 0 & 1 \\ 2 & -5 & 4 \end{bmatrix}\begin{bmatrix} x_1 \\ x_2 \\ x_3 \end{bmatrix} + \begin{bmatrix} 0 \\ 0 \\ 1 \end{bmatrix}u \qquad .....(4.17.7)$$

$$y = \begin{bmatrix} 1 & -7 & 4 \end{bmatrix}\begin{bmatrix} x_1 \\ x_2 \\ x_3 \end{bmatrix} + 4u \qquad .....(4.17.8)$$

The equations (4.17.7) and (4.17.8) is the phase variable form of state model.

b. Jordan Canonical form of state model

Given that, $\dfrac{Y(z)}{U(z)} = \dfrac{4z^3 - 12z^2 + 13z - 7}{(z-1)^2(z-2)}$

$$= \dfrac{4z^3 - 12z^2 + 13z - 7}{z^3 - 4z^2 + 5z - 2}$$

$$= 4 + \dfrac{4z^2 - 7z + 1}{z^3 - 4z^2 + 5z - 2} = 4 + \dfrac{4z^2 - 7z + 1}{(z-1)^2(z-2)} \qquad .....(4.17.9)$$

	4
$z^3 - 4z^2 + 5z - 2$	$4z^3 - 12z^2 + 13z - 7$
	$4z^3 - 16z^2 + 20z - 8$
	$4z^2 - 7z + 1$

By partial fraction expansion the equation (4.17.9) can be written as,

$$\frac{Y(z)}{U(z)} = 4 + \frac{A_1}{(z-1)^2} + \frac{A_2}{z-1} + \frac{A_3}{z-2}$$

$$A_1 = (z-1)^2 \frac{4z^2 - 7z + 1}{(z-1)^2(z-2)}\bigg|_{z=1} = \frac{4z^2 - 7z + 1}{z-2}\bigg|_{z=1} = \frac{4 - 7 + 1}{1 - 2} = 2$$

$$A_2 = \frac{d}{dz}\left[(z-1)^2 \frac{4z^2 - 7z + 1}{(z-1)^2(z-2)}\right]\bigg|_{z=1} = \frac{d}{dz}\left[\frac{4z^2 - 7z + 1}{z-2}\right]\bigg|_{z=1}$$

$$= \frac{(8z - 7)(z-2) - (4z^2 - 7z + 1)}{(z-2)^2}\bigg|_{z=1} = \frac{(8 - 7)(1 - 2) - (4 - 7 + 1)}{(1 - 2)^2} = 1$$

$$A_3 = (z-2)\frac{4z^2 - 7z + 1}{(z-1)^2(z-2)}\bigg|_{z=2} = \frac{4z^2 - 7z + 1}{(z-1)^2}\bigg|_{z=2} = \frac{4 \times 2^2 - 7 \times 2 + 1}{(2-1)^2} = 3$$

$$\therefore \quad \frac{Y(z)}{U(z)} = 4 + \frac{2}{(z-1)^2} + \frac{1}{z-1} + \frac{3}{z-2} \qquad\qquad .....(4.17.10)$$

The equation (4.17.10) can be used to construct the block diagram of the discrete time system shown in fig 4.17.4.

Let us assign state variable at the output of each unit delay element. The state equations are formed by summing up the incoming signals to the unit delay element.

The state equations are,

$$x_1(k+1) = x_1 + x_2$$
$$x_2(k+1) = x_2 + u$$
$$x_3(k+1) = 2x_3 + u$$

The output equation is,

$$y = 2x_1 + x_2 + 3x_3 + 4u$$

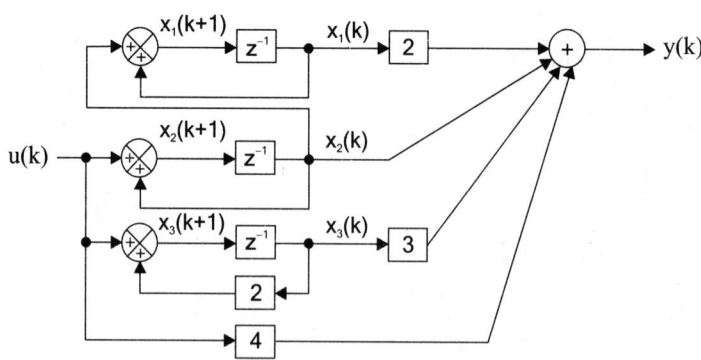

Fig 4.17.4 : Block diagram.

On arranging the state equations and the output equation in the matrix form we get,

$$\begin{bmatrix} x_1(k+1) \\ x_2(k+1) \\ x_3(k+1) \end{bmatrix} = \begin{bmatrix} 1 & 1 & 0 \\ 0 & 1 & 0 \\ 0 & 0 & 2 \end{bmatrix} \begin{bmatrix} x_1 \\ x_2 \\ x_3 \end{bmatrix} + \begin{bmatrix} 0 \\ 1 \\ 1 \end{bmatrix} u \qquad \qquad .....(4.17.11)$$

$$y = \begin{bmatrix} 2 & 1 & 3 \end{bmatrix} \begin{bmatrix} x_1 \\ x_2 \\ x_3 \end{bmatrix} + 4u \qquad \qquad .....(4.17.12)$$

The equations (4.17.11) and (4.17.12) constitute Jordan canonical form of state model of the system.

EXAMPLE 4.18

A discrete time system is described by the difference equation,

$$y(k+2) + 5y(k+1) + 6\,y(k) = u(k)$$

$$y(0) = y(1) = 0; \quad T = 1\text{sec}.$$

(a)   Determine a state model in canonical form.

(b)   Find the state transition matrix

(c)   For input $u(k) = 1$; $k \geq 1$, find the output $y(k)$.

SOLUTION

(a) To determine the canonical form of state model

Given that, $y(k+2) + 5y(k+1) + 6\,y(k) = u(k)$ $\qquad\qquad$ .....(4.18.1)

and $y(0) = y(1) = 0; \quad T = 1\text{sec}.$

On taking Z-transform of equation (4.18.1) with zero initial conditions we get,

$$z^2\,Y(z) + 5z\,Y(z) + 6Y(z) = U(z)$$

$$(z^2 + 5z + 6)\,Y(z) = U(z)$$

$$\therefore \quad \frac{Y(z)}{U(z)} = \frac{1}{z^2 + 5z + 6} = \frac{1}{(z+3)(z+2)} \qquad\qquad .....(4.18.2)$$

By partial fraction expansion the equation (4.18.2) can be expressed as,

$$\frac{Y(z)}{U(z)} = \frac{A_1}{z+3} + \frac{A_2}{z+2}$$

$$A_1 = (z+3)\,\frac{1}{(z+3)(z+2)}\Big|_{z=-3} = \frac{1}{z+2}\Big|_{z=-3} = \frac{1}{-3+2} = -1$$

$$A_2 = (z+2)\,\frac{1}{(z+3)(z+2)}\Big|_{z=-2} = \frac{1}{z+3}\Big|_{z=-2} = \frac{1}{-2+3} = 1$$

$$\therefore \quad \frac{Y(z)}{U(z)} = \frac{-1}{z+3} + \frac{1}{z+2}$$

$$\qquad\qquad .....(4.18.3)$$

The equation (4.18.3) can be used to construct the block diagram shown in fig 1.

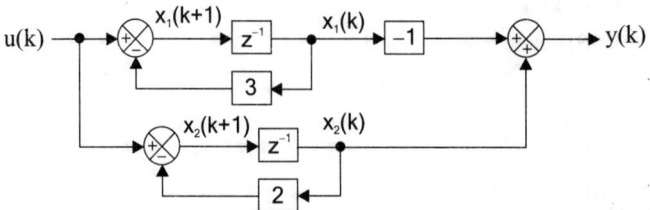

Fig 4.18.1 : Block diagram.

From the block diagram, the state equations are

$$x_1(k+1) = -3x_1 + u$$
$$x_2(k+1) = -2x_2 + u$$

The output equation is

$$y = -x_1 + x_2$$

On arranging the state equations and the output equation in the matrix form, we get,

$$\begin{bmatrix} x_1(k+1) \\ x_2(k+1) \end{bmatrix} = \begin{bmatrix} -3 & 0 \\ 0 & -2 \end{bmatrix}\begin{bmatrix} x_1 \\ x_2 \end{bmatrix} + \begin{bmatrix} 1 \\ 1 \end{bmatrix} u \qquad .....(4.18.4)$$

$$y = \begin{bmatrix} -1 & 1 \end{bmatrix}\begin{bmatrix} x_1 \\ x_2 \end{bmatrix} \qquad .....(4.18.5)$$

The equations (4.18.4) and (4.18.5) constitute the canonical form of state model.

(b) To find state transition matrix A^k

The state transition matrix, $A^k = Z^{-1}\{[zI - A]^{-1}z\}$

From the state equation we get, $A = \begin{bmatrix} -3 & 0 \\ 0 & -2 \end{bmatrix}$

$$\therefore \ zI - A = z\begin{bmatrix} 1 & 0 \\ 0 & 1 \end{bmatrix} - \begin{bmatrix} -3 & 0 \\ 0 & -2 \end{bmatrix} = \begin{bmatrix} z+3 & 0 \\ 0 & z+2 \end{bmatrix}$$

$$|zI - A| = \begin{vmatrix} z+3 & 0 \\ 0 & z+2 \end{vmatrix} = (z+2)(z+3)$$

$$[zI - A]^{-1} = \frac{1}{|zI - A|}\begin{bmatrix} z+2 & 0 \\ 0 & z+3 \end{bmatrix}^T = \frac{1}{(z+2)(z+3)}\begin{bmatrix} z+2 & 0 \\ 0 & z+3 \end{bmatrix}$$

$$= \begin{bmatrix} \dfrac{1}{z+3} & 0 \\ 0 & \dfrac{1}{z+2} \end{bmatrix} \qquad .....(4.18.6)$$

$$A^k = Z^{-1}\{[zI - A]^{-1}z\} = Z^{-1}\begin{bmatrix} \dfrac{z}{z+3} & 0 \\ 0 & \dfrac{z}{z+2} \end{bmatrix}$$

We know that $\mathbf{Z}^{-1}\left\{\dfrac{z}{z-a}\right\} = a^k$

$\therefore \mathbf{Z}^{-1}\left\{\dfrac{z}{z+3}\right\} = (-3)^k$ and $\mathbf{Z}^{-1}\left\{\dfrac{z}{z+2}\right\} = (-2)^k$

$\therefore$ State transition matrix, $\quad \mathbf{A}^k = \begin{bmatrix} (-3)^k & 0 \\ 0 & (-2)^k \end{bmatrix}$ .....(4.18.7)

(c) To find y(k) when the input is unit step

Consider the state equation,

$\mathbf{X}(k+1) = \mathbf{A\,X}(k) + \mathbf{B\,U}(k)$ .....(4.18.8)

On taking Z-transform of equation(6.21.8) we get

$z\,\mathbf{X}(z) - z\,\mathbf{X}(0) = \mathbf{AX}(z) + \mathbf{BU}(z)$

Let us assume that, $\mathbf{X}(0) = 0$

$\therefore\ z\,\mathbf{X}(z) - \mathbf{A\,X}(z) = \mathbf{B\,U}(z)$

$(z\mathbf{I} - \mathbf{A})\,\mathbf{X}(z) = \mathbf{B\,U}(z)$ .....(4.18.9)

On premultiplying both sides of equation (4.18.9) by $(z\mathbf{I} - \mathbf{A})^{-1}$ we get

$\mathbf{X}(z) = (z\mathbf{I} - \mathbf{A})^{-1}\,\mathbf{BU}(z)$ .....(4.18.10)

Given that, u(k) = 1, $\qquad \therefore\ \mathbf{U}(z) = \mathbf{Z}\{u(k)\} = \dfrac{z}{z-1}$ .....(4.18.11)

From equation (4.18.6) we get, $[z\mathbf{I} - \mathbf{A}]^{-1} = \begin{bmatrix} \dfrac{1}{z+3} & 0 \\ 0 & \dfrac{1}{z+2} \end{bmatrix}$

From equation (4.18.4) we get, $\mathbf{B} = \begin{bmatrix} 1 \\ 1 \end{bmatrix}$

$\therefore\ \mathbf{X}(z) = \begin{bmatrix} \dfrac{1}{z+3} & 0 \\ 0 & \dfrac{1}{z+2} \end{bmatrix}\begin{bmatrix} 1 \\ 1 \end{bmatrix}\left[\dfrac{z}{z-1}\right] = \begin{bmatrix} \dfrac{1}{z+3} \\ \dfrac{1}{z+2} \end{bmatrix}\left[\dfrac{z}{z-1}\right] = \begin{bmatrix} \dfrac{z}{(z-1)(z+3)} \\ \dfrac{z}{(z-1)(z+2)} \end{bmatrix}$

We know that $\mathbf{X}(z) = \begin{bmatrix} x_1(z) \\ x_2(z) \end{bmatrix}$ and $\mathbf{X}(k) = \begin{bmatrix} x_1(k) \\ x_2(k) \end{bmatrix}$

Also $\mathbf{X}(k) = \mathbf{Z}^{-1}\{\mathbf{X}(z)\}$

$\therefore\ \mathbf{X}(z) = \begin{bmatrix} X_1(z) \\ X_2(z) \end{bmatrix} = \begin{bmatrix} \dfrac{z}{(z-1)(z+3)} \\ \dfrac{z}{(z-1)(z+2)} \end{bmatrix}$ and $\mathbf{X}(k) = \begin{bmatrix} x_1(z) \\ x_2(z) \end{bmatrix} = \begin{bmatrix} \mathbf{Z}^{-1}\{X_1(z)\} \\ \mathbf{Z}^{-1}\{X_2(z)\} \end{bmatrix}$

$\therefore\ x_1(k) = \mathbf{Z}^{-1}\left\{\dfrac{z}{(z-1)(z+3)}\right\}$ and $x_2(k) = \mathbf{Z}^{-1}\left\{\dfrac{z}{(z-1)(z+2)}\right\}$

By partial fraction expansion,

$$\frac{z}{(z-1)(z+3)} = z\left[\frac{1}{(z-1)(z+3)}\right] = z\left[\frac{A_1}{z-1} + \frac{B_1}{z+3}\right]$$

$$A_1 = \frac{z}{(z-1)(z+3)}(z-1)\bigg|_{z=1} = \frac{1}{z+3}\bigg|_{z=1} = \frac{1}{1+3} = \frac{1}{4}$$

$$B_1 = \frac{1}{(z-1)(z+3)}(z+3)\bigg|_{z=-3} = \frac{1}{z-1}\bigg|_{z=-3} = \frac{1}{-3-1} = -\frac{1}{4}$$

$$\therefore \frac{z}{(z-1)(z+3)} = z\left[\frac{A_1}{z-1} + \frac{B_1}{z+3}\right] = \left[\frac{1}{4}\frac{z}{z-1} - \frac{1}{4}\frac{z}{z+3}\right]$$

$$x_1(k) = \mathcal{Z}^{-1}\left\{\frac{z}{(z-1)(z+3)}\right\} = \mathcal{Z}^{-1}\left[\frac{1}{4}\frac{z}{z-1} - \frac{1}{4}\frac{z}{z+3}\right] = \frac{1}{4}u(k) - \frac{1}{4}(-3)^k$$

By partial fraction expansion,

$$\frac{z}{(z-1)(z+2)} = z\left[\frac{1}{(z-1)(z+2)}\right] = z\left[\frac{A_2}{z-1} + \frac{B_2}{z+2}\right]$$

$$A_2 = \frac{1}{(z-1)(z+2)}(z-1)\bigg|_{z=1} = \frac{1}{z+2}\bigg|_{z=1} = \frac{1}{1+2} = \frac{1}{3}$$

$$B_2 = \frac{1}{(z-1)(z+2)}(z+2)\bigg|_{z=-2} = \frac{1}{z-1}\bigg|_{z=-2} = \frac{1}{-2-1} = -\frac{1}{3}$$

$$\therefore \frac{z}{(z-1)(z+2)} = z\left[\frac{A_2}{z-1} + \frac{B_2}{z+2}\right] = \left[\frac{1}{3}\frac{z}{z-1} - \frac{1}{3}\frac{z}{z+2}\right]$$

$$x_2(k) = \mathcal{Z}^{-1}\left\{\frac{z}{(z-1)(z+2)}\right\} = \mathcal{Z}^{-1}\left[\frac{1}{3}\frac{z}{z-1} - \frac{1}{3}\frac{z}{z+2}\right] = \frac{1}{3}u(k) - \frac{1}{3}(-2)^k$$

From equation (4.18.5) we get,

Response or Output, $y(k) = \begin{bmatrix} -1 & 1 \end{bmatrix}\begin{bmatrix} x_1(k) \\ x_2(k) \end{bmatrix}$

$$\therefore \ y(k) = -x_1(k) + x_2(k)$$

$$= -\frac{1}{4}u(k) + \frac{1}{4}(-3)^k + \frac{1}{3}u(k) - \frac{1}{3}(-2)^k$$

$$= \frac{1}{4}(-3)^k - \frac{1}{3}(-2)^k + \frac{-3+4}{12}u(k)$$

$$= \frac{1}{4}(-3)^k - \frac{1}{3}(-2)^k + \frac{1}{12}u(k)$$

$$= \begin{bmatrix} \dfrac{2(s+2)-(s+1)}{(s+1)(s+2)} & \dfrac{(s+2)-(s+1)}{(s+1)(s+2)} \\[3mm] \dfrac{-2(s+2)+2(s+1)}{(s+1)(s+2)} & \dfrac{-(s+2)+2(s+1)}{(s+1)(s+2)} \end{bmatrix} = \begin{bmatrix} \dfrac{s+3}{(s+1)(s+2)} & \dfrac{1}{(s+1)(s+2)} \\[3mm] \dfrac{-2}{(s+1)(s+2)} & \dfrac{s}{(s+1)(s+2)} \end{bmatrix}$$

Determinant of $\phi(s) = \dfrac{s(s+3)+2}{(s+1)^2(s+2)^2} = \dfrac{s^2+3s+2}{(s+1)(s+2)^2}$

$$= \dfrac{(s+1)(s+2)}{(s+1)^2(s+2)^2} = \dfrac{1}{(s+1)(s+2)}$$

$$\phi(s)^{-1} = \dfrac{1}{(s+1)(s+2)} \begin{bmatrix} \dfrac{s}{(s+1)(s+2)} & \dfrac{-1}{(s+1)(s+2)} \\[3mm] \dfrac{2}{(s+1)(s+2)} & \dfrac{s+3}{(s+1)(s+2)} \end{bmatrix} = \begin{bmatrix} s & -1 \\ 2 & s+3 \end{bmatrix}$$

$$A = sI - \phi(s)^{-1} = s\begin{bmatrix} 1 & 0 \\ 0 & 1 \end{bmatrix} - \begin{bmatrix} s & -1 \\ 2 & s+3 \end{bmatrix} = \begin{bmatrix} s & 0 \\ 0 & s \end{bmatrix} - \begin{bmatrix} s & -1 \\ 2 & s+3 \end{bmatrix} = \begin{bmatrix} 0 & 1 \\ -2 & -3 \end{bmatrix}$$

RESULT

$$A = \begin{bmatrix} 0 & 1 \\ -2 & -3 \end{bmatrix} \; ; \; e^{At} = \begin{bmatrix} 2e^{-t}-e^{-2t} & e^{-t}-e^{-2t} \\ -2e^{-t}+2e^{-2t} & -e^{-t}-2e^{-2t} \end{bmatrix}$$

EXAMPLE 4.16

The state equation and initial condition vector of an linear time-invariant system are given below. Determine the solution of state equation.

$$\begin{bmatrix} \dot{x}_1 \\ \dot{x}_2 \end{bmatrix} = \begin{bmatrix} 1 & 0 \\ 1 & 1 \end{bmatrix} \begin{bmatrix} x_1 \\ x_2 \end{bmatrix} \; ; \; X_0 = \begin{bmatrix} 1 \\ 0 \end{bmatrix}$$

SOLUTION

Here $A = \begin{bmatrix} 1 & 0 \\ 1 & 1 \end{bmatrix}$; $sI - A = s\begin{bmatrix} 1 & 0 \\ 0 & 1 \end{bmatrix} - \begin{bmatrix} 1 & 0 \\ 1 & 1 \end{bmatrix} = \begin{bmatrix} s & 0 \\ 0 & s \end{bmatrix} - \begin{bmatrix} 1 & 0 \\ 1 & 1 \end{bmatrix} = \begin{bmatrix} s-1 & 0 \\ -1 & s-1 \end{bmatrix}$

$$|sI - A| = \begin{vmatrix} s-1 & 0 \\ -1 & s-1 \end{vmatrix} = (s-1)^2 - 0 = (s-1)^2$$

$$(sI - A)^{-1} = \dfrac{1}{(s-1)^2} \begin{bmatrix} s-1 & 0 \\ 1 & s-1 \end{bmatrix} = \begin{bmatrix} \dfrac{1}{s-1} & 0 \\[3mm] \dfrac{1}{(s-1)^2} & \dfrac{1}{s-1} \end{bmatrix}$$

$$e^{At} = \phi(t) = \mathcal{L}^{-1}[\phi(s)] = \mathcal{L}^{-1}[(sI - A)^{-1}] \begin{bmatrix} e^t & 0 \\ te^t & e^t \end{bmatrix}$$

The solution of the state equation is, $X(t) = e^{At} X_0 = \begin{bmatrix} e^t & 0 \\ te^t & e^t \end{bmatrix} \begin{bmatrix} 1 \\ 0 \end{bmatrix} = \begin{bmatrix} e^t \\ te^t \end{bmatrix}$

4.9   STATE SPACE REPRESENTATION OF DISCRETE TIME SYSTEMS

The state variable analysis techniques of continuous time systems can be extended to the discrete-time system. The discrete form of state space representation is quite analogous to the continuous form.

In the state variable formulation of a discrete time system, in general, a system consists of m-inputs, p-outputs and n-state variables. The state space representation of discrete-time system may be visualized as shown in fig 4.6.

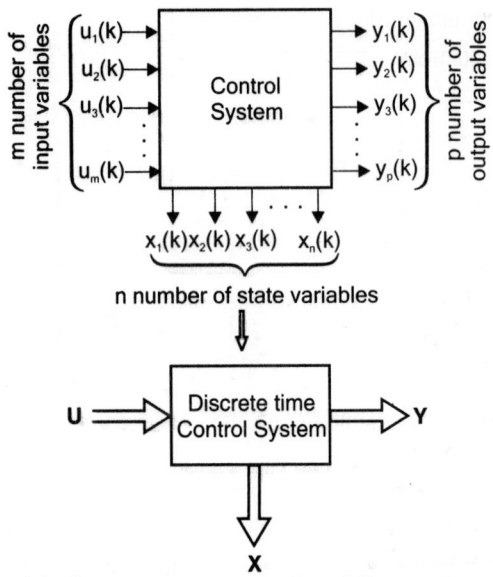

Fig 4.6 : State space representation of discrete time system.

Let,   State variables = $x_1(k)$, $x_2(k)$, $x_3(k)$, ... $x_n(k)$

Input variables = $u_1(k)$, $u_2(k)$, $u_3(k)$, ... $u_m(k)$

Output variables = $y_1(k)$, $y_2(k)$, $y_3(k)$, ... $y_p(k)$

The different variables may be represented by the vectors (column matrix) as shown below.

$$\begin{matrix} \text{Input} \\ \text{vector} \end{matrix} \quad U(k) = \begin{bmatrix} u_1(k) \\ u_2(k) \\ \vdots \\ u_m(k) \end{bmatrix} \quad ; \quad \begin{matrix} \text{Output} \\ \text{vector} \end{matrix} \quad Y(k) = \begin{bmatrix} y_1(k) \\ y_2(k) \\ \vdots \\ y_p(k) \end{bmatrix} \quad ; \quad \begin{matrix} \text{State variable} \\ \text{vector} \end{matrix} \quad X(k) = \begin{bmatrix} x_1(k) \\ x_2(k) \\ \vdots \\ x_n(k) \end{bmatrix}$$

Note : The simplified notation x(k), y(k) and u(k) are used to denote x(kT), y(kT) and u(kT) respectively. Also for convenience the variables are denoted x_1, x_2, x_3, ... ; y_1, y_2, y_3, ... and u_1, u_2, u_3 .....

The state equation of a discrete time system is a set of n-numbers of first order difference equations.

$$x_1(k + 1) = f_1(x_1, x_2, ...x_n ; u_1, u_2, ...u_m)$$

$$x_2(k + 1) = f_2(x_1, x_2, ...x_n ; u_1, u_2, ...u_m)$$

$$\vdots$$

$$x_n(k + 1) = f_n(x_1, x_2, ...x_n ; u_1, u_2, ...u_m)$$

For linear time invariant discrete time systems the above difference equations can be expressed as a linear combination of state variables and inputs.

$$x_1(k+1) = a_{11}x_1 + a_{12}x_2 + \ldots\ldots + a_{1n}x_n + b_{11}u_1 + b_{12}u_2 + \ldots\ldots + b_{1m}u_m$$

$$x_2(k+1) = a_{21}x_1 + a_{22}x_2 + \ldots\ldots + a_{2n}x_n + b_{21}u_1 + b_{22}u_2 + \ldots\ldots + b_{2m}u_m$$

$$\vdots$$

$$x_n(k+1) = a_{n1}x_1 + a_{n2}x_2 + \ldots\ldots + a_{nn}x_n + b_{n1}u_1 + b_{n2}u_2 + \ldots\ldots + b_{nm}u_m$$

where, the coefficients a_{ij} and b_{ij} are constants.

In the matrix form the above equations can be expressed as,

$$\begin{bmatrix} x_1(k+1) \\ x_2(k+1) \\ \vdots \\ x_n(k+1) \end{bmatrix} = \begin{bmatrix} a_{11} & a_{12} & \cdots & a_{1n} \\ a_{21} & a_{22} & \cdots & a_{2n} \\ \vdots & \vdots & & \vdots \\ a_{n1} & a_{n2} & \cdots & a_{nn} \end{bmatrix} \begin{bmatrix} x_1 \\ x_2 \\ \vdots \\ x_n \end{bmatrix} + \begin{bmatrix} b_{11} & b_{12} & \cdots & b_{1m} \\ b_{21} & b_{22} & \cdots & b_{2m} \\ \vdots & \vdots & & \vdots \\ b_{n1} & b_{n2} & \cdots & b_{nm} \end{bmatrix} \begin{bmatrix} u_1 \\ u_2 \\ \vdots \\ u_m \end{bmatrix} \qquad \ldots\ldots(4.82)$$

The matrix equation (4.82) can be written in the vector notation as

$$\mathbf{X}(k+1) = \mathbf{AX}(k) + \mathbf{BU}(k) \qquad \ldots\ldots(4.83)$$

where, $\mathbf{X}(k)$ = State vector of order $(n \times 1)$

$\mathbf{U}(k)$ = Input vector of order $(m \times 1)$

$\mathbf{A}$ = System matrix of order $(n \times n)$

$\mathbf{B}$ = Input matrix of order $(n \times m)$

The equation (4.83) is the state equation of (linear time invariant) discrete time system.

The output at any discrete time instant, k are functions of state variables and inputs. Hence the output variables of linear time invariant system can be expressed as a linear combination of state variables and inputs.

$$y_1 = c_{11}x_1 + c_{12}x_2 + \ldots\ldots + c_{1n}x_n + d_{11}u_1 + d_{12}u_2 + \ldots\ldots + d_{1m}u_m$$

$$y_2 = c_{21}x_1 + c_{22}x_2 + \ldots\ldots + c_{2n}x_n + d_{21}u_1 + d_{22}u_2 + \ldots\ldots + d_{2m}u_m$$

$$\vdots$$

$$y_p = c_{p1}x_1 + c_{p2}x_2 + \ldots\ldots + c_{pn}x_n + d_{p1}u_1 + d_{p2}u_2 + \ldots\ldots + d_{pm}u_m$$

where, the coefficients c_{ij} and d_{ij} are constants.

In the matrix form the above equations can be expressed as,

$$\begin{bmatrix} y_1 \\ y_2 \\ \vdots \\ y_p \end{bmatrix} = \begin{bmatrix} c_{11} & c_{12} & \cdots & c_{1n} \\ c_{21} & c_{22} & \cdots & c_{2n} \\ \vdots & \vdots & & \vdots \\ c_{p1} & c_{p2} & \cdots & c_{pn} \end{bmatrix} \begin{bmatrix} x_1 \\ x_2 \\ \vdots \\ x_n \end{bmatrix} + \begin{bmatrix} d_{11} & d_{12} & \cdots & d_{1m} \\ d_{21} & d_{22} & \cdots & d_{2m} \\ \vdots & \vdots & & \vdots \\ d_{p1} & d_{p2} & \cdots & d_{pm} \end{bmatrix} \begin{bmatrix} u_1 \\ u_2 \\ \vdots \\ u_m \end{bmatrix} \qquad \ldots\ldots(4.84)$$

The matrix equation (4.84) can be written in the vector notation as,

$$\mathbf{Y}(k) = \mathbf{CX}(k) + \mathbf{DU}(k) \qquad \ldots\ldots(4.85)$$

where, $\mathbf{X}(k)$ = State vector of order $(n \times 1)$

$\mathbf{U}(k)$ = Input vector of order $(m \times 1)$

$\mathbf{Y}(k)$ = Output vector of order $(p \times 1)$

$\mathbf{C}$ = Output matrix of order $(p \times n)$

$\mathbf{D}$ = Transmission matrix of order $(p \times m)$

The equation (4.85) is the output equation of (linear time invariant) discrete time system.

The state equation and output equation are together called as state model of the system. Hence the state model of discrete time system is given by the following equations.

$$X(k+1) = A\,X(k) + B\,U(k) \quad \text{........ State equation}$$
$$Y(k) = C\,X(k) + D\,U(k) \quad \text{......... Output equation}$$

STATE DIAGRAM OF DISCRETE TIME SYSTEM

The state diagram of discrete time system can be either in block diagram form or signal flow graph form. The three fundamental elements, scalar, adder and unit delay elements are used to construct the state diagram. These basic elements are shown in table-4.6.

Scalar   :   The scalar is used to multiply a signal by a constant.

Adder   :   The adder is used to add two or more signals.

Unit delay :   The unit delay element will delay the signal passing through it by one sample time.

TABLE-4.6 : Basic Elements of State Diagram of Discrete Time System

Element	Block diagram representation	Signal flow graph representation
Scalar	$x(k)$ → $\boxed{a}$ → $ax(k)$	$x(k)$ ∘———a———∘ $a\,x(k)$
Adder	$x_1(k)$ → ⊕ → $x_1(k)+x_2(k)$ $x_2(k)$ ↑	$x_1(k)$ ∘⟍ 1 $x_2(k)$ ∘⟋ 1 → $x_1(k) + x_2(k)$
Unit delay	$x(k+1)$ → $\boxed{z^{-1}}$ → $x(k)$	$x(k+1)$ ∘——z^{-1}——∘ $x(k)$

The block diagram representation of the state model of discrete time system is shown in fig 4.7. and the signal flow graph representation is shown in fig 4.8.

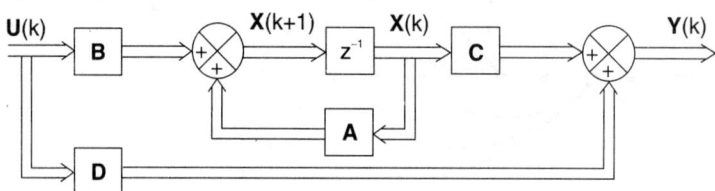

Fig 4.7 : Block diagram representation of discrete time system.

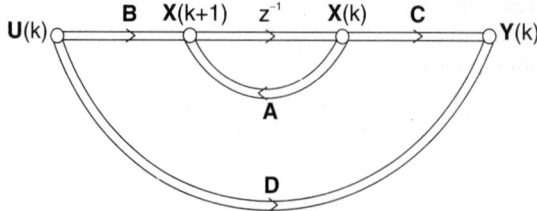

Fig 4.8 : Signal flow graph representation of discrete time system.

PHASE VARIABLE FORM OF STATE MODEL

The discrete time system is governed by n^{th} order difference equation. The general form of nth order difference equation is,

$$y(k) = -\sum_{j=1}^{n} a_j\, y(k-j) + \sum_{j=0}^{m} b_j\, u(k-j) \hspace{2cm} .....(4.86)$$

where, a_j and b_j are constants for time invariant system.

On expanding the summation of equation (4.86) and rearranging we get,

$$y(k) + a_1\, y(k-1) + a_2\, y(k-2) +.........+a_n\, y(k-n) = b_0\, u(k) + b_1\, u(k-1) + b_2\, u(k-2) +......+ b_m u(k-m)$$
$$.....(4.87)$$

On taking $\mathcal{Z}$-transform of equation (4.87) with zero initial conditions we get,

$$y(z) + a_1\, z^{-1} y(z) + a_2\, z^{-2} y(z) +.........+ a_n\, z^{-n} y(z) = b_0\, U(z) + b_1\, z^{-1} U(z) + b_2\, z^{-2} U(z) +......+ b_m z^{-m} U(z)$$

$$Y(z)\,[\, 1 + a_1\, z^{-1} + a_2\, z^{-2}+.......+ a_n\, z^{-n}]\ = [b_0 + b_1\, z^{-1} + b_2\, z^{-2} +........+ b_m\, z^{-m}]\, U(z)$$

When m = n,

$$\frac{Y(z)}{U(z)} = \frac{b_0 + b_1 z^{-1} + b_2 z^{-2} + ... + b_m z^{-m}}{1 + a_1 z^{-1} + a_2 z^{-2} + ... + a_n z^{-n}} \hspace{2cm} .....(4.88)$$

The equation (4.88) is the transfer function of the discrete time system. The equation (4.88) can be expressed as shown below.

$$\frac{Y(z)}{U(z)} = \frac{b_0 + b_1 z^{-1} + b_2 z^{-2} + ... + b_m z^{-m}}{1 - (- a_1 z^{-1} - a_2 z^{-2} - ..... - a_n z^{-n})} \hspace{2cm} .....(4.89)$$

On comparing the equation (4.89) with Mason's gain formula, a signal flow graph can be constructed. Each numerator term of equation (4.89) represent a forward path gain and each denominator term of equation (4.89) represent a individual loop gain. The signal flow graph will not have any non-touching loops. The signal flow graph when n=3 is shown in fig 4.9.

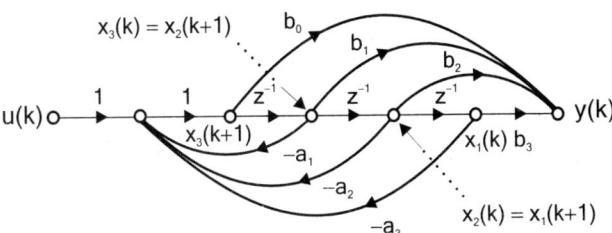

Fig 4.9 : *Signal flow graph representation of discrete time system.*

Let us assign state variables at the output of each unit delay element. Hence the signal at the input of unit delay element will be a signal advanced by one sampling time. The state equations are obtained by summing up the signals at the input of each unit delay element.

In fig 4.9, at node x_2 and x_3 we get,

$$\hspace{10cm} .....(4.90)$$
$$x_1(k+1) = x_2$$

$$\hspace{10cm} .....(4.91)$$
$$x_2(k+1) = x_3$$

By summing up the incoming signals to node $x_3(k+1)$ we get, (refer fig 4.9a)

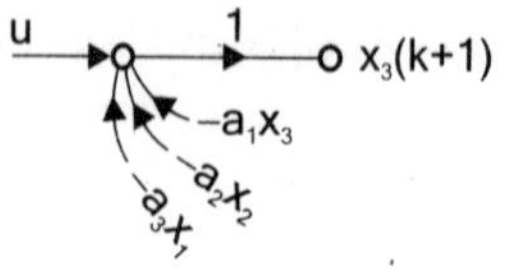

$$x_3(k+1) = -a_3 x_1 - a_2 x_2 - a_1 x_3 + u \qquad .....(4.92)$$

Fig 4.9a.

By summing up the incoming signals to node y we get, (refer fig 4.9b).

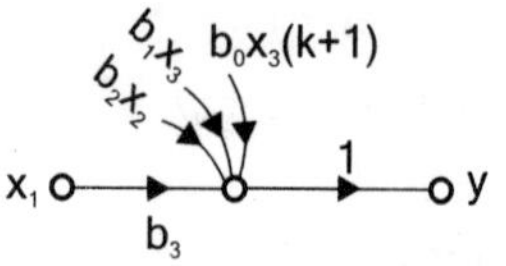

$$y = b_3 x_1 + b_2 x_2 + b_1 x_3 + b_0 x_3(k+1) \qquad .....(4.93)$$

Fig 4.9b.

Substitute for $x_3(k+1)$ from equation (4.92) in equation (4.93).

$$y = b_3 x_1 + b_2 x_2 + b_1 x_3 + b_0 (-a_3 x_1 - a_2 x_2 - a_1 x_3 + u)$$

$$\therefore y = (b_3 - b_0 a_3) x_1 + (b_2 - b_0 a_2) x_2 + (b_1 - b_0 a_1) x_3 + b_0 u \qquad .....(4.94)$$

The equations (4.90), (4.91) and (4.92) are state equations and equation (4.94) is the output equation. On arranging the state equations and output equations in the matrix form we get,

$$\begin{bmatrix} x_1(k+1) \\ x_2(k+1) \\ x_3(k+1) \end{bmatrix} = \begin{bmatrix} 0 & 1 & 0 \\ 0 & 0 & 1 \\ -a_3 & -a_2 & -a_1 \end{bmatrix} \begin{bmatrix} x_1 \\ x_2 \\ x_3 \end{bmatrix} + \begin{bmatrix} 0 \\ 0 \\ 1 \end{bmatrix} u \qquad .....(4.95)$$

$$y = \begin{bmatrix} (b_3 - b_0 a_3)(b_2 - b_0 a_2)(b_1 - b_0 a_1) \end{bmatrix} \begin{bmatrix} x_1 \\ x_2 \\ x_3 \end{bmatrix} + b_0 u \qquad .....(4.96)$$

The equations (4.95) and (4.96) can be extended to n^{th} order system as shown below.

$$\begin{bmatrix} x_1(k+1) \\ x_2(k+1) \\ x_3(k+1) \\ \vdots \\ x_n(k+1) \end{bmatrix} = \begin{bmatrix} 0 & 1 & 0 & \cdots & 0 \\ 0 & 0 & 1 & \cdots & 0 \\ 0 & 0 & 0 & \cdots & 0 \\ \vdots & \vdots & \vdots & & \vdots \\ -a_n & -a_{n-1} & -a_{n-2} & \cdots & -a_1 \end{bmatrix} \begin{bmatrix} x_1 \\ x_2 \\ x_3 \\ \vdots \\ x_n \end{bmatrix} + \begin{bmatrix} 0 \\ 0 \\ 0 \\ \vdots \\ 1 \end{bmatrix} u \qquad .....(4.97)$$

$$y = \begin{bmatrix} (b_n - b_0 a_n)(b_{n-1} - b_0 a_{n-1}) .... (b_1 - b_0 a_1) \end{bmatrix} \begin{bmatrix} x_1 \\ x_2 \\ \vdots \\ x_n \end{bmatrix} + b_0 u \qquad .....(4.98)$$

The equations (4.97) and (4.98) is the phase variable form of state model of discrete time n^{th} order system.

CANONICAL FORM OF STATE MODEL

The transfer function of equation (4.88) can be expressed as a summation using partial fraction technique as shown below.

$$\frac{Y(z)}{U(z)} = b_0 + \frac{C_1}{z + \lambda_1} + \frac{C_2}{z + \lambda_2} + ... + \frac{C_n}{z + \lambda_n}$$

where C_1, C_2 ,...... C_n are residues and λ_1 , λ_2 , ... λ_n are poles of the system.

$$\frac{Y(z)}{U(z)} = b_0 + \frac{C_1}{z\left(1+\dfrac{\lambda_1}{z}\right)} + \frac{C_2}{z\left(1+\dfrac{\lambda_2}{z}\right)} + \text{.......} + \frac{C_n}{z\left(1+\dfrac{\lambda_n}{z}\right)}$$

$$\frac{Y(z)}{U(z)} = b_0 + \frac{z^{-1}C_1}{1+z^{-1}\lambda_1} + \frac{z^{-1}C_2}{1+z^{-1}\lambda_2} + \text{......} + \frac{z^{-1}C_n}{1+z^{-1}\lambda_n} \qquad \text{.....(4.99)}$$

The equation (4.99) can be used to construct the block diagram of state model as shown in fig (4.10).

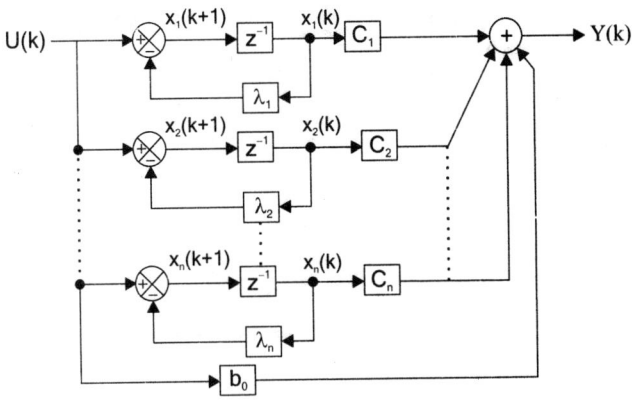

Fig 4.10 : *Block diagram of canonical state model of discrete time system.*

Let us assign the state variables at the output of unit delay elements. Hence the signal at the input of unit delay element will be a signal advanced by one sampling time. The state equations are obtained by summing up the signals at the input of each unit delay element.

The state equations are

$$x_1(k+1) = -\lambda_1 x_1 + u$$
$$x_2(k+1) = -\lambda_2 x_2 + u$$
$$x_3(k+1) = -\lambda_3 x_3 + u$$
$$\vdots \qquad \vdots$$
$$x_n(k+1) = -\lambda_n x_n + u$$

The output equation is

$$y = c_1 x_1 + c_2 x_2 + c_3 x_3 + \text{.....} + c_n x_n + b_0 u$$

The state equations and output equation can be expressed in the matrix form as shown below. The equations (4.100) and (4.101) is the canonical form of state model of discrete time n^{th} order system.

$$\begin{bmatrix} x_1(k+1) \\ x_2(k+1) \\ x_3(k+1) \\ \vdots \\ x_n(k+1) \end{bmatrix} = \begin{bmatrix} -\lambda_1 & 0 & 0 & \cdots & 0 \\ 0 & -\lambda_2 & 0 & \cdots & 0 \\ 0 & 0 & -\lambda_3 & \cdots & 0 \\ \vdots & \vdots & \vdots & & \vdots \\ -a_n & -a_{n-1} & -a_{n-2} & \cdots & -\lambda_n \end{bmatrix} \begin{bmatrix} x_1 \\ x_2 \\ x_3 \\ \vdots \\ x_n \end{bmatrix} + \begin{bmatrix} 1 \\ 1 \\ 1 \\ \vdots \\ 1 \end{bmatrix} u \qquad \text{.....(4.100)}$$

$$y = \begin{bmatrix} C_1 & C_2 & C_3 & \cdots & C_n \end{bmatrix} \begin{bmatrix} x_1 \\ x_2 \\ x_3 \\ \vdots \\ x_n \end{bmatrix} + b_0 u \qquad \text{.....(4.101)}$$

When a pole of the transfer function has multiplicity then the state model will be in a special form called Jordan canonical form. In this form the system matrix **A** will have a Jordan block of size q×q, correspond to a pole of value λ_i with multiplicity q. In the Jordan block the diagonal element will be the poles and the element just above the diagonal is one. Consider a system with poles $\lambda_1, \lambda_2, \lambda_3, \lambda_4, \ldots \lambda_n$, where λ_1 has multiplicity of three. The input matrix and system matrix for this case will be as shown in equation (4.100a). The system matrix is also denoted as **J**. The transfer function of the system for this case is given by equation (4.101) and the block diagram is shown in fig (4.10a).

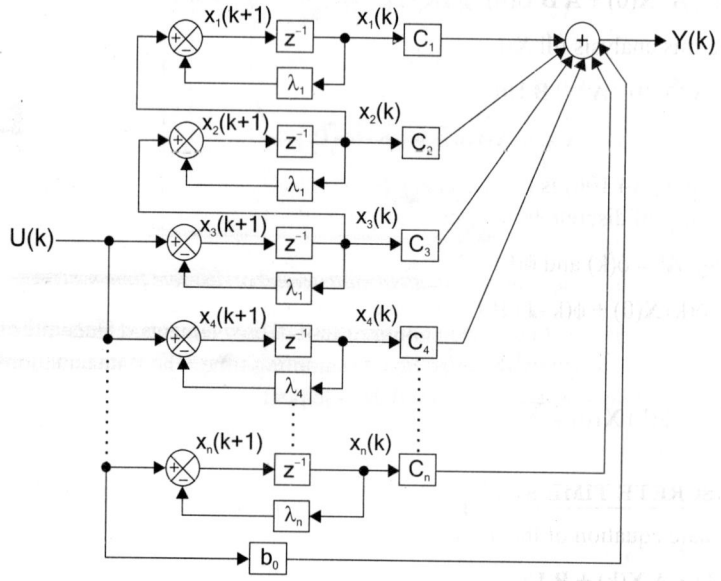

Fig 4.10a : Block diagram of Jordan canonical state model of discrete time system.

Jordan block of size 3×3

$$B = \begin{bmatrix} 0 \\ 0 \\ 1 \\ 1 \\ \vdots \\ 1 \end{bmatrix} \; ; \quad A = J = \begin{bmatrix} -\lambda_1 & 1 & 0 & 0 & \cdots & 0 \\ 0 & -\lambda_1 & 1 & 0 & \cdots & 0 \\ 0 & 0 & -\lambda_1 & 0 & \cdots & 0 \\ 0 & 0 & 0 & -\lambda_4 & \cdots & 0 \\ \vdots & \vdots & \vdots & \vdots & & \vdots \\ 0 & 0 & 0 & 0 & \cdots & -\lambda_n \end{bmatrix} \qquad \text{.....(4.100a)}$$

$$\frac{Y(z)}{U(z)} = b_0 + \frac{C_1}{(z+\lambda_1)^3} + \frac{C_2}{(z+\lambda_1)^2} + \frac{C_3}{z+\lambda_1} + \frac{C_4}{z+\lambda_4} + \ldots + \frac{C_n}{z+\lambda_n} \qquad \text{.....(4.101)}$$

SOLUTION OF DISCRETE TIME STATE EQUATION

The state equation of discrete time system is given by

$$\mathbf{X}(k+1) = \mathbf{A}\,\mathbf{X}(k) + \mathbf{B}\,\mathbf{U}(k) \qquad\qquad .....(4.102)$$

When $k = 0$, the equation (4.102) can be written as

$$\mathbf{X}(1) = \mathbf{A}\,\mathbf{X}(0) + \mathbf{B}\,\mathbf{U}(0) \qquad\qquad .....(4.103)$$

When $k = 1$, the equation (4.102) can be written as

$$\mathbf{X}(2) = \mathbf{A}\,\mathbf{X}(1) + \mathbf{B}\,\mathbf{U}(1) \qquad\qquad .....(4.104)$$

On substituting for $\mathbf{X}(1)$ from equation (4.103) in equation (4.104) we get,

$$\mathbf{X}(2) = \mathbf{A}\,[\mathbf{A}\,\mathbf{X}(0) + \mathbf{B}\,\mathbf{U}(0)] + \mathbf{B}\mathbf{U}(1)$$

$$\therefore \mathbf{X}(2) = \mathbf{A}^2\,\mathbf{X}(0) + \mathbf{A}\,\mathbf{B}\,\mathbf{U}(0) + \mathbf{B}\mathbf{U}(1) \qquad\qquad .....(4.105)$$

On continuing this analysis till $\mathbf{X}(k)$ we get,

$$\mathbf{X}(k) = \mathbf{A}^k\,\mathbf{X}(0) + \mathbf{A}^{(k-1)}\,\mathbf{B}\,\mathbf{U}(0) + \mathbf{A}^{(k-2)}\,\mathbf{B}\mathbf{U}(1) + ...$$

$$......+ \mathbf{A}\mathbf{B}\mathbf{U}(k{-}2) + \mathbf{B}\mathbf{U}(k{-}1) \qquad\qquad .....(4.106)$$

The matrix equation (4.106) is the solution of discrete time state equation. The matrix $\mathbf{A}^k$ is called the state transition matrix of discrete time system and it is also denoted by $\phi(k)$.

On substituting, $\mathbf{A}^k = \phi(k)$ and $\phi(0) = \mathbf{I}$ in equation (4.106) we get

$$\mathbf{X}(k) = \phi(k)\,\mathbf{X}(0) + \phi(k{-}1)\,\mathbf{B}\,\mathbf{U}(0) + \phi(k{-}2)\,\mathbf{B}\mathbf{U}(1) + ...$$

$$.......+ \phi(1)\,\mathbf{B}\mathbf{U}(k{-}2) + \phi(0)\,\mathbf{B}\mathbf{U}(k\,{-}\!-1)$$

$$\therefore \quad \mathbf{X}(k) = \phi(k)\,\mathbf{X}(0) + \sum_{j=0}^{k-1} \phi(k-1-j)\,\mathbf{B}\mathbf{U}(j) \qquad\qquad .....(4.107)$$

SOLUTION OF DISCRETE TIME STATE EQUATION USING Z-TRANSFORM

Consider the state equation of the discrete time system

$$\mathbf{X}(k+1) = \mathbf{A}\,\mathbf{X}(k) + \mathbf{B}\,\mathbf{U}(k) \qquad\qquad .....(4.108)$$

On taking $\mathbb{Z}$-transform of equation (4.108) we get

$$z\,\mathbf{X}(z) - z\,\mathbf{X}(0) = \mathbf{A}\,\mathbf{X}(z) + \mathbf{B}\,\mathbf{U}(z)$$

$$z\,\mathbf{X}(z) - \mathbf{A}\,\mathbf{X}(z) = z\,\mathbf{X}(0) + \mathbf{B}\,\mathbf{U}(z)$$

$$(z\mathbf{I} - \mathbf{A})\,\mathbf{X}(z) = z\,\mathbf{X}(0) + \mathbf{B}\,\mathbf{U}(z) \qquad\qquad .....(4.109)$$

On premultiplying the equation (4.109) by $(z\mathbf{I}{-}\mathbf{A})^{-1}$ we get

$$\mathbf{X}(z) = (z\mathbf{I} - \mathbf{A})^{-1}\,z\,\mathbf{X}(0) + (z\mathbf{I} - \mathbf{A})^{-1}\,\mathbf{B}\,\mathbf{U}(z) \qquad\qquad .....(4.110)$$

On taking inverse $\mathbb{Z}$-transform of equation (4.110) we get $\mathbf{X}(k)$

$$\therefore \quad \mathbf{X}(k) = \mathbb{Z}^{-1}\{\mathbf{X}(z)\} = \mathbb{Z}^{-1}\{(z\mathbf{I} - \mathbf{A})^{-1}\,z\mathbf{X}(0) + (z\mathbf{I} - \mathbf{A})^{-1}\,\mathbf{B}\mathbf{U}(z)\}$$

$$\mathbf{X}(k) = \mathbb{Z}^{-1}\{(z\mathbf{I} - \mathbf{A})^{-1}z\}\mathbf{X}(0) + \mathbb{Z}^{-1}\{(z\mathbf{I} - \mathbf{A})^{-1}\,\mathbf{B}\mathbf{U}(z)\} \qquad\qquad .....(4.111)$$

The equation (4.111) is the solution of discrete time state equation.

PROPERTIES OF STATE TRANSITION MATRIX OF DISCRETE TIME SYSTEM

1. $\phi(0) = \mathbf{I}$

2. $\phi^{-1}(k) = \phi(-k)$

3. $\phi(k, k_0) = \phi(k-k_0) = \mathbf{A}^{(k-k_0)}$;   where, $k > k_0$

COMPUTATION OF STATE TRANSITION MATRIX

The state transition matrix $\mathbf{A}^k$ can be computed by any one of the following methods.

Method 1 : Computation of $\mathbf{A}^k$ using Z-transform

Method 2 : Computation of $\mathbf{A}^k$ by canonical transformation

Method 3 : Computation of $\mathbf{A}^k$ by Cayley - Hamilton theorem

The computation of $\mathbf{A}^k$ using Z-transform have been dealt in this section.

On comparing equations (4.111) and (4.107) [or (4.106)] we can write,

State transition matrix, $\mathbf{A}^k = \mathbf{Z}^{-1}\{(z\mathbf{I} - \mathbf{A})^{-1}z\}$ .....(4.112)

The equation (4.112) can be used to compute the state transition matrix, $\mathbf{A}^k$.

EXAMPLE 4.17

A discrete-time system has the transfer function

$$\frac{Y(z)}{U(z)} = \frac{4z^3 - 12z^2 + 13z - 7}{(z-1)^2(z-2)}$$

Determine the state model of the system in (a) Phase variable form and (b) Jordan canonical form.

SOLUTION

a. Phase variable form of state model

Given that, $\dfrac{Y(z)}{U(z)} = \dfrac{4z^3 - 12z^2 + 13z - 7}{(z-1)^2(z-2)}$

$$= \frac{z^3(4 - 12z^{-1} + 13z^{-2} - 7z^{-3})}{(z^2 - 2z + 1)(z - 2)} = \frac{z^3(4 - 12z^{-1} + 13z^{-2} - 7z^{-3})}{z^3 - 2z^2 - 2z^2 + 4z + z - 2}$$

$$= \frac{z^3(4 - 12z^{-1} + 13z^{-2} - 7z^{-3})}{z^3(1 - 4z^{-1} + 5z^{-2} - 2z^{-3})} = \frac{4 - 12z^{-1} + 13z^{-2} - 7z^{-3}}{1 - (4z^{-1} - 5z^{-2} + 2z^{-3})} \qquad .....(4.17.1)$$

The equation (4.17.1) can be used to construct the signal flow graph of the discrete time system shown in fig 4.17.1.

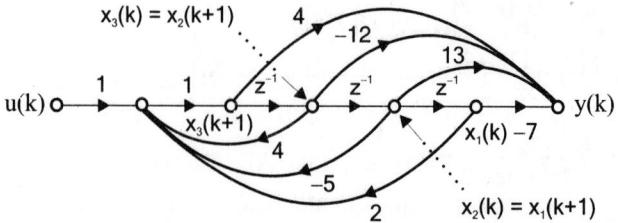

Fig 4.17.1 : Signal flow graph for transfer function of equation (4.17.1).

From node x_2 and x_3 we get,

$$x_1(k+1) = x_2 \qquad .....(4.17.2)$$

$$x_2(k+1) = x_3 \qquad .....(4.17.3)$$

By summing up the incoming signals to node $x_3(k+1)$ we get, (refer fig 4.17.2)

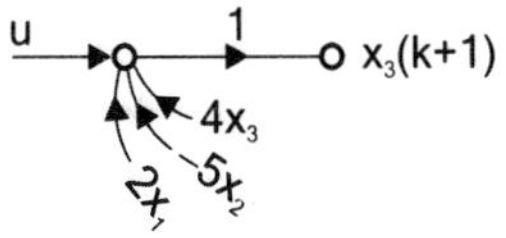

$$x_3(k+1) = 2x_1 - 5x_2 + 4x_3 + u \qquad .....(4.17.4)$$

Fig 4.17.2

The equations (4.17.2) to (4.17.4) are the state equations. The output equation is obtained by summing up the incoming signals to output node (Refer fig 4.17.3)

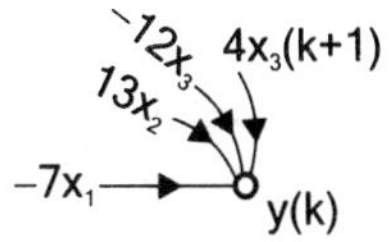

$$y = -7x_1 + 13x_2 - 12x_3 + 4x_3(k+1) \qquad .....(4.17.5)$$

Fig 4.17.3

On substituting for $x_3(k+1)$ from equation (4.17.4) in equation (4.17.5) we get,

$$y = -7x_1 + 13x_2 - 12x_3 + 4\,[2x_1 - 5x_2 + 4x_3 + u]$$

$$y = (-7 + 8)\,x_1 + (13 - 20)\,x_2 + (-12 + 16)\,x_3 + 4u$$

$$y = x_1 - 7x_2 + 4x_3 + 4u \qquad .....(4.17.6)$$

The state equations and output equation can be arranged in the matrix form as shown below.

$$\begin{bmatrix} x_1(k+1) \\ x_2(k+1) \\ x_3(k+1) \end{bmatrix} = \begin{bmatrix} 0 & 1 & 0 \\ 0 & 0 & 1 \\ 2 & -5 & 4 \end{bmatrix}\begin{bmatrix} x_1 \\ x_2 \\ x_3 \end{bmatrix} + \begin{bmatrix} 0 \\ 0 \\ 1 \end{bmatrix} u \qquad .....(4.17.7)$$

$$y = \begin{bmatrix} 1 & -7 & 4 \end{bmatrix}\begin{bmatrix} x_1 \\ x_2 \\ x_3 \end{bmatrix} + 4u \qquad .....(4.17.8)$$

The equations (4.17.7) and (4.17.8) is the phase variable form of state model.

b. Jordan Canonical form of state model

Given that, $\dfrac{Y(z)}{U(z)} = \dfrac{4z^3 - 12z^2 + 13z - 7}{(z-1)^2(z-2)}$

	4
$z^3 - 4z^2 + 5z - 2$	$4z^3 - 12z^2 + 13z - 7$
	$4z^3 - 16z^2 + 20z - 8$
	$4z^2 - 7z + 1$

$$= \frac{4z^3 - 12z^2 + 13z - 7}{z^3 - 4z^2 + 5z - 2}$$

$$= 4 + \frac{4z^2 - 7z + 1}{z^3 - 4z^2 + 5z - 2} = 4 + \frac{4z^2 - 7z + 1}{(z-1)^2(z-2)} \qquad .....(4.17.9)$$

By partial fraction expansion the equation (4.17.9) can be written as,

$$\frac{Y(z)}{U(z)} = 4 + \frac{A_1}{(z-1)^2} + \frac{A_2}{z-1} + \frac{A_3}{z-2}$$

$$A_1 = (z-1)^2 \left.\frac{4z^2 - 7z + 1}{(z-1)^2(z-2)}\right|_{z=1} = \left.\frac{4z^2 - 7z + 1}{z-2}\right|_{z=1} = \frac{4 - 7 + 1}{1 - 2} = 2$$

$$A_2 = \frac{d}{dz}\left[(z-1)^2 \frac{4z^2 - 7z + 1}{(z-1)^2(z-2)}\right]\Bigg|_{z=1} = \frac{d}{dz}\left[\frac{4z^2 - 7z + 1}{z-2}\right]\Bigg|_{z=1}$$

$$= \left.\frac{(8z-7)(z-2) - (4z^2 - 7z + 1)}{(z-2)^2}\right|_{z=1} = \frac{(8-7)(1-2) - (4-7+1)}{(1-2)^2} = 1$$

$$A_3 = (z-2)\left.\frac{4z^2 - 7z + 1}{(z-1)^2(z-2)}\right|_{z=2} = \left.\frac{4z^2 - 7z + 1}{(z-1)^2}\right|_{z=2} = \frac{4 \times 2^2 - 7 \times 2 + 1}{(2-1)^2} = 3$$

$$\therefore \quad \frac{Y(z)}{U(z)} = 4 + \frac{2}{(z-1)^2} + \frac{1}{z-1} + \frac{3}{z-2} \qquad \qquad .....(4.17.10)$$

The equation (4.17.10) can be used to construct the block diagram of the discrete time system shown in fig 4.17.4.

Let us assign state variable at the output of each unit delay element. The state equations are formed by summing up the incoming signals to the unit delay element.

The state equations are,

$$x_1(k+1) = x_1 + x_2$$
$$x_2(k+1) = x_2 + u$$
$$x_3(k+1) = 2x_3 + u$$

The output equation is,

$$y = 2x_1 + x_2 + 3x_3 + 4u$$

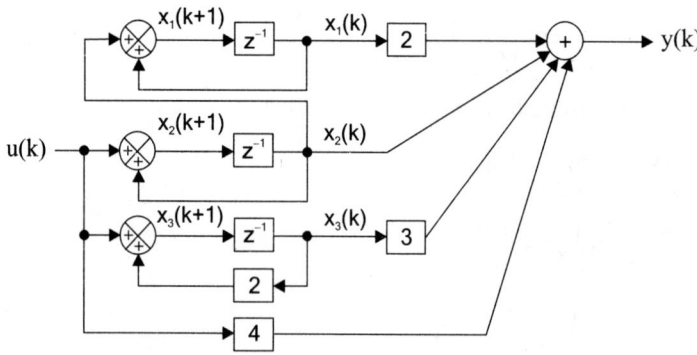

Fig 4.17.4 : Block diagram.

On arranging the state equations and the output equation in the matrix form we get,

$$\begin{bmatrix} x_1(k+1) \\ x_2(k+1) \\ x_3(k+1) \end{bmatrix} = \begin{bmatrix} 1 & 1 & 0 \\ 0 & 1 & 0 \\ 0 & 0 & 2 \end{bmatrix} \begin{bmatrix} x_1 \\ x_2 \\ x_3 \end{bmatrix} + \begin{bmatrix} 0 \\ 1 \\ 1 \end{bmatrix} u \qquad\qquad .....(4.17.11)$$

$$y = \begin{bmatrix} 2 & 1 & 3 \end{bmatrix} \begin{bmatrix} x_1 \\ x_2 \\ x_3 \end{bmatrix} + 4u \qquad\qquad .....(4.17.12)$$

The equations (4.17.11) and (4.17.12) constitute Jordan canonical form of state model of the system.

EXAMPLE 4.18

A discrete time system is described by the difference equation,

$$y(k + 2) + 5y (k + 1) + 6\ y(k) = u(k)$$

$$y(0) = y(1) = 0; \quad T = 1sec.$$

(a)   Determine a state model in canonical form.

(b)   Find the state transition matrix

(c)   For input u(k) = 1 ; k ≥ 1, find the output y(k).

SOLUTION

(a) To determine the canonical form of state model

Given that, $y(k + 2) + 5y (k + 1) + 6\ y(k) = u(k)$ .....(4.18.1)

and $y(0) = y(1) = 0; \quad T = 1sec.$

On taking Z-transform of equation (4.18.1) with zero initial conditions we get,

$$z^2\ Y(z) + 5z\ Y(z) + 6Y(z) = U(z)$$

$$(z^2 + 5z + 6\ ) Y(z) = U(z)$$

$$\therefore\ \frac{Y(z)}{U(z)} = \frac{1}{z^2 + 5z + 6} = \frac{1}{(z+3)(z+2)} \qquad\qquad .....(4.18.2)$$

By partial fraction expansion the equation (4.18.2) can be expressed as,

$$\frac{Y(z)}{U(z)} = \frac{A_1}{z+3} + \frac{A_2}{z+2}$$

$$A_1 = (z+3)\ \frac{1}{(z+3)(z+2)}\Big|_{z=-3} = \frac{1}{z+2}\Big|_{z=-3} = \frac{1}{-3+2} = -1$$

$$A_2 = (z+2)\ \frac{1}{(z+3)(z+2)}\Big|_{z=-2} = \frac{1}{z+3}\Big|_{z=-2} = \frac{1}{-2+3} = 1$$

$$\therefore\ \frac{Y(z)}{U(z)} = \frac{-1}{z+3} + \frac{1}{z+2} \qquad\qquad .....(4.18.3)$$

The equation (4.18.3) can be used to construct the block diagram shown in fig 1.

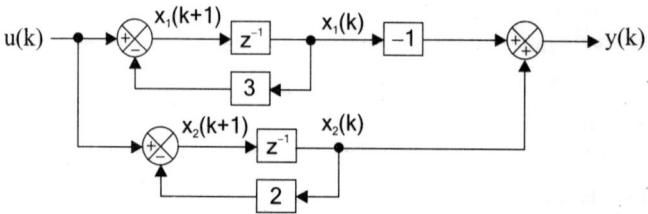

Fig 4.18.1 : *Block diagram.*

From the block diagram, the state equations are

$$x_1(k +1) = -3x_1 + u$$
$$x_2(k +1) = -2x_2 + u$$

The output equation is

$$y = -x_1 + x_2$$

On arranging the state equations and the output equation in the matrix form, we get,

$$\begin{bmatrix} x_1(k+1) \\ x_2(k+1) \end{bmatrix} = \begin{bmatrix} -3 & 0 \\ 0 & -2 \end{bmatrix}\begin{bmatrix} x_1 \\ x_2 \end{bmatrix} + \begin{bmatrix} 1 \\ 1 \end{bmatrix} u \qquad\qquad .....(4.18.4)$$

$$y = \begin{bmatrix} -1 & 1 \end{bmatrix}\begin{bmatrix} x_1 \\ x_2 \end{bmatrix} \qquad\qquad .....(4.18.5)$$

The equations (4.18.4) and (4.18.5) constitute the canonical form of state model.

(b) To find state transition matrix A^k

The state transition matrix, $A^k = \mathbf{Z}^{-1}\{[z\mathbf{I} - \mathbf{A}]^{-1}z\}$

From the state equation we get, $A = \begin{bmatrix} -3 & 0 \\ 0 & -2 \end{bmatrix}$

$$\therefore\ z\mathbf{I} - \mathbf{A} = z\begin{bmatrix} 1 & 0 \\ 0 & 1 \end{bmatrix} - \begin{bmatrix} -3 & 0 \\ 0 & -2 \end{bmatrix} = \begin{bmatrix} z+3 & 0 \\ 0 & z+2 \end{bmatrix}$$

$$|z\mathbf{I} - \mathbf{A}| = \begin{vmatrix} z+3 & 0 \\ 0 & z+2 \end{vmatrix} = (z+2)(z+3)$$

$$[z\mathbf{I} - \mathbf{A}]^{-1} = \frac{1}{|z\mathbf{I} - \mathbf{A}|}\begin{bmatrix} z+2 & 0 \\ 0 & z+3 \end{bmatrix}^{T} = \frac{1}{(z+2)(z+3)}\begin{bmatrix} z+2 & 0 \\ 0 & z+3 \end{bmatrix}$$

$$= \begin{bmatrix} \dfrac{1}{z+3} & 0 \\ 0 & \dfrac{1}{z+2} \end{bmatrix} \qquad\qquad .....(4.18.6)$$

$$A^k = \mathbf{Z}^{-1}\{[z\mathbf{I} - \mathbf{A}]^{-1}z\} = \mathbf{Z}^{-1}\begin{bmatrix} \dfrac{z}{z+3} & 0 \\ 0 & \dfrac{z}{z+2} \end{bmatrix}$$

We know that $\mathbb{Z}^{-1}\left\{\dfrac{z}{z-a}\right\} = a^k$

$\therefore \ \mathbb{Z}^{-1}\left\{\dfrac{z}{z+3}\right\} = (-3)^k$ and $\mathbb{Z}^{-1}\left\{\dfrac{z}{z+2}\right\} = (-2)^k$

$\therefore$ State transition matrix, $\quad \mathbf{A}^k = \begin{bmatrix} (-3)^k & 0 \\ 0 & (-2)^k \end{bmatrix}$ .....(4.18.7)

(c) To find y(k) when the input is unit step

Consider the state equation,

$$X(k+1) = A\,X(k) + B\,U(k)$$ .....(4.18.8)

On taking Z-transform of equation(6.21.8) we get

$$z\,X(z) - z\,X(0) = AX(z) + BU(z)$$

Let us assume that, $X(0) = 0$

$\therefore \ z\,X(z) - A\,X(z) = B\,U(z)$

$$(zI - A)\,X(z) = B\,U(z)$$ .....(4.18.9)

On premultiplying both sides of equation (4.18.9) by $(zI - A)^{-1}$ we get

$$X(z) = (zI - A)^{-1}\,BU(z)$$ .....(4.18.10)

Given that, $u(k) = 1,$ $\qquad \therefore \ U(z) = \mathbb{Z}\{u(k)\} = \dfrac{z}{z-1}$ .....(4.18.11)

From equation (4.18.6) we get, $[zI - A]^{-1} = \begin{bmatrix} \dfrac{1}{z+3} & 0 \\ 0 & \dfrac{1}{z+2} \end{bmatrix}$

From equation (4.18.4) we get, $\mathbf{B} = \begin{bmatrix} 1 \\ 1 \end{bmatrix}$

$\therefore \ X(z) = \begin{bmatrix} \dfrac{1}{z+3} & 0 \\ 0 & \dfrac{1}{z+2} \end{bmatrix}\begin{bmatrix} 1 \\ 1 \end{bmatrix}\left[\dfrac{z}{z-1}\right] = \begin{bmatrix} \dfrac{1}{z+3} \\ \dfrac{1}{z+2} \end{bmatrix}\left[\dfrac{z}{z-1}\right] = \begin{bmatrix} \dfrac{z}{(z-1)(z+3)} \\ \dfrac{z}{(z-1)(z+2)} \end{bmatrix}$

We know that $X(z) = \begin{bmatrix} x_1(z) \\ x_2(z) \end{bmatrix}$ and $X(k) = \begin{bmatrix} x_1(k) \\ x_2(k) \end{bmatrix}$

Also $X(k) = \mathbb{Z}^{-1}\{X(z)\}$

$\therefore \ X(z) = \begin{bmatrix} X_1(z) \\ X_2(z) \end{bmatrix} = \begin{bmatrix} \dfrac{z}{(z-1)(z+3)} \\ \dfrac{z}{(z-1)(z+2)} \end{bmatrix}$ and $X(k) = \begin{bmatrix} x_1(z) \\ x_2(z) \end{bmatrix} = \begin{bmatrix} \mathbb{Z}^{-1}\{X_1(z)\} \\ \mathbb{Z}^{-1}\{X_2(z)\} \end{bmatrix}$

$\therefore \ x_1(k) = \mathbb{Z}^{-1}\left\{\dfrac{z}{(z-1)(z+3)}\right\}$ and $x_2(k) = \mathbb{Z}^{-1}\left\{\dfrac{z}{(z-1)(z+2)}\right\}$

By partial fraction expansion,

$$\frac{z}{(z-1)(z+3)} = z\left[\frac{1}{(z-1)(z+3)}\right] = z\left[\frac{A_1}{z-1} + \frac{B_1}{z+3}\right]$$

$$A_1 = \frac{z}{(z-1)(z+3)}(z-1)\bigg|_{z=1} = \frac{1}{z+3}\bigg|_{z=1} = \frac{1}{1+3} = \frac{1}{4}$$

$$B_1 = \frac{1}{(z-1)(z+3)}(z+3)\bigg|_{z=-3} = \frac{1}{z-1}\bigg|_{z=-3} = \frac{1}{-3-1} = -\frac{1}{4}$$

$$\therefore \quad \frac{z}{(z-1)(z+3)} = z\left[\frac{A_1}{z-1} + \frac{B_1}{z+3}\right] = \left[\frac{1}{4}\frac{z}{z-1} - \frac{1}{4}\frac{z}{z+3}\right]$$

$$x_1(k) = \mathbf{Z}^{-1}\left\{\frac{z}{(z-1)(z+3)}\right\} = \mathbf{Z}^{-1}\left[\frac{1}{4}\frac{z}{z-1} - \frac{1}{4}\frac{z}{z+3}\right] = \frac{1}{4}u(k) - \frac{1}{4}(-3)^k$$

By partial fraction expansion,

$$\frac{z}{(z-1)(z+2)} = z\left[\frac{1}{(z-1)(z+2)}\right] = z\left[\frac{A_2}{z-1} + \frac{B_2}{z+2}\right]$$

$$A_2 = \frac{1}{(z-1)(z+2)}(z-1)\bigg|_{z=1} = \frac{1}{z+2}\bigg|_{z=1} = \frac{1}{1+2} = \frac{1}{3}$$

$$B_2 = \frac{1}{(z-1)(z+2)}(z+2)\bigg|_{z=-2} = \frac{1}{z-1}\bigg|_{z=-2} = \frac{1}{-2-1} = -\frac{1}{3}$$

$$\therefore \quad \frac{z}{(z-1)(z+2)} = z\left[\frac{A_2}{z-1} + \frac{B_2}{z+2}\right] = \left[\frac{1}{3}\frac{z}{z-1} - \frac{1}{3}\frac{z}{z+2}\right]$$

$$x_2(k) = \mathbf{Z}^{-1}\left\{\frac{z}{(z-1)(z+2)}\right\} = \mathbf{Z}^{-1}\left[\frac{1}{3}\frac{z}{z-1} - \frac{1}{3}\frac{z}{z+2}\right] = \frac{1}{3}u(k) - \frac{1}{3}(-2)^k$$

From equation (4.18.5) we get,

Response or Output, $y(k) = \begin{bmatrix} -1 & 1 \end{bmatrix}\begin{bmatrix} x_1(k) \\ x_2(k) \end{bmatrix}$

$$\therefore \quad y(k) = -x_1(k) + x_2(k)$$

$$= -\frac{1}{4}u(k) + \frac{1}{4}(-3)^k + \frac{1}{3}u(k) - \frac{1}{3}(-2)^k$$

$$= \frac{1}{4}(-3)^k - \frac{1}{3}(-2)^k + \frac{-3+4}{12}u(k)$$

$$= \frac{1}{4}(-3)^k - \frac{1}{3}(-2)^k + \frac{1}{12}u(k)$$

4.10 SHORT-ANSWER QUESTIONS

Q4.1   *What are the advantages of state space analysis?*

1. The state space analysis is applicable to any type of systems. They can be used for modelling and analysis of linear & non-linear systems, time invariant & time variant systems and multiple input & multiple output systems.

2. The state space analysis can be performed with initial conditions.

3. The variables used to represent the system can be any variables in the system.

4. Using this analysis the internal states of the system at any time instant can be predicted.

Q4.2   *What are the drawbacks in transfer function model analysis?*

1. Transfer function is defined under zero initial conditions.

2. Transfer function is applicable to linear time invariant systems.

3. Transfer function analysis is restricted to single input and output systems.

4. Does not provides information regarding the internal state of the system.

Q4.3   *What is state and state variable?*

The state is the condition of a system at any time instant, t. A set of variable which describes the state of the system at any time instant are called state variables.

Q4.4   *What is a state vector?*

The state vector is a $(n \times 1)$ column matrix (or vector) whose elements are state variables of the system, (where n is the order of the system). It is denoted by X(t).

Q4.5   *Write the state model of nth order system?*

The state model of a system consists of state equation and output equation. The state model of a nth order system with m-inputs and p-outputs are

$$\dot{X}(t) = AX(t) + BU(t)   ..... \text{state equation}$$

$$Y(t) = CX(t) + DU(t)   ..... \text{output equation}$$

where X(t) = State vector of order $(n \times 1)$;   U(t) = Input vector of order $(m \times 1)$

  A = System matrix of order $(n \times n)$;   **B** = Input matrix of order $(n \times m)$

  Y(t) = Output vector of order $(p \times 1)$;   **C** = Output matrix of order $(p \times n)$.

  D = Transmission matrix of order $(p \times m)$.

Q4.6   *What is state space?*

The set of all possible values which the state vector **X**(t) can have (or assume) at time t forms the state space of the system.

Q4.7   *What is input and output space?*

The set of all possible values which the input vector **U**(t) can have (or assume) at time t forms the input space of the system.

The set of all possible values which the output vector **Y**(t) can have (or assume) at time t forms the output space.

Q4.8   *The state model of a linear time invariant system is given by*

$$\dot{X}(t) = AX(t) + BU(t)$$

$$Y(t) = CX(t) + DU(t)$$

Obtain the expression for transfer function of the system.

Solution : Given that $\dot{X}(t) = AX(t) + BU(t)$ .....(1)

and $Y(t) = CX(t) + DU(t)$ .....(2)

On taking Laplace transform of equation (1) with zero initial conditions we get,

$s\,X(s) = A\,X(s) + B\,U(s)$

$s\,X(s) - A\,X(s) = B\,U(s)$

$(sI - A)\,X(s) = B\,U(s)$ .....(3)

On premultiplying equation (3) by $(sI-A)^{-1}$ we get

$X(s) = (sI - A)^{-1}\,B\,U(s)$ .....(4)

On taking Laplace transform of equation (2) we get,

$Y(s) = C\,X\,(s) + D\,U(s)$ .....(5)

Substitute for $X(s)$ from equation (4) in equation (5) we get,

$Y(s) = C\,[(sI - A)^{-1}\,B\,U(s)] + D\,U\,(s)$

$\quad\;\; = [C\,(sI - A)^{-1}\,B + D]\,U\,(s)$

$\therefore\; \dfrac{Y(s)}{U(s)} = C\,(sI - A)^{-1}B + D$ .....(6)

The equation (6) is the transfer function of the system.

Q4.9   ***What is state diagram?***

The pictorial representation of the state model of the system is called state diagram. The state diagram of the system can be either in block diagram or in signal flow graph form.

Q4.10   ***Draw the block diagram representation of state model?***

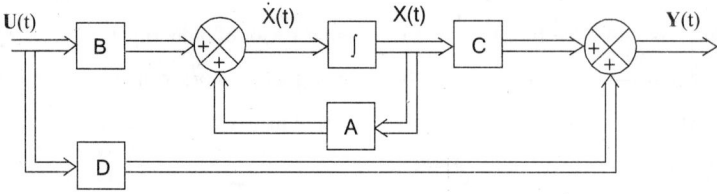

Fig Q4.10 : Block diagram of state model.

Q4.11   ***Draw the signal flow graph representation of state model?***

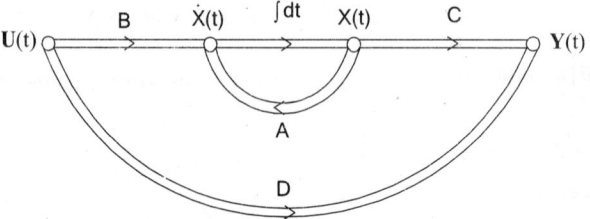

Fig Q4.11 : Signal flow graph of state model.

Q4.12   What are the basic elements used to construct the state diagram?

The basic elements used to construct the state diagram are Scalar, Adder and Integrator.

Q4.13   Sketch the basic elements used to construct the block diagram of a state model.

The basic elements used to construct block diagram of a state model are shown in the following table.

Elements	Time domain	s-domain
Scalar	$x(t)$ → $\boxed{a}$ → $a\,x(t)$	$X(s)$ → $\boxed{a}$ → $a\,X(s)$
Adder	$x(t)$ → ⊕ → $x(t)+x(t)$, with $x(t)$ ↑	$X(s)$ → ⊕ → $X(s)+X(s)$, with $X(s)$ ↑
Integrator	$\dot{x}(t)$ → $\boxed{\int}$ → ⊕ → $\int_{0}^{t}\dot{x}(T)dT+x(0)$, with $x(0)$ ↓	$X(s)$ → $\boxed{1/s}$ → ⊕ → $\dfrac{X(s)}{s}+\dfrac{X(0)}{s}$, with $X(0)/s$ ↓

Q4.14   Sketch the basic elements used to construct the signal flow graph of a state model.

The basic elements used to construct the signal flow graph of a state model are shown in the following table.

Elements	Time domain	s-domain
Scalar	$x(t)$ —a→ $a\,x(t)$	$X(s)$ —a→ $a\,X(s)$
Adder	$x(t)$ —1→, $x(t)$ —1→ : $x(t)+x(t)$	$X(s)$ —1→, $X(s)$ —1→ : $X(s)+X(s)$
Integrator	$\dot{x}(t)$ —$\int dt$→ $x(t)+x(0)$, with $x(0)$ —1→	$X(s)$ —$\frac{1}{s}$→ $\dfrac{X(s)}{s}+\dfrac{X(0)}{s}$, with $\dfrac{X(0)}{s}$ —1→

Q4.15   Draw the block diagram of the system described by the state model,

$$\begin{bmatrix}\dot{x}_1\\\dot{x}_2\\\dot{x}_3\end{bmatrix}=\begin{bmatrix}0 & 1 & 0\\0 & 0 & 1\\0 & a_2 & a_3\end{bmatrix}\begin{bmatrix}x_1\\x_1\\x_1\end{bmatrix}+\begin{bmatrix}0\\0\\1\end{bmatrix}u \quad and \quad y=x_1$$

Solution

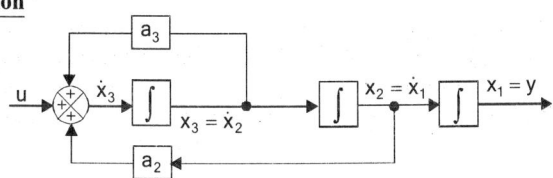

Q4.16 *Draw the signal flow graph of the system described by the state model.*

$$\begin{bmatrix} \dot{x}_1 \\ \dot{x}_2 \\ \dot{x}_3 \end{bmatrix} = \begin{bmatrix} a_1 & a_2 & 0 \\ 1 & 0 & 1 \\ 0 & 1 & 0 \end{bmatrix} \begin{bmatrix} x_1 \\ x_2 \\ x_3 \end{bmatrix} + \begin{bmatrix} 1 \\ 0 \\ 0 \end{bmatrix} u \quad and \quad y = x_3$$

Solution

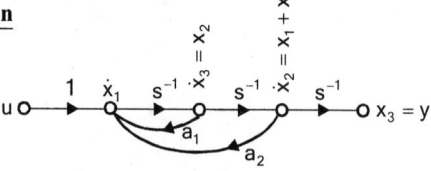

Q4.17 *Determine the state model of the system represented by the block diagram of fig Q4.17.*

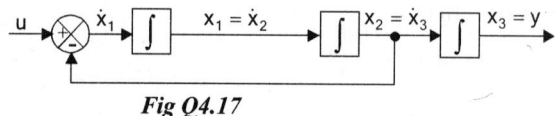

Fig Q4.17

Solution

$$\begin{bmatrix} \dot{x}_1 \\ \dot{x}_2 \\ \dot{x}_3 \end{bmatrix} = \begin{bmatrix} 0 & 1 & 0 \\ 1 & 0 & 0 \\ 0 & 1 & 0 \end{bmatrix} \begin{bmatrix} x_1 \\ x_2 \\ x_3 \end{bmatrix} + \begin{bmatrix} 1 \\ 0 \\ 0 \end{bmatrix} u \quad and \quad y = x_3$$

Q4.18 *Determine the state model of the system represented by the signal flow graph of fig Q4.18.*

Fig Q4.18

Solution

$$\begin{bmatrix} \dot{x}_1 \\ \dot{x}_2 \\ \dot{x}_3 \end{bmatrix} = \begin{bmatrix} 0 & 1 & 0 \\ -2 & 0 & 1 \\ 0 & 0 & 0 \end{bmatrix} \begin{bmatrix} x_1 \\ x_2 \\ x_3 \end{bmatrix} + \begin{bmatrix} 0 \\ 0 \\ 1 \end{bmatrix} u \quad and \quad y = x_1$$

Q4.19 *A system is characterized by the differential equation,*

$$\frac{d^2 y}{dt^2} + 10\frac{dy}{dt} + 7y - u = 0$$

Determine its transfer function.

Solution

Given that, $\dfrac{d^2 y}{dt^2} + 10\dfrac{dy}{dt} + 7y - u = 0$ .....(1)

On taking Laplace transform of equation (1) with zero initial conditions we get,

$$s^2\, Y(s) + 10\, s\, Y(s) + 7\, Y(s) - U(s) = 0$$

$$(s^2 + 10\, s + 7)\, Y(s) = U(s)$$

$$\therefore \frac{Y(s)}{U(s)} = \frac{1}{s^2 + 10s + 7} \qquad .....(2)$$

The equation (2) is the transfer function of the system.

Q4.20 *The transfer function of a system is given by* $\dfrac{Y(s)}{U(s)} = \dfrac{10}{4s^2 + 2s + 1}$ *. Determine the differential equation governing the system.*

Solution

Given that, $\dfrac{Y(s)}{U(s)} = \dfrac{10}{4s^2 + 2s + 1}$

$[4s^2 + 2s + 1] Y(s) = 10 \, U(s)$

$4s^2 \, Y(s) + 2s \, Y(s) + Y(s) = 10 \, U(s) \qquad .....(1)$

On taking inverse Laplace transform of equation (1) we get,

$$4 \frac{d^2 y}{dt^2} + 2 \frac{dy}{dt} + y = 10u$$

$$4 \frac{d^2 y}{dt^2} + 2 \frac{dy}{dt} + y - 10u = 0 \qquad .....(2)$$

The equation (2) is the differential equation governing the system.

Q4.21 *What are the advantages of state space modelling using physical variable?*

The advantages of choosing the physical variable are the following,

1. The state variable can be utilized for the purpose of feedback
2. The implementation of design with state variable feedback becomes straight forward.
3. The solution of state equation gives time variation of variables which have direct relevance to the physical system.

Q4.22 *What are phase variables?*

The phase variables are defined as those particular state variables which are obtained from one of the system variable and its derivatives. Usually the variable used is the system output and the remaining state variables are then derivatives of the output.

Q4.23 *What is bush form or companion form of state model?*

In bush form or companion form of state model, the system matrix, A has all 1's in the upper off-diagonal and its last row is comprised of the negative of the coefficients of the original differential equation and all other elements are zero. The companion form of state model is shown below

$$\begin{bmatrix} \dot{x}_1 \\ \dot{x}_2 \\ \dot{x}_3 \\ \vdots \\ \dot{x}_{n-1} \\ \dot{x}_n \end{bmatrix} = \begin{bmatrix} 0 & 1 & 0 & 0 & \cdots & 0 \\ 0 & 0 & 1 & 0 & \cdots & 0 \\ 0 & 0 & 0 & 1 & \cdots & 0 \\ \vdots & \vdots & \vdots & \vdots & & \vdots \\ 0 & 0 & 0 & 0 & \cdots & 1 \\ -a_n & -a_{n-1} & -a_{n-2} & -a_{n-3} & \cdots & -a_1 \end{bmatrix} \begin{bmatrix} x_1 \\ x_2 \\ x_3 \\ \vdots \\ x_{n-1} \\ x_n \end{bmatrix} + \begin{bmatrix} 0 \\ 0 \\ 0 \\ \vdots \\ 0 \\ b \end{bmatrix} [u]$$

$$y = \begin{bmatrix} 1 & 0 & 0 & \cdots & 0 \end{bmatrix} \begin{bmatrix} x_1 \\ x_2 \\ x_3 \\ \vdots \\ x_n \end{bmatrix}$$

Q4.24   What are the advantages in choosing phase variables for state space modelling?

1. Using phase variables the system state model can be written directly by inspection from the differential equation governing the system.

2. The phase variables provides a link between the transfer function design approach and time-domain design approach.

Q4.25   What is the disadvantage in choosing phase variable for state-space modelling?

The disadvantage in choosing phase variables is that the phase variables are not physical variables of the system and therefore are not available for measurement and control purposes.

Q4.26   Write the canonical form of state model of nth order system.

In canonical form of state model, the system matrix, A will be a diagonal matrix. The canonical form of state model in the matrix form is given below.

$$\begin{bmatrix} \dot{x}_1 \\ \dot{x}_2 \\ \dot{x}_3 \\ \vdots \\ \dot{x}_n \end{bmatrix} = \begin{bmatrix} \lambda_1 & 0 & 0 & \cdots & 0 \\ 0 & \lambda_2 & 0 & \cdots & 0 \\ 0 & 0 & \lambda_3 & \cdots & 0 \\ \vdots & \vdots & \vdots & & \vdots \\ 0 & 0 & 0 & \cdots & \lambda_n \end{bmatrix} \begin{bmatrix} x_1 \\ x_2 \\ x_3 \\ \vdots \\ x_n \end{bmatrix} + \begin{bmatrix} 1 \\ 1 \\ 1 \\ \vdots \\ 1 \end{bmatrix} [u]$$

$$y = [c_1 \quad c_2 \quad c_3 \quad \cdots \quad c_n] \begin{bmatrix} x_1 \\ x_2 \\ x_3 \\ \vdots \\ x_n \end{bmatrix} + [b_o] [u]$$

Q4.27   What is the advantage and the disadvantage in canonical form of state model.

The advantage of canonical form is that the state equations are independent of each other. The disadvantage is that the canonical variables are not physical variables and so they are not available for measurement and control.

Q4.28   What is state transition matrix and how it is related to state of a system?

The matrix exponential e^{At} is called state transition matrix. In the expanded form,

$$e^{At} = I + At + \frac{1}{2!} A^2 t^2 + \frac{1}{3!} A^3 t^3 + \cdots + \frac{1}{i!} A^i t^i + \cdots$$

The state transition matrix is used to find the state of the system, at any time instant t, from the knowledge of the state at time, t_0.

When the input is zero, $X(t) = e^{At} X(t_0)$

When the input vector is U(t), $X(t) = e^{A(t - t_0)} X(t_0) + \int_{t_0}^{t} e^{A(t - t)} B U(\tau) dt$

where, $X(t)$ = State vector at time, t and $X(t_0)$ = State vector at time, t_0.

Q4.29   Write the properties of state transition matrix.

The following are the properties of state transition matrix.

1. $\phi(0) = e^{A \times 0} = I$ (Unit matrix)

2. $\phi(t) = e^{At} = (e^{-At})^{-1} = [\phi(-t)]^{-1}$

3. $\phi(t_1 + t_2) = e^{A(t1 + t2)} = e^{At1} + e^{At2} = \phi(t_1) \phi(t_2) = \phi(t_2) \phi(t_1)$

Q4.30   *Write the solution of homogeneous state equations.*

The solution of homogeneous state equation is, $X(t) = e^{At} X_0$

where, $X(t)$ = State vector at time, t

e^{At} = State transition matrix

and X_0 = Initial condition vector at t = 0.

Q4.31   *Write the solution of non-homogeneous state equations.*

The solution of non-homogeneous state equation is

$$X(t) = e^{A(t-t_0)} X(t_0) + \int_{t_0}^{t} e^{A(t-\tau)} B\, U(\tau)\, dt$$

where, $X(t)$ = State vector of time, t ; $X(t_0)$ = Initial condition vector at t = t_0.

B = Input matrix   and $U(t)$ = Input vector

If initial conditions are known at t = 0, then put t_0 = 0.

Q4.32   *What is resolvant matrix?*

The Laplace transform of state transition matrix is called resolvant matrix.

Resolvant matrix, $\phi(s) = \mathcal{L}[\phi(t)] = \mathcal{L}[e^{At}]$

Also, $\phi(s) = [sI - A]^{-1}$

Q4.33   *What are the different methods available for computing e^{At}.*

The following four methods are available for computing e^{At}.

1. Computation of e^{At} using matrix exponential.
2. Computation of e^{At} using laplace transform.
3. Computation of e^{At} by canonical transformation.
4. Computation of e^{At} using Cayley-Hamilton Theorem.

Q4.34   *Write the state model of nth order discrete time system.*

The state model of a system consists of state equation and output equation. The state model of a nth order discrete time system with m-inputs and p-outputs are

$X(k +1) = A\, X(k) + B\, U(k)$ ..... state equation

$Y(k) = C\, X(k) + D\, U(k)$ ..... output equation

where, $X(k)$ $=$ State vector of order (n ×1)

$Y(k)$ $=$ Output vector of order (p×1)

$U(k)$ $=$ Input vector of order (m×1)

A $=$ System matrix of order (n×n)

B $=$ Input matrix of order (n × m)

C $=$ Output matrix of order (p×n)

D $=$ Transmission matrix of order (p×m)

Q4.35   *What are the fundamental elements used to construct the state diagram of discrete time system?*

The fundamental elements used to construct the state diagram of discrete time system are scalar, adder and unit delay element.

Q4.36   *Sketch the basic elements used to construct the block diagram of discrete time system.*

The basic elements used to construct the block diagram of discrete time system are shown below.

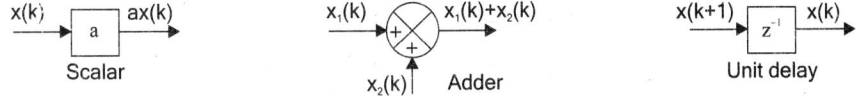

Q4.37   *Draw the basic elements used to construct the signal flow graph of discrete time system.*

The basic elements used to construct the signal flow graph of discrete time system are shown below.

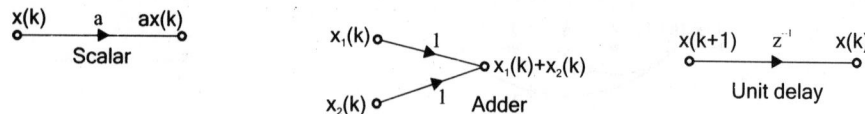

Q4.38   *Draw the block diagram representation of the state model of discrete time system.*

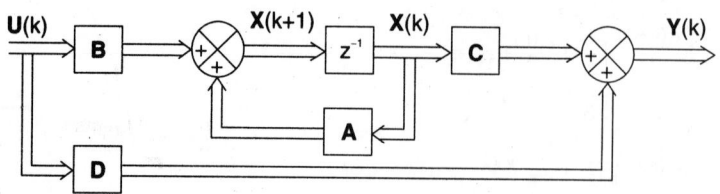

Fig Q4.38 : *Block diagram representation of discrete time system.*

Q4.39   *Draw the block diagram representation of the state model of discrete time system.*

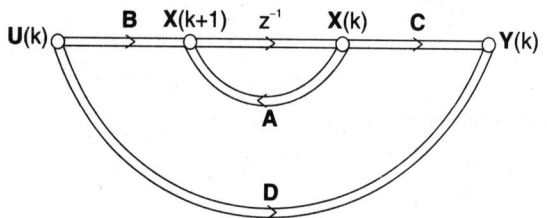

Fig Q4.39 : *Signal flow graph representation of discrete time system.*

Q4.40   *Draw the block diagram of the discrete time system described by the state model,*

$$\begin{bmatrix} x_1(k+1) \\ x_2(k+1) \end{bmatrix} = \begin{bmatrix} 4 & 0 \\ 0 & 2 \end{bmatrix} \begin{bmatrix} x_1(k) \\ x_2(k) \end{bmatrix} + \begin{bmatrix} 1 \\ 1 \end{bmatrix} u \quad ; \quad y(k) = \begin{bmatrix} 3 & 5 \end{bmatrix} \begin{bmatrix} x_1(k) \\ x_2(k) \end{bmatrix} + 10\, u(k)$$

Solution

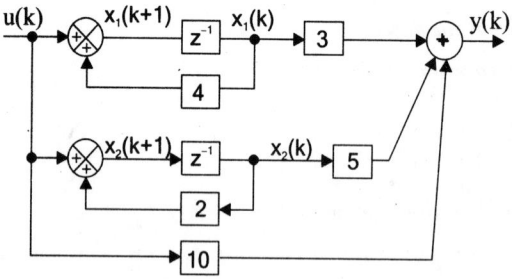

Q4.41   *Draw the signal flow graph of the discrete time system described by the state model.*

$$\begin{bmatrix} x_1(k+1) \\ x_2(k+1) \end{bmatrix} = \begin{bmatrix} 0 & 1 \\ 2 & 3 \end{bmatrix} \begin{bmatrix} x_1(k) \\ x_2(k) \end{bmatrix} + \begin{bmatrix} 0 \\ 1 \end{bmatrix} u \quad ; \quad y(k) = x_1(k)$$

Solution

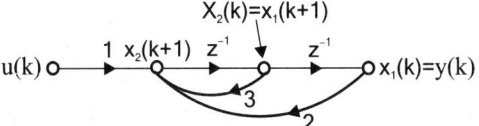

Q4.42   *Determine the state model of the discrete time system represented by the block diagram of fig Q4.42.*

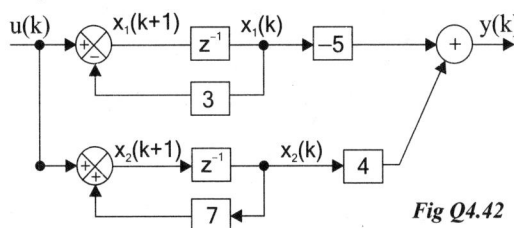

Fig Q4.42

Solution

$$\begin{bmatrix} x_1(k+1) \\ x_2(k+1) \end{bmatrix} = \begin{bmatrix} -3 & 0 \\ 0 & 7 \end{bmatrix} \begin{bmatrix} x_1(k) \\ x_2(k) \end{bmatrix} + \begin{bmatrix} 1 \\ 1 \end{bmatrix} u \quad ; \quad y(k) = \begin{bmatrix} -5 & 4 \end{bmatrix} \begin{bmatrix} x_1(k) \\ x_2(k) \end{bmatrix}$$

Q4.43   *Determine the state model of the discrete time system represented by the signal flow graph of fig Q4.43.*

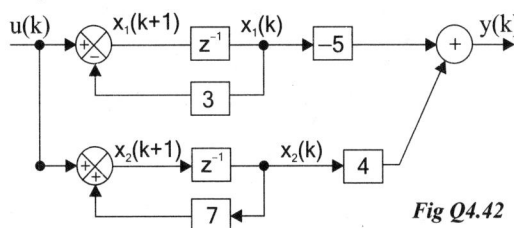

Fig Q4.43

Solution

$$\begin{bmatrix} x_1(k+1) \\ x_2(k+1) \end{bmatrix} = \begin{bmatrix} 0 & 0 \\ 1 & -2 \end{bmatrix} \begin{bmatrix} x_1(k) \\ x_2(k) \end{bmatrix} + \begin{bmatrix} 1 \\ 1 \end{bmatrix} u(k) \quad ; \quad y(k) = x_2(k)$$

Q4.44   *A discrete time system is described by the difference equation,*

y(k+2) + 3 y(k+1) + 5 y(k) = u(k). Determine the transfer function of the system.

Solution

Given that, $y(k+2) + 3 y(k+1) + 5 y(k) = u(k)$ .....(1)

On taking Z-transform of equation (1) with zero initial conditions we get,

$z^2 Y(z) + 3z\,Y(z) + 6\,Y(z) = U(z)$

$(z^2 + 3z + 6)\,Y(z) = U(z)$

$$\therefore \frac{Y(z)}{U(z)} = \frac{1}{z^2 + 3z + 6}$$ .....(2)

The equation(2) is the transfer function of the system.

Q4.45   *What is state transition matrix of discrete time system ?*

The matrix $\mathbf{A}^k$ (or $\phi(k)$) is called the state transition matrix of discrete time system.

It is given by, $\mathbf{A}^k = \mathbf{Z}^{-1}\{(z\mathbf{I} - \mathbf{A})^{-1}z\}$

The state transition matrix is used to find the state of the system, at any discrete time instant k.

Q4.46   *Write the properties of the state transition matrix of discrete time system.*

The properties of the state transition matrix of discrete time system are given below.

1. $\phi(0) = \mathbf{I}$

2. $\phi^{-1}(k) = \phi(-k)$

3. $\phi(k, k_0) = \phi(k-k_0) = \mathbf{A}^{(k-k0)}$;   where $k > k_0$.

Q4.47   *What are the different methods available for computing $\mathbf{A}^k$?*

The various methods are available for computing $\mathbf{A}^k$ are,

1.   Computation of $\mathbf{A}^k$ using z-transform.

2.   Computation of $\mathbf{A}^k$ by canonical transformation.

3.   Computation of $\mathbf{A}^k$ using Cayley-Hamilton Theorem.

Q4.48   *Write the solution of discrete time state equation.*

The solution of discrete time state equation is

$$X(k) = \phi(k)X(0) + \sum_{j=0}^{k-1} \phi(k-1-j)\,B\,U(j)$$

Q4.49   *Write the expression to determine the solution of discrete time state equation using z-transform.*

The solution of discrete time state equation is given by

$$X(k) = \mathbf{Z}^{-1}\{(z\mathbf{I}-\mathbf{A})^{-1}z\}\,X(0) + \mathbf{Z}^{-1}\{(z\mathbf{I}-\mathbf{A})^{-1}\,B\,U(z)\}$$

Q4.50   *The state model of a discrete time system is given by,*

$$X(k+1) = A\,X(k) + B\,U(k)$$

$$Y(k)\ \ = C\,X(k) + D\,U(k)$$

Determine its transfer function.

Solution

The state equation is, $X(k+1) = A\,X(k) + B\,U(k)$ .....(1)

On taking Z-transform of equation (1) with zero initial conditions we get,

$z\,X(z) = A\,X(z) + B\,U(z)$

$z\,X(z) - A\,X(z) = B\,U(z)$

$(z\mathbf{I} - A)\,X(z) = B\,U(z)$ .....(2)

On premultiplying equation (2) by $(z\mathbf{I} - A)^{-1}$ we get,

$X(z) = (z\mathbf{I} - A)^{-1}\,B\,U(z)$ .....(3)

The output equation is, $Y(k) = C\,X(k) + D\,U(k)$ .....(4)

On taking Z-transform of equation (4) we get,

$Y(z) = C\,X(z) + D\,U(z)$ .....(5)

On substituting for $X(z)$ from equation (3) in equation (5) we get,

$Y(z) = C\,(z\mathbf{I} - A)^{-1}\,BU(z) + DU(z)$

$Y(z) = [C\,(z\mathbf{I} - A)^{-1}\,B + D]\,U(z)$

$\therefore\ \dfrac{Y(z)}{U(z)} = C(z\mathbf{I} - A)^{-1}B + D$ .....(6)

The equation (6) is the transform function of discrete time system.

4.11 EXERCISES

I. FILL IN THE BLANKS

1.   A set of variables which describes the system at any time instant are called ....

2.   The set of all possible values which the state X(t) can assume at time t forms the .... o the system.

3.   The .... of a system consist of the state equation and output equation.

4.   The pictorial representation of the state model of the system is called ....

5.   In vector notation notation the state equation of the system is ....

6.   In vector notation notation the output equation of the system is ....

7.   In state modelling the number of state variables will decide the .... of the system.

8.   The number of .... in a state diagram is equal to the number of ....

9.   The .... are obtained from one of the system variables and its derivatives.

10.   In .... of state model the last row is comprised of the negative of the coefficients of the differential equation governing the system.

11.   In canonical form of state model, the system matrix will be a ....

12.   In a diagonalized system matrix the elements on the diagonal are .... of the transfer function of the system.

13.   When a pole of the transfer function has multiplicity the canonical state model will be in a special form called ....

14.   In the Jordan block the diagonal elements are .... and the element just above the diagonal is ....

15.   The matrix .... is called state transition matrix.

16.   In vector notation the state equation of the discrete time system is ....

17.   In vector notation the output equation of the discrete time system is ....

18.   The number of .... in a state diagram of discrete time system is equal to number of state variables.

19.   The matrix .... is called the state transition matrix of discrete time system.

20.   The state transition matrix of discrete time system A^k = ...

ANSWERS		
1. state variables	8. integrator, state variables	14. poles, one
2. state space	9. phase variables	15. e^{At}
3. state model	10. bush or companion form	16. $X(k + 1) = AX(t) + BU(t)$
4. state diagram	11. diagonal matrix	17. $Y(k) = C\,X(k) + D\,U(k)$
5. $X(t) = A\,X(t) + B\,U(t)$	12. poles	18. unit delay
6. $Y(t) = C\,X(t) + D\,U(t)$	13. Jordan canonical form	19. A^k
7. order	14. poles, one	20. $\mathbf{Z}^{-1}\{(zI - A)^{-1}z\}$

II. State whether the following statements are TRUE / FALSE

1. The state variable analysis can be applied for any type (linear/nonlinear and time varying invarying) of system.

2. The transfer function anlysis is applicable to nonlinear systems.

3. The transfer function anlysis can be carried with initial conditions.

4. The state space analysis can be carried with initial conditions and on multiple input and output systems.

5. The state model of a system is nonunique but the transfer function is unique.

6. The state and output equations of a system are functions of state variables and inputs.

7. The diagram can not be used for simulation of the system in analog computers.

8. The state diagram provides a direct relation between the time domain and s-domain.

9. The Bush or Companion form of state model can be written dirctely by directly by inspection from the differential equation governing the system.

10. In a system matrix of the companion form of state model the elements just above the main diagonal are always one.

11. The phase variables are physical variables of the system.

12. In canonical form of state model, the state equations are independent of each other.

13. The canonical variables are physical variables of the system.

14. In canonical form of state model, when a pole has multiplicity, q the system matrix will have a Jordan bolck of $(q \times q)$.

15. The discrete time form of state space representation is quite analogous to the continuous tome form.

16. The state equation of a discrete time system is a set of n-numbers of first order differential equations.

17. The state diagram of discrete time system can be used for simulation of the system on digital computers.

18. The state diagram of discrete time system provides a direct relation between time domain and z-domain.

19. In discrete time system the value of state variables at $(k + 1)^{th}$ time instant are functions of state variables and inputs at k^{th} time instant.

20. In discrete time system the value of output at $(k + 1)^{th}$ time instant are functions of state variables and inputs at k^{th} time instant.

ANSWERS

1. True	6. True	11. False	16. False
2. False	7. False	12. True	17. True
3. False	8. True	13. False	18. True
4. True	9. True	14. True	19. True
5. True	10. True	15. True	20. False

III. UNSOLVED PROBLEMS

E4.1   *Obtain the state model of the electrical network shown below by choosing minimal number of state variables.*

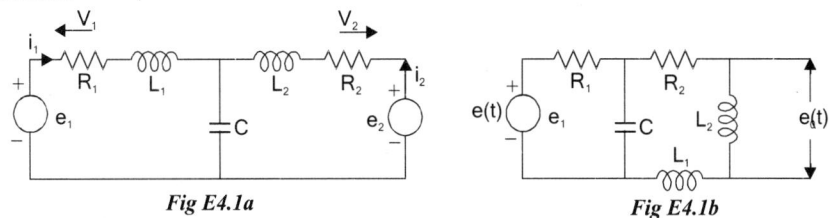

Fig E4.1a   Fig E4.1b

E4.2   *Construct the state model of mechanical systems shown in fig E4.2a & E4.2b.*

E4.3   *Construct a state model for the systems characterized by the following differential equations.*

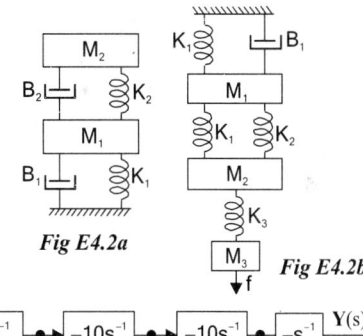

    (i) $\dddot{y} + 3\ddot{y} + 2\dot{y} = \dot{u} + u$

    (ii) $\dddot{y} + 6\ddot{y} + 11\dot{y} + 6y = u$

    (iii) $\ddot{y} + 2\dot{y} + y = \dot{u} + u$

                       Fig E4.2a   Fig E4.2b

E4.4   *In the state diagrams of systems shown in figE4.4a and figE4.4b. Assign state variables and obtain state model of the system.*

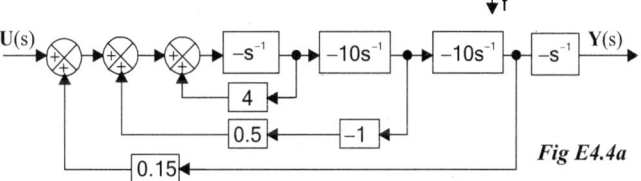

Fig E4.4a

E4.5   *The state diagram of a linear system is shown in fig E4.5. Assign state variables and obtain the state model.*

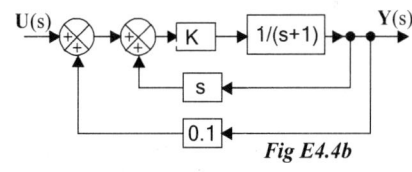

Fig E4.4b

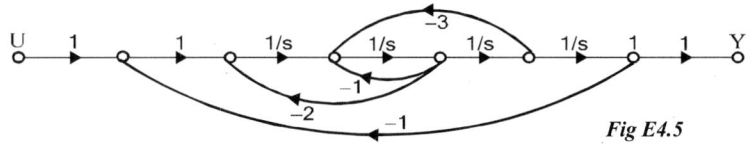

Fig E4.5

E4.6   *Obtain the state model of the system whose transfer function is given as*

    (i) $\dfrac{1}{s^2 + 3s + 2}$ **(ii)** $\dfrac{s+3}{s^2 + 3s + 2}$ **(iii)** $\dfrac{s^2 + 4s + 3}{s^3 + 9s + 20}$ **(iv)** $\dfrac{s^2 + 6s + 8}{(s+3)(s^2 + 2s + 2)}$

E4.7   *For the system matrices given below compute e^{At}*

a) $A = \begin{bmatrix} 0 & 1 \\ 1 & 1 \end{bmatrix}$ b) $A = \begin{bmatrix} -2 & 1 \\ 1 & -2 \end{bmatrix}$ c) $A = \begin{bmatrix} -2 & 0 \\ 0 & -1 \end{bmatrix}$ d) $A = \begin{bmatrix} -1 & 0 \\ 0 & -3 \end{bmatrix}$

E4.8   *The following facts are known about the linear system, $\dot{x}(t) = A\,x(t)$.*

$$If \ x(0) = \begin{bmatrix} 2 \\ -1 \end{bmatrix} \ then \ x(t) = \begin{bmatrix} e^{-2t} \\ -4e^{-2t} \end{bmatrix} \ ; \ If \ x(0) = \begin{bmatrix} 1 \\ -2 \end{bmatrix} \ then \ x(t) = \begin{bmatrix} e^{-t} \\ -2e^{-t} \end{bmatrix}$$

Find e^{At} and hence A.

E4.9   *The figure E4.9 shows the block diagram of a control system with state variable feedback and integral control.*

Derive the state model of the system.

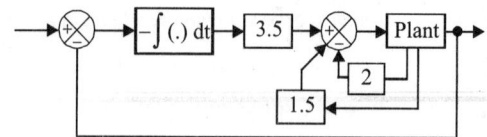

Fig E4.9

E4.10   *A system is described by the state equation.*

$$\dot{X} = \begin{bmatrix} 0 & 1 & 0 \\ 0 & 0 & 1 \\ -1 & 0 & -3 \end{bmatrix} X + \begin{bmatrix} 0 \\ 0 \\ 1 \end{bmatrix} u \ ; \ X(0) = x_0$$

Using Laplace transform technique, transform the state equation into a set of linear algebraic equations in the form $X(s) = \phi(s)\,X(0) + \phi(s)\,B\,U(s)$

E4.11   *Consider the state equation*

$$\begin{bmatrix} \dot{x}_1 \\ \dot{x}_2 \end{bmatrix} = \begin{bmatrix} 0 & 1 \\ -1 & -2 \end{bmatrix} \begin{bmatrix} x_1 \\ x_2 \end{bmatrix}$$

Compute the solution of homogeneous equation, assuming $x_1(2) = 2$.

E4.12   *Compute the solution of following state equations.*

a) $\dot{X} = \begin{bmatrix} 0 & 0 & -2 \\ 0 & 1 & 0 \\ 1 & 0 & 3 \end{bmatrix} X \ ; \ X(0) = \begin{bmatrix} 0 \\ 1 \\ 0 \end{bmatrix}$ b) $\begin{bmatrix} \dot{x}_1 \\ \dot{x}_2 \end{bmatrix} = \begin{bmatrix} 1 & 0 \\ 1 & 1 \end{bmatrix} \begin{bmatrix} x_1 \\ x_2 \end{bmatrix} + \begin{bmatrix} 1 \\ 1 \end{bmatrix} u(t) \ with \ X_0 = \begin{bmatrix} 1 \\ 1 \end{bmatrix}$

Chapter 5

Analysis and Design of Control System in State Space

5.1 DEFINITIONS INVOLVING MATRICES

Matrix : A matrix is an ordered array of elements which may be real numbers, complex numbers, functions or operators. In general the array consists of m rows and n columns. When m = n, the matrix is called square matrix. When n = 1, the matrix is called column matrix or vector. When m = 1, the matrix is called row matrix or vector.

Diagonal matrix : It is a square matrix whose elements other than main diagonal are all zeros.

Unit Matrix : It is a diagonal matrix whose diagonal elements are all equal to unity. The elements other than diagonal are all zeros. It is denoted by $\mathbf{I}$.

Transpose : If the rows and columns of an $m \times n$ matrix $\mathbf{A}$ are interchanged, then the resulting $n \times m$ matrix is called the transpose of $\mathbf{A}$. The transpose of $\mathbf{A}$ is denoted by $\mathbf{A}^T$.

Determinant : A determinant consisting of the elements of a square matrix (in the order given in the matrix) is called the determinant of the matrix.

Symmetric matrix : A square matrix is symmetric if it is equal to its transpose, i.e., $\mathbf{A}^T = \mathbf{A}$. If $\mathbf{A}$ is a square matrix, then $\mathbf{A} + \mathbf{A}^T$ is a symmetric matrix.

Skew-symmetric matrix : A square matrix is skew-symmetric if it is equal to the negative of its transpose, i.e., $\mathbf{A}^T = -\mathbf{A}$. If $\mathbf{A}$ is a square matrix then $\mathbf{A} - \mathbf{A}^T$ is a skew symmetric matrix.

Orthogonal Matrix : A matrix $\mathbf{A}$ is called an orthogonal matrix if it is real and satisfies the relationship $\mathbf{A}^T \mathbf{A} = \mathbf{A}\mathbf{A}^T = \mathbf{I}$.

Minor : If the ith row and j^{th} column of determinant $\mathbf{A}$ are deleted, the remaining (n–1) rows and columns form a determinant $\mathbf{M}_{ij}$. This determinant is called the minor of the element a_{ij}.

Cofactor : The cofactor C_{ij} of element a_{ij} of the matrix $\mathbf{A}$ is defined as $C_{ij} = (-1)^{(i+j)} \mathbf{M}_{ij}$, where $\mathbf{M}_{ij}$ is the minor of a_{ij}.

Adjoint matrix : The adjoint matrix of a square matrix $\mathbf{A}$ is found by replacing each element a_{ij} of matrix $\mathbf{A}$ by its cofactor C_{ij} and then transposing.

Singular matrix : A square matrix is called singular if its associated determinant is zero. If the determinant of the matrix is nonzero then the matrix is nonsingular.

Rank of matrix : A matrix **A** is said to have a rank r if there exists an $r \times r$ submatrix of **A** which is nonsingular and all other $q \times q$ submatrices are singular, where $q \geq (r+1)$.

Conjugate matrix : The conjugate of a matrix **A** is the matrix in which each element is the complex conjugate of the corresponding element of **A**. The conjugate of **A** is denoted by **A***.

Real matrix : If all the elements of a matrix are real then the matrix is called real matrix. A real matrix is equal to its conjugate.

5.2   EIGENVALUES AND EIGENVECTORS

A nonzero column vector **X** is an eigenvector of a square matrix **A**, if there exists a scalar λ such that **AX** = λ**X**, then λ is a eigenvalue (or characteristic value) of **A**. Eigenvalue may be zero but the corresponding vector may not be a zero vector.

The characteristic equation of n × n matrix **A** is the nth degree polynomial of equation, $|\lambda \mathbf{I} - \mathbf{A}| =$ 0, where **I** is the unit matrix. Solving the characteristic equation for λ gives the eigenvalues of **A**. The eigenvalues may be real, complex or multiples of each other.

Once an eigenvalue is determined it may be substituted into **AX** = λ**X** and then that equation may be solved for the corresponding eigenvector.

PROPERTIES OF EIGENVALUES AND EIGENVECTORS

1.   *The sum of the eigenvalues of a matrix is equal to its trace, which is the sum of the elements on its main diagonal.*

2.   *Eigenvectors corresponding to different eigenvalues are linearly independent.*

3.   *A matrix is singular if and only if it has a zero eigenvalue.*

4.   *If X is an eigenvector of A corresponding to the eigenvalue of λ and A is invertible, then X is an eigenvector of A^{-1} corresponding to its eigenvalue $1/\lambda$.*

5.   *If X is an eigenvector of a matrix then KX is also an eigenvector for any nonzero constant K. Here both X and KX correspond to the same eigenvalue.*

6.   *A matrix and its transpose have the same eigenvalues.*

7.   *The eigenvalues of an upper or lower triangular matrix are the elements on its main diagonal.*

8.   *The product of the eigenvalues (counting multiplicities) of the matrix equals the determinant of the matrix.*

9.   *If X is an eigenvector of A corresponding to eigenvalue of λ , then X is an eigenvector of A-CI corresponding to the eigenvalue $\lambda - C$ for any scalar C.*

DETERMINATION OF EIGENVECTORS

Case i : **Distinct eigenvalues**

If the eigenvalues of **A** are all distinct, then we have only one independent eigenvector corresponding to any particular eigenvalue λ_i. The eigenvector corresponding to λ_i may be obtained by taking cofactors of matrix $[\lambda_i \mathbf{I} - \mathbf{A}]$ along any row.

Let, $\mathbf{m}_i$ = Eigenvector corresponding to λ_i

Now the eigenvector $\mathbf{m}_i$ is given by

$$\mathbf{m}_i = \begin{bmatrix} C_{k1} \\ C_{k2} \\ \vdots \\ C_{kn} \end{bmatrix} \; ; \; k = 1 \text{ or } 2 \text{ or }, ...n \qquad .....(5.1)$$

where $C_{k1}, C_{k2}, ... C_{kn}$ are cofactors of matrix $[\lambda_i I - A]$ along k^{th} row.

Case ii : **Multiple eigenvalues**

In this case the eigenvectors corresponding to the distinct eigenvalues are evaluated as mentioned in case (i).

If the matrix has repeated eigenvalues with multiplicity "q", then there exists only one independent eigenvector corresponding to that repeated eigenvalue. If λ_i is a repeated eigenvalue, then the independent vector corresponding to λ_i can be evaluated by taking the cofactor of matrix $[\lambda_i I - A]$ along any row as mentioned in case (1). The remaining (q−1) eigenvectors can be obtained as shown in equation (5.2).

Let, $\mathbf{m}_p$ = p^{th} eigenvector corresponding to repeated eigenvalue λ_1.

$$\mathbf{m}_p = \begin{bmatrix} \dfrac{1}{p!} \dfrac{d^p}{d\lambda_i^p} C_{k1} \\[2mm] \dfrac{1}{p!} \dfrac{d^p}{d\lambda_i^p} C_{k2} \\[2mm] \vdots \\[2mm] \dfrac{1}{p!} \dfrac{d^p}{d\lambda_i^p} C_{kn} \end{bmatrix} \; ; \; p = 1, 2, 3, ...(q-1) \qquad .....(5.2)$$

where $C_{k1}, C_{k2}, C_{k3} .....C_{kn}$ are cofactors of matrix $[\lambda_i I - A]$ along k^{th} row.

5.3 SIMILARITY TRANSFORMATION

The square matrices $\mathbf{A}$ and $\mathbf{B}$ are said to be similar if a nonsingular matrix $\mathbf{P}$ exists such that

$$\mathbf{P}^{-1}\mathbf{A}\mathbf{P} = \mathbf{B} \qquad .....(5.3)$$

The process of transformation is called similarity transformation and it is a linear transformation. The matrix $\mathbf{P}$ is called transformation matrix. Also the matrix, $\mathbf{A}$ can be obtained from $\mathbf{B}$ by a similarity transformation with a transformation matrix $\mathbf{P}^{-1}$,

i.e., $\mathbf{A} = \mathbf{P}\,\mathbf{B}\,\mathbf{P}^{-1}$ $\qquad .....(5.4)$

The similarity transformation can be used for diagonalization of a square matrix. If an $n \times n$ matrix has n linearly independent eigenvectors (i.e., with distinct eigenvalues) then it can be diagonalized by a similarity transformation. If a matrix has multiple eigenvalues then it will not have a complete set of n linearly independent eigenvectors and so it cannot be diagonalized. However such a matrix can be transformed into a Jordan matrix (Jordan canonical form).

The transformation matrix for diagonalization or converting to Jordan form can be obtained from eigenvectors. For a system with n state variables we can find n numbers of eigenvectors $\mathbf{m}_1$, $\mathbf{m}_2$, $\mathbf{m}_3$,........, $\mathbf{m}_n$. The eigenvectors are column vectors of order ($n \times 1$). The transformation matrix is obtained by arranging the eigenvectors columnwise as shown in equation (5.5). This transformation matrix is also called *Modal matrix* and denoted by M.

$$\textbf{Modal matrix, } \mathbf{M} = [\, \mathbf{m}_1 \ \ \mathbf{m}_2 \ \ \mathbf{m}_3 \ \ .... \ \ \mathbf{m}_n \,] \quad\quad\quad .....(5.5)$$

The similarity transformation will not alter certain properties of the matrix. A property of a matrix is said to be invariant if it is possessed by all similar matrices. The determinant, characteristic equation and trace of a matrix are invariant under a similarity transformation. Since the characteristic equation is invariant the eigenvalues are also invariant under a linear or similarity transformation.

PROOF FOR INVARIANCE OF DETERMINANT

Let $\mathbf{A}$ and $\mathbf{B}$ are similar matrices and $\mathbf{P}$ be the transformation matrix which transforms $\mathbf{A}$ to $\mathbf{B}$ by a similarity transformation, $\mathbf{P}^{-1}\mathbf{AP} = \mathbf{B}$

$$\therefore \mathbf{B} = \mathbf{P}^{-1}\mathbf{AP} \quad\quad\quad .....(5.6)$$

On taking determinant of equation (5.6) we get,

$$|\mathbf{B}| = |\mathbf{P}^{-1}\mathbf{AP}| \quad\quad\quad .....(5.7)$$

Since the determinant of a product of two or more square matrices is equal to the product of their individual determinants, the equation (5.7) can be written as,

$$|\mathbf{B}| = |\mathbf{P}^{-1}|\,|\mathbf{A}|\,|\mathbf{P}| = |\mathbf{A}|\,|\mathbf{P}^{-1}|\,|\mathbf{P}|$$

$$= |\mathbf{A}|\,|\mathbf{P}^{-1}\,\mathbf{P}| = |\mathbf{A}|\,|\mathbf{I}| \quad\quad\quad\quad (\because \ \mathbf{P}^{-1}\mathbf{P} = 1)$$

$$= |\mathbf{A}| \quad\quad\quad\quad\quad\quad\quad\quad\quad\quad\quad\quad (\because \ |\mathbf{I}| = 1)$$

From the above analysis it is evident that the determinant of a matrix is invariant under a similarity transformation.

PROOF FOR INVARIANCE OF CHARACTERISTIC EQUATION AND EIGENVALUES

Let $\mathbf{A}$ and $\mathbf{B}$ are similar matrices and $\mathbf{P}$ be the transformation matrix which transforms $\mathbf{A}$ to $\mathbf{B}$ by a similarity transformation, $\mathbf{P}^{-1}\mathbf{AP} = \mathbf{B}$.

The characteristic equation of matrix $\mathbf{B}$ is given by

$$|\lambda \mathbf{I} - \mathbf{B}| = 0 \quad\quad\quad .....(5.8)$$

On substituting $\mathbf{B} = \mathbf{P}^{-1}\,\mathbf{AP}$ in equation (5.8) we get,

$$|\lambda \mathbf{I} - \mathbf{B}| = |\lambda \mathbf{I} - \mathbf{P}^{-1}\,\mathbf{AP}|$$

$$= |\lambda \mathbf{P}^{-1}\,\mathbf{P} - \mathbf{P}^{-1}\mathbf{AP}| \quad\quad\quad (\because \ \mathbf{P}^{-1}\mathbf{P} = 1)$$

$$= |\mathbf{P}^{-1}\,(\lambda \mathbf{I} - \mathbf{A})\mathbf{P}| \quad\quad\quad\quad\quad\quad .....(5.9)$$

Since the determinant of a product is the product of the determinant, the equation (5.9) can be written as,

$$|\lambda I - B| = |P^{-1}| \, |\lambda I - A| \, |P| = |\lambda I - A| \, |P^{-1}| \, |P|$$

$$= |\lambda I - A| \, |P^{-1} \, P|$$

$$= |\lambda I - A| \, |I| \qquad (\because P^{-1}P = 1)$$

$$= |\lambda I - A| \qquad (\because |I| = 1)$$

From the above analysis it is clear that the characteristic equations of **A** and **B** are identical. Since the characteristic equations are identical, the eigenvalues of **A** and **B** are identical. Hence the eigenvalues are invariant under a similarity (linearity) transformation.

PROOF FOR INVARIANCE OF TRACE OF A MATRIX

Let **A** and **B** are similar matrices and **P** be the transformation matrix which transforms **A** to **B** by a similarity transformation, $P^{-1}AP = B$.

$$\therefore \text{tr } B = \text{tr } P^{-1}AP \qquad\qquad\qquad .....(5.10)$$

For an $n \times m$ matrix **C** and $m \times n$ matrix **D**, regardless of whether $CD = DC$ or $CD \neq DC$, we have,

$$\text{tr } (CD) = \text{tr } (DC) \qquad\qquad\qquad .....(5.11)$$

Using the property of equation (5.11), the equation (5.10) can be written as,

$$\text{tr } B = \text{tr } APP^{-1}$$

$$= \text{tr } AI = \text{tr } A \qquad (\because PP^{-1} = I \text{ and } AI = A)$$

From the above analysis it is clear that the trace of a matrix is invariant under a similarity transformation.

5.4 CAYLEY-HAMILTON THEOREM

The Cayley-Hamilton theorem states that every square matrix satisfies its own characteristic equation.

Consider an $n \times n$ matrix **A** and its characteristic equation [equation (5.12)].

$$|\lambda I - A| = \lambda^n + a_1 \lambda^{n-1} + \ldots\ldots + a_{n-1} \lambda + a_n = 0 \qquad\qquad .....(5.12)$$

By Cayley-Hamilton theorem, the matrix **A** has to satisfy its characteristic equation, hence equation (5.12) can be written as,

$$A^n + a_1 A^{n-1} + \ldots\ldots + a_{n-1} A + a_n I = 0 \qquad\qquad .....(5.13)$$

PROOF OF CAYLEY - HAMILTON THEOREM

Let **A** be a square matrix. The characteristic equation of **A** is given by

$$|\lambda I - A| = \lambda^n + a_1 \lambda^{n-1} + a_2 \lambda^{n-2} + \ldots\ldots + a_{n-2} \lambda^2 + a_{n-1} \lambda + a_n = 0 \qquad .....(5.12)$$

We have to prove that **A** satisfies the characteristic equation,

i.e., $A^n + a_1 A^{n-1} + a_2 A^{n-2} + \ldots\ldots + a_{n-2} A^2 + a_{n-1} A + a_n I = 0 \qquad\qquad .....(5.13)$

where **I** is the unit matrix of order $(n \times n)$.

Consider the matrix $[\lambda I - A]$. Let the matrix $\mathbf{B}$ be adjoint of $[\lambda I - A]$.

$$\therefore \mathbf{B} = \text{adj } [\lambda I - A] \qquad \qquad \qquad \qquad \qquad \dots(5.14)$$

The elements of adj $[\lambda I - A]$ are the cofactors of the elements of $[\lambda I - A]$. Therefore each element of $\mathbf{B}$ will be a polynomial in λ of degree $(n-1)$ or less. We know that every matrix whose elements are ordinary polynomials can be written as matrix polynomial. Hence the matrix $\mathbf{B}$ can be written as a matrix polynomial as shown in equation (5.15).

$$\therefore \mathbf{B} = \mathbf{B}_1 \lambda^{n-1} + \mathbf{B}_2 \lambda^{n-2} + \mathbf{B}_3 \lambda^{n-3} + \dots + \mathbf{B}_{n-2} \lambda^2 + \mathbf{B}_{n-1} \lambda + \mathbf{B}_n \qquad \dots(5.15)$$

From equations (5.14) and (5.15) we get,

$$\text{adj } [\lambda I - A] = \mathbf{B}_1 \lambda^{n-1} + \mathbf{B}_2 \lambda^{n-2} + \mathbf{B}_3 \lambda^{n-3} + \dots + \mathbf{B}_{n-2} \lambda^2 + \mathbf{B}_{n-1} \lambda + \mathbf{B}_n \qquad \dots(5.16)$$

We know that, $[\lambda I - A] [\lambda I - A]^{-1} = I$

But, $(\lambda I - A)^{-1} = \dfrac{\text{adj} (\lambda I - A)}{|\lambda I - A|}$

$$\therefore (\lambda I - A) = \dfrac{\text{adj} (\lambda I - A)}{|\lambda I - A|} = I$$

$$(\lambda I - A) \text{ adj } (\lambda I - A) = |\lambda I - A| I \qquad \qquad \qquad \dots(5.17)$$

Using equations (5.12) and (5.16), the equation (5.17) can be written as,

$$[\lambda I - A] (\mathbf{B}_1 \lambda^{n-1} + \mathbf{B}_2 \lambda^{n-2} + \mathbf{B}_3 \lambda^{n-3} + \dots + \mathbf{B}_{n-2} \lambda^2 + \mathbf{B}_{n-1} \lambda + \mathbf{B}_n)$$
$$= (\lambda^n + a_1 \lambda^{n-1} + a_2 \lambda^{n-2} + \dots + a_{n-2} \lambda^2 + a_{n-1} \lambda + a_n)I$$

$$(\mathbf{B}_1 \lambda^n + \mathbf{B}_2 \lambda^{n-1} + \mathbf{B}_3 \lambda^{n-2} + \dots + \mathbf{B}_{n-2} \lambda^3 + \mathbf{B}_{n-1} \lambda^2 + \mathbf{B}_n \lambda)$$
$$- A\mathbf{B}_1 \lambda^{n-1} - A\mathbf{B}_2 \lambda^{n-2} - A\mathbf{B}_3 \lambda^{n-3} - \dots - A\mathbf{B}_{n-2} \lambda^2 - A\mathbf{B}_{n-1} \lambda - A\mathbf{B}_n$$
$$= I\lambda^n + a_1 I \lambda^{n-1} + a_2 I \lambda^{n-2} + \dots + a_{n-2} I \lambda^2 + a_{n-1} I\lambda + a_n I \qquad \dots(5.18)$$

On equating the coefficients of like powers of λ in equation (5.18) we get the following $(n + 1)$ equations

$$
\left.
\begin{aligned}
\mathbf{B}_1 &= I & \dots(1) \\
\mathbf{B}_2 - A\mathbf{B}_1 &= a_1 I & \dots(2) \\
\mathbf{B}_3 - A\mathbf{B}_2 &= a_2 I & \dots(3) \\
&\vdots & \\
\mathbf{B}_{n-1} - A\mathbf{B}_{n-2} &= a_{n-2} I & \dots(n-1) \\
\mathbf{B}_n - A\mathbf{B}_{n-1} &= a_{n-1} I & \dots(n) \\
- A\mathbf{B}_n &= a_{n-2} I & \dots(n+1)
\end{aligned}
\right\} \quad (n+1) \text{ equations}
$$

On premultiplying both sides of equations (1), (2), (3), ... (n - 1), (n) and (n + 1) by A^n, A^{n+1}, A^{n-2}, ...A^2, A and I respectively we get the following $(n + 1)$ equations.

$$
\begin{aligned}
\mathbf{A}^n \times \text{equ}(1) &\Rightarrow & \mathbf{A}^n\mathbf{B}_1 &= \mathbf{A}^n\mathbf{I} \\
\mathbf{A}^{n-1} \times \text{equ}(2) &\Rightarrow & \mathbf{A}^{n-1}\mathbf{B}_2 - \mathbf{A}^n\mathbf{B}_1 &= a_1\mathbf{A}^{n-1}\mathbf{I} \\
\mathbf{A}^{n-2} \times \text{equ}(3) &\Rightarrow & \mathbf{A}^{n-2}\mathbf{B}_3 - \mathbf{A}^{n-1}\mathbf{B}_2 &= a_2\mathbf{A}^{n-2}\mathbf{I} \\
&\vdots & \vdots \quad &\quad \vdots \\
\mathbf{A}^2 \times \text{equ}(n-1) &\Rightarrow & \mathbf{A}^2\mathbf{B}_{n-1} - \mathbf{A}^3\mathbf{B}_{n-2} &= a_{n-2}\mathbf{A}^2\mathbf{I} \\
\mathbf{A} \times \text{equ}(n) &\Rightarrow & \mathbf{A}\mathbf{B}_n - \mathbf{A}^2\mathbf{B}_{n-1} &= a_{n-1}\mathbf{A}\,\mathbf{I} \\
1 \times \text{equ}(n+1) &\Rightarrow & -\mathbf{A}\,\mathbf{B}_n &= a_n\mathbf{I}
\end{aligned}
\right\} (n+1) \text{ equations}
$$

On adding the above (n+1) equations we get, (i.e., all the left hand side terms gets cancelled and becomes zero),

$$0 = \mathbf{A}^n\mathbf{I} + a_1\,\mathbf{A}^{n-1}\mathbf{I} + a_2\,\mathbf{A}^{n-2}\mathbf{I} + \ldots + a_{n-2}\,\mathbf{A}^2\mathbf{I} + a_{n-1}\,\mathbf{A}\mathbf{I} + a_n\mathbf{I}$$

$$\therefore \ \mathbf{A}^n + a_1\,\mathbf{A}^{n-1} + a_2\,\mathbf{A}^{n-2} + \ldots + a_{n-2}\,\mathbf{A}^2 + a_{n-1}\,\mathbf{A} + a_n\mathbf{I} = 0 \qquad \qquad \ldots(5.19)$$

The equation (5.19) shows that the matrix **A** satisfies its characteristic equation. Thus Cayley-Hamilton theorem is proved.

COMPUTATION OF THE FUNCTION OF A MATRIX USING

CAYLEY-HAMILTON THEOREM

The Cayley-Hamilton theorem provides a simple procedure for evaluating the function of a matrix. Consider a matrix **A** of order (n × n) with eigenvalues $\lambda_1, \lambda_2, \lambda_3, \ldots \lambda_n$. The characteristic equation q (λ) of matrix **A** will be as shown in equation (5.20)

$$\therefore q(\lambda) = |\lambda\mathbf{I} - \mathbf{A}| = \lambda^n + a_1\,\lambda^{n-1} + \ldots\ldots + a_{n-1}\,\lambda + a_n = 0 \qquad \qquad \ldots(5.20)$$

Let f(**A**) be a function of matrix **A** and f(**A**) can be expressed as a matrix polynomial. Let f(λ) be a scalar polynomial obtained from f(**A**) after substituting **A** by λ .

On dividing f(λ) by q(λ), we get

$$\frac{f(\lambda)}{q(\lambda)} = Q(\lambda) + \frac{R(\lambda)}{q(\lambda)} \qquad \qquad \ldots(5.21)$$

where   Q(λ) = Quotient polynomial

and   R(λ) = Remainder polynomial

$$\therefore \ \ f(\lambda) = Q(\lambda)\,q(\lambda) + R(\lambda) \qquad \qquad \ldots(5.22)$$

If we evaluate the equation (5.22) using the eigenvalues $\lambda_1, \lambda_2, \lambda_3, \ldots \lambda_n$ then from equation (5.20) we get, q(λ) = 0 and we have,

$$f(\lambda_i) = R(\lambda_i) \qquad \qquad \ldots(5.23)$$

where i = 1, 2, 3, ...n

The remainder polynomial R(λ) will be in the form of equation (5.24) shown below.

$$R(\lambda) = \alpha_0 + \alpha_1\,\lambda + \alpha_2\,\lambda^2 + \ldots\ldots + \alpha_{n-1}\,\lambda^{n-1} \qquad \qquad \ldots(5.24)$$

where $\alpha_0, \alpha_1, \alpha_2, \ldots\ldots, \alpha_{n-1}$ are constants.

From equations (5.23) and (5.24) when $\lambda = \lambda_i$ we get,

$$f(\lambda_i) = \alpha_0 + \alpha_1 \lambda_i + \alpha_2 \lambda_i^2 + \ldots + \alpha_{n-1} \lambda_i^{n-1} \qquad \ldots(5.25)$$

where i = 1, 2, 3, ... , n.

On substituting the n number of eigenvalues in equation (5.25), one by one, we get n number of equations. There equations can be solved to find the constants α_0, α_1, ..., α_{n-1}.

In equation (5.22) on substituting **A** for the variable λ we get,

$$f(\mathbf{A}) = Q(\mathbf{A}) \, q(\mathbf{A}) + R(\mathbf{A}) \qquad \ldots(5.26)$$

The Cayley-Hamilton theorem says that every square matrix satisfies its characteristic equation and so $q(\mathbf{A}) = 0$. Therefore the equation (5.26) can be written as,

$$f(\mathbf{A}) = R(\mathbf{A}) \qquad \ldots(5.27)$$

From equation (5.24) we get,

$$R(\mathbf{A}) = \alpha_0 \mathbf{I} + \alpha_1 \mathbf{A} + \alpha_2 \mathbf{A}^2 + \ldots + \alpha_{n-1} \mathbf{A}^{n-1} \qquad \ldots(5.28)$$

From equations (5.27) and (5.28) we get,

$$f(\mathbf{A}) = \alpha_0 \mathbf{I} + \alpha_1 \mathbf{A} + \alpha_2 \mathbf{A}^2 + \ldots + \alpha_{n-1} \mathbf{A}^{n-1} \qquad \ldots(5.29)$$

The equation (5.29) can be used to evaluate the function $f(\mathbf{A})$. On substituting the eigenvalues in equation (5.25) we get n-number of linear equations. The constants α_0, α_1, α_2, α_3, ... , α_{n-1}, are obtained by solving these n-number of linear equations.

The equation (5.25) can be used to form n-number of independent equations only when all the eigenvalues are distinct. If the matrix **A** have a multiple eigenvalue with multiplicity m then only one independent equation can be obtained by substituting the multiple eigenvalue in equation (5.25). The remaining (m−1) equations are obtained by differentiating equation (5.25) after replacing λ_i by λ and then evaluating with $\lambda = \lambda_p$ where λ_p is the multiple eigenvalue, as shown in equation (5.30). [The equations corresponding to distinct eigenvalues are obtained by substituting the eigenvalues in equation (5.25)].

$$\left. \frac{\mathbf{d}^j}{\mathbf{d}\lambda^j}[f(\lambda)] \right|_{\lambda = \lambda_p} = \left. \frac{\mathbf{d}^j}{\mathbf{d}\lambda^j}[\alpha_0 + \alpha_1 \lambda + \alpha_2 \lambda^2 + \ldots + \alpha_{n-1}\lambda^{n-1}] \right|_{\lambda = \lambda_p} \qquad \ldots(5.30)$$

where j = 1, 2, 3, ... , (n−1).

The equation (5.29) can also be used to compute the state transition matrix of continuous time system $e^{\mathbf{A}t}$ by taking $f(\mathbf{A}) = e^{\mathbf{A}t}$ and the state transition matrix of discrete time system $\mathbf{A}^k$ by taking $f(\mathbf{A}) = \mathbf{A}^k$.

Note : In order to solve $f(\mathbf{A}) = e^{\mathbf{A}t}$, when the eigenvalues are distinct the equations (5.25) and (5.29) can also be obtained by using the sylvester's interpolation formula given below.

$$\begin{vmatrix} 1 & \lambda_1 & \lambda_1^2 & \cdots & \lambda_1^{n-1} & e^{\lambda_1 t} \\ 1 & \lambda_2 & \lambda_2^2 & \cdots & \lambda_2^{n-1} & e^{\lambda_2 t} \\ \vdots & \vdots & \vdots & & \vdots & \vdots \\ 1 & \lambda_n & \lambda_n^2 & \cdots & \lambda_n^{n-1} & e^{\lambda_n t} \\ 1 & \mathbf{A} & \mathbf{A}^2 & \cdots & \mathbf{A}^{n-1} & e^{\mathbf{A}t} \end{vmatrix} = 0$$

EXAMPLE 5.1

Find $f(\mathbf{A}) = \mathbf{A}^7$ for $\mathbf{A} = \begin{bmatrix} 0 & 3 \\ -2 & -5 \end{bmatrix}$

SOLUTION

$$[\lambda\mathbf{I} - \mathbf{A}] = \lambda\begin{bmatrix} 1 & 0 \\ 0 & 1 \end{bmatrix} - \begin{bmatrix} 0 & 3 \\ -2 & -5 \end{bmatrix} = \begin{bmatrix} \lambda & -3 \\ 2 & \lambda+5 \end{bmatrix}$$

$$|\lambda\mathbf{I} - \mathbf{A}| = \begin{vmatrix} \lambda & -3 \\ 2 & \lambda+5 \end{vmatrix} = \lambda(\lambda+5) + 3 \times 2 = \lambda^2 + 5\lambda + 6 = (\lambda+2)(\lambda+3)$$

The characteristic equation is given by

$$|\lambda\mathbf{I} - \mathbf{A}| = 0, \qquad\qquad \therefore (\lambda+2)(\lambda+3) = 0$$

The eigenvalues λ_1, λ_2 are roots of characteristic equation.

$$\therefore \lambda_1 = -2, \qquad \lambda_2 = -3,$$

Given that, $f(\mathbf{A}) = \mathbf{A}^7$, $\qquad \therefore f(\lambda) = \lambda^7$

when $\lambda_i = \lambda_1 = -2$; $\quad f(\lambda_1) = f(-2) = (-2)^7 = -128$ .....(5.1.1)

when $\lambda_i = \lambda_2 = -3$; $\quad f(\lambda_2) = f(-3) = (-3)^7 = -2187$ .....(5.1.2)

We know that , $f(\lambda_i) = \alpha_0 + \alpha_1\lambda_i + \alpha_2\lambda_i^2 + \ldots\ldots + \alpha_{n-1}\lambda_i^{n-1}$

Here $n = 2$, $\qquad \therefore f(\lambda_i) = \alpha_0 + \alpha_1\lambda_i$ .....(5.1.3)

From equations (5.1.1) and (5.1.3) when $\lambda_i = \lambda_1 = -2$, we get,

$$f(\lambda_1) = \alpha_0 + \alpha_1\lambda_1$$

$$-128 = \alpha_0 + \alpha_1(-2)$$

$$\therefore \alpha_0 = 2\alpha_1 - 128$$ .....(5.1.4)

From equations (5.1.2) and (5.1.3) when $\lambda_i = \lambda_2 = -3$, we get,

$$f(\lambda_2) = \alpha_0 + \alpha_1\lambda_2$$

$$-2187 = \alpha_0 + \alpha_1(-3)$$

$$3\alpha_1 = \alpha_0 + 2187$$

$$\therefore \alpha_1 = \frac{\alpha_0}{3} + \frac{2187}{3} = \frac{1}{3}\alpha_0 + 729$$ .....(5.1.5)

On substituting for α_1 from equation (5.1.5) in equation (5.1.4) we get,

$$\alpha_0 = 2\,((1/3)\,\alpha_0 + 729) - 128$$

$$\alpha_0 = (2/3)\,\alpha_0 + 1458 - 128$$

$$\alpha_0 - (2/3)\, \alpha_0 = 1330$$

$$(1/3)\, \alpha_0 = 1330 \qquad ; \qquad \therefore\ \alpha_0 = 3 \times 1330 = 3990$$

On substituting the value of α_0 in equation (5.1.5) we get,

$$\alpha_1 = (1/3)\,(3990) + 729 = 1330 + 729 = 2059$$

We know that, $f(\mathbf{A}) = \alpha_0 \mathbf{I} + \alpha_1 \mathbf{A}$

$$\therefore f(\mathbf{A}) = 3990 \begin{bmatrix} 1 & 0 \\ 0 & 1 \end{bmatrix} + 2059 \begin{bmatrix} 0 & 3 \\ -2 & -5 \end{bmatrix}$$

$$= \begin{bmatrix} 3990 & 0 \\ 0 & 3990 \end{bmatrix} + \begin{bmatrix} 0 & 6177 \\ -4118 & -10295 \end{bmatrix} = \begin{bmatrix} 3990 & 6177 \\ -4118 & -6305 \end{bmatrix}$$

$$\therefore \mathbf{A}^7 = \begin{bmatrix} 3990 & 6177 \\ -4118 & -6305 \end{bmatrix}$$

ALTERNATE METHOD

$$\mathbf{A}^2 = \mathbf{AA} = \begin{bmatrix} 0 & 3 \\ -2 & -5 \end{bmatrix} \begin{bmatrix} 0 & 3 \\ -2 & -5 \end{bmatrix} = \begin{bmatrix} -6 & -15 \\ 10 & 19 \end{bmatrix}$$

$$\mathbf{A}^3 = \mathbf{A}^2\mathbf{A} = \begin{bmatrix} -6 & -15 \\ 10 & 19 \end{bmatrix} \begin{bmatrix} 0 & 3 \\ -2 & -5 \end{bmatrix} = \begin{bmatrix} 30 & 57 \\ -38 & -65 \end{bmatrix}$$

$$\mathbf{A}^4 = \mathbf{A}^3\mathbf{A} = \begin{bmatrix} 30 & 57 \\ -38 & -65 \end{bmatrix} \begin{bmatrix} 0 & 3 \\ -2 & -5 \end{bmatrix} = \begin{bmatrix} -114 & -195 \\ 130 & 211 \end{bmatrix}$$

$$\therefore \mathbf{A}^7 = \mathbf{A}^4\mathbf{A}^3 = \begin{bmatrix} -114 & -195 \\ 130 & 211 \end{bmatrix} \begin{bmatrix} 30 & 57 \\ -38 & -65 \end{bmatrix} = \begin{bmatrix} 3990 & 6177 \\ -4118 & -6305 \end{bmatrix}$$

EXAMPLE 5.2

For $\mathbf{A} = \begin{bmatrix} 0 & 1 \\ -2 & -3 \end{bmatrix}$

Compute the state transition matrix $e^{\mathbf{A}t}$ using Cayley-Hamilton theorem.

SOLUTION

Given that, $\mathbf{A} = \begin{bmatrix} 0 & 1 \\ -2 & -3 \end{bmatrix}$

$$[\lambda\mathbf{I} - \mathbf{A}] = \lambda\begin{bmatrix} 1 & 0 \\ 0 & 1 \end{bmatrix} - \begin{bmatrix} 0 & 1 \\ -2 & -3 \end{bmatrix} = \begin{bmatrix} \lambda & -1 \\ 2 & \lambda+3 \end{bmatrix}$$

$$|\lambda\mathbf{I} - \mathbf{A}| = \begin{vmatrix} \lambda & -1 \\ 2 & \lambda+3 \end{vmatrix} = \lambda(\lambda+3) + 1 \times 2 = \lambda^2 + 3\lambda + 2$$

$$= (\lambda + 1)(\lambda + 2)$$

The characteristic equation is given by

$$|\lambda\mathbf{I} - \mathbf{A}| = 0, \qquad\qquad \therefore (\lambda + 1)(\lambda + 2) = 0$$

The eigenvalues λ_1, λ_2 are roots of characteristic equation.

$$\therefore \lambda_1 = -1, \qquad\qquad \lambda_2 = -2,$$

Let $f(\mathbf{A}) = e^{\mathbf{A}t}$; $\therefore f(\lambda) = e^{\lambda t}$

when $\lambda_i = \lambda_1 = -1$; $f(\lambda_1) = f(-1) = e^{-t}$ .....(5.2.1)

when $\lambda_i = \lambda_2 = -2$; $f(\lambda_2) = f(-2) = e^{-2t}$ .....(5.2.2)

We know that , $f(\lambda_i) = \alpha_0 + \alpha_1 \lambda_i + \alpha_2 \lambda_i^2 + \ldots\ldots + \alpha_{n-1} \lambda_i^{n-1}$

Here n = 2, $\therefore f(\lambda_i) = \alpha_0 + \alpha_1 \lambda_i$ .....(5.2.3)

From equations (5.2.1) and (5.2.3) when $\lambda_i = \lambda_1 = -1$, we get,

$$f(\lambda_1) = \alpha_0 + \alpha_1 \lambda_1$$

$$e^{-t} = \alpha_0 + \alpha_1(-1)$$

$$\therefore \alpha_0 = \alpha_1 + e^{-t} \qquad\qquad .....(5.2.4)$$

From equations (5.2.2) and (5.2.3) when $\lambda_i = \lambda_2 = -2$, we get,

$$f(\lambda_2) = \alpha_0 + \alpha_1 \lambda_2$$

$$e^{-2t} = \alpha_0 + \alpha_1(-2)$$

$$2\alpha_1 = \alpha_0 - e^{-2t}$$

$$\therefore \alpha_1 = \frac{1}{2}\alpha_0 - \frac{1}{2}e^{-2t} \qquad\qquad .....(5.2.5)$$

On substituting for α_1 from equation (5.2.5) in equation (5.2.4) we get,

$$\alpha_0 = \frac{1}{2}\alpha_0 - \frac{1}{2}e^{-2t} + e^{-t}$$

$$\alpha_0 - \frac{1}{2}\alpha_0 = -\frac{1}{2}e^{-2t} + e^{-t}$$

$$\frac{1}{2}\alpha_0 = e^{-t} - \frac{1}{2}e^{-2t}$$

$$\therefore \alpha_0 = 2e^{-t} - e^{-2t}$$

On substituting the value of α_0 in equation (5.2.5) we get,

$$\alpha_1 = \frac{1}{2}(2e^{-t} - e^{-2t}) - \frac{1}{2}e^{-2t}$$

$$\alpha_1 = e^{-t} - \frac{1}{2}e^{-2t} - \frac{1}{2}e^{-2t} = e^{-t} - e^{-2t}$$

By Cayley-Hamilton theorem,

$$f(\mathbf{A}) = \alpha_0 \mathbf{I} + \alpha_1 \mathbf{A}$$

$$\therefore \ f(\mathbf{A}) = (2e^{-t} - e^{-2t})\begin{bmatrix} 1 & 0 \\ 0 & 1 \end{bmatrix} + (e^{-t} - e^{-2t})\begin{bmatrix} 0 & 1 \\ -2 & -3 \end{bmatrix}$$

$$= \begin{bmatrix} 2e^{-t} - e^{-2t} & 0 \\ 0 & 2e^{-t} - e^{-2t} \end{bmatrix} + \begin{bmatrix} 0 & e^{-t} - e^{-2t} \\ -2e^{-t} + 2e^{-2t} & -3e^{-t} + 3e^{-2t} \end{bmatrix}$$

$$= \begin{bmatrix} 2e^{-t} - e^{-2t} & e^{-t} - e^{-2t} \\ -2e^{-t} + 2e^{-2t} & -e^{-t} + 2e^{-2t} \end{bmatrix}$$

$$\therefore \text{State transition matrix, } e^{\mathbf{A}t} = \begin{bmatrix} 2e^{-t} - e^{-2t} & e^{-t} - e^{-2t} \\ -2e^{-t} + 2e^{-2t} & -e^{-t} + 2e^{-2t} \end{bmatrix}$$

EXAMPLE 5.3

The system matrix $\mathbf{A}$ of a discrete time system is given by $\mathbf{A} = \begin{bmatrix} 0 & 1 \\ -2 & -3 \end{bmatrix}$

Compute the state transition matrix $\mathbf{A}^k$ using the Cayley-Hamilton theorem.

SOLUTION

Given that, $\mathbf{A} = \begin{bmatrix} 0 & 1 \\ -2 & -3 \end{bmatrix}$

$$[\lambda \mathbf{I} - \mathbf{A}] = \lambda\begin{bmatrix} 1 & 0 \\ 0 & 1 \end{bmatrix} - \begin{bmatrix} 0 & 1 \\ -2 & -3 \end{bmatrix} = \begin{bmatrix} \lambda & -1 \\ 2 & \lambda + 3 \end{bmatrix}$$

$$|\lambda \mathbf{I} - \mathbf{A}| = \begin{vmatrix} \lambda & -1 \\ 2 & \lambda + 3 \end{vmatrix} = \lambda(\lambda + 3) + 1 \times 2 = \lambda^2 + 3\lambda + 2$$

$$= (\lambda + 1)(\lambda + 2)$$

The characteristic equation is given by

$$|\lambda \mathbf{I} - \mathbf{A}| = 0, \qquad\qquad \therefore (\lambda + 1)(\lambda + 2) = 0$$

The eigenvalues λ_1, λ_2 are roots of characteristic equation.

$$\therefore \lambda_1 = -1, \qquad\qquad \lambda_2 = -2,$$

Let $f(\mathbf{A}) = \mathbf{A}^k \qquad ; \quad \therefore f(\lambda) = \lambda^k$

when $\lambda_i = \lambda_1 = -1$; $f(\lambda_1) = f(-1) = (-1)^k$ .....(5.3.1)

when $\lambda_i = \lambda_2 = -2$; $f(\lambda_2) = f(-2) = (-2)^k$ .....(5.3.2)

We know that, $f(\lambda_i) = \alpha_0 + \alpha_1\lambda_i + \alpha_2\lambda_i^2 + \ldots\ldots + \alpha_{n-1}\lambda_i^{n-1}$

Here n = 2, $\therefore f(\lambda_i) = \alpha_0 + \alpha_1\lambda_i$ .....(5.3.3)

From equations (5.3.1) and (5.3.3) when $\lambda_i = \lambda_1 = -1$, we get,

$$f(\lambda_1) = \alpha_0 + \alpha_1\lambda_1$$

$$(-1)^k = \alpha_0 + \alpha_1(-1)$$

$$\therefore \alpha_0 = \alpha_1 + (-1)^k$$ .....(5.3.4)

From equations (5.3.2) and (5.3.3) when $\lambda_i = \lambda_2 = -2$, we get,

$$f(\lambda_2) = \alpha_0 + \alpha_1\lambda_2$$

$$(-2)^k = \alpha_0 + \alpha_1(-2)$$

$$2\alpha_1 = \alpha_0 - (-2)^k$$

$$\therefore \alpha_1 = \frac{1}{2}\alpha_0 - \frac{1}{2}(-2)^k$$ .....(5.3.5)

On substituting for α_1 from equation (5.3.5) in equation (5.3.4) we get,

$$\alpha_0 = \frac{1}{2}\alpha_0 - \frac{1}{2}(-2)^k + (-1)^k$$

$$\alpha_0 - \frac{1}{2}\alpha_0 = (-1)^k - \frac{1}{2}(-2)^k$$

$$\frac{1}{2}\alpha_0 = (-1)^k - \frac{1}{2}(-2)^k$$

$$\therefore \alpha_0 = 2(-1)^k - (-2)^k$$

On substituting the value of α_0 in equation (5.3.5) we get,

$$\alpha_1 = \frac{1}{2}\left(2(-1)^k - (-2)^k\right) - \frac{1}{2}(-2)^k$$

$$\alpha_1 = (-1)^k - \frac{1}{2}(-2)^k - \frac{1}{2}(-2)^k$$

$$\therefore \alpha_1 = (-1)^k - (-2)^k$$

By Cayley-Hamilton theorem,

$$f(A) = \alpha_0 I + \alpha_1 A$$

$$\therefore f(A) = \left(2(-1)^k - (-2)^k\right)\begin{bmatrix} 1 & 0 \\ 0 & 1 \end{bmatrix} + \left((-1)^k - (-2)^k\right)\begin{bmatrix} 0 & 1 \\ -2 & -3 \end{bmatrix}$$

$$= \begin{bmatrix} 2(-1)^k - (-2)^k & 0 \\ 0 & 2(-1)^k - (-2)^k \end{bmatrix} + \begin{bmatrix} 0 & (-1)^k - (-2)^k \\ -2(-1)^k + 2(-2)^k & -3(-1)^k + 3(-2)^k \end{bmatrix}$$

$$= \begin{bmatrix} 2(-1)^k - (-2)^k & (-1)^k - (-2)^k \\ -2(-1)^k + 2(-2)^k & -(-1)^k + 2(-2)^k \end{bmatrix}$$

$$\left.\begin{array}{l} \therefore \text{ State transition matrix} \\ \text{of discreten time system} \end{array}\right\} \mathbf{A}^k = \begin{bmatrix} 2(-1)^k - (-2)^k & (-1)^k - (-2)^k \\ -2(-1)^k + 2(-2)^k & -(-1)^k + 2(-2)^k \end{bmatrix}$$

5.5   TRANSFORMATION OF STATE MODEL

The state model of a system is not unique and it can be formed using physical variables, phase variables or canonical variables. The physical variables are useful from application point of view, because they can be measured and used for control purposes. However, the state model using physical variables is not convenient for investigation of system properties and evaluation of time response. But the canonical state model is most convenient for time domain analysis. In canonical model the system matrix $\mathbf{A}$ will be a diagonal matrix. Therefore each component state variable equation is a first order equation and is decoupled from all other component state variable equation.

When a non-diagonal system matrix $\mathbf{A}$ has distinct eigenvalues, it can be converted to diagonal matrix by a similarity transformation using modal matrix, $\mathbf{M}$. Due to this the state model is transformed to canonical form.

When a non-diagonal system matrix has multiple eigenvalues, it can be converted to Jordan matrix by a similarity transformation using modal matrix, $\mathbf{M}$. Due to this the state model is transformed to Jordan canonical form.

CANONICAL FORM OF STATE MODEL

Consider the state equation of a system, $\dot{\mathbf{X}} = \mathbf{AX} + \mathbf{BU}$, where $\mathbf{X}$ is the state variable vector of order $n \times 1$. Let us define a new state variable vector $\mathbf{Z}$, such that $\mathbf{X} = \mathbf{MZ}$, where $\mathbf{M}$ is the Modal matrix or Diagonalization matrix.

The state model of the nth order system is given by

$$\dot{\mathbf{X}} = \mathbf{AX} + \mathbf{BU}$$

$$\mathbf{Y} = \mathbf{CX} + \mathbf{DU}$$

On substituting $\mathbf{X} = \mathbf{MZ}$ in the state model of the system, we get

$$\dot{\mathbf{X}} = \mathbf{AMZ} + \mathbf{BU} \qquad\qquad .....(5.31)$$

$$\mathbf{Y} = \mathbf{CMZ} + \mathbf{DU} \qquad\qquad .....(5.32)$$

Premultiply equation (5.31) by $\mathbf{M}^{-1}$

$$\therefore \mathbf{M}^{-1}\mathbf{X} = \mathbf{M}^{-1}\mathbf{AMZ} + \mathbf{M}^{-1}\mathbf{BU} \qquad\qquad .....(5.33)$$

The relation governing $\mathbf{X}$ and $\mathbf{Z}$ is, $\mathbf{X} = \mathbf{MZ}$. $\qquad\qquad .....(5.34)$

On differentiating equation (5.34), we get, $\dot{\mathbf{X}} = \mathbf{M}\dot{\mathbf{Z}}$. $\qquad\qquad .....(5.35)$

On premultiplying the equation (5.35) by $\mathbf{M}^{-1}$ we get

$$\mathbf{M}^{-1}\mathbf{X} = \dot{\mathbf{Z}} \qquad\qquad .....(5.36)$$

From equations (5.33) and (5.36), we get,

$$\dot{\mathbf{Z}} = \mathbf{M}^{-1}\mathbf{AMZ} + \mathbf{M}^{-1}\mathbf{BU} \qquad\qquad .....(5.37)$$

Let, $M^{-1}AM = \Lambda$ (called grammian matrix) .....(5.38)

$$M^{-1}B = \tilde{B}$$ .....(5.39)

$$CM = \tilde{C}$$ .....(5.40)

From equations (5.32) and (5.37) to (5.40) the transformed state model is obtained as shown below.

$$\dot{Z} = \Lambda Z + \tilde{B}U$$ .....(5.41)

$$Y = \tilde{C}Z + DU$$ .....(5.42)

In the transformed state model [equations (5.41) and (5.42)] the grammian matrix Λ, will be a diagonal matrix and the transformed state model is called canonical form of state model. The model matrix M is obtained from Eigenvectors. When the system matrix A is in the companion or Bush form then the modal matrix is given by a special matrix called vander monde matrix, V, shown below.

i.e., If $A = \begin{bmatrix} 0 & 1 & 0 & \cdots & 0 \\ 0 & 0 & 1 & \cdots & 0 \\ \vdots & \vdots & \vdots & & \vdots \\ 0 & 0 & 0 & \cdots & 1 \\ -a_n & -a_{n-1} & -a_{n-2} & \cdots & -a_1 \end{bmatrix}$ then $M = V = \begin{bmatrix} 1 & 1 & \cdots\cdots & 1 \\ \lambda_1 & \lambda_2 & \cdots\cdots & \lambda_n \\ \lambda_1^2 & \lambda_2^2 & & \lambda_n^2 \\ \vdots & & & \\ \lambda_1^{n-1} & \lambda_1^{n-1} & \cdots\cdots & \lambda_n^{n-1} \end{bmatrix}$

Note : *When the state model is available in canonical form then $M = M^{-1} = I$.*

JORDAN CANONICAL FORM OF STATE MODEL

If the eigenvalues has multiplicity then the system matrix cannot be diagonalized. In this case, the transformation as explained in previous section will give a Jordan matrix, J where $J = M^{-1}AM$. The transformed state model is called Jordan canonical form and it is given by equation (5.43) and (5.44).

$$\dot{Z} = JZ + \tilde{B}U \quad .....(5.43)$$

$$Y = \tilde{C}Z + DU \quad .....(5.44)$$

where $J = M^{-1}AM$
$\tilde{B} = M^{-1}B$
$\tilde{C} = CM$

The Jordan matrix J will have a Jordan block of size $q \times q$ correspond to a eigenvalue of λ_i with multiplicity q. In the Jordan block the diagonal elements will be eigenvalue and the element just above the diagonal is one.

Consider a system matrix with eigenvalues $\lambda_1, \lambda_1, \lambda_1, \lambda_4, \lambda_5........\lambda_n$ where λ_1 has multiplicity of three. The Jordan matrix for this case will be as shown in equation (5.45)

$$J = \begin{bmatrix} \lambda_1 & 1 & 0 & 0 & \cdots & 0 \\ 0 & \lambda_1 & 1 & 0 & \cdots & 0 \\ 0 & 0 & \lambda_1 & 0 & \cdots & 0 \\ 0 & 0 & 0 & \lambda_4 & \cdots & 0 \\ \vdots & \vdots & \vdots & \vdots & & \vdots \\ 0 & 0 & 0 & 0 & \cdots & \lambda_n \end{bmatrix}$$ ——— Jordan Block of size 3×3

.....(5.45)

If the system matrix **A** is in the companion form and has multiple eigen values, then the modal matrix is given by modified vander monde matrix shown in equation (5.46).

Let λ_1 has multiplicity of q and the other eigen values are $\lambda_{q+1}, \lambda_{q+2}, \ldots \lambda_n$.

$$\mathbf{M} = \mathbf{V} = \begin{bmatrix} 1 & 0 & 0 & \cdots & 1 & \cdots & 1 \\ \lambda_1 & 1 & 0 & \cdots & \lambda_{q+1} & \cdots & \lambda_n \\ \lambda_1^2 & 2\lambda_1 & 1 & \cdots & \lambda_{q+1}^2 & \cdots & \lambda_n^2 \\ \lambda_1^3 & 3\lambda_1^2 & 3\lambda_1 & \cdots & \lambda_{q+1}^3 & \cdots & \lambda_n^3 \\ \vdots & \vdots & \vdots & & \vdots & \cdots & \vdots \\ \vdots & \vdots & \vdots & & \vdots & \cdots & \vdots \\ \lambda_1^{(n-1)} & \dfrac{d(\lambda_1^{n-1})}{d\lambda_1} & \dfrac{1}{2!}\dfrac{d^2(\lambda_1^{n-1})}{d\lambda_1^2} & \cdots & \lambda_{q+1}^{n-1} & \cdots & \lambda_n^{n-1} \end{bmatrix} \quad \ldots(5.46)$$

COMPUTATION OF STATE TRANSITION MATRIX BY CANONICAL TRANSFORMATION

Consider the state equation without input,

$$\dot{\mathbf{X}}(t) = \mathbf{A}\,\mathbf{X}(t) \qquad \ldots(5.47)$$

The solution of state equation is

$$\mathbf{X}(t) = e^{\mathbf{A}t}\,\mathbf{X}(0) \qquad \ldots(5.48)$$

Let us assume that the system matrix **A** is non-diagonal and its eigen values are distinct. Now the system matrix **A** can be diagonalized by a similarity transformation using modal matrix, **M**. Due to this transformation we get a new state vector **Z**(t). The relation governing the old and new state vector is $\mathbf{X}(t) = \mathbf{M}\,\mathbf{Z}(t)$. Also, the state equation (5.47) modifies to the form shown in equation (5.49).

$$\dot{\mathbf{Z}}(t) = \mathbf{M}^{-1}\mathbf{A}\,\mathbf{M}\,\mathbf{Z}(t) = \Lambda\,\mathbf{Z}(t) \qquad \ldots(5.49)$$

$$\text{where, } \Lambda = \mathbf{M}^{-1}\mathbf{A}\mathbf{M} = \begin{bmatrix} \lambda_1 & 0 & 0 & \cdots & 0 \\ 0 & \lambda_2 & 0 & \cdots & 0 \\ \vdots & \vdots & \vdots & & \vdots \\ 0 & 0 & 0 & \cdots & \lambda_n \end{bmatrix}$$

The solution of the modified state equation (equation (5.49)) is given by equation (5.50).

$$\mathbf{Z}(t) = e^{\Lambda t}\,\mathbf{Z}(0) \qquad \ldots(5.50)$$

The matrix $e^{\Lambda t}$ can be expressed as an infinite series shown below.

$$e^{\Lambda t} = \mathbf{I} + \Lambda t + \frac{1}{2!}\Lambda^2 t^2 + \frac{1}{3!}\Lambda^3 t^3 + \ldots\ldots$$

$$e^{\Lambda t} = \begin{bmatrix} 1 & 0 & \cdots & 0 \\ 0 & 1 & \cdots & 0 \\ \vdots & \vdots & & \vdots \\ 0 & 0 & \cdots & 1 \end{bmatrix} + \begin{bmatrix} \lambda_1 & 0 & \cdots & 0 \\ 0 & \lambda_2 & \cdots & 0 \\ \vdots & \vdots & & \vdots \\ 0 & 0 & \cdots & \lambda_n \end{bmatrix} t + \frac{1}{2!}\begin{bmatrix} \lambda_1^2 & 0 & \cdots & 0 \\ 0 & \lambda_2^2 & \cdots & 0 \\ \vdots & \vdots & & \vdots \\ 0 & 0 & \cdots & \lambda_n^2 \end{bmatrix} t^2 + \frac{1}{3!}\begin{bmatrix} \lambda_1^3 & 0 & \cdots & 0 \\ 0 & \lambda_2^3 & \cdots & 0 \\ \vdots & \vdots & & \vdots \\ 0 & 0 & \cdots & \lambda_n^3 \end{bmatrix} t^3 + \ldots\ldots$$

$$e^{\Lambda t} = \begin{bmatrix} 1+\lambda_1 t+\frac{1}{2!}\lambda_1^2 t^2+\frac{1}{3!}\lambda_1^3 t^3+..... & & ....0 \\ \vdots & 1+\lambda_2 t+\frac{1}{2!}\lambda_2^2 t^2+\frac{1}{3!}\lambda_2^3 t^3+..... & \\ 0 & & 1+\lambda_n t+\frac{1}{2!}\lambda_n^2 t^2+\frac{1}{3!}\lambda_n^3 t^3+..... \end{bmatrix}$$

$$\therefore e^{\Lambda t} = \begin{bmatrix} e^{\lambda_1 t} & 0 & 0 \\ 0 & e^{\lambda_2 t} & 0 \\ \vdots & \vdots & \vdots \\ 0 & 0 & e^{\lambda_n t} \end{bmatrix} \qquad \qquad .....(5.51)$$

We know that, $\mathbf{M}\ \mathbf{Z}(t) = \mathbf{X}(t)$ .....(5.52)

On premultiplying the equation (5.52) by $\mathbf{M}^{-1}$ we get,

$\mathbf{Z}(t) = \mathbf{M}^{-1}\ \mathbf{X}(t)$ .....(5.53)

From equation (5.53) when $t = 0$, we get, $\mathbf{Z}(0) = \mathbf{M}^{-1}\ \mathbf{X}(0)$ .....(5.54)

Using equations (5.53) and (5.54) the equation (5.50) can be written as,

$\mathbf{M}^{-1}\ \mathbf{X}(t) = e^{\Lambda t}\ \mathbf{M}^{-1}\ \mathbf{X}(0)$ .....(5.55)

On premultiplying, the equation (5.55) by M we get,

$\mathbf{X}(t) = \mathbf{M}\ e^{\Lambda t}\ \mathbf{M}^{-1}\ \mathbf{X}(0)$ .....(5.56)

On comparing equations (5.48) and (5.56) we get,

The state transition matrix, $e^{\mathbf{A}t} = \mathbf{M}\ e^{\Lambda t}\ \mathbf{M}^{-1}$ .....(5.57)

When the eigenvalues are distinct the equation (5.57) can be used to compute the state transition matrix $e^{\mathbf{A}t}$.

When the eigenvalues have multiplicity the system matrix cannot be diagonalized but can be transformed to Jordan matrix. When one of the eigenvalues λ_1 repeats q times the solution of state equation and state transition matrix are given by equations (5.58) and (5.59) respectively.

$\mathbf{X}(t) = \mathbf{M}\ \mathbf{Q}(t)\ e^{\Lambda t}\ \mathbf{M}^{-1}\ \mathbf{X}(0)$ .....(5.58)

$e^{\Lambda t} = \mathbf{M}\ \mathbf{Q}(t)\ e^{\Lambda t}\ \mathbf{M}^{-1}$ .....(5.59)

$$\text{where, } Q(t) = \begin{bmatrix} 1 & t & \frac{1}{2!}t^2 & \cdots & \frac{1}{(q-1)!}t^{q-1} & 0 & \cdots & 0 \\ 0 & 1 & t & \cdots & \frac{1}{(q-2)!}t^{q-2} & 0 & \cdots & 0 \\ 0 & 0 & 1 & \cdots & \frac{1}{(q-3)!}t^{q-3} & 0 & \cdots & 0 \\ \vdots & \vdots & \vdots & & \vdots & \vdots & & \vdots \\ 0 & 0 & 0 & \cdots & t & 0 & \cdots & 0 \\ 0 & 0 & 0 & \cdots & 1 & 0 & \cdots & 0 \\ 0 & 0 & 0 & \cdots & 0 & 1 & \cdots & 0 \\ \vdots & \vdots & \vdots & & \vdots & \vdots & & \vdots \\ 0 & 0 & 0 & \cdots & 0 & 0 & \cdots & 1 \end{bmatrix}$$ .....(5.60)

> Note : In this case the matrix e^{At} is same as equation (5.51)

The state transition matrix of discrete time system can be computed by an analysis similar to that discussed above.

When the eigenvalues are distinct the transition matrix of discrte time system is given by equ (5.61).

$$A^k = M \wedge^k M^{-1}$$ .....(5.61)

$$\text{where, } \wedge^k = \begin{bmatrix} \lambda_1^k & 0 & \cdots & 0 \\ 0 & \lambda_2^k & \cdots & 0 \\ \vdots & \vdots & & \vdots \\ 0 & 0 & & \lambda_n^k \end{bmatrix}$$ .....(5.62)

When one of the eigenvalues $\lambda 1$ has a multiplicity of q, the state transition matrix of discrete time system is given by equ (5.63).

$$A^k = M \, Q(k) \wedge^k M^{-1}$$ .....(5.63)

where Q(k) is obtained from Q(t) [equ (5.60)] after replacing t by kT and $\wedge^k$ is given by equ (5.62).

EXAMPLE 5.4

A linear time invariant system is described by the following state model.

$$\begin{bmatrix} \dot{x}_1 \\ \dot{x}_2 \\ \dot{x}_3 \end{bmatrix} = \begin{bmatrix} 0 & 1 & 0 \\ 0 & 0 & 1 \\ -6 & -11 & -6 \end{bmatrix} \begin{bmatrix} x_1 \\ x_2 \\ x_3 \end{bmatrix} + \begin{bmatrix} 0 \\ 0 \\ 2 \end{bmatrix} [u] \quad \text{and} \quad y = \begin{bmatrix} 1 & 0 & 0 \end{bmatrix} \begin{bmatrix} x_1 \\ x_2 \\ x_3 \end{bmatrix}$$

Transform this state model into a canonical state model. Also compute the state transition matrix, e^{At}.

SOLUTION

To find eigenvalues

$$(\lambda I - A) = \lambda \begin{bmatrix} 1 & 0 & 0 \\ 0 & 1 & 0 \\ 0 & 0 & 1 \end{bmatrix} - \begin{bmatrix} 0 & 1 & 0 \\ 0 & 0 & 1 \\ -6 & -11 & -6 \end{bmatrix}$$

$$= \begin{bmatrix} \lambda & 0 & 0 \\ 0 & \lambda & 0 \\ 0 & 0 & \lambda \end{bmatrix} - \begin{bmatrix} 0 & 1 & 0 \\ 0 & 0 & 1 \\ -6 & -11 & -6 \end{bmatrix} = \begin{bmatrix} \lambda & -1 & 0 \\ 0 & \lambda & -1 \\ 6 & 11 & \lambda+6 \end{bmatrix}$$

The characteristic equation is $|\lambda I - A| = 0$

$$|\lambda I - A| = \begin{vmatrix} \lambda & -1 & 0 \\ 0 & \lambda & -1 \\ 6 & 11 & \lambda+6 \end{vmatrix} = \lambda[\lambda(\lambda+6)+11] + 1 \times 6 = 0$$

$\therefore \lambda(\lambda^2 + 6\lambda + 11) + 6 = 0$

$\lambda^3 + 6\lambda^2 + 11\lambda + 6 = 0$

$\lambda = -1$ is one of the root of the above equation,

$\lambda = -1$	1	6	11	6
$\downarrow$		-1	-5	-6
	1	5	6	0

$\therefore (\lambda + 1)(\lambda^2 + 5\lambda + 6) = 0$

$(\lambda + 1)(\lambda + 2)(\lambda + 3) = 0$

The eigenvalues are $\lambda = -1, -2, -3$; i.e., $\lambda_1 = -1, \lambda_2 = -2$, and $\lambda_3 = -3$.

To find modal matrix

The system matrix is given in the companion form and therefore modal matrix, **M** is given by vander monde matrix, **V**.

$$M = V = \begin{bmatrix} 1 & 1 & 1 \\ \lambda_1 & \lambda_2 & \lambda_3 \\ \lambda_1^2 & \lambda_2^2 & \lambda_3^2 \end{bmatrix} = \begin{bmatrix} 1 & 1 & 1 \\ -1 & -2 & -3 \\ 1 & 4 & 9 \end{bmatrix}$$

Alternatively modal matrix can be found from eigenvectors. The alternate method is shown below.

$$(\lambda_1 I - A) = (-1)\begin{bmatrix} 1 & 0 & 0 \\ 0 & 1 & 0 \\ 0 & 0 & 1 \end{bmatrix} - \begin{bmatrix} 0 & 1 & 0 \\ 0 & 0 & 1 \\ -6 & -11 & -6 \end{bmatrix} = \begin{bmatrix} -1 & -1 & 0 \\ 0 & -1 & -1 \\ 6 & 11 & 5 \end{bmatrix}$$

Let the cofactors of $[\lambda_1 I - A]$ along Ist row be C_{11}, C_{12} and C_{13}

$$C_{11} = (+1)\begin{vmatrix} -1 & -1 \\ 11 & 5 \end{vmatrix} = 6 \ ; \ C_{12} = (-1)\begin{vmatrix} 0 & -1 \\ 6 & 5 \end{vmatrix} = -6 \ ; \ C_{13} = (+1)\begin{vmatrix} 0 & -1 \\ 6 & 11 \end{vmatrix} = 6$$

$\therefore$ Eigenvector corresponding to $\lambda_1 = m_1 = \begin{bmatrix} C_{11} \\ C_{12} \\ C_{13} \end{bmatrix} = \begin{bmatrix} 6 \\ -6 \\ 6 \end{bmatrix} = \begin{bmatrix} 1 \\ -1 \\ 1 \end{bmatrix}$

> **Note : The elements of eigenvector can be divided by a constant.**

$$(\lambda_2 I - A) = (-2)\begin{bmatrix} 1 & 0 & 0 \\ 0 & 1 & 0 \\ 0 & 0 & 1 \end{bmatrix} - \begin{bmatrix} 0 & 1 & 0 \\ 0 & 0 & 1 \\ -6 & -11 & -6 \end{bmatrix} = \begin{bmatrix} -2 & -1 & 0 \\ 0 & -2 & -1 \\ 6 & 11 & 4 \end{bmatrix}$$

Let the cofactors of $[\lambda_2 I - A]$ along Ist row be C_{11}, C_{12} and C_{13}

$$\therefore\ C_{11} = (+1)\begin{vmatrix} -2 & -1 \\ 11 & 4 \end{vmatrix} = 3\ ;\ C_{12} = (-1)\begin{vmatrix} 0 & -1 \\ 6 & 4 \end{vmatrix} = -6\ ;\ C_{13} = (+1)\begin{vmatrix} 0 & -2 \\ 6 & 11 \end{vmatrix} = 12$$

$$\therefore\ \text{Eigenvector corresponding to } \lambda_2 = \mathbf{m}_2 = \begin{bmatrix} C_{11} \\ C_{12} \\ C_{13} \end{bmatrix} = \begin{bmatrix} 3 \\ -6 \\ 12 \end{bmatrix} = \begin{bmatrix} 1 \\ -2 \\ 4 \end{bmatrix}$$

$$(\lambda_3 I - A) = (-3)\begin{bmatrix} 1 & 0 & 0 \\ 0 & 1 & 0 \\ 0 & 0 & 1 \end{bmatrix} - \begin{bmatrix} 0 & 1 & 0 \\ 0 & 0 & 1 \\ -6 & -11 & -6 \end{bmatrix} = \begin{bmatrix} -3 & -1 & 0 \\ 0 & -3 & -1 \\ 6 & 11 & 3 \end{bmatrix}$$

Let the cofactors of $[\lambda_3 I - A]$ along Ist row be C_{11}, C_{12} and C_{13}

$$C_{11} = (+1)\begin{vmatrix} -3 & -1 \\ 11 & 3 \end{vmatrix} = 2\ ;\ C_{12} = (-1)\begin{vmatrix} 0 & -1 \\ 6 & 3 \end{vmatrix} = -6\ ;\ C_{13} = (+1)\begin{vmatrix} 0 & -3 \\ 6 & 11 \end{vmatrix} = 18$$

$$\therefore\ \text{Eigenvector corresponding to } \lambda_3 = \mathbf{m}_3 = \begin{bmatrix} C_{11} \\ C_{12} \\ C_{13} \end{bmatrix} = \begin{bmatrix} 2 \\ -6 \\ 18 \end{bmatrix} = \begin{bmatrix} 1 \\ -3 \\ 9 \end{bmatrix}$$

$$\text{The modal matrix, } \mathbf{M} = [\mathbf{m}_1 \quad \mathbf{m}_2 \quad \mathbf{m}_3] = \begin{bmatrix} 1 & 1 & 1 \\ -1 & -2 & -3 \\ 1 & 4 & 9 \end{bmatrix}$$

It is observed that the modal matrix obtained from vander monde matrix and from eigenvectors are same.

To find M^{-1}

$$\mathbf{M}^{-1} = \frac{(\text{Cofactor of } \mathbf{M})^T}{\text{Determinant of } \mathbf{M}} = \frac{\mathbf{M}^T_{cof}}{\Delta_M}$$

$$\Delta_M = \begin{vmatrix} 1 & 1 & 1 \\ -1 & -2 & -3 \\ 1 & 4 & 9 \end{vmatrix} = 1(-18+12) - 1(-9+3) + 1(-4+2) = -2$$

$$\mathbf{M}^T_{cof} = \begin{bmatrix} -6 & 6 & -2 \\ -5 & 8 & -3 \\ -1 & 2 & -1 \end{bmatrix}^T = \begin{bmatrix} -6 & -5 & -1 \\ 6 & 8 & 2 \\ -2 & -3 & -1 \end{bmatrix} \qquad \therefore\ \mathbf{M}^{-1} = \frac{1}{-2}\begin{bmatrix} -6 & -5 & -1 \\ 6 & 8 & 2 \\ -2 & -3 & -1 \end{bmatrix}$$

To find canonical form of state model

$$\left.\begin{array}{l}\text{Grammian matrix} \\ \Lambda = \mathbf{M}^{-1}\mathbf{A}\mathbf{M}\end{array}\right\} = \frac{1}{-2}\begin{bmatrix} -6 & -5 & -1 \\ 6 & 8 & 2 \\ -2 & -3 & -1 \end{bmatrix}\begin{bmatrix} 0 & 1 & 0 \\ 0 & 0 & 1 \\ -6 & -11 & -6 \end{bmatrix}\begin{bmatrix} 1 & 1 & 1 \\ -1 & -2 & -3 \\ 1 & 4 & 9 \end{bmatrix}$$

$$= \frac{1}{-2}\begin{bmatrix} 6 & 5 & 1 \\ -12 & -16 & -4 \\ 6 & 9 & 3 \end{bmatrix}\begin{bmatrix} 1 & 1 & 1 \\ -1 & -2 & -3 \\ 1 & 4 & 9 \end{bmatrix} = \begin{bmatrix} -1 & 0 & 0 \\ 0 & -2 & 0 \\ 0 & 0 & -3 \end{bmatrix}$$

$$\tilde{B} = M^{-1}B = -\frac{1}{2}\begin{bmatrix} -6 & -5 & -1 \\ 6 & 8 & 2 \\ -2 & -3 & -1 \end{bmatrix}\begin{bmatrix} 0 \\ 0 \\ 2 \end{bmatrix} = \begin{bmatrix} 1 \\ -2 \\ 1 \end{bmatrix}$$

$$\tilde{C} = CM = \begin{bmatrix} 1 & 0 & 0 \end{bmatrix}\begin{bmatrix} 1 & 1 & 1 \\ -1 & -2 & -3 \\ 1 & 4 & 9 \end{bmatrix} = \begin{bmatrix} 1 & 1 & 1 \end{bmatrix}$$

The canonical form of state model is,

$$\dot{Z} = \Lambda Z + \tilde{B}U$$

$$Y = \tilde{C}Z + DU \qquad \text{(Here } DU \text{ is not defined)}$$

$$\begin{bmatrix} \dot{z}_1 \\ \dot{z}_2 \\ \dot{z}_3 \end{bmatrix} = \begin{bmatrix} -1 & 0 & 0 \\ 0 & -2 & 0 \\ 0 & 0 & -3 \end{bmatrix}\begin{bmatrix} z_1 \\ z_2 \\ z_3 \end{bmatrix} + \begin{bmatrix} 1 \\ -2 \\ 1 \end{bmatrix}[u] \quad ; \quad Y = \begin{bmatrix} 1 & 1 & 1 \end{bmatrix}\begin{bmatrix} z_1 \\ z_2 \\ z_3 \end{bmatrix}$$

To compute state transition matrix, $e^{\Lambda t}$

The state transition matrix, $e^{\Lambda t} = M\, e^{\Lambda t}\, M^{-1}$

$$e^{\Lambda} = \begin{bmatrix} e^{\lambda_1 t} & 0 & 0 \\ 0 & e^{\lambda_2 t} & 0 \\ 0 & 0 & e^{\lambda_3 t} \end{bmatrix} = \begin{bmatrix} e^{-t} & 0 & 0 \\ 0 & e^{-2t} & 0 \\ 0 & 0 & e^{-3t} \end{bmatrix}$$

$$\therefore e^{At} = \begin{bmatrix} 1 & 1 & 1 \\ -1 & -2 & -3 \\ 1 & 4 & 9 \end{bmatrix}\begin{bmatrix} e^{-t} & 0 & 0 \\ 0 & e^{-2t} & 0 \\ 0 & 0 & e^{-3t} \end{bmatrix}\left(-\frac{1}{2}\right)\begin{bmatrix} -6 & -5 & -1 \\ 6 & 8 & 2 \\ -2 & -3 & -1 \end{bmatrix}$$

$$= \begin{bmatrix} e^{-t} & e^{-2t} & e^{-3t} \\ -e^{-t} & -2e^{-2t} & -3e^{-3t} \\ e^{-t} & 4e^{-2t} & 9e^{-3t} \end{bmatrix}\begin{bmatrix} 3 & 2.5 & 0.5 \\ -3 & -4 & -1 \\ 1 & 1.5 & 0.5 \end{bmatrix}$$

$$= \begin{bmatrix} 3e^{-t} - 3e^{-2t} + e^{-3t} & 2.5e^{-t} - 4e^{-2t} + 1.5e^{-3t} & 0.5e^{-t} - e^{-2t} + 0.5e^{-3t} \\ -3e^{-t} + 6e^{-2t} - 3e^{-3t} & -2.5e^{-t} + 8e^{-2t} - 4.5e^{-3t} & -0.5e^{-t} + 2e^{-2t} - 1.5e^{-3t} \\ 3e^{-t} - 12e^{-2t} + 9e^{-3t} & 2.5e^{-t} - 16e^{-2t} + 13.5e^{-3t} & 0.5e^{-t} - 4e^{-2t} + 4.5e^{-3t} \end{bmatrix}$$

EXAMPLE 5.5

Convert the following system matrix to canonical form and hence calculate the state transition matrix e^{At}

$$A = \begin{bmatrix} 4 & 1 & -2 \\ 1 & 0 & 2 \\ 1 & -1 & 3 \end{bmatrix}$$

SOLUTION

To find eigenvalues

$$(\lambda \mathbf{I} - \mathbf{A}) = \lambda \begin{bmatrix} 1 & 0 & 0 \\ 0 & 1 & 0 \\ 0 & 0 & 1 \end{bmatrix} - \begin{bmatrix} 4 & 1 & -2 \\ 1 & 0 & 2 \\ 1 & -1 & 3 \end{bmatrix} = \begin{bmatrix} \lambda-4 & -1 & 2 \\ -1 & \lambda & -2 \\ -1 & 1 & \lambda-3 \end{bmatrix}$$

$$|\lambda \mathbf{I} - \mathbf{A}| = \begin{vmatrix} \lambda-4 & -1 & 2 \\ -1 & \lambda & -2 \\ -1 & 1 & -3 \end{vmatrix} = (\lambda-4)\left[\lambda(\lambda-3)+2\right]+1(-\lambda+3-2)+2(-1+\lambda)$$

$$= (\lambda-4)(\lambda^2-3\lambda+2)-\lambda+1-2+2\lambda = (\lambda-4)(\lambda-1)(\lambda-2)+(\lambda-1)$$

$$= (\lambda-1)(\lambda-4)(\lambda-2)+1] = (\lambda-1)[\lambda^2-6\lambda+8+1]$$

$$= (\lambda-1)(\lambda^2-6\lambda+9) = (\lambda-1)(\lambda-3)^2$$

The eigenvalues are $\lambda_1 = 1, \lambda_2 = 3, \lambda_3 = 3$. Since one of the eigenvalue has a multiplicity of 2 (repeats two times) the canonical form will be a Jordan canonical form.

To find eigenvectors

$$[\lambda_1 \mathbf{I} - \mathbf{A}] = (1)\begin{bmatrix} 1 & 0 & 0 \\ 0 & 1 & 0 \\ 0 & 0 & 1 \end{bmatrix} - \begin{bmatrix} 4 & 1 & -2 \\ 1 & 0 & 2 \\ 1 & -1 & 3 \end{bmatrix} = \begin{bmatrix} -3 & -1 & 2 \\ -1 & 1 & -2 \\ -1 & 1 & -2 \end{bmatrix}.$$

Let the cofactors of $[\lambda_1 \mathbf{I} - \mathbf{A}]$ along Ist row be C_{11}, C_{12} and C_{13}.

$$C_{11} = (+1)\begin{vmatrix} 1 & -2 \\ 1 & -2 \end{vmatrix} = 0 \ ; \ C_{12} = (-1)\begin{vmatrix} -1 & -2 \\ -1 & -2 \end{vmatrix} = 0 \ ; \ C_{13} = (+1)\begin{vmatrix} -1 & 1 \\ -1 & 1 \end{vmatrix} = 0$$

$$\therefore \ \mathbf{m}_1 = \begin{bmatrix} C_{11} \\ C_{12} \\ C_{13} \end{bmatrix} = \begin{bmatrix} 0 \\ 0 \\ 0 \end{bmatrix} \quad \text{The cofactors along 1st row gives null solution.}$$

Let C_{21}, C_{22} and C_{23} be cofactors of $[\lambda_1 \mathbf{I} - \mathbf{A}]$ along IInd row.

$$C_{21} = (-1)\begin{vmatrix} -1 & 2 \\ 1 & -2 \end{vmatrix} = 0 \ ; \ C_{22} = (+1)\begin{vmatrix} -3 & 2 \\ -1 & -2 \end{vmatrix} = 8 \ ; \ C_{23} = (-1)\begin{vmatrix} -3 & -1 \\ -1 & 1 \end{vmatrix} = 4$$

$$\therefore \ \mathbf{m}_1 = \begin{bmatrix} C_{21} \\ C_{22} \\ C_{23} \end{bmatrix} = \begin{bmatrix} 0 \\ 8 \\ 4 \end{bmatrix}$$

$$(\lambda_2 \mathbf{I} - \mathbf{A}) = \lambda_2 \begin{bmatrix} 1 & 0 & 0 \\ 0 & 1 & 0 \\ 0 & 0 & 1 \end{bmatrix} - \begin{bmatrix} 4 & 1 & -2 \\ 1 & 0 & 2 \\ 1 & -1 & 3 \end{bmatrix} = \begin{bmatrix} \lambda_2-4 & -1 & 2 \\ -1 & \lambda_2 & -2 \\ -1 & 1 & \lambda_2-3 \end{bmatrix}$$

The cofactor of $[\lambda_2 I - A]$ along Ist row be C_{11}, C_{12} and C_{13}.

$$C_{11} = (+1) \begin{vmatrix} \lambda_2 & -2 \\ 1 & \lambda_2 - 3 \end{vmatrix} = \lambda_2(\lambda_2 - 3) + 2 = \lambda_2^2 - 3\lambda_2 + 2$$

$$C_{12} = (-1) \begin{vmatrix} -1 & -2 \\ -1 & \lambda_2 - 3 \end{vmatrix} = (-1)[(-1)(\lambda_2 - 3) - 2] = \lambda_2 - 1$$

$$C_{13} = (+1) \begin{vmatrix} -1 & \lambda_2 \\ -1 & 1 \end{vmatrix} = -1 + \lambda_2 = \lambda_2 - 1$$

Let $\mathbf{m}_2$ be the independent eigenvector corresponding to $\lambda_2 = 3$

$$\text{Now, } \mathbf{m}_2 = \begin{bmatrix} C_{11} \\ C_{12} \\ C_{13} \end{bmatrix} = \begin{bmatrix} \lambda_2^2 - 3\lambda_2 + 2 \\ \lambda_2 - 1 \\ \lambda_2 - 1 \end{bmatrix} = \begin{bmatrix} 3^2 - (3 \times 3) + 2 \\ 3 - 1 \\ 3 - 1 \end{bmatrix} = \begin{bmatrix} 2 \\ 2 \\ 2 \end{bmatrix}$$

The eigenvector $\mathbf{m}_3$ is given by

$$\mathbf{m}_3 = \begin{bmatrix} \dfrac{d}{d\lambda_2} C_{11} \\[2mm] \dfrac{d}{d\lambda_2} C_{12} \\[2mm] \dfrac{d}{d\lambda_2} C_{13} \end{bmatrix} = \begin{bmatrix} \dfrac{d}{d\lambda_2}(\lambda_2^2 - 3\lambda_2 + 2) \\[2mm] \dfrac{d}{d\lambda_2}(\lambda_2 - 1) \\[2mm] \dfrac{d}{d\lambda_2}(\lambda_2 - 1) \end{bmatrix} = \begin{bmatrix} 2\lambda_2 - 3 \\ 1 \\ 1 \end{bmatrix} = \begin{bmatrix} 2 \times 3 - 3 \\ 1 \\ 1 \end{bmatrix} = \begin{bmatrix} 3 \\ 1 \\ 1 \end{bmatrix}$$

To find canonical form of system matrix

The modal matrix is given by, $\mathbf{M} = [\mathbf{m}_1 \ \ \mathbf{m}_2 \ \ \mathbf{m}_3] = \begin{bmatrix} 0 & 2 & 3 \\ 8 & 2 & 1 \\ 4 & 2 & 1 \end{bmatrix}$

$$\mathbf{M}^{-1} = \frac{[\text{Cofactor of } \mathbf{M}]^T}{\text{Determinant of } \mathbf{M}} = \frac{\mathbf{M}_{cof}^T}{\Delta_M} \ ; \ \ \Delta_M = \begin{vmatrix} 0 & 2 & 3 \\ 8 & 2 & 1 \\ 4 & 2 & 1 \end{vmatrix} = -8 + 24 = 16$$

$$\mathbf{M}_{cof}^T = \begin{bmatrix} 0 & -4 & 8 \\ 4 & -12 & 8 \\ -4 & 24 & -16 \end{bmatrix}^T = \begin{bmatrix} 0 & 4 & -4 \\ -4 & -12 & 24 \\ 8 & 8 & -16 \end{bmatrix}$$

$$\therefore \mathbf{M}^{-1} = \frac{1}{\Delta_M} \mathbf{M}_{cof}^T = \frac{1}{16} \begin{bmatrix} 0 & 4 & -4 \\ -4 & -12 & 24 \\ 8 & 8 & -16 \end{bmatrix} = \frac{1}{4} \begin{bmatrix} 0 & 1 & -1 \\ -1 & -3 & 6 \\ 2 & 2 & -4 \end{bmatrix}$$

$$\mathbf{M}^{-1} \mathbf{A} \mathbf{M} = \frac{1}{4} \begin{bmatrix} 0 & 1 & -1 \\ -1 & -3 & 6 \\ 2 & 2 & -4 \end{bmatrix} \begin{bmatrix} 4 & 1 & -2 \\ 1 & 0 & 2 \\ 1 & -1 & 3 \end{bmatrix} \begin{bmatrix} 0 & 2 & 3 \\ 8 & 2 & 1 \\ 4 & 2 & 1 \end{bmatrix}$$

$$= \frac{1}{4} \begin{bmatrix} 0 & 1 & -1 \\ -1 & -7 & 14 \\ 6 & 6 & -12 \end{bmatrix} \begin{bmatrix} 0 & 2 & 3 \\ 8 & 2 & 1 \\ 4 & 2 & 1 \end{bmatrix} = \frac{1}{14} \begin{bmatrix} 4 & 0 & 0 \\ 0 & 12 & 4 \\ 0 & 0 & 12 \end{bmatrix} = \begin{bmatrix} 1 & 0 & 0 \\ 0 & 3 & 1 \\ 0 & 0 & 3 \end{bmatrix}$$

$$\therefore \; J = M^{-1}AM = \begin{bmatrix} 1 & 0 & 0 \\ 0 & 3 & 1 \\ 0 & 0 & 3 \end{bmatrix} \quad \text{—— Jordan Block}$$

To compute state transition matrix, e^{At}

The state transition matrix, $e^{At} = M \, Q(t) \, e^{\Lambda t} \, M^{-1}$

$$Q(t) = \begin{bmatrix} 1 & t & \dfrac{t^2}{2} \\ 0 & 1 & t \\ 0 & 0 & 1 \end{bmatrix} ; \; e^{\Lambda t} = \begin{bmatrix} e^{\lambda_1 t} & 0 & 0 \\ 0 & e^{\lambda_2 t} & 0 \\ 0 & 0 & e^{\lambda_3 t} \end{bmatrix} = \begin{bmatrix} e^t & 0 & 0 \\ 0 & e^{3t} & 0 \\ 0 & 0 & e^{3t} \end{bmatrix}$$

$$\therefore e^{At} = \begin{bmatrix} 0 & 2 & 3 \\ 8 & 2 & 1 \\ 4 & 2 & 1 \end{bmatrix} \begin{bmatrix} 1 & t & t^2/2 \\ 0 & 1 & t \\ 0 & 0 & 1 \end{bmatrix} \begin{bmatrix} e^t & 0 & 0 \\ 0 & e^{2t} & 0 \\ 0 & 0 & e^{3t} \end{bmatrix} \begin{bmatrix} 0 & 1 & -1 \\ -1 & -3 & 6 \\ 2 & 2 & -4 \end{bmatrix} \times \dfrac{1}{4}$$

$$= \begin{bmatrix} 0 & 2 & 2t+3 \\ 8 & 8t+2 & 4t^2+2t+1 \\ 4 & 4t+2 & 2t^2+2t+1 \end{bmatrix} \begin{bmatrix} e^t & 0 & 0 \\ 0 & e^{3t} & 0 \\ 0 & 0 & e^{3t} \end{bmatrix} \begin{bmatrix} 0 & 0.25 & -0.25 \\ -0.25 & -0.75 & 1.5 \\ 0.5 & 0.5 & -1 \end{bmatrix}$$

$$= \begin{bmatrix} 0 & 2e^{3t} & (2t+3)e^{3t} \\ 8e^t & (8t+2)e^{3t} & (4t^2+2t+1)e^{3t} \\ 4e^t & (4t+2)e^{3t} & (2t^2+2t+1)e^{3t} \end{bmatrix} \begin{bmatrix} 0 & 0.25 & -0.25 \\ -0.25 & -0.75 & 1.5 \\ 0.5 & 0.5 & -1.0 \end{bmatrix}$$

$$= \begin{bmatrix} -0.5e^{3t}+(t+1.5)e^{3t} & -1.5e^{3t}+(t+1.5)e^{3t} & 3e^{3t}-(2t+3)e^{3t} \\ (-2t-0.5)e^{3t} & 2e^t+(-6t-1.5)e^{3t} & -2e^{-t}+(12t+3)e^{3t} \\ +(2t^2+t+0.5)e^{3t} & +(2t^2+t+0.5)e^{3t} & -(4t^2+2t+1)e^{3t} \\ (-t-0.5)e^{3t} & e^{-t}+(-3t-1.5)e^{3t} & -e^{-t}+(6t+3)e^{3t} \\ +(t^2+t+0.5)e^{3t} & +(t^2+t+0.5)e^{3t} & -(2t^2+2t+1)e^{3t} \end{bmatrix}$$

$$= \begin{bmatrix} (t+1)e^{3t} & te^{3t} & -2te^{3t} \\ (2t^2-t)e^{3t} & 2e^t+(2t^2-5t-1)e^{3t} & -2e^t-(4t^2-10t-2)e^{3t} \\ t^2e^{3t} & e^t+(t^2-2t-1)e^{3t} & -e^t-(2t^2-4t-2)e^{3t} \end{bmatrix}$$

5.6   CONCEPTS OF CONTROLLABILITY AND OBSERVABILITY

CONTROLLABILITY

The controllability verifies the usefulness of a state variable. In the controllability test we can find, whether the state variable can be controlled to achieve the desired output. The choice of state variables is arbitrary while forming the state model. After determining the state model, the controllability of the state variable is verified. If the state variable is not controllable then we have to go for another choice of state variable.

Definition of controllability

    A system is said to be completely state controllable if it is possible to transfer the system state from any initial state $X(t_o)$ to any other desired state $X(t_f)$ in specified finite time by a control vector U(t).

    The controllability of a state model can be tested by Kalman's test or Gilbert's test.

Gilbert's method of testing controllability

Case(i) : When the system matrix has distinct eigenvalues

    In this case the system matrix can be diagonalized and the state model can be converted to canonical form.

    Consider the state model of the system,

$$\dot{X} = AX + BU$$
$$Y = CX + DU$$

    The state model can be converted to canonical form by a transformation, $X = MZ$, where M is the modal matrix and Z is the transformed state variable vector.

    The transformed state model is given by

$$\dot{Z} = \Lambda Z + \tilde{B}U$$
$$Y = \tilde{C}Z + DU$$

    where $\Lambda = M^{-1}AM$
$$\tilde{B} = M^{-1}B$$
$$\tilde{C} = CM$$

    In this case the necessary and sufficient condition for complete controllability is that, the matrix $\tilde{B}$ must have no rows with all zeros. If any row of the matrix $\tilde{B}$ is zero then the corresponding state variable is uncontrollable.

Case(ii) : When the system matrix has repeated eigenvalues

    In this case, the system matrix cannot be diagonalized but can be transformed to Jordan canonical form.

    Consider the state model of the system,

$$\dot{X} = AX + BU$$
$$Y = CX + DU$$

    The state model can be transformed to Jordan canonical form by a transformation, $X = MZ$, where M is modal matrix and Z is the transformed state variable vector.

    The transformed state model is given by,

$$\dot{Z} = JZ + \tilde{B}U$$
$$Y = \tilde{C}Z + DU$$

    where $J = M^{-1}AM$
$$\tilde{B} = M^{-1}B$$
$$\tilde{C} = CM$$

In this case, the system is completely controllable if the elements of any row of $\tilde{\mathbf{B}}$ that correspond to the last row of each Jordan block are not all zero and the rows corresponding to other state variables must not have all zeros.

Kalman's method of testing controllability

Consider a system with state equation, $\dot{\mathbf{X}} = \mathbf{AX} + \mathbf{BU}$. For this system, a composite matrix, $\mathbf{Q}_c$ can be formed such that,

$$\mathbf{Q}_c = [\mathbf{B} \quad \mathbf{AB} \quad \mathbf{A}^2\mathbf{B} \quad .... \quad \mathbf{A}^{n-1}\mathbf{B}] \qquad\qquad .....(5.64)$$

where n is the order of the system (n is also equal to number of state variables).

In this case the system is completely state controllable if the rank of the composite matrix, $\mathbf{Q}_c$ is n.

The rank of the matrix is n, if the determinant of (n $\times$ n) composite matrix $\mathbf{Q}_c$ is non-zero. i.e, if $|\mathbf{Q}_c| \neq 0$, then rank of $\mathbf{Q}_c = n$ and the system is completely state controllable.

The advantage in kalman's test is that the calculations are simpler. But the disadvantage in kalman's test is that, we can't find the state variable which is uncontrollable. But in Gilbert's method we can find the uncontrollable state variable which is the state variable corresponding to the row of which has all zeros.

Condition for complete state controllability in the s-plane

A necessary and sufficient condition for complete state controllability is that no cancellation of poles and zeros occurs in the transfer function of the system. If cancellation occurs then the system cannot be controlled in the direction of the cancelled mode.

OBSERVABILITY

In observability test we can find whether the state variable is observable or measurable. The concept of observability is useful in solving the problem of reconstructing unmeasurable state variables from measurable ones in the minimum possible length of time. In state feedback control the estimation of unmeasurable state variables is essential in order to construct the control signals.

Definition of observability

A system is said to be completely observable if every state X(t) can be completely identified by measurements of the output Y(t) over a finite time interval.

The observability of a system can be tested by either Gilbert's method or Kalman's method.

Gilbert's method of testing observability

Consider a state model of nth order system, $\dot{\mathbf{X}} = \mathbf{AX} + \mathbf{BU}$; $\mathbf{Y} = \mathbf{CX} + \mathbf{DU}$

The state model can be transformed to a canonical or Jordan canonical form by a transformation, $\mathbf{X} = \mathbf{MZ}$, where $\mathbf{M}$ is the modal matrix and $\mathbf{Z}$ is the transformed state variable vector.

The transformed state model is,

$$\dot{\mathbf{Z}} = \Lambda\mathbf{Z} + \tilde{\mathbf{B}}\mathbf{U} \quad \text{(or)} \quad \dot{\mathbf{Z}} = \mathbf{JZ} + \tilde{\mathbf{B}}\mathbf{U}$$
$$\mathbf{Y} = \tilde{\mathbf{C}}\mathbf{Z} + \mathbf{DU} \qquad\quad \mathbf{Y} = \tilde{\mathbf{C}}\mathbf{Z} + \mathbf{DU}$$

where $\Lambda = M^{-1}AM$; if eigenvalues are distinct ; $\tilde{B} = M^{-1}B$

$J = M^{-1}AM$; if eigenvalues have multiplicity ; $\tilde{C} = CM$

The necessary and sufficient condition for complete observability is that none of the columns of the matrix $\tilde{C}$ be zero. If any of the column's of $\tilde{C}$ has all zeros then the corresponding state variable is not observable.

Kalman's Test for observability

Consider a system with state model, $\dot{X} = AX + BU$; $Y = CX + DU$

For this system, a composite matrix, Q_0 can be formed such that,

$$Q_0 = \begin{bmatrix} C^T & A^T C^T & (A^T)^2 C^T & (A^T)^3 C^T & \text{.........} & (A^T)^{n-1} C^T \end{bmatrix} \qquad \text{.....(5.65)}$$

where n is the order of the system (n is also equal to number of state variables)

In this case, the system is completely observable if the rank of composite matrix, Q_0 is n.

The rank of the matrix is n, if the determinant of $n \times n$ composite matrix Q_0 is non-zero. The disadvantage in Kalman's test is that, the nonobservable state variables cannot be determined.

Condition for complete observability in the s-plane

The necessary and sufficient condition for complete observability is that no cancellation of poles and zeros occurs in the transfer function. If cancellation occurs, the cancelled mode cannot be observed in the output.

RELATIONSHIPS BETWEEN CONTROLLABILITY, OBSERVABILITY & TRANSFER FUNCTIONS

The concepts of controllability and observability play an important role in the design of control systems in state space. They govern the existence of a complete solution to the control system design problem. The solution to this problem may not exist if the system considered is not controllable.

It is important to note that all physical systems are controllable and observable. However, the mathematical models of these systems may not posses the property of the controllability or observability. Then it is necessary to know the conditions under which a system is controllable and observable and the designer can seek another state model which is controllable and observable.

Duality property

The concepts of controllability and observability are dual concepts and it is proposed by Kalman as principle of duality.

The principle of duality states that a system is completely state controllable if and only if its dual system is completely observable or viceversa. [i.e., if the system is observable then its dual is controllable]. Using the principle of duality, the observability of a given system can be checked by testing the state controllability of its dual or vice-versa.

Consider the system S_1, described by the state model shown below.

$\dot{X} = AX + BU$

$Y = CX$

Let the dual of system S_1 be denoted as S_2 and the dual system S_2 is described by the following state model.

$$\dot{Z} = A^T Z + C^T V$$

$$N = B^T Z$$

where, Z = State vector of dual system

V = Input vector of dual system   ; N = Output vector of dual system

For the system S_1 the composite matrix, Q_{C1} for controllability is given by equation (5.66) and the composite matrix, Q_{O1} for observability is given by equation (5.67).

$$Q_{C1} = [B \quad AB \quad A^2 B \dots\dots A^{n-1} B] \qquad \qquad \dots\dots(5.66)$$

$$Q_{O1} = [C^T \quad A^T C^T \quad (A^T)^2 C^T \dots\dots (A^T)^{n-1} C^T] \qquad \qquad \dots\dots(5.67)$$

For the dual system S_2 the composite matrix, Q_{C2} for controllability is given by equation (5.68) and the composite matrix Q_{O2} for observability is given by equation (5.69)

$$Q_{C2} = [C^T \quad A^T C^T \quad (A^T)^2 C^T \dots\dots (A^T)^{n-1} C^T] \qquad \qquad \dots\dots(5.68)$$

$$Q_{O2} = [A \quad AB \quad A^2 B \dots\dots A^{n-1} B] \qquad \qquad \dots\dots(5.69)$$

From equations (5.66) and (5.69) we get $Q_{C1} = Q_{O2}$, hence if the system S_1 is controllable then its dual system S_2 is observable.

From equations (5.67) and (5.68) we get $Q_{O1} = Q_{C2}$, hence if the system S_1 is observable then its dual system S_2 is controllable.

Effect of pole-zero cancellation in transfer function

The concepts of controllability and observability are closely related to the properties of the transfer function. Consider an nth order system with distinct eigenvalues. The transfer function of the system can be expressed as a ratio of two polynomials as shown in equation (5.70).

$$T(s) = \frac{Y(s)}{U(s)} = \frac{b_0 s^m + b_1 s^{m-1} + \dots\dots + b_{m-1} s_1 + b_m}{s^n + a_1 s^{n-1} + \dots\dots + a_{n-1} s + a_n} \; ; \; m < n \qquad \dots\dots(5.70)$$

$$= \frac{K(s + \beta_1)(s + \beta_2) \dots\dots (s + \beta_m)}{(s + \lambda_1)(s + \lambda_2) \dots\dots (s + \lambda_n)} \qquad \qquad \dots\dots(5.71)$$

By partial fraction expansion technique the equation (5.71) can be written as,

$$\frac{Y(s)}{U(s)} = \frac{C_1}{s_1 + \lambda_1} + \frac{C_2}{s + \lambda_2} + \dots\dots + \frac{C_i}{s + \lambda_i} + \dots\dots + \frac{C_1}{s + \lambda_n} \qquad \dots\dots(5.72)$$

where $C_1, C_2, C_3, \dots\dots C_n$ are residues.

If the transfer function has identical pair of pole and zero at $\beta_i = \lambda_i$, then $C_i = 0$. The effect of this cancellation on controllability and observability properties depends on the choice of state variables [or depends on the method of forming state model].

In one method of state space modelling using canonical of variables, the $C_i = 0$, will appear in input (control) vector B and the the state x_i is uncontrollable. In another method of state space modelling using canonical variables, the $C_i = 0$, will appear in output vector C and the state x_i is shielded from observation.

From the above discussion we can conclude that if cancellation of pole-zero occurs in the transfer function of a system, then the system will be either not state controllable or unobservable, depending on how the state variables are defined (or chosen). If the transfer function does not have pole-zero cancellation, the system can always be represented by completely controllable and observable state model.

EXAMPLE 5.6

Write the state equations for the system shown in fig 5.6.1 in which x_1, x_2 and x_3 constitute the state vector. Determine whether the system is ncompletely controllable and observable.

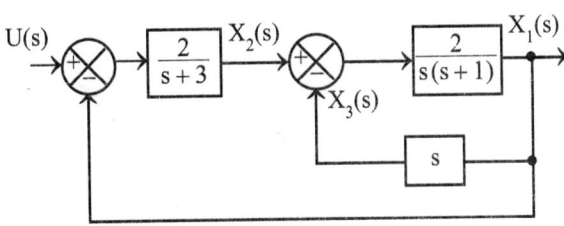

Fig 5.6.1

SOLUTION

To find state model

The state equations are obtained by writing equations for the output of each block and then taking inverse Laplace transform.

With reference to fig 5.6.2 we can write,

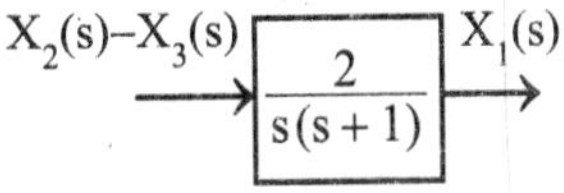

$$X_1(s) = [X_2(s) - X_3(s)] \left[\frac{2}{s(s+1)}\right]$$

$$s(s+1) X_1(s) = 2X_2(s) - 2X_3(s)$$

$$s^2 X_1(s) + s X_1(s) = 2X_2(s) - 2X_3(s)$$

On taking inverse laplace transform,

$$\ddot{x}_1 + \dot{x}_1 = 2x_2 - 2x_3 \qquad .....(5.6.1)$$

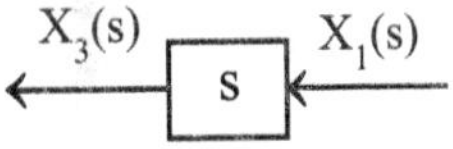

With reference to fig 5.6.3, we can write,

$$X_3(s) = s X_1(s)$$

On taking inverse Laplace transform

$$x_3 = \dot{x}_1 \qquad .....(5.6.2)$$

With reference to fig 5.6.4 we can write

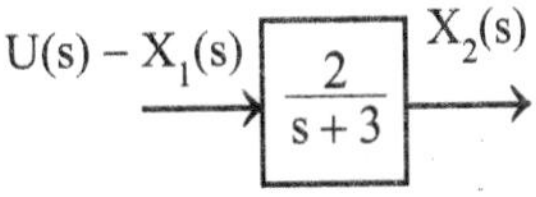

$$X_2(s) = [U(s) - X_1(s)] \left[\frac{2}{s+3}\right]$$

$$X_2(s) (s+3) = 2U(s) - 2X_1(s)$$

$$s X_2(s) + 3 X_2(s) = 2U(s) - 2X_1(s)$$

On taking inverse laplace transform

$$\dot{x}_2 + 3x_2 = 2u - 2x_1$$

$$\dot{x}_2 = -2x_1 = 3x_2 + 2u \qquad .....(5.6.3)$$

Fig 5.6.2: $X_2(s) - X_3(s) \rightarrow \boxed{\dfrac{2}{s(s+1)}} \rightarrow X_1(s)$

Fig 5.6.2

Fig 5.6.3: $X_3(s) \leftarrow \boxed{s} \leftarrow X_1(s)$

Fig 5.6.3

Fig 5.6.4: $U(s) - X_1(s) \rightarrow \boxed{\dfrac{2}{s+3}} \rightarrow X_2(s)$

Fig 5.6.4

From equation (5.6.2) we get, $\dot{x}_1 = x_3$; $\therefore \ddot{x}_1 = \dot{x}_3$

Put $\dot{x}_1 = x_3$ and $\ddot{x}_1 = \dot{x}_3$ in equation (5.6.1)

$$\therefore \dot{x}_3 + x_3 = 2x_2 - 2x_3$$

$$\dot{x}_3 = 2x_2 - 2x_3 - x_3$$

$$\dot{x}_3 = 2x_2 - 3x_3 \qquad \qquad \dots\dots(5.6.4)$$

The state equation are given by equations (5.6.2), (5.6.3) and (5.6.4)

$$\dot{x}_1 = x_3$$

$$\dot{x}_2 = -2x_1 - 3x_2 + 2u$$

$$\dot{x}_3 = 2x_2 - 3x_3$$

The output equation is $y = x_1$

The state model in the matrix form is

$$\begin{bmatrix} \dot{x}_1 \\ \dot{x}_2 \\ \dot{x}_3 \end{bmatrix} = \begin{bmatrix} 0 & 0 & 1 \\ -2 & -3 & 0 \\ 0 & 2 & -3 \end{bmatrix} \begin{bmatrix} x_1 \\ x_2 \\ x_3 \end{bmatrix} + \begin{bmatrix} 0 \\ 2 \\ 0 \end{bmatrix} u \quad ; \quad y = \begin{bmatrix} 1 & 0 & 0 \end{bmatrix} \begin{bmatrix} x_1 \\ x_2 \\ x_3 \end{bmatrix}$$

To find eigenvalues

Here the system matrix, $A = \begin{bmatrix} 0 & 0 & 1 \\ -2 & -3 & 0 \\ 0 & 2 & -3 \end{bmatrix}$

The characteristic equation is $|\lambda I - A| = 0$

$$(\lambda I - A) = \lambda \begin{bmatrix} 1 & 0 & 0 \\ 0 & 1 & 0 \\ 0 & 0 & 1 \end{bmatrix} - \begin{bmatrix} 0 & 0 & 1 \\ -2 & -3 & 0 \\ 0 & 2 & -3 \end{bmatrix} = \begin{bmatrix} \lambda & 0 & -1 \\ 2 & \lambda+3 & 0 \\ 0 & -2 & \lambda+3 \end{bmatrix}$$

$$|\lambda I - A| = \begin{vmatrix} \lambda & 0 & -1 \\ 2 & \lambda+3 & 0 \\ 0 & -2 & \lambda+3 \end{vmatrix} = \lambda(\lambda+3)^2 - 1(-4) = \lambda(\lambda^2 + 6\lambda + 9) + 4$$

$$= \lambda^3 + 6\lambda^2 + 9\lambda + 4 = (\lambda + 1)(\lambda^2 + 5\lambda + 4)$$

$$= (\lambda + 1)(\lambda + 1)(\lambda + 4) = (\lambda + 1)^2 (\lambda + 4)$$

$$\begin{array}{r|rrr|r} \lambda = -1 & 1 & 6 & 9 & 4 \\ \downarrow & & -1 & -5 & -4 \\ \hline & 1 & 5 & 4 & 0 \end{array}$$

The eigenvalues are $\lambda_1 = -1$, $\lambda_2 = -1$, and $\lambda_3 = -4$

To find eigenvectors

$$(\lambda_1 I - A) = \lambda_1 \begin{bmatrix} 1 & 0 & 0 \\ 0 & 1 & 0 \\ 0 & 0 & 1 \end{bmatrix} - \begin{bmatrix} 0 & 0 & 1 \\ -2 & -3 & 0 \\ 0 & 2 & -3 \end{bmatrix} = \begin{bmatrix} \lambda_1 & 0 & -1 \\ 2 & \lambda_1+3 & 0 \\ 0 & -2 & \lambda_1+3 \end{bmatrix}$$

Let C_{11}, C_{12} and C_{13} be cofactors along Ist row of the matrix $[\lambda_1 I - A]$

$$C_{11} = (+1) \begin{vmatrix} \lambda_1 + 3 & 0 \\ -2 & \lambda_1 + 3 \end{vmatrix} = (\lambda_1 + 3)^2 = \lambda_1^2 + 6\lambda_1 + 9$$

$$C_{12} = (-1) \begin{vmatrix} 2 & 0 \\ 0 & \lambda_1 + 3 \end{vmatrix} = -(2(\lambda_1 + 3)) = -2\lambda_1 - 6$$

$$C_{13} = (+1) \begin{vmatrix} 2 & \lambda_1 + 3 \\ 0 & -2 \end{vmatrix} = -4$$

$$\mathbf{m}_1 = \begin{bmatrix} C_{11} \\ C_{12} \\ C_{13} \end{bmatrix} = \begin{bmatrix} \lambda_1^2 + 6\lambda_1 + 9 \\ -2\lambda_1 - 6 \\ -4 \end{bmatrix} = \begin{bmatrix} 1 - 6 + 9 \\ 2 - 6 \\ -4 \end{bmatrix} = \begin{bmatrix} 4 \\ -4 \\ -4 \end{bmatrix}$$

$$\mathbf{m}_2 = \begin{bmatrix} \dfrac{d}{d\lambda_1} C_{11} \\ \dfrac{d}{d\lambda_1} C_{12} \\ \dfrac{d}{d\lambda_1} C_{13} \end{bmatrix} = \begin{bmatrix} 2\lambda_1 + 6 \\ -2 \\ 0 \end{bmatrix} = \begin{bmatrix} -2 + 6 \\ -2 \\ 0 \end{bmatrix} = \begin{bmatrix} 4 \\ -2 \\ 0 \end{bmatrix}$$

$$(\lambda_3 \mathbf{I} - \mathbf{A}) = (-4) \begin{bmatrix} 1 & 0 & 0 \\ 0 & 1 & 0 \\ 0 & 0 & 1 \end{bmatrix} - \begin{bmatrix} 0 & 0 & 1 \\ -2 & -3 & 0 \\ 0 & 2 & -3 \end{bmatrix} = \begin{bmatrix} -4 & 0 & -1 \\ 2 & -1 & 0 \\ 0 & -2 & -1 \end{bmatrix}$$

Let C_{11}, C_{12} and C_{13} be the cofactors along Ist row of the matrix $[\lambda_3 \mathbf{I} - \mathbf{A}]$.

$$C_{11} = (+1) \begin{vmatrix} -1 & 0 \\ -2 & -1 \end{vmatrix} = 1 \; ; \; C_{12} = (-1) \begin{vmatrix} 2 & 0 \\ 0 & -1 \end{vmatrix} = 2 \; ; \; C_{13} = (+1) \begin{vmatrix} 2 & -1 \\ 0 & -2 \end{vmatrix} = -4$$

$$\therefore \; \mathbf{m}_3 = \begin{bmatrix} C_{11} \\ C_{12} \\ C_{13} \end{bmatrix} = \begin{bmatrix} 1 \\ 2 \\ -4 \end{bmatrix}$$

To find canonical form of state model

The modal matrix, **M** is given by

$$\mathbf{M} = [\mathbf{m}_1 \; \mathbf{m}_2 \; \mathbf{m}_3] = \begin{bmatrix} 4 & 4 & 1 \\ -4 & -2 & 2 \\ -4 & 0 & -4 \end{bmatrix}$$

$$\mathbf{M}^{-1} = \frac{[\text{Cofactor of } \mathbf{M}]^T}{\text{Determinant of } \mathbf{M}} = \frac{\mathbf{M}_{cof}^T}{\Delta_M}$$

$$\Delta_M = \begin{vmatrix} 4 & 4 & 1 \\ -4 & -2 & 2 \\ -4 & 0 & -4 \end{vmatrix} = 4(8) - 4(24) + 1(-8) = 32 - 96 - 8 = -72$$

$$\mathbf{M}_{cof}^T = \begin{bmatrix} 8 & -24 & -8 \\ 16 & -12 & -16 \\ 10 & -12 & 8 \end{bmatrix}^T = \begin{bmatrix} 8 & 16 & 10 \\ -24 & -12 & -12 \\ -8 & -16 & 8 \end{bmatrix}$$

$$M^{-1} = \frac{1}{-72}\begin{bmatrix} 8 & 16 & 10 \\ -24 & -12 & -12 \\ -8 & -16 & 8 \end{bmatrix} = \frac{1}{18}\begin{bmatrix} -2 & -4 & -2.5 \\ 6 & 3 & 3 \\ 2 & 4 & -2 \end{bmatrix}$$

$$J = M^{-1}AM = \frac{1}{18}\begin{bmatrix} -2 & -4 & -2.5 \\ 6 & 3 & 3 \\ 2 & 4 & -2 \end{bmatrix}\begin{bmatrix} 0 & 0 & 1 \\ -2 & -3 & 0 \\ 0 & 2 & -3 \end{bmatrix}\begin{bmatrix} 4 & 4 & 1 \\ -4 & -2 & 2 \\ -4 & 0 & -4 \end{bmatrix}$$

$$= \frac{1}{18}\begin{bmatrix} 8 & 7 & 5.5 \\ -6 & -3 & -3 \\ -8 & -16 & 8 \end{bmatrix}\begin{bmatrix} 4 & 4 & 1 \\ -4 & -2 & 2 \\ -4 & 0 & -4 \end{bmatrix}$$

$$= \frac{1}{18}\begin{bmatrix} -18 & 18 & 0 \\ 0 & -18 & 0 \\ 0 & 0 & -72 \end{bmatrix} = \begin{bmatrix} \boxed{\begin{matrix} -1 & 1 \\ 0 & -1 \end{matrix}} & \begin{matrix} 0 \\ 0 \end{matrix} \\ \begin{matrix} 0 & 0 \end{matrix} & -4 \end{bmatrix} \leftarrow \text{Jordan block}$$

$$\tilde{B} = M^{-1}B = \frac{1}{18}\begin{bmatrix} -2 & -4 & -2.5 \\ 6 & 3 & 3 \\ 2 & 4 & -2 \end{bmatrix}\begin{bmatrix} 0 \\ 2 \\ 0 \end{bmatrix} = \begin{bmatrix} -8/18 \\ 6/18 \\ 8/18 \end{bmatrix} = \begin{bmatrix} -4/9 \\ 3/9 \\ 4/9 \end{bmatrix}$$

$$\tilde{C} = CM = \begin{bmatrix} 1 & 0 & 0 \end{bmatrix}\begin{bmatrix} 4 & 4 & 1 \\ -4 & -2 & 2 \\ -4 & 0 & -4 \end{bmatrix} = \begin{bmatrix} 4 & 4 & 1 \end{bmatrix}$$

The Jordan canonical form of state model is shown below.

$$\dot{Z} = JZ + \tilde{B}U \; ; \; Y = \tilde{C}Z + DU \qquad \text{(Here DU is not defined)}$$

$$\begin{bmatrix} \dot{z}_1 \\ \dot{z}_2 \\ \dot{z}_3 \end{bmatrix} = \begin{bmatrix} -1 & 1 & 0 \\ 0 & -1 & 0 \\ 0 & 0 & -4 \end{bmatrix}\begin{bmatrix} z_1 \\ z_2 \\ z_3 \end{bmatrix} + \begin{bmatrix} -4/9 \\ 3/9 \\ 4/9 \end{bmatrix}[u] \; ; \; Y = \begin{bmatrix} 4 & 4 & 1 \end{bmatrix}\begin{bmatrix} z_1 \\ z_2 \\ z_3 \end{bmatrix}$$

CONCLUSION

It is observed that the elements of the rows of $\tilde{B}$ are not all zeros. Hence the system is completely controllable (or state controllable).

It is observed that the elements of the columns of $\tilde{C}$ are not all zeros. Hence the system is completely observable [i.e, all the state variables are observable].

ALTERNATE METHOD

KALMAN'S TEST FOR CONTROLLABILITY

$$A^2 = A.A = \begin{bmatrix} 0 & 0 & 1 \\ -2 & -3 & 0 \\ 0 & 2 & -3 \end{bmatrix}\begin{bmatrix} 0 & 0 & 1 \\ -2 & -3 & 0 \\ 0 & 2 & -3 \end{bmatrix} = \begin{bmatrix} 0 & 2 & -3 \\ 6 & 9 & -2 \\ -4 & -12 & 9 \end{bmatrix}$$

$$A.B = \begin{bmatrix} 0 & 0 & 1 \\ -2 & -3 & 0 \\ 0 & 2 & -3 \end{bmatrix}\begin{bmatrix} 0 \\ 2 \\ 0 \end{bmatrix} = \begin{bmatrix} 0 \\ -6 \\ 4 \end{bmatrix}$$

$$\mathbf{A^2B} = \begin{bmatrix} 0 & 2 & -3 \\ 6 & 9 & -2 \\ -4 & -12 & 9 \end{bmatrix} \begin{bmatrix} 0 \\ 2 \\ 0 \end{bmatrix} = \begin{bmatrix} 4 \\ 18 \\ -24 \end{bmatrix}$$

The composite matrix for controllability, $\mathbf{Q_c} = \begin{bmatrix} \mathbf{B} & \mathbf{AB} & \mathbf{A^2B} \end{bmatrix}$

$$= \begin{bmatrix} 0 & 0 & 4 \\ 2 & -6 & 18 \\ 0 & 4 & -24 \end{bmatrix}$$

Determinant of $\mathbf{Q_c} = \begin{vmatrix} 0 & 0 & 4 \\ 2 & -6 & 18 \\ 0 & 4 & -24 \end{vmatrix} = 4 \times 8 = 32$; Since $|\mathbf{Q_c}| \neq 0$, the rank of $\mathbf{Q_c} = 3$.

Hence the system is completely state controllable.

KALMAN'S TEST FOR OBSERVABILITY

$$\mathbf{A^T} = \begin{bmatrix} 0 & 0 & 1 \\ -2 & -3 & 0 \\ 0 & 2 & -3 \end{bmatrix}^T = \begin{bmatrix} 0 & -2 & 0 \\ 0 & -3 & 2 \\ 1 & 0 & -3 \end{bmatrix}$$

$$\mathbf{C^T} = \begin{bmatrix} 1 & 0 & 0 \end{bmatrix}^T = \begin{bmatrix} 1 \\ 0 \\ 0 \end{bmatrix}$$

$$\left(\mathbf{A^T}\right)^2 = \begin{bmatrix} 0 & -2 & 0 \\ 0 & -3 & 2 \\ 1 & 0 & -3 \end{bmatrix} \begin{bmatrix} 0 & -2 & 0 \\ 0 & -3 & 2 \\ 1 & 0 & -3 \end{bmatrix} = \begin{bmatrix} 0 & 6 & -4 \\ 2 & 9 & -12 \\ -3 & -2 & 9 \end{bmatrix}$$

$$\mathbf{A^T C^T} = \begin{bmatrix} 0 & -2 & 0 \\ 0 & -3 & 2 \\ 1 & 0 & -3 \end{bmatrix} \begin{bmatrix} 1 \\ 0 \\ 0 \end{bmatrix} = \begin{bmatrix} 0 \\ 0 \\ 1 \end{bmatrix}$$

$$\left(\mathbf{A^T}\right)^2 \mathbf{C^T} = \begin{bmatrix} 0 & 6 & -4 \\ 2 & 9 & -12 \\ -3 & -2 & 9 \end{bmatrix} \begin{bmatrix} 1 \\ 0 \\ 0 \end{bmatrix} = \begin{bmatrix} 0 \\ 2 \\ -3 \end{bmatrix}$$

The composite matrix for observability } $\mathbf{Q_0} = \begin{bmatrix} \mathbf{C^T} & \mathbf{A^T C^T} & \left(\mathbf{A^T}\right)^2 \mathbf{C^T} \end{bmatrix} = \begin{bmatrix} 1 & 0 & 0 \\ 0 & 0 & 2 \\ 0 & 1 & -3 \end{bmatrix}$

Determinant of $\mathbf{Q_0} = \begin{vmatrix} 1 & 0 & 0 \\ 0 & 0 & 2 \\ 0 & 1 & -3 \end{vmatrix} = 1 \times -2 = -2$; Since $|\mathbf{Q_0}| \neq 0$, the rank of $\mathbf{Q_0} = 3$

Hence the system is completely observable. (or all the state variables of the system are observable).

5.7   CONTROLLABLE PHASE VARIABLE FORM OF STATE MODEL

A controllable system can be represented by a modified state model called controllable phase variable form by transforming the system matrix, A into phase variable form (Bush form or companion form).

Consider the state model of nth order system with single-input and single output as shown below.

$$\dot{X} = AX + Bu \qquad\qquad .....(5.73)$$

$$y = CX + Du \qquad\qquad .....(5.74)$$

Let us choose a transformation, $Z = P_C X$ to transform the state model to controllable phase variable form.

Here Z = Transformed state vector of order $(n \times 1)$

P_C = Transformation matrix of order $(n \times n)$

$$Z = \begin{bmatrix} z_1 \\ z_2 \\ z_3 \\ \vdots \\ z_n \end{bmatrix} \quad \text{and} \quad P_c = \begin{bmatrix} p_{11} & p_{12} & p_{13} & \cdots & p_{1n} \\ p_{21} & p_{22} & p_{23} & \cdots & p_{21} \\ p_{31} & p_{32} & p_{33} & \cdots & p_{3n} \\ \vdots & \vdots & \vdots & & \vdots \\ p_{n1} & p_{n2} & p_{n3} & \cdots & p_{nn} \end{bmatrix}$$

On premultiplying the equation $Z = P_C X$ by P_c^{-1} we get

$$P_c^{-1} Z = P_c^{-1} P_c X$$

$$\therefore X = P_c^{-1} Z$$

On differentiating the equation $X = P_c^{-1} Z$ we get,

$$\dot{X} = P_c^{-1} \dot{Z}$$

On substituting $X = P_c^{-1} Z$ and $\dot{X} = P_c^{-1} \dot{Z}$ in the state model (equations (5.73) and (5.74)) of the system we get,

$$P_c^{-1} \dot{Z} = A P_c^{-1} Z + Bu \qquad\qquad .....(5.75)$$

$$y = C P_c^{-1} Z + Du \qquad\qquad .....(5.76)$$

On premultiplying the equations (5.75) by P_C we get,

$$\dot{Z} = P_c A P_c^{-1} Z + P_c Bu$$

$$y = C P_c^{-1} Z + Du$$

Let, $P_c A P_c^{-1} = A_C$; $P_c B = B_C$ and $C P_c^{-1} = C_C$

$$\therefore \dot{Z} = A_C z + B_C u \qquad\qquad .....(5.77)$$

$$y = C_C Z + Du \qquad\qquad .....(5.78)$$

The equations (5.77) and (5.78) are called the controllable phase variable form of state model of the system.

Note : In controllable phase variable form of state model the matrices A_C, B_C *and* C_C *will be as shown below.*

$$A_c = \begin{bmatrix} 0 & 1 & 0 & \cdots & 0 \\ 0 & 0 & 1 & \cdots & 0 \\ 0 & 0 & 0 & \cdots & 0 \\ \vdots & \vdots & \vdots & & \vdots \\ 0 & 0 & 0 & \cdots & 1 \\ -a_n & -a_{n-1} & -a_{n-2} & \cdots & -a_1 \end{bmatrix} \; ; \; B_c = \begin{bmatrix} 0 \\ 0 \\ 0 \\ \vdots \\ 0 \\ 1 \end{bmatrix} \; ; \; C_c = \begin{bmatrix} c_{11} & c_{12} & c_{13} & \cdots & c_{1n} \end{bmatrix}$$

Determination of transformation matrix, P_c

The $n \times n$ transformation matrix, P_c and be expressed as n-numbers of row vectors (matrices) as shown below.

$$P_c = \begin{bmatrix} p_{11} & p_{12} & p_{13} & \cdots & p_{1n} \\ p_{21} & p_{22} & p_{23} & \cdots & p_{21} \\ p_{31} & p_{32} & p_{33} & \cdots & p_{3n} \\ \vdots & \vdots & \vdots & & \vdots \\ p_{n1} & p_{n2} & p_{n3} & \cdots & p_{nn} \end{bmatrix} = \begin{bmatrix} P_1 \\ P_2 \\ P_3 \\ \vdots \\ P_n \end{bmatrix}$$ .....(5.79)

$$
\begin{aligned}
P_1 &= \begin{bmatrix} p_{11} & p_{12} & p_{13} & \cdots & p_{1n} \end{bmatrix} \\
P_2 &= \begin{bmatrix} p_{21} & p_{22} & p_{23} & \cdots & p_{2n} \end{bmatrix} \\
\text{where } P_1 &= \begin{bmatrix} p_{31} & p_{32} & p_{33} & \cdots & p_{3n} \end{bmatrix} \\
&\;\vdots \qquad \vdots \qquad\qquad\qquad \vdots \\
P_n &= \begin{bmatrix} p_{n1} & p_{n2} & p_{n3} & \cdots & p_{nn} \end{bmatrix}
\end{aligned}
$$

The transformation $Z = P_c X$ can be written in the expanded form as,

$$\begin{bmatrix} z_1 \\ z_2 \\ z_3 \\ \vdots \\ z_n \end{bmatrix} \begin{bmatrix} p_{11} & p_{12} & p_{13} & \cdots & p_{1n} \\ p_{21} & p_{22} & p_{23} & \cdots & p_{2n} \\ p_{31} & p_{32} & p_{33} & \cdots & p_{3n} \\ \vdots & \vdots & \vdots & \vdots & \vdots \\ p_{n1} & p_{n2} & p_{n3} & \cdots & p_{nn} \end{bmatrix} = \begin{bmatrix} x_1 \\ x_2 \\ x_3 \\ \vdots \\ x_n \end{bmatrix}$$ .....(5.80)

From equation (5.80) we get,

$$z_1 = p_{11} x_1 + p_{12} x_2 + p_{13} x_3 + \ldots + p_{1n} x_n$$

$$= \begin{bmatrix} p_{11} & p_{12} & p_{13} & \cdots & p_{1n} \end{bmatrix} \begin{bmatrix} x_1 \\ x_2 \\ x_3 \\ \vdots \\ x_n \end{bmatrix}$$

$$\therefore z_1 = P_1 X$$ ..... (5.81)

On differentiating equation (5.81) we get

$$\dot{z}_1 = P_1 \dot{X}$$ ..... (5.82)

On substituting for $\dot{X}$ from equation (5.73) in equation (5.82) we get

$$\dot{z}_1 = P_1\dot{X} = P_1(AX + Bu) = P_1AX + P_1Bu$$

Since the transformed state variables are functions of state variables alone, the term P_1B will be zero (i.e, $P_1B = 0$)

$$\therefore \dot{z}_1 = P_1AX$$

We know that, $\dot{z}_1 = z_2$

$$\therefore z_2 = \dot{z}_1 = P_1AX \qquad\qquad\qquad ..... (5.83)$$

On differentiating equation (5.83) we get

$$\dot{z}_2 = P_1A\dot{X} \qquad\qquad\qquad ..... (5.84)$$

On substituting for $\dot{X}$ from equation (5.73) in equation (5.84) we get,

$$\dot{z}_2 = P_1A(AX + Bu) = P_1A^2X + P_1ABu$$

$$= P_1A^2X \qquad\qquad\qquad (\because P_1AB = 0)$$

We know that, $\dot{z}_2 = z_3$

$$\therefore z_3 = P_1A^2X \qquad\qquad\qquad ..... (5.85)$$

Similarly the kth transformed state variable z_k can be expressed as

$$z_k = P_1A^{(k-1)}X \quad \text{and} \quad P_1A^{(k-2)}B = 0$$

Hence the n-numbers of transformed state variables can be expressed as shown below.

$$\therefore z_1 = P_1X$$
$$z_2 = P_2AX$$
$$z_3 = P_1A^2X$$
$$\vdots$$
$$z_k = P_1A^{(k-1)}X$$
$$\vdots$$
$$z_n = P_1A^{(n-1)}X$$

On arranging the above equations in the matrix form we get

$$\begin{bmatrix} z_1 \\ z_2 \\ z_3 \\ \vdots \\ z_n \end{bmatrix} = \begin{bmatrix} P_1 \\ P_1A \\ P_1A^2 \\ \vdots \\ P_1A^{(n-1)} \end{bmatrix} X \qquad\qquad .....(5.86)$$

Provided $P_1B = P_1AB = ... = P_1A^{(n-2)}B = 0$ and $P_1A^{(n-1)}B = 1$.

The equation (5.86) is same as $\mathbf{Z} = \mathbf{P_C} \, \mathbf{X}$ we can write,

$$\mathbf{P_c} = \begin{bmatrix} \mathbf{P_1} \\ \mathbf{P_1A} \\ \mathbf{P_1A^2} \\ \vdots \\ \mathbf{P_1A^{(n-1)}} \end{bmatrix} \qquad\qquad \text{.....(5.87)}$$

On arranging the elements $\mathbf{P_1B}$, $\mathbf{P_1AB}$, $\mathbf{P_1A^2B}$, ... , $\mathbf{P_1A^{(n-1)}} \mathbf{B}$ as column vector we get,

$$\begin{bmatrix} \mathbf{P_1B} & \mathbf{P_1AB} & \mathbf{P_1A^2B} & \cdots & \mathbf{P_1A^{(n-2)}B} & \mathbf{P_1A^{(n-1)}B} \end{bmatrix} = \begin{bmatrix} 0 & 0 & 0 & \cdots & 0 & 1 \end{bmatrix}$$

$$\mathbf{P_1} \begin{bmatrix} \mathbf{B} & \mathbf{AB} & \mathbf{A^2B} & \cdots & \mathbf{A^{(n-2)}B} & \mathbf{A^{(n-1)}B} \end{bmatrix} = \begin{bmatrix} 0 & 0 & 0 & \cdots & 0 & 1 \end{bmatrix} \qquad \text{.....(5.88)}$$

$$\mathbf{P_1Q_c} = \begin{bmatrix} 0 & 0 & 0 & \cdots & 0 & 1 \end{bmatrix}$$

where, $\mathbf{Q_C} = \begin{bmatrix} \mathbf{B} & \mathbf{AB} & \mathbf{A^2B} & ... & \mathbf{A^{(n-2)}} & \mathbf{B} & \mathbf{A^{(n-1)}} & \mathbf{B} \end{bmatrix}$ $\mathbf{P_1} = \begin{bmatrix} 0 & 0 & 0 & \cdots & 0 & 1 \end{bmatrix} \mathbf{Q_c^{-1}}$ .....(5.89)

Using the equation (5.87), (5.88) and (5.89), the transformation matrix, $\mathbf{P_C}$ can be evaluated.

Alternate method to find transformation matrix, $\mathbf{P_C}$

Let $\mathbf{A}$ be the system matrix of original state model. Now the characteristic equation governing the system is given by equation (5.90).

$$|\lambda \mathbf{I} - \mathbf{A}| = \lambda^n + a_1 \lambda^{n-1} + a_2 \lambda^{n-2} + ... + a_{n-1} \lambda + a_n = 0 \qquad \text{.....(5.90)}$$

Using the coefficients a_1, a_2, ... a_{n-2}, a_{n-1} of characteristic equation [equation(5.90)] we can form a matrix, $\mathbf{W}$ as shown in equation (5.91)

$$\mathbf{W} = \begin{bmatrix} a_{n-1} & a_{n-2} & \cdots & a_2 & a_1 & 0 \\ a_{n-2} & a_{n-3} & \cdots & a_1 & 1 & 0 \\ \vdots & \vdots & & \vdots & \vdots & \vdots \\ a_1 & 1 & \cdots & 0 & 0 & 0 \\ 1 & 0 & \cdots & 0 & 0 & 0 \end{bmatrix} \qquad \text{.....(5.91)}$$

Now the transformation matrix, $\mathbf{P_c}$ is given by

$$\mathbf{P_C} = (\mathbf{Q_C} \, \mathbf{W})^{-1} \qquad\qquad \text{.....(5.92)}$$

(or) $\mathbf{P_C^{-1}} = (\mathbf{Q_C} \, \mathbf{W})$ .....(5.93)

where, $\mathbf{Q_C} = \begin{bmatrix} \mathbf{B} & \mathbf{AB} & \mathbf{A^2B} & ..... & \mathbf{A^{(n-2)}} \, \mathbf{B} & \mathbf{A^{(n-1)}} \, \mathbf{B} \end{bmatrix}$

EXAMPLE 5.7

The state model of a system is given by

$$\begin{bmatrix} \dot{x}_1 \\ \dot{x}_2 \\ \dot{x}_3 \end{bmatrix} = \begin{bmatrix} 0 & 0 & 1 \\ -2 & -3 & 0 \\ 0 & 2 & -3 \end{bmatrix} \begin{bmatrix} x_1 \\ x_2 \\ x_3 \end{bmatrix} + \begin{bmatrix} 0 \\ 2 \\ 0 \end{bmatrix} [u] \;\; ; \;\; y = \begin{bmatrix} 1 & 0 & 0 \end{bmatrix} \begin{bmatrix} x_1 \\ x_2 \\ x_3 \end{bmatrix}$$

Convert the state model to controllable phase variable form.

SOLUTION

The given state model can be transformed to controllable phase variable form, only if the system is completely state controllable. Hence check for controllability.

Kalman's test for controllability

From the given state model we get,

$$A = \begin{bmatrix} 0 & 0 & 1 \\ -2 & -3 & 0 \\ 0 & 2 & -3 \end{bmatrix} \text{ and } B = \begin{bmatrix} 0 \\ 2 \\ 0 \end{bmatrix}$$

$$A^2 = AA = \begin{bmatrix} 0 & 0 & 1 \\ -2 & -3 & 0 \\ 0 & 2 & -3 \end{bmatrix}\begin{bmatrix} 0 & 0 & 1 \\ -2 & -3 & 0 \\ 0 & 2 & -3 \end{bmatrix} = \begin{bmatrix} 0 & 2 & -3 \\ 6 & 9 & -2 \\ -4 & -12 & 9 \end{bmatrix}$$

$$A.B = \begin{bmatrix} 0 & 0 & 1 \\ -2 & -3 & 0 \\ 0 & -2 & -3 \end{bmatrix}\begin{bmatrix} 0 \\ 2 \\ 0 \end{bmatrix} = \begin{bmatrix} 0 \\ -6 \\ 4 \end{bmatrix}$$

$$A^2B = \begin{bmatrix} 0 & 2 & -3 \\ 6 & 9 & -2 \\ -4 & -12 & 9 \end{bmatrix}\begin{bmatrix} 0 \\ 2 \\ 0 \end{bmatrix} = \begin{bmatrix} 4 \\ 18 \\ -24 \end{bmatrix}$$

The composite matrix for controllability, $Q_C = [B \quad AB \quad A^2B] = \begin{bmatrix} 0 & 0 & 4 \\ 2 & -6 & 18 \\ 0 & 4 & -24 \end{bmatrix}$

Determinant of $Q_C = \Delta_{QC} = \begin{vmatrix} 0 & 0 & 4 \\ 2 & -6 & 18 \\ 0 & 4 & -24 \end{vmatrix} = 4 \times 8 = 324 \times 8 = 32$

Since, $\Delta_{QC} \neq 0$, the rank of $Q_C = 3$. Hence the system is completely state controllable.

To find transformation matrix, P_C

The system state model can be converted to controllable phase variable form by choosing a transformation matrix, P_C.

where $P_C = \begin{bmatrix} P_1 \\ P_1A \\ P_1A^2 \end{bmatrix}$ and $P_1 = [0 \quad 0 \quad 1] Q_C^{-1}$

$$Q_C^{-1} = \frac{[\text{Cofactor of } Q_c]}{\text{Determinant of } Q_c} = \frac{Q_{c,cof}^T}{\Delta_{QC}}$$

$$Q_{c,cof}^T = \begin{bmatrix} 72 & 48 & 8 \\ 16 & 0 & 0 \\ 24 & 8 & 0 \end{bmatrix} = \begin{bmatrix} 72 & 16 & 24 \\ 48 & 0 & 8 \\ 8 & 0 & 0 \end{bmatrix}$$

$$Q_{c,cof}^{-1} = \frac{1}{32}\begin{bmatrix} 72 & 16 & 24 \\ 48 & 0 & 8 \\ 8 & 0 & 0 \end{bmatrix} = \begin{bmatrix} 2.25 & 0.5 & 0.75 \\ 1.5 & 0 & 0.25 \\ 0.25 & 0 & 0 \end{bmatrix}$$

$$P_1 = \begin{bmatrix} 0 & 0 & 1 \end{bmatrix} Q_c^{-1} = \begin{bmatrix} 0 & 0 & 1 \end{bmatrix} \begin{bmatrix} 2.25 & 0.5 & 0.75 \\ 1.5 & 0 & 0.25 \\ 0.25 & 0 & 0 \end{bmatrix} = \begin{bmatrix} 0.25 & 0 & 0 \end{bmatrix}$$

$$P_1 A = \begin{bmatrix} 0.25 & 0 & 0 \end{bmatrix} \begin{bmatrix} 0 & 0 & 1 \\ -2 & -3 & 0 \\ 0 & 2 & -3 \end{bmatrix} = \begin{bmatrix} 0 & 0 & 0.25 \end{bmatrix}$$

$$P_1 A^2 = \begin{bmatrix} 0.25 & 0 & 0 \end{bmatrix} \begin{bmatrix} 0 & 2 & -3 \\ 6 & 9 & -2 \\ -4 & -12 & 9 \end{bmatrix} = \begin{bmatrix} 0 & 0.5 & -0.75 \end{bmatrix}$$

$$\therefore \text{ Transformation matrix, } P_c = \begin{bmatrix} P_1 \\ P_1 A \\ P_1 A^2 \end{bmatrix} = \begin{bmatrix} 0.25 & 0 & 0 \\ 0 & 0 & 0.25 \\ 0 & 0.5 & -0.75 \end{bmatrix}$$

To determine the controllable phase variable form of state model

The controllable phase variable form of state model is given by,

$$\dot{Z} = A_c Z + B_c u$$
$$y = C_c Z \qquad \text{(Here D is not given)}$$

where $A_c = P_c A P_c^{-1}$; $\quad B_c = P_c B$ and $C_c = C P_c^{-1}$

$$\text{The transformation matrix, } P_c = \begin{bmatrix} 0.25 & 0 & 0 \\ 0 & 0 & 0.25 \\ 0 & 0.5 & -0.75 \end{bmatrix}$$

$$\therefore \ P_c^{-1} = \frac{[\text{Cofactor of } P_c]^T}{\text{Determinant of } P_c} = \frac{P_{c,cof}^T}{\Delta_{P_c}}$$

$$\Delta_{P_c} = \begin{vmatrix} 0.25 & 0 & 0 \\ 0 & 0 & 0.25 \\ 0 & 0.5 & -0.75 \end{vmatrix} = 0.25 \times (-0.5 \times 0.25) = -0.03125$$

$$P_{c,cof}^T = \begin{bmatrix} -0.125 & 0 & 0 \\ 0 & -0.1875 & -0.125 \\ 0 & -0.0625 & 0 \end{bmatrix}^T = \begin{bmatrix} -0.125 & 0 & 0 \\ 0 & -0.1875 & -0.0625 \\ 0 & -0.125 & 0 \end{bmatrix}$$

$$\therefore \ P_c^{-1} = \frac{1}{-0.03125} \begin{bmatrix} -0.125 & 0 & 0 \\ 0 & -0.1875 & -0.0625 \\ 0 & -0.125 & 0 \end{bmatrix} = \begin{bmatrix} 4 & 0 & 0 \\ 0 & 6 & 2 \\ 0 & 4 & 0 \end{bmatrix}$$

$$A_C = P_C A P_C^{-1} = \begin{bmatrix} 0.25 & 0 & 0 \\ 0 & 0 & 0.25 \\ 0 & 0.5 & -0.75 \end{bmatrix} \begin{bmatrix} 0 & 0 & 1 \\ -2 & -3 & 0 \\ 0 & 2 & -3 \end{bmatrix} \begin{bmatrix} 4 & 0 & 0 \\ 0 & 6 & 2 \\ 0 & 4 & 0 \end{bmatrix}$$

$$= \begin{bmatrix} 0 & 0 & 0.25 \\ 0 & 0.5 & -0.75 \\ -1 & -3 & 2.25 \end{bmatrix} \begin{bmatrix} 4 & 0 & 0 \\ 0 & 6 & 2 \\ 0 & 4 & 0 \end{bmatrix} = \begin{bmatrix} 0 & 1 & 0 \\ 0 & 0 & 1 \\ -4 & -9 & -6 \end{bmatrix}$$

$$B_C = P_C B = \begin{bmatrix} 0.25 & 0 & 0 \\ 0 & 0 & -0.25 \\ 0 & 0.5 & -0.75 \end{bmatrix} \begin{bmatrix} 0 \\ 2 \\ 0 \end{bmatrix} = \begin{bmatrix} 0 \\ 0 \\ 1 \end{bmatrix}$$

$$C_C = C P_C^{-1} = \begin{bmatrix} 1 & 0 & 0 \end{bmatrix} \begin{bmatrix} 4 & 0 & 0 \\ 0 & 6 & 2 \\ 0 & 4 & 0 \end{bmatrix} = \begin{bmatrix} 4 & 0 & 0 \end{bmatrix}$$

The controllable phase variable form of state model is given by,

$$\begin{bmatrix} \dot{z}_1 \\ \dot{z}_2 \\ \dot{z}_3 \end{bmatrix} = \begin{bmatrix} 0 & 1 & 0 \\ 0 & 0 & 1 \\ -4 & -9 & -6 \end{bmatrix} \begin{bmatrix} z_1 \\ z_2 \\ z_3 \end{bmatrix} + \begin{bmatrix} 0 \\ 0 \\ 1 \end{bmatrix} u$$

$$y = \begin{bmatrix} 4 & 0 & 0 \end{bmatrix} \begin{bmatrix} z_1 \\ z_2 \\ z_3 \end{bmatrix}$$

Alternate method to find P_C

From the given state model we get,

$$A = \begin{bmatrix} 0 & 0 & 1 \\ -2 & -3 & 0 \\ 0 & 2 & -3 \end{bmatrix}$$

$$[\lambda I - A] = \lambda \begin{bmatrix} 1 & 0 & 0 \\ 0 & 1 & 0 \\ 0 & 0 & 1 \end{bmatrix} - \begin{bmatrix} 0 & 0 & 1 \\ -2 & -3 & 0 \\ 0 & 2 & -3 \end{bmatrix} = \begin{bmatrix} \lambda & 0 & -1 \\ 2 & \lambda+3 & 0 \\ 0 & -2 & \lambda+3 \end{bmatrix}$$

$$|\lambda I - A| = \begin{vmatrix} \lambda & 0 & -1 \\ 2 & \lambda+3 & 0 \\ 0 & -2 & \lambda+3 \end{vmatrix} \quad \begin{aligned} &= \lambda(\lambda+3)^2 - 1(-4) = \lambda(\lambda^2 + 6\lambda + 9) + 4 \\ &= \lambda^3 + 6\lambda^2 + 9\lambda + 4 \end{aligned}$$

The characteristic equation is, $\lambda^3 + 6\lambda^2 + 9\lambda + 4 = 0$

The standard form of characteristic equation when n = 3 is given by

$$\lambda^3 + a_1\lambda^2 + a_2\lambda + a_3 = 0$$

On comparing the characteristic equation of the system with standard form we get,

$$a_1 = 6, \qquad a_2 = 9 \text{ and } a_3 = 4$$

$$\therefore \ \mathbf{W} = \begin{bmatrix} a_2 & a_1 & 1 \\ a_1 & 1 & 0 \\ 1 & 0 & 0 \end{bmatrix} = \begin{bmatrix} 9 & 6 & 1 \\ 6 & 1 & 0 \\ 1 & 0 & 0 \end{bmatrix}$$

$$\therefore \ \mathbf{P_C^{-1}} = \mathbf{Q_C W} = \begin{bmatrix} 0 & 0 & 4 \\ 2 & -6 & 18 \\ 0 & 4 & -24 \end{bmatrix}\begin{bmatrix} 9 & 6 & 1 \\ 6 & 1 & 0 \\ 1 & 0 & 0 \end{bmatrix} = \begin{bmatrix} 4 & 0 & 0 \\ 0 & 6 & 2 \\ 0 & 4 & 0 \end{bmatrix}$$

$$\therefore \ \mathbf{P_C} = (\mathbf{P_C^{-1}})^{-1} = \begin{bmatrix} 4 & 0 & 0 \\ 0 & 6 & 2 \\ 0 & 4 & 0 \end{bmatrix}^{-1}$$

$$\text{Let } \mathbf{T} = \begin{bmatrix} 4 & 0 & 0 \\ 0 & 6 & 2 \\ 0 & 4 & 0 \end{bmatrix} \ ; \ \therefore \ \mathbf{P_C} = \mathbf{T^{-1}}$$

$$\mathbf{T^{-1}} = \frac{[\text{Cofactor of } \mathbf{T}]^T}{\text{Determinant of } \mathbf{T}} = \frac{\mathbf{T_{cof}^T}}{\Delta_T} \ ; \quad \Delta_T = \begin{vmatrix} 4 & 0 & 0 \\ 0 & 6 & 2 \\ 0 & 4 & 0 \end{vmatrix} = 4(-4 \times 2) = -32$$

$$\mathbf{T_{cof}^T} = \begin{bmatrix} -8 & 0 & 0 \\ 0 & 0 & -16 \\ 0 & -8 & 24 \end{bmatrix}^T = \begin{bmatrix} -8 & 0 & 0 \\ 0 & 0 & -8 \\ 0 & -16 & 24 \end{bmatrix}$$

$$\therefore \ \mathbf{P_C} = \mathbf{T^{-1}} = \frac{\mathbf{T_{cof}^T}}{\Delta_T} = \frac{1}{-32}\begin{bmatrix} -8 & 0 & 0 \\ 0 & 0 & -8 \\ 0 & -16 & 24 \end{bmatrix} = \begin{bmatrix} 0.25 & 0 & 0 \\ 0 & 0 & 0.25 \\ 0 & 0.5 & -0.75 \end{bmatrix}$$

5.8 CONTROL SYSTEM DESIGN VIA POLE PLACEMENT BY STATE FEEDBACK

In the conventional approach to the design of a single-input, single-output control system, a controller or compensator is designed such that the dominant closed-loop poles have a desired damping ratio, ζ and undamped natural frequency, ω_n. In the compensated system the output alone is used as feedback signal to achieve desired performance. In state space design any inner parameter or variable of a system can be used for feedback. If the state variables (inner parameters or variables of the system) are used for feedback, then the system can be optimized for satisfying a desired performance index.

In control system design by pole placement or pole assignment technique, the state variables are used for feedback, to achieve desired closed loop poles. The advantage in this system is that the closed loop poles may be placed at any desired locations by means of state feedback through an appropriate state feedback gain matrix, **K**. The necessary and sufficient condition to be satisfied by the system for arbitrary pole placement is that the system be completely state controllable.

Consider the nth order single-input single-output system with and without state variable feedback as shown in fig 5.1. The state model of the system without state feedback is given by,

$$\dot{X} = AX + Bu \qquad\qquad .....(5.94)$$
$$y = CX \qquad\qquad .....(5.95)$$

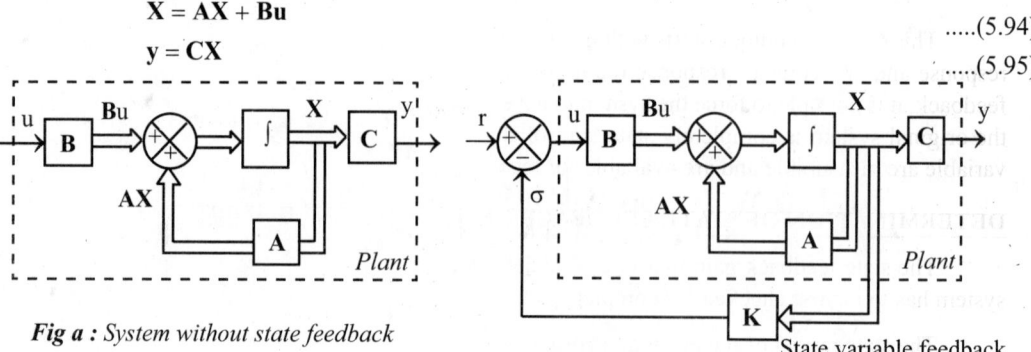

Fig a : *System without state feedback*

State variable feedback

Fig b : *System with state variable feedback*

Fig 5.1 : *The nth order single - input single - output system*

Let, r = System input when state variable feedback is employed.

σ = Feedback signal obtained from state variables.

u = Plant input.

The feedback signal, σ is obtained from state feedback and it is related to the state variables by the equation,

$$\sigma = KX \qquad\qquad .....(5.96)$$

where **K** = State feedback gain matrix of order (1 × n) and

$$K = [k_1 \ k_2 \ k_3 \ ... \ k_n] \qquad\qquad .....(5.97)$$

In system employing state variable feedback, the plant input, u is the difference between system input, r and feedback input, σ.

∴ Plant input, u = r − σ $\qquad\qquad .....(5.98)$

On substituting, σ = **KX** in equation (5.98) we get,

$$u = r - KX \qquad\qquad .....(5.99)$$

The equation (5.99) is called control law.

The state equation of the system with state variable feedback is obtained by substituting the expression for u, from equation (5.99) in equation (5.94).

$$\therefore \ \dot{X} = AX + Bu = AX + B\,(r - KX)$$

$$= AX + Br - BKX = (A - BK)\,X + Br$$

Therefore, the state model of the system with state variable feedback is given by the following equations [equation (5.100) and (5.101)]

$$\dot{X} = (A - BK)\,X + Br \qquad\qquad .....(5.100)$$
$$y = CX \qquad\qquad .....(5.101)$$

where, $\mathbf{K} = [k_1 \ k_2 \ k_3 \ \ldots \ k_n]$

and $r = u + \mathbf{KX}$

This design technique starts with the determination of desired closed-loop poles to satisfy transient response and/or frequency response requirements. By choosing an appropriate gain matrix, $\mathbf{K}$ for state feedback, it is possible to force the system to have closed loop poles at the desired locations, provided that the original system is completely state controllable. In this design technique it is assumed that all state variable are measurable and are available for feedback.

DETERMINATION OF STATE FEEDBACK GAIN MATRIX, K

The state feedback gain matrix can be determined by three methods. In all the three methods, the system has to be first checked for complete state controllability.

The state model of the original n^{th} order system is given by

$$\dot{\mathbf{X}} = \mathbf{AX} + \mathbf{B}u$$

$$y = \mathbf{CX}$$

To check for controllability of original system, determine the composite matrix for controllability, $\mathbf{Q}_C$.

where, $\mathbf{Q}_C = [\mathbf{B} \ \mathbf{AB} \ \mathbf{A^2B} \ \ldots \ \mathbf{A^{n-1}B}]$

Then calculate the determinant of $\mathbf{Q}_C$. If the determinant of $\mathbf{Q}_C$ is not equal to zero, then the rank of $\mathbf{Q}_C$ is n and so the system is completely state controllable. (Here n is the order of the system). If the rank is not equal to n then arbitrary pole placement is not possible. When the system is completely state controllable any one of the following method can be used to find $\mathbf{K}$.

METHOD - I

1. Determine the characteristic polynomial of original system. The characteristic polynomial is given by $|\lambda\mathbf{I}-\mathbf{A}| = 0$

    Let, $|\lambda\mathbf{I}-\mathbf{A}| = \lambda^n + a_1\lambda^{n-1} + a_2\lambda^{n-2} + \ldots + a_{n-1}\lambda + a_n$.

2. Determine the desired characteristic polynomial from the specified closed loop poles.

    Let the specified or desired closed loop poles be $\mu_1, \mu_2, \mu_3, \ldots \mu_n$.

    Now the desired characteristic polynomial is given by

    $$(\lambda-\mu_1)(\lambda-\mu_2)(\lambda-\mu_3) \ldots (\lambda-\mu_n) = \lambda^n + \alpha_1\lambda^{n-1} + \alpha_2\lambda^{n-2} + \ldots + \alpha_{n-1}\lambda + \alpha_n$$

3. Determine the transformation matrix, $\mathbf{P}_C$ which transforms the original state model to controllable phase variable form.

    The transformation matrix, $\mathbf{P}_C = \begin{bmatrix} P_1 \\ P_1A \\ \vdots \\ P_1A^{n-1} \end{bmatrix}$

    and, $\mathbf{P}_1 = [0 \ 0 \ \ldots \ 0 \ 1] \mathbf{Q}_C^{-1}$

4.   Determine the state feedback gain matrix, from the following equation.

$$K = [\alpha_n - a_n \quad \alpha_{n-1} - a_{n-1} \quad ... \quad \alpha_2 - a_2 \quad \alpha_1 - a_1] \, P_C$$

> *Note : If the given system state model is in controllable phase variable form then* $P_C = I$, *unit matrix.*

METHOD - II

1.   Determine the characteristic polynomial of the system with state feedback, which is given by, $|\lambda I - (A - BK)| = 0$.

Here take, $K = [k_1 \quad k_2 \quad k_3 \quad ... \quad k_n]$

Let $|\lambda I - (A - BK)| = |\lambda I - A + BK| = \lambda^n + b_1 \lambda^{n-1} + b_2 \lambda^{n-2} + ... + b_{n-1} \lambda + b_n$.

The coefficients of this polynomial $b_1, b_2, b_3, ... b_n$ will be functions of $k_1, k_2, k_3, ... k_n$.

2.   Determine the desired characteristic polynomial from the specified closed loop poles.

Let the specified or desired closed loop poles be $\mu_1, \mu_2, \mu_3, ... \mu_n$.
Now the desired characteristic polynomial is given by,

$$(\lambda - \mu_1)(\lambda - \mu_2)(\lambda - \mu_3)...(\lambda - \mu_n) = \lambda^n + \alpha_1 \lambda^{n-1} + \alpha_2 \lambda^{n-2} + ... + \alpha_{n-1} \lambda + \alpha_n.$$

3.   By equating the coefficients of polynomials obtained in step-1 and step-2, we get n-number of equations.

i.e, $b_1 = \alpha_1$; $b_2 = \alpha_2$; ... ; $b_{n-1} = \alpha_{n-1}$ and $b_n = a_n$.

On solving these equations we get the elements $k_1, k_2, ... k_n$ of state feedback gain matrix, **K**.

> *Note : Method - II is suitable only for low values of n (i.e, for 2nd and 3rd order systems) otherwise calculations will be tedious.*

METHOD - III

1.   Determine the desired characteristic polynomial from the specified closed loop poles. Let the specified or desired closed loop poles be $\mu_1, \mu_2, \mu_3, ... \mu_n$.

Now the desired characteristic polynomial is given by,

$$(\lambda - \mu_1)(\lambda - \mu_2)(\lambda - \mu_3)... (\lambda - \mu_n) = \lambda^n + \alpha_1 \lambda^{n-1} + \alpha_2 \lambda^{n-2} + ... + a_{n-1} \lambda + \alpha_n.$$

2.   Determine the matrix $\phi(A)$ using the coefficients of desired characteristic polynomial.

$$\phi(A) = A^n + \alpha_1 A^{n-1} + \alpha_2 A^{n-2} + ... + \alpha_{n-1} A + \alpha_n I$$

3.   Calculate the state feedback gain matrix, **K** using the Ackermann's formula given below.

$$K = [0 \quad 0 \quad .... \quad 0 \quad 1] \, Q_C^{-1} \phi(A)$$

where, $Q_C = [B \quad AB \quad A^2B \quad ... \quad A^{n-1}B]$

EXAMPLE 5.8

Consider a linear system described by the transfer function

$$\frac{Y(s)}{U(s)} = \frac{10}{s(s+1)(s+2)}$$

Design a feedback controller with a state feedback so that the closed loop poles are placed at $-2, -1 \pm j1$

SOLUTION

To determine the state equation of the system

Given that, $\dfrac{Y(s)}{U(s)} = \dfrac{10}{s(s+1)(s+2)}$ ..... (5.8.1)

On cross multiplying the equation (5.8.1) we get,

$$Y(s)\,[s(s+1)(s+2)] \;=\; 10\,U(s)$$
$$Y(s)\,[s(s^2+3s+2)] \;=\; 10\,U(s)$$
$$Y(s)\,[s^3+3s^2+2s] \;=\; 10\,U(s)$$
$$\therefore s^3 Y(s) + 3s^2\,Y(s) + 2sY(s) = 10\,U(s) \qquad\qquad .....(5.8.2)$$

On taking inverse Laplace transform of equation (5.8.2) we get,

$$\dddot{y} + 3\ddot{y} + 2\dot{y} = 10u \qquad\qquad .....(5.8.3)$$

Let us define state variables as follows,

$$x_1 = y \;;\; x_2 = \dot{y} \;;\; x_3 = \ddot{y}$$

Put $\dddot{y} = \dot{x}_3 \;;\; \ddot{y} = x_3 \;;\; \dot{y} = x_2$ and $y = x_1$ in equation (5.8.3)

$$\therefore \; \dot{x}_3 + 3x_3 + 2x_2 = 10u$$

or $\dot{x}_3 = -2x_2 - 3x_3 + 10u$

The state equations governing the system are

$$\dot{x}_1 = x_2$$
$$\dot{x}_2 = x_3$$
$$\dot{x}_3 = -2x_2 - 3x_3 + 10u$$

The state equation in the matrix form is

$$\begin{bmatrix} \dot{x}_1 \\ \dot{x}_2 \\ \dot{x}_3 \end{bmatrix} = \begin{bmatrix} 0 & 1 & 0 \\ 0 & 0 & 1 \\ 0 & -2 & -3 \end{bmatrix} \begin{bmatrix} x_1 \\ x_2 \\ x_3 \end{bmatrix} + \begin{bmatrix} 0 \\ 0 \\ 10 \end{bmatrix} u \qquad\qquad ......(5.8.4)$$

Check for controllability

Given that, $\mathbf{A} = \begin{bmatrix} 0 & 1 & 0 \\ 0 & 0 & 1 \\ 0 & -2 & -3 \end{bmatrix}$ and $\mathbf{B} = \begin{bmatrix} 0 \\ 0 \\ 10 \end{bmatrix}$

$$A^2 = AA = \begin{bmatrix} 0 & 1 & 0 \\ 0 & 0 & 1 \\ 0 & -2 & -3 \end{bmatrix} \begin{bmatrix} 0 & 1 & 0 \\ 0 & 0 & 1 \\ 0 & -2 & -3 \end{bmatrix} = \begin{bmatrix} 0 & 0 & 1 \\ 0 & -2 & -3 \\ 0 & 6 & 7 \end{bmatrix} \qquad .....(5.8.5)$$

$$AB = \begin{bmatrix} 0 & 1 & 0 \\ 0 & 0 & 1 \\ 0 & -2 & -3 \end{bmatrix} \begin{bmatrix} 0 \\ 0 \\ 10 \end{bmatrix} = \begin{bmatrix} 0 \\ 10 \\ -30 \end{bmatrix}$$

$$A^2B = \begin{bmatrix} 0 & 0 & 1 \\ 0 & -2 & -3 \\ 0 & 6 & 7 \end{bmatrix} \begin{bmatrix} 0 \\ 0 \\ 10 \end{bmatrix} = \begin{bmatrix} 10 \\ -30 \\ 70 \end{bmatrix}$$

Composite matrix for controllability, $Q_C = [B \ \ AB \ \ A^2B] = \begin{bmatrix} 0 & 0 & 10 \\ 0 & 10 & -30 \\ 10 & -30 & 70 \end{bmatrix}$ ....(5.8.6)

Determinant of $\ \ Q_C = \Delta_{QC} = \begin{bmatrix} 0 & 0 & 10 \\ 0 & 10 & -30 \\ 10 & -30 & 70 \end{bmatrix} = 10(-10 \times 10) = -1000$ ....(5.8.7)

Since, $\Delta_{Qc} \neq 0$, the system is completely state controllable.

To find Q_C^{-1}

From equation (5.8.6) and (5.8.7) we get

$$Q_C = \begin{bmatrix} 0 & 0 & 10 \\ 0 & 10 & -30 \\ 10 & -30 & 70 \end{bmatrix} \text{ and } \Delta_{QC} = -1000.$$

$$Q_C^{-1} = \frac{[\text{Cofactor of } Q_C]^T}{\text{Determinant of } Q_C} = \frac{1}{\Delta_{QC}} \begin{bmatrix} 610 & -300 & -100 \\ 300 & -100 & 0 \\ -100 & 0 & 0 \end{bmatrix}^T$$

$$= \frac{1}{-1000} \begin{bmatrix} 610 & 300 & -100 \\ 300 & -100 & 0 \\ -100 & 0 & 0 \end{bmatrix} = \begin{bmatrix} -0.61 & 0.3 & 0.1 \\ -0.3 & 0.1 & 0 \\ 0.1 & 0 & 0 \end{bmatrix} \qquad .....(5.8.8)$$

To find desired characteristic polynomial

The desired closed loop poles are

$$\mu_1 = -2, \ \mu_2 = -1 + j1 \text{ and } \mu_3 = -1 - j1$$

Hence the desired characteristic polynomial is

$$(\lambda - \mu_1)(\lambda - \mu_2)(\lambda - \mu_3) = (\lambda + 2)(\lambda + 1 - j1)(\lambda + 1 + j1)$$

$$= (\lambda + 2)((\lambda + 1)^2 - (j1)^2)$$

$$= (\lambda + 2)(\lambda^2 + 2\lambda + 1 + 1)$$

$$= (\lambda + 2)(\lambda^2 + 2\lambda + 2)$$

$$= \lambda^3 + 2\lambda^2 + 2\lambda + 2\lambda^2 + 4\lambda + 4$$

$$= \lambda^3 + 4\lambda^2 + 6\lambda + 4$$

The desired characteristic polynomial is

$$\lambda^3 + 4\lambda^2 + 6\lambda + 4 = 0 \qquad\qquad .....(5.8.9)$$

To determine the state variable feedback matrix, K

Method - I

The characteristic polynomial of original system is given by $|\lambda I - A| = 0$

$$[\lambda I - A] = \lambda \begin{bmatrix} 1 & 0 & 0 \\ 0 & 1 & 0 \\ 0 & 0 & 1 \end{bmatrix} - \begin{bmatrix} 0 & 1 & 0 \\ 0 & 0 & 1 \\ 0 & -2 & -3 \end{bmatrix} = \begin{bmatrix} \lambda & -1 & 0 \\ 0 & \lambda & -1 \\ 0 & 2 & \lambda+3 \end{bmatrix}$$

$$|\lambda I - A| = \begin{vmatrix} \lambda & -1 & 0 \\ 0 & \lambda & -1 \\ 0 & 2 & \lambda+3 \end{vmatrix} = \lambda[\lambda(\lambda+3)+2] = \lambda^2(\lambda+3) + 2\lambda = \lambda^3 + 3\lambda^2 + 2\lambda$$

The characteristic polynomial of original system is,

$$\lambda^3 + 3\lambda^2 + 2\lambda = 0 \qquad\qquad .....(5.8.10)$$

From equation (5.8.9) we get the desired characteristic polynomial as

$$\lambda^3 + 4\lambda^2 + 6\lambda + 4 = 0 \qquad\qquad .....(5.8.11)$$

From equation (5.8.8) we get,

$$Q_C^{-1} = \begin{bmatrix} -0.61 & 0.3 & 0.1 \\ -0.3 & 0.1 & 0 \\ 0.1 & 0 & 0 \end{bmatrix}$$

$$P_1 = [0\ 0\ 1] Q_C^{-1} = [0\ 0\ 1] \begin{bmatrix} -0.61 & 0.3 & 0.1 \\ -0.3 & 0.1 & 0 \\ 0.1 & 0 & 0 \end{bmatrix} = [0.1\ 0\ 0]$$

$$P_1 A = [0.1\ 0\ 0] \begin{bmatrix} 0 & 1 & 0 \\ 0 & 0 & 1 \\ 0 & -2 & -3 \end{bmatrix} = [0\ 0.1\ 0]$$

$$P_1 A^2 = [0.1\ 0\ 0] \begin{bmatrix} 0 & 0 & 1 \\ 0 & -2 & -3 \\ 0 & 6 & 7 \end{bmatrix} = [0\ 0\ 0.1]$$

$$\therefore P_C = \begin{bmatrix} P_1 \\ P_1 A \\ P_1 A^2 \end{bmatrix} = \begin{bmatrix} 0.1 & 0 & 0 \\ 0 & 0.1 & 0 \\ 0 & 0 & 0.1 \end{bmatrix}$$

The state feedback gain matrix, $K = [\alpha_3-a_3\ \ \alpha_2-a_2\ \ \alpha_1-a_1] P_C$

From equation (5.8.11) we get, $\alpha_3 = 4$; $\alpha_2 = 6$; $\alpha_1 = 4$

From equation (5.8.10) we get, $a_3 = 0$; $a_2 = 2$; $a_1 = 3$

$$\therefore \mathbf{K} = [4-0 \quad 6-2 \quad 4-3]\, \mathbf{P}_C$$

$$= [4 \ \ 4 \ \ 1] \begin{bmatrix} 0.1 & 0 & 0 \\ 0 & 0.1 & 0 \\ 0 & 0 & 0.1 \end{bmatrix} = [0.4 \ \ 0.4 \ \ 0.1]$$

Method - II

From the given state model we get,

$$\mathbf{A} = \begin{bmatrix} 0 & 1 & 0 \\ 0 & 0 & 1 \\ 0 & -2 & -3 \end{bmatrix} \text{ and } \mathbf{B} = \begin{bmatrix} 0 \\ 0 \\ 10 \end{bmatrix}$$

Let, $\mathbf{K} = [k_1 \ \ k_2 \ \ k_3]$

The characteristic polynomial of the system with state feedback is given by,

$$|\lambda \mathbf{I} - (\mathbf{A} - \mathbf{BK})| = |\lambda \mathbf{I} - \mathbf{A} + \mathbf{BK}| = 0$$

$$[\lambda \mathbf{I} - \mathbf{A} + \mathbf{BK}] = \lambda \begin{bmatrix} 1 & 0 & 0 \\ 0 & 1 & 0 \\ 0 & 0 & 1 \end{bmatrix} - \begin{bmatrix} 0 & 1 & 0 \\ 0 & 0 & 1 \\ 0 & -2 & -3 \end{bmatrix} + \begin{bmatrix} 0 \\ 0 \\ 10 \end{bmatrix} [k_1 \ \ k_2 \ \ k_3]$$

$$= \begin{bmatrix} \lambda & 0 & 0 \\ 0 & \lambda & 0 \\ 0 & 0 & \lambda \end{bmatrix} - \begin{bmatrix} 0 & 1 & 0 \\ 0 & 0 & 1 \\ 0 & -2 & -3 \end{bmatrix} + \begin{bmatrix} 0 & 0 & 0 \\ 0 & 0 & 0 \\ 10k_1 & 10k_2 & 10k_3 \end{bmatrix}$$

$$= \begin{bmatrix} \lambda & -1 & 0 \\ 0 & \lambda & -1 \\ 10k_1 & 2+10k_2 & \lambda+3+10k_3 \end{bmatrix}$$

$$|\lambda \mathbf{I} - \mathbf{A} + \mathbf{BK}| = \begin{vmatrix} \lambda & -1 & 0 \\ 0 & \lambda & -1 \\ 10k_1 & 2+10k_2 & \lambda+3+10k_3 \end{vmatrix}$$

$$= \lambda[\lambda(\lambda + 3 + 10k_3) + 2 + 10k_2] + I[10k_1]$$

$$= \lambda^2(\lambda + 3 + 10k_3) + (2 + 10k_2)\,\lambda + 10k_1$$

$$= \lambda^3 + (3 + 10k_3)\lambda^2 + (2 + 10k_2)\lambda + 10k_1$$

The characteristic polynomial of the system with state feedback is

$$\lambda^3 + (3 + 10k_3)\,\lambda^2 + (2 + 10k_2)\,\lambda + 10k_1 = 0 \qquad \qquad .....(5.8.12)$$

From equation (5.8.9) we get the desired characteristic polynomial as,

$$\lambda_3 + 4\lambda^2 + 6\lambda + 4 = 0 \qquad\qquad\qquad .....(5.8.13)$$

On equating the coefficients of λ^0 term (constant) in equations (5.8.12) and (5.8.13) we get,

$$10k_1 = 4 \ ; \ \therefore k_1 = \frac{4}{10} = 0.4$$

On equating the coefficients of λ^1 term in equations (5.8.12) and (5.8.13) we get,

$$2 + 10k_3 = 6 \ ; \ \therefore k_2 = \frac{6-2}{10} = 0.4$$

On equating the coefficients of λ^2 term in equations (5.8.12) and (5.8.13) we get,

$$3 + 10k_3 = 4 \ ; \ \therefore k_3 = \frac{4-3}{10} = 0.1$$

The state feedback gain matrix, $\mathbf{K} = [k_1 \quad k_2 \quad k_3] = [0.4 \quad 0.4 \quad 0.1]$

Method - III

From equation (5.8.9) we get the desired characteristic polynomial as,

$$\lambda^3 + 4\lambda^2 + 6\lambda + 4 = 0 \qquad\qquad\qquad .....(5.8.14)$$

Here, $\phi(\mathbf{A}) = \mathbf{A}^3 + \alpha_1\mathbf{A}^2 + \alpha_2\mathbf{A} + \alpha_3\mathbf{I}$

From equation (5.8.14) we get, $\alpha_1 = 4; \ \alpha_2 = 6 \ ; \ \alpha_3 = 4$

From the given state equation and equation (5.8.5) we get,

$$\mathbf{A} = \begin{bmatrix} 0 & 1 & 0 \\ 0 & 0 & 1 \\ 0 & -2 & -3 \end{bmatrix} \text{ and } \mathbf{A}^2 = \begin{bmatrix} 0 & 0 & 1 \\ 0 & -2 & -3 \\ 0 & 6 & 7 \end{bmatrix}$$

$$\therefore \mathbf{A}^3 = \mathbf{A}^2\mathbf{A} = \begin{bmatrix} 0 & 0 & 1 \\ 0 & -2 & -3 \\ 0 & 6 & 7 \end{bmatrix}\begin{bmatrix} 0 & 1 & 0 \\ 0 & 0 & 1 \\ 0 & -2 & -3 \end{bmatrix} = \begin{bmatrix} 0 & -2 & -3 \\ 0 & 6 & 7 \\ 0 & -14 & -15 \end{bmatrix}$$

$$\therefore \ \phi(\mathbf{A}) = \mathbf{A}^3 + \alpha_1\mathbf{A}^2 + \alpha_2\mathbf{A} + \alpha_3\mathbf{I}$$

$$= \begin{bmatrix} 0 & -2 & -3 \\ 0 & 6 & 7 \\ 0 & -14 & -15 \end{bmatrix} + 4\begin{bmatrix} 0 & 0 & 0 \\ 0 & -2 & -3 \\ 0 & 6 & 7 \end{bmatrix} + 6\begin{bmatrix} 0 & 1 & 0 \\ 0 & 0 & 1 \\ 0 & -2 & -3 \end{bmatrix} + 4\begin{bmatrix} 1 & 0 & 0 \\ 0 & 1 & 0 \\ 0 & 0 & 1 \end{bmatrix}$$

$$= \begin{bmatrix} 0 & -2 & -3 \\ 0 & 6 & 7 \\ 0 & -14 & -15 \end{bmatrix} + 4\begin{bmatrix} 0 & 0 & 0 \\ 0 & -8 & -12 \\ 0 & 24 & 28 \end{bmatrix} + \begin{bmatrix} 0 & 6 & 0 \\ 0 & 0 & 6 \\ 0 & -12 & -18 \end{bmatrix} + \begin{bmatrix} 4 & 0 & 0 \\ 0 & 4 & 0 \\ 0 & 0 & 4 \end{bmatrix} = \begin{bmatrix} 4 & 4 & 1 \\ 0 & 2 & 1 \\ 0 & -2 & -1 \end{bmatrix}$$

From equation (5.8.8) we get, $\mathbf{Q}_C^{-1} = \begin{bmatrix} -0.61 & 0.3 & 0.1 \\ -0.3 & 0.1 & 0 \\ 0.1 & 0 & 0 \end{bmatrix}$

From Ackermann's formula we get,

$$\mathbf{K} = \begin{bmatrix} 0 & 0 & 1 \end{bmatrix} \mathbf{Q}_C^{-1} \phi(\mathbf{A})$$

$$= \begin{bmatrix} 0 & 0 & 1 \end{bmatrix} \begin{bmatrix} -0.61 & 0.3 & 0.1 \\ -0.3 & 0.1 & 0 \\ 0.1 & 0 & 0 \end{bmatrix} \begin{bmatrix} 4 & 4 & 1 \\ 0 & 2 & 1 \\ 0 & -2 & -1 \end{bmatrix}$$

$$= \begin{bmatrix} 0.1 & 0 & 0 \end{bmatrix} \begin{bmatrix} 4 & 4 & 1 \\ 0 & 2 & 1 \\ 0 & -2 & -1 \end{bmatrix} = \begin{bmatrix} 0.4 & 0.4 & 0.1 \end{bmatrix}$$

The state feedback gain matrix $\mathbf{K} = \begin{bmatrix} 0.4 & 0.4 & 0.1 \end{bmatrix}$

Note : *It is observed that the values of k_1, k_2, k_3 obtained by all the three methods are same. Because for a given set of poles the values of k_1, k_2, k_3, ... will be unique.*

EXAMPLE 5.9

A single-input system is described by the following state equation.

$$\begin{bmatrix} \dot{x}_1 \\ \dot{x}_2 \\ \dot{x}_3 \end{bmatrix} = \begin{bmatrix} -1 & 0 & 0 \\ 1 & -2 & 0 \\ 2 & 1 & -3 \end{bmatrix} \begin{bmatrix} x_1 \\ x_2 \\ x_3 \end{bmatrix} + \begin{bmatrix} 10 \\ 1 \\ 0 \end{bmatrix} u$$

Design a state feedback controller which will give closed-loop poles at $-1 \pm j2$, -6.

SOLUTION

Check for controllability

Given that $\mathbf{A} = \begin{bmatrix} -1 & 0 & 0 \\ 1 & -2 & 0 \\ 2 & 1 & -3 \end{bmatrix}$ and $\mathbf{B} = \begin{bmatrix} 10 \\ 1 \\ 0 \end{bmatrix}$

$$\mathbf{A}^2 = \mathbf{A}\mathbf{A} = \begin{bmatrix} -1 & 0 & 0 \\ 1 & -2 & 0 \\ 2 & 1 & -3 \end{bmatrix} \begin{bmatrix} -1 & 0 & 0 \\ 1 & -2 & 0 \\ 2 & 1 & -3 \end{bmatrix} = \begin{bmatrix} 1 & 0 & 0 \\ -3 & 4 & 0 \\ -7 & -5 & 9 \end{bmatrix} \qquad .....(5.91)$$

$$\mathbf{A}\mathbf{B} = \begin{bmatrix} -1 & 0 & 0 \\ 1 & -2 & 0 \\ 2 & 1 & -3 \end{bmatrix} \begin{bmatrix} 10 \\ 1 \\ 0 \end{bmatrix} = \begin{bmatrix} -10 \\ 8 \\ 21 \end{bmatrix}$$

$$\mathbf{A}^2\mathbf{B} = \begin{bmatrix} 1 & 0 & 0 \\ -3 & 4 & 0 \\ -7 & -5 & 9 \end{bmatrix} \begin{bmatrix} 10 \\ 1 \\ 0 \end{bmatrix} = \begin{bmatrix} 10 \\ -26 \\ -75 \end{bmatrix}$$

$$\left.\begin{matrix} \text{Composite matrix} \\ \text{for controllability} \end{matrix}\right\} \mathbf{Q}_C = \begin{bmatrix} \mathbf{B} & \mathbf{A}\mathbf{B} & \mathbf{A}^2\mathbf{B} \end{bmatrix} = \begin{bmatrix} 10 & -10 & 10 \\ 1 & 8 & -26 \\ 0 & 21 & -75 \end{bmatrix} \qquad .....(5.9.2)$$

$$\text{Determinant of } \mathbf{Q}_C = \Delta_{QC} = \begin{vmatrix} 10 & -10 & 10 \\ 1 & 8 & -26 \\ 0 & 21 & -75 \end{vmatrix}$$

$$\therefore \Delta_{QC} = 10\,[\,8 \times (-75) - 21 \times (-26)\,] + 10\,[-75] + 10\,[21]$$

$$= -540 - 750 + 210$$

$$= -1080 \qquad\qquad\qquad .....(5.9.3)$$

Since, $\Delta_{QC} \neq 0$, The system is completely state controllable.

To find $\mathbf{Q}_C^{-1}$

From equations (5.9.2) and (5.9.3) we get,

$$\mathbf{Q}_C = \begin{bmatrix} 10 & -10 & 10 \\ 1 & 8 & -26 \\ 0 & 21 & -75 \end{bmatrix} \text{ and } \Delta_{QC} = -1080$$

$$\mathbf{Q}_C^{-1} = \frac{[\text{Cofactor of } \mathbf{Q}_C]^T}{\text{Determinant of } \mathbf{Q}_C} = \frac{1}{\Delta_{QC}} \begin{bmatrix} -54 & 75 & 21 \\ -540 & -750 & -210 \\ 180 & 270 & 90 \end{bmatrix}^T$$

$$= \frac{1}{-1080} \begin{bmatrix} -54 & -540 & 180 \\ 75 & -750 & 270 \\ 21 & -210 & 90 \end{bmatrix} = \begin{bmatrix} 0.05 & 0.5 & -0.1667 \\ -0.0694 & 0.6944 & -0.25 \\ -0.0194 & 0.1944 & -0.0833 \end{bmatrix} \qquad .....(5.9.4)$$

To find desired characteristic polynomial

The desired closed loop poles are,

$$\mu_1 = -1 + j2, \ \mu_2 = -1 - j2 \text{ and } \mu_3 = -6$$

Hence the desired characteristic polynomial is,

$$(\lambda - \mu_1)(\lambda - \mu_2)(\lambda - \mu_3) = (\lambda + 1 - j2)(\lambda + 1 + j2)(\lambda + 6)$$

$$= ((\lambda + 1)^2 - (j2)^2)(\lambda + 6)$$

$$= (\lambda^2 + 2\lambda + 1 + 4)(\lambda + 6)$$

$$= (\lambda^2 + 2\lambda + 5)(\lambda + 6)$$

$$= \lambda^3 + 2\lambda^2 + 5\lambda + 6\lambda^2 + 12\lambda + 30$$

$$= \lambda^3 + 8\lambda^2 + 17\lambda + 30$$

The desired characteristic polynomial is $\lambda^3 + 8\lambda^2 + 17\lambda + 30 = 0$ .....(5.9.5)

To determine the state variable feedback matrix, K

Method - I

The characteristic equation of original system is given by,

$$[\lambda I - A] = \lambda \begin{bmatrix} 1 & 0 & 0 \\ 0 & 1 & 0 \\ 0 & 0 & 1 \end{bmatrix} - \begin{bmatrix} -1 & 0 & 0 \\ 1 & -2 & 0 \\ 2 & 1 & -3 \end{bmatrix} = \begin{bmatrix} \lambda+1 & 0 & 0 \\ -1 & \lambda+2 & 0 \\ -2 & -1 & \lambda+3 \end{bmatrix}$$

$$|\lambda I - A| = \begin{vmatrix} \lambda+1 & 0 & 0 \\ -1 & \lambda+2 & 0 \\ -2 & -1 & \lambda+3 \end{vmatrix} = (\lambda+1)(\lambda+2)(\lambda+3)$$

$$= (\lambda+1)(\lambda^2 + 5\lambda + 6) = \lambda^3 + 5\lambda^2 + 6\lambda + \lambda^2 + 5\lambda + 6$$

$$= \lambda^3 + 6\lambda^2 + 11\lambda + 6$$

The characteristic polynomial of original system is,

$$\lambda^3 + 6\lambda^2 + 11\lambda + 6 = 0 \qquad\qquad\qquad .....(5.9.6)$$

From equation (5.9.5) we get the desired characteristic polynomial as

$$\lambda^3 + 8\lambda^2 + 17\lambda + 30 = 0 \qquad\qquad\qquad .....(5.9.7)$$

From equation (5.9.4) we get,

$$Q_C^{-1} = \begin{bmatrix} 0.05 & 0.5 & -0.1667 \\ -0.0694 & 0.6944 & -0.25 \\ -0.0194 & 0.1944 & -0.0833 \end{bmatrix}$$

$$P_1 = \begin{bmatrix} 0 & 0 & 1 \end{bmatrix} Q_C^{-1} = \begin{bmatrix} 0 & 0 & 1 \end{bmatrix} \begin{bmatrix} 0.05 & 0.5 & -0.1667 \\ -0.0694 & 0.6944 & -0.25 \\ -0.0194 & 0.1944 & -0.0833 \end{bmatrix}$$

$$= \begin{bmatrix} -0.0194 & 0.1944 & -0.0833 \end{bmatrix}$$

$$P_1 A = \begin{bmatrix} -0.0194 & 0.1944 & -0.0833 \end{bmatrix} \begin{bmatrix} -1 & 0 & 0 \\ 1 & -2 & 0 \\ 2 & 1 & -3 \end{bmatrix}$$

$$= \begin{bmatrix} 0.0472 & -0.4721 & 0.2499 \end{bmatrix}$$

$$P_1 A^2 = \begin{bmatrix} -0.0194 & 0.1944 & -0.0833 \end{bmatrix} \begin{bmatrix} 1 & 0 & 0 \\ -3 & 4 & 0 \\ -7 & -5 & 9 \end{bmatrix}$$

$$= \begin{bmatrix} -0.0195 & 1.1941 & -0.7497 \end{bmatrix}$$

$$P_C = \begin{bmatrix} P_1 \\ P_1A \\ P_1A^2 \end{bmatrix} = \begin{bmatrix} -0.0194 & 0.1944 & -0.0833 \\ 0.0472 & -0.4721 & 0.2499 \\ -0.0195 & 1.1941 & -0.7497 \end{bmatrix}$$

The state feedback gain matrix, $K = [\alpha_3 - a_3 \quad \alpha_2 - a_2 \quad \alpha_1 - a_1]P_C$

From equation (5.9.7) we get, $\alpha_3 = 30$; $\alpha_2 = 17$; $\alpha_1 = 8$

From equation (5.9.6) we get, $a_3 = 6$; $a_2 = 11$; $a_1 = 6$

$$\therefore K = [30 - 6 \quad 17 - 11 \quad 8 - 6]P_C$$

$$= [24 \ 6 \ 2]\begin{bmatrix} -0.0194 & 0.1944 & -0.0833 \\ 0.0472 & -0.4721 & 0.2499 \\ -0.0195 & 1.1941 & -0.7497 \end{bmatrix}$$

$$= [-0.22 \quad 4.22 \quad -2]$$

Method - II

From the given state model we get

$$A = \begin{bmatrix} -1 & 0 & 0 \\ 1 & -2 & 0 \\ 2 & 1 & -3 \end{bmatrix} \text{ and } B = \begin{bmatrix} 10 \\ 1 \\ 0 \end{bmatrix}$$

Let, $K = [k_1 \quad k_2 \quad k_3]$

The characteristic polynomial of the systems with state feedback is given by,

$$|\lambda I - (A - BK)| = |\lambda I - A + BK| = 0$$

$$[\lambda I - A + BK] = \lambda\begin{bmatrix} 1 & 0 & 0 \\ 0 & 1 & 0 \\ 0 & 0 & 1 \end{bmatrix} - \begin{bmatrix} -1 & 0 & 0 \\ 1 & -2 & 0 \\ 2 & 1 & -3 \end{bmatrix} + \begin{bmatrix} 10 \\ 1 \\ 0 \end{bmatrix}[k_1 \ k_2 \ k_3]$$

$$= \begin{bmatrix} \lambda & 0 & 0 \\ 0 & \lambda & 0 \\ 0 & 0 & \lambda \end{bmatrix} - \begin{bmatrix} -1 & 0 & 0 \\ 1 & -2 & 0 \\ 2 & 1 & -3 \end{bmatrix} + \begin{bmatrix} 10k_1 & 10k_2 & 10k_3 \\ k_1 & k_2 & k_3 \\ 0 & 0 & 0 \end{bmatrix}$$

$$= \begin{bmatrix} \lambda + 1 + 10k_1 & 10k_2 & 10k_3 \\ -1 + k_1 & \lambda + 2 + k_2 & k_3 \\ -2 & -1 & \lambda + 3 \end{bmatrix}$$

$$|\lambda I - A + BK| = \begin{bmatrix} \lambda + 1 + 10k_1 & 10k_2 & 10k_3 \\ -1 + k_1 & \lambda + 2 + k_2 & k_3 \\ -2 & -1 & \lambda + 3 \end{bmatrix}$$

$$= (\lambda + 1 + 10k_1)[(\lambda + 2 + k_2)(\lambda + 3) + k_3] - 10k_2[(-1 + k_1)(\lambda + 3) + 2k_3]$$
$$+ 10k_3[-(-1 + k_1) + 2(\lambda + 2 + k_2)]$$

$$= [\lambda + (1 + 10k_1)] [\lambda^2 + 3\lambda + 2\lambda + 6 + \lambda k_2 + 3k_2 + k_3]$$
$$- 10k_2[-\lambda - 3 + \lambda k_1 + 3k_1 + 2k_3] + 10k_3[1 - k_1 + 2\lambda + 4 + 2k_2]$$

$$= [\lambda + (1 + 10k_1)] [\lambda^2 + (5 + k_2)\lambda + (6 + 3k_2 + k_3)]$$
$$- 10k_2[(-1 + k_1)\lambda + (-3 + 3k_1 + 2k_3)] + 10k_3[2\lambda + (5 - k_1 + 2k_2)]$$

$$= \lambda^3 + (5 + k_2)\lambda^2 + (6 + 3k_2 + k_3)\lambda$$
$$+ (1 + 10k_1)\lambda^2 + (1 + 10k_1)(5 + k_2)\lambda + (1 + 10k_1)(6 + 3k_2 + k_3)$$
$$- 10k_2(-1 + k_1)\lambda - 10k_2(-3 + 3k_1 + 2k_3) + 20k_3\lambda + 10k_3(5 - k_1 + 2k_2)$$

$$= \lambda^3 + (6 + 10k_1 + k_2)\lambda^2 + (6 + 3k_2 + k_3 + 5 + k_2 + 50k_1 + 10k_1k_2$$
$$+ 10k_2 - 10k_1k_2 + 20k_3)\lambda + (6 + 3k_2 + k_3 + 60k_1 + 30k_1k_2$$
$$+ 10k_1k_3 + 30k_2 - 30k_1k_2 - 20k_2k_3 + 50k_3 - 10k_1k_3 + 20k_2k_3)$$

$$= \lambda^3 + (6 + 10k_1 + k_2)\lambda^2 + (11 + 50k_1 + 14k_2 + 21k_3)\lambda + (6 + 60k_1 + 33k_2 + 51k_3)$$

The characteristic polynomial of system with state feedback is

$$\lambda^3 + (6 + 10k_1 + k_2)\lambda^2 + (11 + 50k_1 + 14k_2 + 21k_3)\lambda + (6 + 60k_1 + 33k_2 + 51k_3) = 0 \qquad ...(5.9.8)$$

From equation (5.9.5) we get the desired characteristic polynomial as,

$$\lambda^3 + 8\lambda^2 + 17\lambda + 30 = 0 \qquad .....(5.9.9)$$

On equating the coefficients of λ^2 term in equations (5.9.8) and (5.9.9) we get,

$$6 + 10k_1 + k_2 = 8$$
$$10k_1 + k_2 = 8 - 6$$
$$\therefore 10k_1 + k_2 = 2 \qquad .....(5.9.10)$$

On equation the coefficients of λ^1 term in equations (5.9.8) and (5.9.9) we get,

$$11 + 50k_1 + 14k_2 + 21k_3 = 17$$
$$50k_1 + 14k_2 + 21k_3 = 17 - 11$$
$$\therefore 50k_1 + 14k_2 + 21k_3 = 6 \qquad .....(5.9.11)$$

On equating the coefficients of λ^0 term (constant) in equations (5.9.8) and (5.9.9) we get,

$$6 + 60k_1 + 33k_2 + 51k_3 = 30$$
$$60k_1 + 33k_2 + 51k_3 = 30 - 6$$
$$\therefore 60k_1 + 33k_2 + 51k_3 = 24 \qquad .....(5.9.12)$$

The equations (5.9.10), (5.9.11) and (5.9.12) can be arranged in the matrix form and k_1, k_2 and k_3 are solved using cramer's rule as shown below.

$$\begin{bmatrix} 10 & 1 & 0 \\ 50 & 14 & 21 \\ 60 & 33 & 51 \end{bmatrix} \begin{bmatrix} k_1 \\ k_2 \\ k_3 \end{bmatrix} = \begin{bmatrix} 2 \\ 6 \\ 24 \end{bmatrix}$$

$$\Delta = \begin{vmatrix} 10 & 1 & 0 \\ 50 & 14 & 21 \\ 60 & 33 & 51 \end{vmatrix} \begin{aligned} &= 10(14 \times 51 - 33 \times 21) - 1(50 \times 51 - 60 \times 21) \\ &= 210 - 1290 = -1080 \end{aligned}$$

$$\Delta_1 = \begin{vmatrix} 2 & 1 & 0 \\ 6 & 14 & 21 \\ 24 & 33 & 51 \end{vmatrix} \begin{aligned} &= 2(14 \times 51 - 33 \times 21) - 1(6 \times 51 - 24 \times 21) \\ &= 42 + 198 = 240 \end{aligned}$$

$$\Delta_2 = \begin{vmatrix} 10 & 2 & 0 \\ 50 & 6 & 21 \\ 60 & 24 & 51 \end{vmatrix} \begin{aligned} &= 10(6 \times 51 - 24 \times 21) - 2(50 \times 51 - 60 \times 21) \\ &= -1980 - 2580 = -4560 \end{aligned}$$

$$\Delta_3 = \begin{vmatrix} 10 & 1 & 2 \\ 50 & 14 & 6 \\ 60 & 33 & 24 \end{vmatrix} \begin{aligned} &= 10(14 \times 24 - 33 \times 6) - 1(50 \times 24 - 60 \times 6) + 2(50 \times 33 - 60 \times 14) \\ &= 1380 - 840 + 1620 = 2160 \end{aligned}$$

$$k_1 = \frac{\Delta_1}{\Delta} = \frac{240}{-1080} = -0.22$$

$$k_2 = \frac{\Delta_2}{\Delta} = \frac{-4560}{-1080} = 4.22$$

$$k_3 = \frac{\Delta_3}{\Delta} = \frac{2160}{-1080} = -2$$

The state feedback gain matrix, $\mathbf{K} = [k_1 \quad k_2 \quad k_3] = [-0.22 \quad 4.22 \quad -2]$

Method III

From equation (5.9.5) we get the desired characteristic polynomial as,

$$\lambda^3 + 8\lambda^2 + 17\lambda + 30 = 0 \qquad\qquad\qquad .....(5.9.13)$$

Here, $\phi(\mathbf{A}) = \mathbf{A}^3 + \alpha_1 \mathbf{A}^2 + \alpha_2 \mathbf{A} + \alpha_3 \mathbf{I}$

From equation (5.9.13) we get, $\alpha_1 = 8$; $\alpha_2 = 17$; $\alpha_3 = 30$

From the given state equation and equation (5.9.1) we get,

$$\mathbf{A} = \begin{bmatrix} -1 & 0 & 0 \\ 1 & -2 & 0 \\ 2 & 1 & -3 \end{bmatrix} \text{ and } \mathbf{A}^2 = \begin{bmatrix} 1 & 0 & 0 \\ -3 & 4 & 0 \\ -7 & -5 & 9 \end{bmatrix}$$

$$\therefore \ \mathbf{A}^3 = \mathbf{A}^2 \mathbf{A} = \begin{bmatrix} 1 & 0 & 0 \\ -3 & 4 & 0 \\ -7 & -5 & 9 \end{bmatrix} \begin{bmatrix} -1 & 0 & 0 \\ 1 & -2 & 0 \\ 2 & 1 & -3 \end{bmatrix} = \begin{bmatrix} -1 & 0 & 0 \\ 7 & -8 & 0 \\ 20 & 19 & -27 \end{bmatrix}$$

$$\therefore \ \phi(\mathbf{A}) = \mathbf{A}^3 + \alpha_1 \mathbf{A}^2 + \alpha_2 \mathbf{A} + \alpha_2 \mathbf{I}$$

$$= \begin{bmatrix} -1 & 0 & 0 \\ 7 & -8 & 0 \\ 20 & 19 & -27 \end{bmatrix} + 8\begin{bmatrix} 1 & 0 & 0 \\ -3 & 4 & 0 \\ -7 & -5 & 9 \end{bmatrix} + 17\begin{bmatrix} -1 & 0 & 0 \\ 1 & -2 & 0 \\ 2 & 1 & -3 \end{bmatrix} + 30\begin{bmatrix} 1 & 0 & 0 \\ 0 & 1 & 0 \\ 0 & 0 & 1 \end{bmatrix}$$

$$= \begin{bmatrix} -1 & 0 & 0 \\ 7 & -8 & 0 \\ 20 & 19 & -27 \end{bmatrix} + \begin{bmatrix} 8 & 0 & 0 \\ -24 & 32 & 0 \\ -56 & -40 & 72 \end{bmatrix} + \begin{bmatrix} -17 & 0 & 0 \\ 17 & -34 & 0 \\ 34 & 17 & -51 \end{bmatrix} + \begin{bmatrix} 30 & 0 & 0 \\ 0 & 30 & 0 \\ 0 & 0 & 30 \end{bmatrix}$$

$$= \begin{bmatrix} 20 & 0 & 0 \\ 0 & 20 & 0 \\ -2 & -4 & 24 \end{bmatrix}$$

From equation (5.9.4) we get,

$$Q_C^{-1} = \begin{bmatrix} 0.05 & 0.5 & -0.1667 \\ -0.0694 & 0.6944 & -0.25 \\ -0.0194 & 0.1944 & -0.0833 \end{bmatrix}$$

From Ackermann's formula we get,

$$K = [0 \ 0 \ 1] \, Q_C^{-1} \phi(A)$$

$$= [0 \ 0 \ 1] \begin{bmatrix} 0.05 & 0.5 & -0.1667 \\ -0.0694 & 0.6944 & -0.25 \\ -0.0194 & 0.1944 & -0.0833 \end{bmatrix} \begin{bmatrix} 20 & 0 & 0 \\ 0 & 20 & 0 \\ -2 & -4 & 24 \end{bmatrix}$$

$$= [-0.0194 \ 0.1944 \ -0.0833] \begin{bmatrix} 20 & 0 & 0 \\ 0 & 20 & 0 \\ -2 & -4 & 24 \end{bmatrix} = [-0.22 \ 4.22 \ -2]$$

The state feedback gain matrix, $K = [-0.22 \quad 4.22 \quad -2]$

Note : *The result obtained from all the three methods are same.*

5.9   OBSERVABLE PHASE VARIABLE FORM OF STATE MODEL

An observable system can be represented by a modified state model called observable phase variable form by transforming the system matrix A into the transpose of bush or companion form as shown in equation (5.102).

$$A_o = \begin{bmatrix} 0 & 0 & \cdots & 0 & -a_n \\ 1 & 0 & \cdots & 0 & -a_{n-1} \\ \vdots & \vdots & & \vdots & \vdots \\ 0 & 0 & \cdots & 1 & -a_1 \end{bmatrix}$$ .....(5.102)

Consider the state model of a nth order system with single-input and single-output as shown below.

$$\dot{X} = AX + Bu$$ .....(5.103)

$$y = CX + Du$$ .....(5.104)

Let us choose a transformation $Z = P_o X$ to transform the state model to observable phase variable form.

Here, Z = Transformed state vector of order $(n \times 1)$

P_o = Transformation matrix of order $(n \times n)$

On premultiplying the equation, $Z = P_o X$ by P_o^{-1} we get,

$$P_o^{-1} Z = P_o^{-1} P_o X$$

$$\therefore X = P_o^{-1} Z$$

On differentiating the equation $X = P_0^{-1} Z$ we get,

$$\dot{X} = P_0^{-1} \dot{Z}$$

On substituting $X = P_0^{-1} Z$ and $\dot{X} = P_0^{-1} \dot{Z}$ in the state model [equations (5.103) and (5.104)] of the system we get,

$$P_0^{-1} \dot{Z} = A P_0^{-1} Z + B u \qquad \qquad ....(5.105)$$

$$y = C P_0^{-1} Z + D u \qquad \qquad .....(5.106)$$

On premultiplying the equation (5.75) by P_0 we get,

$$\dot{Z} = P_0 A P_0^{-1} Z + P_0 B u$$

$$y = C P_0^{-1} Z + D u$$

Let $P_0 A P_0^{-1} = A_0$; $P_0 B = B_0$ and $C P_0^{-1} = C_0$

$$\therefore \dot{Z} = A_0 Z + B_0 u \qquad \qquad .....(5.107)$$

$$y = C_0 Z + D u \qquad \qquad .....(5.108)$$

The equations(5.107) and (5.108) are called observable phase variable form of state model of the system.

Note : In observable phase variable form of state model the matrices A_0 will be as shown in equation (5.102) and the B_0 and C_0 will in the form shown below.

$$B_0 = \begin{bmatrix} \beta_n \\ \beta_{n-1} \\ \vdots \\ \beta_1 \end{bmatrix} \quad ; \quad C_0 = \begin{bmatrix} 0 & 0 & \cdots & 0 & 1 \end{bmatrix}$$

DETERMINATION OF TRANSFORMATION MATRIX, P_0

Let A be the system matrix of original state model. Now the characteristic equation governing the system is given by equation (5.109).

$$|\lambda I - A| = \lambda^n + a_1 \lambda^{n-1} + a_2 \lambda^{n-2} + \ldots + a_{n-1} \lambda + a_n = 0 \qquad \qquad .....(5.109)$$

Using the coefficients $a_1, a_2, \ldots a_{n-2}, a_{n-1}$ of characteristic equation [equation (5.109)] we can form a matrix, W as shown in equation (5.110).

$$W = \begin{bmatrix} a_{n-1} & a_{n-2} & \cdots & a_2 & a_1 & 0 \\ a_{n-2} & a_{n-3} & \cdots & a_1 & 1 & 0 \\ \vdots & \vdots & & \vdots & \vdots & \vdots \\ a_1 & 1 & \cdots & 0 & 0 & 0 \\ 1 & 0 & \cdots & 0 & 0 & 0 \end{bmatrix} \qquad \qquad .....(5.110)$$

Now the transformation matrix P_0 is given by

$$P_0 = W Q_0^T \qquad \qquad .....(5.111)$$

where, $Q_0 = \begin{bmatrix} C^T & A^T C^T & (A^T)^2 C^T & \ldots & (A^T)^{n-1} C^T \end{bmatrix}$

EXAMPLE 5.10

The state model of a system is given by

$$\begin{bmatrix} \dot{x}_1 \\ \dot{x}_2 \\ \dot{x}_3 \end{bmatrix} = \begin{bmatrix} 0 & 0 & 1 \\ -2 & -3 & 0 \\ 0 & 2 & -3 \end{bmatrix}\begin{bmatrix} x_1 \\ x_2 \\ x_3 \end{bmatrix} + \begin{bmatrix} 0 \\ 2 \\ 0 \end{bmatrix}[u] \ ; \ y = [1 \ 0 \ 0]\begin{bmatrix} x_1 \\ x_2 \\ x_3 \end{bmatrix}$$

Convert the state model to observable phase variable form.

SOLUTION

The given state model can be transformed to observable phase variable form, only if the system is completely observable. Hence check for observability.

Kalman's test for observability

From the given state model we get,

$$A = \begin{bmatrix} 0 & 0 & 1 \\ -2 & -3 & 0 \\ 0 & 2 & -3 \end{bmatrix} \text{ and } C = [1 \ 0 \ 0]$$

$$A^T = \begin{bmatrix} 0 & 0 & 1 \\ -2 & -3 & 0 \\ 0 & 2 & -3 \end{bmatrix}^T = \begin{bmatrix} 0 & -2 & 0 \\ 0 & -3 & 2 \\ 1 & 0 & -3 \end{bmatrix}$$

$$(A^T)^2 = \begin{bmatrix} 0 & -2 & 0 \\ 0 & -3 & 2 \\ 1 & 0 & -3 \end{bmatrix}\begin{bmatrix} 0 & -2 & 0 \\ 0 & -3 & 2 \\ 1 & 0 & -3 \end{bmatrix} = \begin{bmatrix} 0 & 6 & -4 \\ 2 & 9 & -12 \\ -3 & -2 & 9 \end{bmatrix}$$

$$C^T = [1 \ 0 \ 0]^T = \begin{bmatrix} 1 \\ 0 \\ 0 \end{bmatrix}$$

$$A^TC^T = \begin{bmatrix} 0 & -2 & 0 \\ 0 & -3 & 2 \\ 1 & 0 & -3 \end{bmatrix}\begin{bmatrix} 1 \\ 0 \\ 0 \end{bmatrix} = \begin{bmatrix} 0 \\ 0 \\ 1 \end{bmatrix} \ ; \ (A^T)^2C^T = \begin{bmatrix} 0 & 6 & -4 \\ 2 & 9 & -12 \\ -3 & -2 & 9 \end{bmatrix}\begin{bmatrix} 1 \\ 0 \\ 0 \end{bmatrix} = \begin{bmatrix} 0 \\ 2 \\ -3 \end{bmatrix}$$

The composite matrix for observability $\Big\}$ $Q_o = [C^T \ A^TC^T \ (A^T)^2C^T] = \begin{bmatrix} 1 & 0 & 0 \\ 0 & 0 & 2 \\ 0 & 1 & -3 \end{bmatrix}$

Determinant of $Q_o = \Delta_{Qo}$ $\begin{vmatrix} 1 & 0 & 0 \\ 0 & 0 & 2 \\ 0 & 1 & -3 \end{vmatrix} = 1(-2) = -2$

Since, $\Delta_{Qo} \neq$, the rank of $Q_o = 3$. Hence the system is completely observable.

To find transformation matrix, P_o

From the given state model we get, $A = \begin{bmatrix} 0 & 0 & 1 \\ -2 & -3 & 0 \\ 0 & 2 & -3 \end{bmatrix}$

$$[\lambda I - A] = \lambda \begin{bmatrix} 1 & 0 & 0 \\ 0 & 1 & 0 \\ 0 & 0 & 1 \end{bmatrix} - \begin{bmatrix} 0 & 0 & 1 \\ -2 & -3 & 0 \\ 0 & 2 & -3 \end{bmatrix} = \begin{bmatrix} \lambda & 0 & -1 \\ 2 & \lambda+3 & 0 \\ 0 & -2 & \lambda+3 \end{bmatrix}$$

$$|\lambda I - A| = \begin{vmatrix} \lambda & 0 & -1 \\ 2 & \lambda+3 & 0 \\ 0 & -2 & \lambda+3 \end{vmatrix} = \lambda(\lambda+3)^2 - 1(-4) = \lambda(\lambda^2 + 6\lambda + 9) + 4 = \lambda^3 + 6\lambda^2 + 9\lambda + 4$$

The characteristic equation is,

$$\lambda^3 + 6\lambda^2 + 9\lambda + 4 = 0$$

The standard form of characteristic equation when n = 3 is given by,

$$\lambda^3 + a_1\lambda^2 + a_2\lambda + a_3 = 0$$

On comparing the characteristic equation of the system with standard form we get,

$$a_1 = 6, \qquad a_2 = 9 \quad \text{and} \quad a_3 = 4$$

$$\therefore W = \begin{bmatrix} a_2 & a_1 & 1 \\ a_1 & 1 & 0 \\ 1 & 0 & 0 \end{bmatrix} = \begin{bmatrix} 9 & 6 & 1 \\ 6 & 1 & 0 \\ 1 & 0 & 0 \end{bmatrix}$$

$$\therefore P_o = WQ_o^T = \begin{bmatrix} 9 & 6 & 1 \\ 6 & 1 & 0 \\ 1 & 0 & 0 \end{bmatrix} \begin{bmatrix} 1 & 0 & 0 \\ 0 & 0 & 2 \\ 0 & 1 & -3 \end{bmatrix}^T = \begin{bmatrix} 9 & 6 & 1 \\ 6 & 1 & 0 \\ 1 & 0 & 0 \end{bmatrix} \begin{bmatrix} 1 & 0 & 0 \\ 0 & 0 & 1 \\ 0 & 2 & -3 \end{bmatrix} = \begin{bmatrix} 9 & 2 & 3 \\ 6 & 0 & 1 \\ 1 & 0 & 0 \end{bmatrix}$$

$$P_o^{-1} = \frac{[\text{Cofactor of } P_o]^T}{\text{Determinant of } P_o} = \frac{P_{o,cof}^T}{\Delta_{P_o}}$$

$$\Delta_{P_o} = \begin{vmatrix} 9 & 2 & 3 \\ 6 & 0 & 1 \\ 1 & 0 & 0 \end{vmatrix} = -2(-1) = 2$$

$$P_{o,cof}^T = \begin{bmatrix} 0 & 1 & 0 \\ 0 & -3 & 2 \\ 2 & 9 & -12 \end{bmatrix}^T = \begin{bmatrix} 0 & 0 & 2 \\ 1 & -3 & 9 \\ 0 & 2 & -12 \end{bmatrix}$$

$$P_o^{-1} = \frac{1}{\Delta_{P_o}} P_{o,cof}^T = \frac{1}{2} \begin{bmatrix} 0 & 0 & 2 \\ 1 & -3 & 9 \\ 0 & 2 & -12 \end{bmatrix} = \begin{bmatrix} 0 & 0 & 1 \\ 0.5 & -1.5 & 4.5 \\ 0 & 1 & -6 \end{bmatrix}$$

To determine the observable phase variable form of state model

The observable phase variable form of state model is given by,

$$\dot{Z} = A_0 Z + B_0 u$$

$$y = C_0 Z \qquad\qquad \text{(Here } \mathbf{D} \text{ is not given)}$$

where, $A_0 = P_0 A P_0^{-1}$; $B_0 = P_0 B$ and $C_0 = C P_0^{-1}$

$$P_0 A P_0^{-1} = \begin{bmatrix} 9 & 2 & 3 \\ 6 & 0 & 1 \\ 1 & 0 & 0 \end{bmatrix} \begin{bmatrix} 0 & 0 & 1 \\ -2 & -3 & 0 \\ 0 & 2 & -3 \end{bmatrix} \begin{bmatrix} 0 & 0 & 1 \\ 0.5 & -1.5 & 4.5 \\ 0 & 1 & -6 \end{bmatrix}$$

$$= \begin{bmatrix} -4 & 0 & 0 \\ 0 & 2 & 3 \\ 0 & 0 & 1 \end{bmatrix} \begin{bmatrix} 0 & 0 & 1 \\ 0.5 & -1.5 & 4.5 \\ 0 & 1 & -6 \end{bmatrix}$$

$$= \begin{bmatrix} 0 & 0 & -4 \\ 1 & 0 & -9 \\ 0 & 1 & -6 \end{bmatrix}$$

$$B_0 = P_0 B = \begin{bmatrix} 9 & 2 & 3 \\ 6 & 0 & 1 \\ 1 & 0 & 0 \end{bmatrix} \begin{bmatrix} 0 \\ 2 \\ 0 \end{bmatrix} = \begin{bmatrix} 4 \\ 0 \\ 0 \end{bmatrix}$$

$$C_0 = C P_0^{-1} = \begin{bmatrix} 1 & 0 & 0 \end{bmatrix} \begin{bmatrix} 0 & 0 & 1 \\ 0.5 & -1.5 & 4.5 \\ 0 & 1 & -6 \end{bmatrix} = \begin{bmatrix} 0 & 0 & 1 \end{bmatrix}$$

The observable phase variable form of state model is given by,

$$\begin{bmatrix} \dot{z}_1 \\ \dot{z}_2 \\ \dot{z}_3 \end{bmatrix} = \begin{bmatrix} 0 & 0 & -4 \\ 1 & 0 & -9 \\ 0 & 1 & -6 \end{bmatrix} \begin{bmatrix} z_1 \\ z_2 \\ z_3 \end{bmatrix} + \begin{bmatrix} 4 \\ 0 \\ 0 \end{bmatrix} u \ ; \ y = \begin{bmatrix} 0 & 0 & 1 \end{bmatrix} \begin{bmatrix} z_1 \\ z_2 \\ z_3 \end{bmatrix}$$

5.10 STATE OBSERVERS

The control system design problem of pole placement by state feedback requires feedback of all state variables. In practice, however, not all state variables are available for measurement and feedback. In such situations we need to estimate unmeasurable state variables.

The estimation of unmeasurable state variables is commonly called observation. A device (or a computer program) that estimates or observes the state variables is called state observer. If the state observer, observes all the state variables of the system (regardless of whether some state variables are available for direct measurement), it is called a full-order state observer.

In certain systems some of the state variables may be accurately measured. In this case all the n state variables of the system need not be observed and so a reduced-order observer is sufficient. An observer that estimates m number of state variable, where m is less than n is called a reduced-order observer. If the order of the reduced order observer is the minimum possible, then the observer is called a minimum order observer.

Consider an nth order system with single-input and single-output described by the state model

$$\dot{X} = AX + Bu \qquad\qquad .....(5.112)$$

$$y = CX \qquad\qquad .....(5.113)$$

Let us assume that the state variables of the system are not available for feedback and so an observer system has to be designed to measure or observe the state variables. The observed state variables can be used to implement the feedback control low.

Let us construct a full-order state observer as shown in fig 5.2, with y and u as inputs and $\hat{X}$ and $\hat{y}$ as outputs,

where , $\hat{X}$ = Observed state vector (or Estimated state vector)

$\hat{y}$ = Output of observer system (or Estimated output)

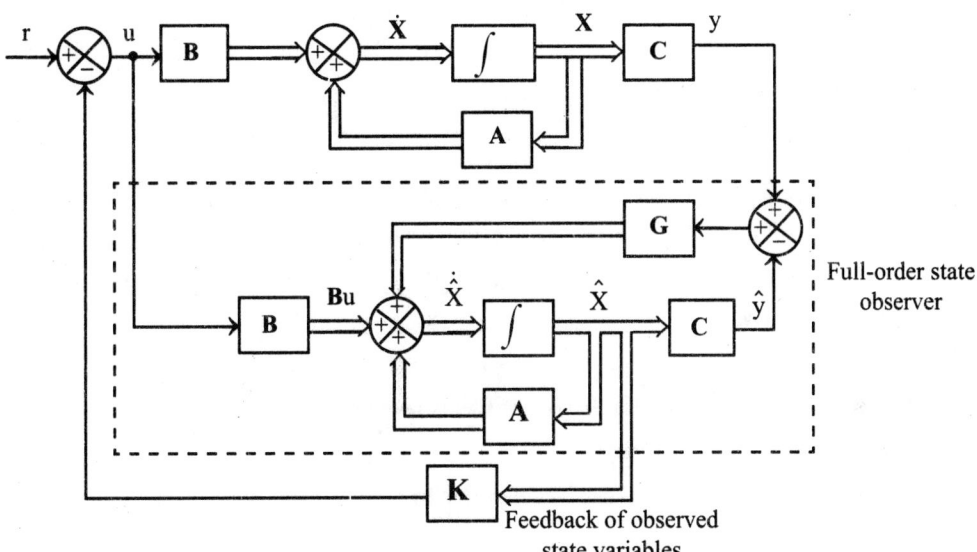

Fig 5.2 : *A linear system with full-order state observer*

Logically the observer system should have the same state equations as the original system. The criterion for state observer design is to minimize the difference between the system output (y) and estimated output ($\hat{y}$). Therefore the dynamic model of the state observer can be written as,

$$\dot{\hat{X}} = A\hat{X} + Bu + G(y - \hat{y}) \qquad\qquad .....(5.114)$$

The equation (5.114) is used to construct the state observer shown in fig 5.2. In state observer the difference (y – $\hat{y}$) is multiplied by (n × 1) observer gain matrix or vector **G** and fed into the input of the integrators of the observer. Now the problem is to design **G** so that (y – $\hat{y}$) is minimized.

From fig(5.2) we get y = **C X** and $\hat{y}$ = **C$\hat{X}$**. On substituting y = **CX** and $\hat{y}$= **C$\hat{X}$** in equation (5.114) we get

$$\dot{\hat{X}} = A\hat{X} + Bu + G(CX - C\hat{X})$$

$$\dot{\hat{X}} = A\hat{X} + GCX - GC\hat{X} + Bu$$

$$\dot{\hat{X}} = (A - GC)\hat{X} + GCX + Bu \qquad\qquad .....(5.115)$$

On subtracting the eqution (5.115) from eqution (5.112) we get,

$$\dot{X} - \dot{\hat{X}} = AX - (A - GC)\hat{X} - GCX$$

$$= (A - GC)X - (A - GC)\hat{X}$$

$$= (A - GC)(X - \hat{X}) \qquad\qquad .....(5.116)$$

Let, Error vector, $E = X - \hat{X}$ $\qquad\qquad .....(5.117)$

On differentiating equation (5.117) we get,

$$\dot{E} = \dot{X} - \dot{\hat{X}} \qquad\qquad .....(5.118)$$

Using equations (5.114) and (5.118) the equation (5.116) can be written as,

$$\dot{E} = (A - GC)E \qquad\qquad .....(5.119)$$

From equation (5.118) we can say that the dynamic behaviour of error vector is determined by the eigenvalues of matrix $(A-GC)$. If the matrix $(A-GC)$ is a stable matrix, the error vector will converge to zero for any initial error vector $E(0)$. (i.e., $\hat{X}(t)$ will converge to $X(t)$ regardless of the values of $X(0)$ and $\hat{X}(0)$).

It can be proved that if the system is completely observable then it is possible to choose the observable gain matrix G such that $(A-GC)$ has arbitrarily desired eigenvalues.

Determination of observer gain matrix, G

The state observer gain matrix, G can be determined by three methods. In all the three methods, the system has to be checked for observability.

The state model of the original n^{th} order system is given by

$$\dot{X} = AX + Bu$$
$$y = CX$$

To check for observability of original system, determine the composite matrix for observability, Q_o.

where, $Q_o = [C^T \quad A^T C^T \quad (A^T)^2 C^T ...... (A^T)^{n-1} C^T]$

Then calculate the determinant of Q_o. If the determinant of Q_o is not equal to zero then the rank of Q_o is n and so the system is completely state controllable. (Here n is the order of the system). If the rank is not equal to n then the matrix $(A-GC)$ cannot have arbitrary eigenvalues. When the system is completely observable any one of the following method can be used to determine the observer gain matrix G. The matrix G will have an order of $(n \times 1)$ and the elements of G are denoted as $g_1, g_2, ... g_n$ as shown in equation (5.120)

$$G = \begin{bmatrix} g_1 \\ g_2 \\ g_3 \\ \vdots \\ g_n \end{bmatrix}$$ .....(5.120)

METHOD I

1. Determine the characteristic polynomial of original system. The characteristic polynomial is given by $|\lambda I - A| = 0$

    Let, $|\lambda I - A| = \lambda^n + a_1 \lambda^{n-1} + a_2 \lambda^{n-2} + ... + a_{n-1}\lambda + a_n$.

2. Determine the desired characteristic polynomial from the specified eigenvalues.

    Let the specified eigenvalues be $\mu_1, \mu_2, \mu_3, ..., \mu_n$.

    Now the desired characteristic polynomial is given by,

    $(\lambda - \mu_1)(\lambda - \mu_2)(\lambda - \mu_3) ..... (\lambda - \mu_n) = \lambda^n + \alpha_1 \lambda^{n-1} + \alpha_2 \lambda^{n-2} + ...+ \alpha_{n-1}\lambda + \alpha_n.$

3. Determine the transformation matrix P_o which transforms the original state model to observable phase variable form.

    The transformation matrix, $P_o = W Q_o^T$

    where $Q_o = [C^T \quad A^T C^T \quad (A^T)^2 C^T .... (A^T)^{n-1} C^T]$

    and $W = \begin{bmatrix} a_{n-1} & a_{n-2} & \cdots & a_2 & a_1 & 0 \\ a_{n-2} & a_{n-3} & \cdots & a_1 & 1 & 0 \\ \vdots & \vdots & & \vdots & \vdots & \vdots \\ a_1 & 1 & \cdots & 0 & 0 & 0 \\ 1 & 0 & \cdots & 0 & 0 & 0 \end{bmatrix}$

    (Here $a_{n-1}, a_{n-2}, ...a_1$ are coefficients of characteristic equation of original system).

4. Determine the observer gain matrix from the following equation.

    $$G = P_o^{-1} \begin{bmatrix} \alpha_n - a_n \\ \alpha_{n-1} - a_{n-1} \\ \vdots \\ \alpha_2 - a_2 \\ \alpha_1 - a_1 \end{bmatrix}$$

> Note : If the given state model is in observable phase variable form then $P_o = I$, unit matrix.

METHOD II

1. Determine the characteristic polynomial of the system with state observer, which is given by, $|\lambda I - (A - GC)| = 0$

$$G = \begin{bmatrix} g_1 \\ g_2 \\ \vdots \\ g_n \end{bmatrix}$$

Let, $|\lambda I - (A-GC)| = |\lambda I - A + GC| = \lambda^n + b_1 \lambda^{n-1} + b_2 \lambda^{n-2} + ... + b_{n-1} \lambda + b_n$.

The coefficients of this polynomial $b_1, b_2, ... b_n$ will be functions of $g_1, g_2, ... g_n$.

2.   Determine the desired characteristic polynomial from the specified eigenvalues.

Let the specified eigenvalues be $\mu_1, \mu_2, \mu_3, ..., \mu_n$.

Now the desired characteristic polynomial is given by,

$(\lambda - \mu_1)(\lambda - \mu_2)(\lambda - \mu_3) ..... (\lambda - \mu_n) = \lambda^n + \alpha_1 \lambda^{n-1} + \alpha_2 \lambda^{n-2} + ... + \alpha_{n-1} \lambda + \alpha_n$.

3.   By equating the coefficients of polynomials obtained in step-1 and step-2 we get n number of equations.

i.e., $b_1 = \alpha_1$; $b_2 = \alpha_2$;   .... ; $b_{n-1} = \alpha_{n-1}$ and $b_n = \alpha_n$.

On solving these equations we get the elements g_1, g_2 ..... g_n of state observer gain matrix, **G**.

Note : Method-II is suitable only for low values of n (i.e., for 2nd and 3rd order systems otherwise calculations will be tedious)

METHOD III

1.   Determine the desired characteristic polynomial from the specified eigenvalues.

Let the specified eigenvalues be $\mu_1, \mu_2, \mu_3, ..., \mu_n$.

Now the desired characteristic polynomial is given by,

$(\lambda - \mu_1)(\lambda - \mu_2)(\lambda - \mu_3) ..... (\lambda - \mu_n) = \lambda^n + \alpha_1 \lambda^{n-1} + \alpha_2 \lambda^{n-2} + ... + \alpha_{n-1} \lambda + \alpha_n$.

2.   Determine the matrix $\phi(A)$ using the coefficients of desired characteristic polynomial.

$\phi(A) = A^n + \alpha_1 A^{n-1} + \alpha_2 A^{n-2} + ... + \alpha_{n-1} A + \alpha_n I$

3.   Calculate the state observer gain matrix **G** using the Ackermann's formula given below.

$$G = \phi(A) \begin{bmatrix} C \\ CA \\ \vdots \\ CA^{n-2} \\ CA^{n-1} \end{bmatrix}^{-1} \begin{bmatrix} 0 \\ 0 \\ \vdots \\ 0 \\ 1 \end{bmatrix}$$

EXAMPLE 5.11

Consider the system described by the state model

$$\dot{X} = AX \qquad \text{where } A = \begin{bmatrix} -1 & 1 \\ 1 & -2 \end{bmatrix} ; \ C = \begin{bmatrix} 1 & 0 \end{bmatrix}$$
$$y = CX$$

Design a full-order state observer. The desired eigenvalues for the observer matrix are $\mu_1 = -5$; $\mu_2 = -5$.

SOLUTION

Check for observability

Given that, $A = \begin{bmatrix} -1 & 1 \\ 1 & -2 \end{bmatrix}$; $C = \begin{bmatrix} 1 & 0 \end{bmatrix}$

$$C^T = \begin{bmatrix} 1 & 0 \end{bmatrix}^T = \begin{bmatrix} 1 \\ 0 \end{bmatrix}$$

$$A^T = \begin{bmatrix} -1 & 1 \\ 1 & -2 \end{bmatrix}^T = \begin{bmatrix} -1 & 1 \\ 1 & -2 \end{bmatrix} \quad ; \quad A^T C^T = \begin{bmatrix} -1 & 1 \\ 1 & -2 \end{bmatrix}\begin{bmatrix} 1 \\ 0 \end{bmatrix} = \begin{bmatrix} -1 \\ 1 \end{bmatrix}$$

Composite matrix for observability $\left.\right\}$ $Q_o = \begin{bmatrix} C^T & A^T C^T \end{bmatrix} = \begin{bmatrix} 1 & -1 \\ 0 & 1 \end{bmatrix}$ .....(5.11.1)

Determinant of $Q_0 = \Delta_{Qo} = \begin{vmatrix} 1 & -1 \\ 0 & 1 \end{vmatrix} = 1$ .....(5.11.2)

Since, $\Delta_{Qo} \neq 0$, the system is observable.

To determine the characteristic polynomial of original system

The characteristic polynomial of the system is given by $|\lambda I - A| = 0$.

$$[\lambda I - A] = \lambda \begin{bmatrix} 1 & 0 \\ 0 & 1 \end{bmatrix} - \begin{bmatrix} -1 & 1 \\ 1 & -2 \end{bmatrix} = \begin{bmatrix} \lambda + 1 & -1 \\ -1 & \lambda + 2 \end{bmatrix}$$

$$|\lambda I - A| = (\lambda + 1)(\lambda + 2) - 1 = \lambda^2 + 3\lambda + 2 - 1 = \lambda^2 + 3\lambda + 1$$

Hence the characteristic polynomial of original system is

$$\lambda^3 + 3\lambda + 1 = 0 \qquad\qquad\qquad .....(5.11.2)$$

To determine the desired characteristic polunomial

The desired eigenvalues are $\mu_1 = -5$ and $\mu_2 = -5$

$$\therefore (\lambda - \mu_1)(\lambda - \mu_2) = (\lambda + 5)(\lambda + 5) = \lambda^2 + 10\lambda + 25$$

The desired characteristic polynomial is

$$= \lambda^2 + 10\lambda + 25 = 0 \qquad\qquad\qquad .....(5.11.3)$$

To determine the state observer gain matrix, G

METHOD-I

From equation (5.11.2) we get the characteristic polynomial of original system as

$$\lambda^3 + 3\lambda + 1 = 0$$

$$\therefore a_1 = 3 \ ; a_2 = 1$$

From equation (5.11.3) we get the desired characteristic polynomial as,

$$\lambda^2 + 10\lambda + 25 = 0$$

$$\therefore \alpha_1 = 10 \ ; \ \alpha_2 = 25$$

Using the coefficients of characteristic polynomial of original system the matrix **W** is constructed as shown below.

$$\mathbf{W} = \begin{bmatrix} a_1 & 1 \\ 1 & 0 \end{bmatrix} \begin{bmatrix} 3 & 1 \\ 1 & 0 \end{bmatrix}$$

From equation (5.11.1) we get

$$\mathbf{Q}_o = \begin{bmatrix} 1 & -1 \\ 0 & 1 \end{bmatrix} ; \quad \mathbf{Q}_o^T = \begin{bmatrix} 1 & -1 \\ 0 & 1 \end{bmatrix}^T = \begin{bmatrix} 1 & 0 \\ -1 & 1 \end{bmatrix}$$

The transformation matrix, **P**$_o$ which transforms the state model to observable phase variables form is given by,

$$\mathbf{P}_o = \mathbf{W}\,\mathbf{Q}_o^T = \begin{bmatrix} 3 & 1 \\ 1 & 0 \end{bmatrix} \begin{bmatrix} 1 & 0 \\ -1 & 1 \end{bmatrix} = \begin{bmatrix} 2 & 1 \\ 1 & 0 \end{bmatrix}$$

$$\mathbf{P}_o^{-1} = \frac{[\text{Cofactor of } \mathbf{P}_o]^T}{\text{Determinant of } \mathbf{P}_o} = \frac{\mathbf{P}_{o,cof}^T}{\Delta_{P_o}}$$

$$\Delta_{P_o} = \begin{vmatrix} 2 & 1 \\ 1 & 0 \end{vmatrix} = -1 ; \quad \mathbf{P}_{o,cof}^T = \begin{bmatrix} 0 & -1 \\ -1 & 2 \end{bmatrix}^T = \begin{bmatrix} 0 & -1 \\ -1 & 2 \end{bmatrix}$$

$$\therefore \mathbf{P}_o^{-1} = \frac{\mathbf{P}_{o,cof}^T}{\Delta_{P_o}} = (-1) \begin{bmatrix} 0 & -1 \\ -1 & 2 \end{bmatrix} = \begin{bmatrix} 0 & -1 \\ 1 & -2 \end{bmatrix}$$

The observer gain matrix
$$\left.\begin{array}{l}\text{The observer}\\ \text{gain matrix}\end{array}\right\} \mathbf{G} = \mathbf{P}_o^{-1} \begin{bmatrix} \alpha_2 - a_2 \\ \alpha_1 - a_1 \end{bmatrix} = \begin{bmatrix} 0 & 1 \\ 1 & -2 \end{bmatrix} \begin{bmatrix} 25 & -1 \\ 10 & -3 \end{bmatrix}$$

$$= \begin{bmatrix} 0 & 1 \\ 1 & -2 \end{bmatrix} \begin{bmatrix} 24 \\ 7 \end{bmatrix} = \begin{bmatrix} 7 \\ 10 \end{bmatrix}$$

$$\therefore \mathbf{G} = \begin{bmatrix} g_1 \\ g_2 \end{bmatrix} = \begin{bmatrix} 7 \\ 10 \end{bmatrix}$$

METHOD - II

The characteristic polynomial of the system with state observer is given by

$$|\lambda \mathbf{I} - (\mathbf{A} - \mathbf{GC})| = 0$$

Here, $\mathbf{G} = \begin{bmatrix} g_1 \\ g_2 \end{bmatrix}$

$$[\lambda \mathbf{I} - (\mathbf{A} - \mathbf{GC})] = [\lambda \mathbf{I} - \mathbf{A} + \mathbf{GC}]$$

$$= \lambda \begin{bmatrix} 1 & 0 \\ 0 & 1 \end{bmatrix} - \begin{bmatrix} -1 & 1 \\ 1 & -2 \end{bmatrix} + \begin{bmatrix} g_1 \\ g_2 \end{bmatrix} \begin{bmatrix} 1 & 0 \end{bmatrix}$$

$$= \begin{bmatrix} \lambda & 0 \\ 0 & \lambda \end{bmatrix} - \begin{bmatrix} -1 & 1 \\ 1 & -2 \end{bmatrix} + \begin{bmatrix} g_1 & 0 \\ g_2 & 0 \end{bmatrix} = \begin{bmatrix} \lambda + 1 + g_1 & -1 \\ -1 + g_2 & \lambda + 2 \end{bmatrix}$$

$$\left| \lambda I - A + GC \right| = \begin{vmatrix} \lambda + 1 + g_1 & -1 \\ -1 + g_2 & \lambda + 2 \end{vmatrix} = (\lambda + 1 + g_1)(\lambda + 2) - (-1 + g_2)(-1)$$

$$= \lambda^2 + 2\lambda + \lambda + 2 + g_1\lambda + 2g_1 - 1 + g_2$$

$$= \lambda^2 + (3 + g_1)\lambda + (2g_1 + g_2 + 1)$$

The characteristic polynomial of the system with state observer is

$$\lambda^2 + (3 + g_1)\lambda + (2g_1 + g_2 + 1) = 0 \qquad\qquad .....(5.11.4)$$

From equation (5.11.3) we get the desired characteristic polynomial of the system as,

$$\lambda^2 + 10\lambda + 25 = 0 \qquad\qquad .....(5.11.5)$$

On comparing the coefficients of λ in equations (5.11.4) and (5.11.5) we get,

$$3 + g_1 = 0$$

$$\therefore g_1 = 10 - 3 = 7$$

On comparing the coefficients of λ^0 in equations (5.11.4) and (5.11.5) we get,

$$2g_1 + g_2 + 1 = 25$$

$$\therefore g_2 = 25 - 1 - 2g_1 = 24 - 2(7) = 10$$

$\therefore$ The observer gain matrix, $G = \begin{bmatrix} g_1 \\ g_2 \end{bmatrix} = \begin{bmatrix} 7 \\ 10 \end{bmatrix}$

METHOD-III

From equation (5.11.3) we get the desired characteristic polynomial of the system as,

$$\lambda^2 + 10\lambda + 25 = 0 \qquad\qquad .....(5.11.6)$$

$$\therefore \alpha_1 = 10 \quad ; \quad \alpha_2 = 25$$

Here, $\phi(A) = A^2 + \alpha_1 A + \alpha_2 I$

$$\therefore \phi(A) = \begin{bmatrix} -1 & 1 \\ 1 & -2 \end{bmatrix}\begin{bmatrix} -1 & 1 \\ 1 & -2 \end{bmatrix} + 10\begin{bmatrix} -1 & 1 \\ 1 & -2 \end{bmatrix} + 25\begin{bmatrix} 1 & 0 \\ 0 & 1 \end{bmatrix}$$

$$= \begin{bmatrix} 2 & -3 \\ -3 & 5 \end{bmatrix} + \begin{bmatrix} -10 & 10 \\ 10 & -20 \end{bmatrix} + \begin{bmatrix} 25 & 0 \\ 0 & 25 \end{bmatrix}$$

$$= \begin{bmatrix} 17 & 7 \\ 7 & 10 \end{bmatrix}$$

$$CA = \begin{bmatrix} 1 & 0 \end{bmatrix} \begin{bmatrix} -1 & 1 \\ 1 & -2 \end{bmatrix} = \begin{bmatrix} -1 & 1 \end{bmatrix}$$

$$\begin{bmatrix} C \\ CA \end{bmatrix}^{-1} = \begin{bmatrix} 1 & 0 \\ -1 & 1 \end{bmatrix}^{-1} = 1 \times \begin{bmatrix} 1 & 1 \\ 0 & 1 \end{bmatrix}^{T} = \begin{bmatrix} 1 & 1 \\ 0 & 1 \end{bmatrix}$$

From Ackermann's formula we get,

$$\mathbf{G} = \begin{bmatrix} g_1 \\ g_2 \end{bmatrix} = \phi(\mathbf{A}) \begin{bmatrix} C \\ CA \end{bmatrix}^{-1} \begin{bmatrix} 0 \\ 1 \end{bmatrix} = \begin{bmatrix} 17 & 7 \\ 7 & 10 \end{bmatrix} \begin{bmatrix} 1 & 0 \\ 1 & 1 \end{bmatrix} \begin{bmatrix} 0 \\ 1 \end{bmatrix} = \begin{bmatrix} 24 & 7 \\ 17 & 10 \end{bmatrix} \begin{bmatrix} 0 \\ 1 \end{bmatrix}$$

$$= \begin{bmatrix} 7 \\ 10 \end{bmatrix}$$

> **Note :** *The results obtained from all the three methods are same.*

5.11   SHORT-ANSWER QUESTIONS

Q5.1   *Define the characteristic equation of a matrix.*

The characteristic equation of a n x n matrix **A** is the nth degree polynomial of equation, $|\lambda \mathbf{I} - \mathbf{A}| = 0$, where **I** is the unit matrix.

Q5.2   *What are eigenvalues and eigenvector?*

A non-zero column vector **X** is an eigenvector of a square matrix **A**, if there exists a scalar λ such that **AX** = λ**X**, then λ is an eigenvalue of **A**.

Q5.3   *How the eigenvalues are calculated?*

The eigenvalues of a matrix **A** are the roots of characteristic equation of **A**. The characteristic equation is $|\lambda \mathbf{I} - \mathbf{A}| = 0$. Solving the characteristic equation for λ gives the eigenvalues of **A**.

Q5.4   *Write any two properties of eigenvalues.*

1.   A matrix and its transpose have the same eigenvalues.

2.   The product of the eigenvalues (counting multiplicities) of the matrix equals the determinant of the matrix.

Q5.5   *How the eigenvectors are calculated, when the eigenvalues are distinct?*

When the eigenvalues are distinct, then there exists only one independent eigenvector corresponding to each eigenvalue. The independent eigenvector corresponding to eigenvalue λ_i may be obtained by taking cofactors of matrix $[\lambda_i \mathbf{I} - \mathbf{A}]$ along any row.

If $C_{k1}, C_{k2}, \dots\dots C_{kn}$ are cofactors of matrix $[\lambda_i \mathbf{I} - \mathbf{A}]$ along k^{th} row then the eigenvector $\mathbf{m}_i$ corresponding to λ_i is given by

$$\mathbf{m}_i = \begin{bmatrix} C_{k1} \\ C_{k2} \\ \vdots \\ C_{k3} \end{bmatrix} \quad ; \quad k = 1 \text{ or } 2 \text{ or } \dots\dots n$$

Q5.6   *What is similarity transformation?*

The process of transforming a square matrix **A** to another similar matrix **B** by a transformation $\mathbf{P}^{-1} \mathbf{A} \mathbf{P} = \mathbf{B}$ is called similarity transformation. The matrix **P** is called transformation matrix.

Q5.7   *What is meant by diagonalization?*

The process of converting the system matrix **A** into a diagonal matrix by a similarity transformation using the modal matrix **M** is called diagonalization.

Q5.8   *What are the characteristics or property that are invariant under a similarity transformation?*

The determinant, characteristic equation, eigenvalues and trace of a matrix are invariant under a similarity transformation.

Q5.9   *What is modal matrix?*

The modal matrix is a matrix used to diagonalize the system matrix. It is also called diagonalization matrix.

If   **A** = System matrix

   M = Modal matrix

and **M**$^{-1}$ = Inverse of modal matrix

then **M**$^{-1}$ **AM** will be a diagonalized system matrix.

Q5.10   *How the modal matrix is determined?*

The modal matrix **M** can be formed from eigenvectors. Let m_1, m_2, m_3, ... m_n be the eigenvectors of a nth order system. Now the modal matrix **M** is obtained by arranging all the eigenvectors columnwise as shown below

Modal matrix, $\mathbf{M} = [m_1 \quad m_2 \quad m_3 \quad .... \quad m_n]$

Q5.11   *State Cayley-Hamilton theorem.*

The Cayley-Hamilton theorem states that every square matrix satisfies its own characteristic equation.

Consider an n × n square matrix **A**. Its characteristic equation is given by $|\lambda \mathbf{I} - \mathbf{A}| = 0$

Let, $|\lambda \mathbf{I} - \mathbf{A}| = \lambda^n + a_1 \lambda^{n-1} + a_2 \lambda^{n-2} + ... + a_{n-1} \lambda + a_n = 0$

By Cayley-Hamilton theorem, the matrix **A** has to satisfy its characteristic equation, hence in the above equation on replacing λ by A we get,

$\mathbf{A}^n + a_1 \mathbf{A}^{n-1} + a_2 \mathbf{A}^{n-2} + ... + a_{n-1} \mathbf{A} + a_n \mathbf{I} = 0$

Q5.12   *How the state transition matrix e^{At} is computed using Cayley-Hamilton theorem?*

1.   Let A be the system matrix and $f(\mathbf{A}) = e^{At}$

2.   Compute the eigenvalues of **A**

3.   On substituting the eigenvalues λ_1, λ_2, λ_3, ..., λ_n in the following equation, we can form n-number of linear equations.

$f(\lambda_i) = \alpha_0 + \alpha_1 \lambda_i + \alpha_2 \lambda_i^2 + ... + \alpha_{n-1} \lambda_i^{n-1}$

The constants α_0, α_1, α_2, ..., α_{n-1} are obtained by solving these n-number of equations.

4.   Now $f(\mathbf{A}) = e^{At} = \alpha_0 \mathbf{I} + \alpha_1 \mathbf{A} + \alpha_2 \mathbf{A}^2 + ... + \alpha_{n-1}\mathbf{A}^{n-1}$.

Q5.13   *How the state transition matrix of discrete time system A^k is computed using Cayley-Hamilton theorem.*

1. Let **A** be the system matrix and $f(\mathbf{A}) = \mathbf{A}^k$

2. Compute the eigenvalues of **A**

3. On substituting the eigenvalues $\lambda_1, \lambda_2, \lambda_3, \ldots\ldots, \lambda_n$ in the following equation, we can form n-number of linear equations.

$$f(\lambda_i) = \alpha_0 + \alpha_1\lambda_i + \alpha_2\lambda_i^2 + \ldots\ldots + \alpha_{n-1}\lambda_i^{n-1}$$

The constants $\alpha_0, \alpha_1, \alpha_2, \ldots\ldots, \alpha_{n-1}$ are obtained by solving these n-number of equations.

4. Now $f(\mathbf{A}) = \mathbf{A}^k = \alpha_0\mathbf{I} + \alpha_1\mathbf{I} + \alpha_2\mathbf{A}^2 + \ldots\ldots + \alpha_{n-1}\mathbf{A}^{n-1}$.

Q5.14.   *What is canonical form of state model?*

If the system matrix, **A** is in the form of diagonal matrix then the state model is called canonical form [In diagonal matrix, we have the eigenvalues on the main diagonal and all other elements are zero].

Q5.15   *What is the transformation used to diagonalize a system matrix?*

The transformation used to diagonalize a system matrix is $\mathbf{X} = \mathbf{MZ}$

where   **X** = Original state vector

   Z = Transformed state vector   ;   **M** = Modal matrix.

Q5.16   *Write the transformed canonical state model of a system.*

The transformed state model of a system is

$$\dot{\mathbf{Z}} = \Lambda\mathbf{Z} + \tilde{\mathbf{B}}\mathbf{U}$$

$$\mathbf{Y} = \tilde{\mathbf{C}}\mathbf{U} + \mathbf{DU}$$

where, **Z** = Transformed state vector ; $\tilde{\mathbf{B}} = \mathbf{M}^{-1}\mathbf{B}$

   M = Modal matrix   ; $\tilde{\mathbf{C}} = \mathbf{CM}$

   $\Lambda = \mathbf{M}^{-1}\mathbf{AM}$

Q5.17   *When modal matrix is called vander monde matrix.*

When the system matrix **A** is in the companion or bush form then the modal matrix is given by a special matrix called vander monde matrix, **V**, shown below.

$$\mathbf{M} = \mathbf{V} = \begin{bmatrix} 1 & 1 & \cdots & 1 \\ \lambda_1 & \lambda_2 & \cdots & \lambda_n \\ \lambda_1^2 & \lambda_2^2 & \cdots & \lambda_n^2 \\ \vdots & \vdots & & \vdots \\ \lambda_1^{n-1} & \lambda_2^{n-2} & \cdots & \lambda_n^{n-1} \end{bmatrix}$$

Q5.18   *What is Jordan canonical form?*

When the eigenvalues have multiplicity the system matrix cannot be diagonalized. But the transformation, $\mathbf{X} = \mathbf{MZ}$ will transform the system matrix to a form called Jordan matrix, **J**, where **J** is $\mathbf{M}^{-1}\mathbf{AM}$. The transformed state model in this case is called Jordan canonical form.

Q5.19   ***What is Jordan matrix?***

The Jordan matrix is the transformed system matrix using the transformation, $\mathbf{X} = \mathbf{MZ}$. The Jordan matrix $\mathbf{J}$, will have a Jordan block of size q x q correspond to a eigenvalue of λ_1 with multiplicity q. In the Jordan block the diagonal elements will be eigenvalues and elements just above the diagonal is one. A typical Jordan matrix is shown below.

$$\mathbf{J} = \begin{bmatrix} \lambda_1 & 1 & 0 & \cdots & 0 \\ 0 & \lambda_1 & 1 & \cdots & 0 \\ 0 & 0 & \lambda_1 & \cdots & 0 \\ 0 & 0 & 0 & \lambda_4 \cdots & 0 \\ \vdots & \vdots & \vdots & \vdots & \vdots \\ 0 & 0 & 0 & 0 & \lambda_n \end{bmatrix} \quad \longleftarrow \text{Jordan block of size } 3 \times 3$$

Q5.20   ***How the state transition matrix e^{At} is computed by canonical transformation.***

When the eigenvalues are distinct the following procedure can be used to compute e^{At}.

1. Compute the eigenvalues $\lambda_1, \lambda_2, \ldots\ldots, \lambda_n$ and eigenvector $m_1, m_2, \ldots\ldots, m_n$ of the system matrix $\mathbf{A}$.

2. Determine the modal matrix $\mathbf{M}$ and $\mathbf{M}^{-1}$. Modal matrix, $\mathbf{M} = [m_1 \ m_2 \ \ldots\ldots \ m_n]$.

3. The state transition matrix e^{At} is given by $e^{At} = \mathbf{M} \ e^{At} \ \mathbf{M}^{-1}$

$$\text{where, } e^{At} = \begin{bmatrix} e^{\lambda_1 t} & 0 & \cdots & 0 \\ 0 & e^{\lambda_2 t} & \cdots & 0 \\ \vdots & \vdots & & \vdots \\ 0 & 0 & \cdots & e^{\lambda_n t} \end{bmatrix}$$

Q5.21   ***How the state transition matrix of discrete time system A^k is computed by canonical transformation.***

When the eigenvalues are distinct, the following procedure can be used to compute A^k.

1. Compute the eigenvalues $\lambda_1, \lambda_2, \ldots.., \lambda_n$ and eigenvectors $m_1, m_2, \ldots.., m_n$ of the system matrix $\mathbf{A}$.

2. Determine the model matrix $\mathbf{M}$ and $\mathbf{M}^{-1}$

Modal matrix, $\mathbf{M} = [m_1 \quad m_2 \quad \ldots \quad m_n]$

3. The state transition matrix of discrete time system $\mathbf{A}^k$ is given by

$$\mathbf{A}^k = \mathbf{M} \ \mathbf{L}^k \ \mathbf{M}^{-1}$$

$$\text{where, } \Lambda^k = \begin{bmatrix} \lambda_1^k & 0 & \cdots & 0 \\ 0 & \lambda_2^k & \cdots & 0 \\ \vdots & \vdots & & \vdots \\ 0 & 0 & \cdots & \lambda_n^k \end{bmatrix}$$

Q5.22   ***Define controllability.***

A system is said to be completely state controllable if it is possible to transfer the system state from any initial state $\mathbf{X}(t_0)$ at any other desired state $\mathbf{X}(t)$, in specified finite time by a control vector $\mathbf{U}(t)$.

Q5.23   ***What is the need for controllability test?***

The controllability test is necessary to find the usefulness of a state variable. If the state variables are controllable then by controlling (i.e., varying) the state variables the desired outputs of the system are achieved.

Q5.24   *State the condition for controllability by Gilbert's method.*

Case(i) : When eigenvalues are distinct

Consider the canonical form of state model shown below which is obtained by using the transformation **X = MZ.**

$$\dot{Z} = \Lambda Z + \tilde{B}U$$

$$Y = \tilde{C}Z + DU$$

where, $\Lambda = M^{-1}AM$; $\tilde{C} = CM$

$\tilde{B} = M^{-1}B$ and   M = Modal matrix

In this case the necessary and sufficient condition for complete controllability is that, the matrix $\tilde{B}$ must have no rows with all zeros. If any row of the matrix $\tilde{B}$ is zero then the corresponding state variable is uncontrollable.

Case (ii) : When eigenvalues have multiplicity

In this case the state model can be converted to Jordan canonical form shown below

$$\dot{Z} = JZ + \tilde{B}U$$

$$Y = \tilde{C}Z + DU   ;   \text{where } J = M^{-1}AM$$

In this case the system is completely controllable, if the elements of any row of $\tilde{B}$ that correspond to the last row of each Jordan block are not all zero and all rows corresponding to other state variable must not have all zeros.

Q5.25   *State the condition for controllability by Kalman's method.*

For a nth order system described by state equation, $\dot{X} = AX + BU$, we can form a composite matrix, Q_C where

$$Q_C = [B\quad AB\quad A^2B\ ....\ A^{n-1}B]$$

In this case the system is completely state controllable if the rank of the composite matrix Q_C is n.

Q5.26   *What is the advantage and the disadvantage in Kalman's test for controllability?*

The advantage in Kalman's test is that the calculations are simpler. But the disadvantage in Kalman's test is that, we can't find the state variable which is uncontrollable. But in Gilbert's method we can find the uncontrollable state variable which is the state variable corresponding to the row of $\tilde{B}$ which has all zeros.

Q5.27   *Define observability.*

A system is said to be completely observable if every state **X**(t) can be completely identified by measurements of the output **Y**(t) over a finite time interval.

Q5.28   *What is the need for observability test?*

The observability test is necessary to find whether the state variables are measurable or not. If the state variables are measurable then the state of the system can be determined by practical measurements of the state variables.

Q5.29   *State the condition for observability by Gilbert's method.*

Consider the transformed canonical or Jordan canonical form of the state model shown below which is obtained by using the transformation , $X = MZ$

$$\dot{Z} = \Lambda Z + \tilde{B}U \qquad \dot{Z} = JZ + \tilde{B}U$$
$$Y = \tilde{C}Z + DU \qquad\quad Y = \tilde{C}Z + DU$$

(or)

where $\tilde{C} = CM$ and M = Modal matrix.

The necessary and sufficient condition for complete observability is that none of the columns of the matrix $\tilde{C}$ be zero. If any of the column is of $\tilde{C}$ has all zeros then the corresponding state variable is not observable.

Q5.30   *State the condition for observability by Kalman's method?*

For a nth order system described by state model

$$\dot{X} = AX + BU$$
$$Y = CX + DU$$

we can form a composite matrix Q_0, where

$$Q_0 = [C^T \quad A^T C^T \quad (A^T)^2 C^T \quad (AT)^3 C^T \quad ..... \quad (A^T)^{n-1} C^T]$$

In this case the system is completely observable if the rank of composite matrix, Q_0 is n.

Q5.31   *What is the advantage and the disadvantage in Kalman's test for observability?*

The advantage in Kalman's test is that the calculations are simpler. The disadvantage in Kalman's test is that the nonobservable state variables cannot be determined.

Q5.32   *State the duality between controllability and observability.*

The concepts of controllability and observability are dual concepts and it is proposed by Kalman as principle of duality.

The principle of duality states that a system is completely state controllable if and only if its dual system is completely observable or viceversa (i.e., if the system is observable then its dual is controllable).

Q5.33   *What is the effect of pole-zero cancellation in transfer function.*

If cancellation of pole-zero occurs in the transfer function of a system, then the system will be either not state controllable or unobservable depending on how the state variables are defined (or chosen).

Q5.34   *Write the controllable phase variable form of state model.*

The controllable phase variable form of state model is given by the following equation.

$$\dot{Z} = A_c Z + B_c u$$
$$Y = C_c X + D\,u$$

$$A_c = \begin{bmatrix} 0 & 1 & 0 & \cdots & 0 \\ 0 & 0 & 1 & \cdots & 0 \\ 0 & 0 & 0 & \cdots & 0 \\ \vdots & \vdots & \vdots & & \vdots \\ 0 & 0 & 0 & \cdots & 1 \\ -a_n & -a_{n-1} & -a_{n-2} & \cdots & -a_1 \end{bmatrix} \;;\; B_c = \begin{bmatrix} 0 \\ 0 \\ 0 \\ \vdots \\ 0 \\ 1 \end{bmatrix} \;;\; C_c = [C_{11} \quad C_{12} \quad \cdots \quad C_{1n}]$$

Q5.35   *How will you find the transformation matrix, P_c for transforming the state model to controllable canonical form?*

    1.   Compute the composite matrix for controllability, $\mathbf{Q_C}$ and $\mathbf{Q_C}^{-1}$.

$$\mathbf{Q_C} = [\mathbf{B} \quad \mathbf{AB} \quad \mathbf{A^2B} \; .... \; \mathbf{A^{(n-2)} B} \quad \mathbf{A^{(n-1)} B}]$$

    2.   Determine the matrix, $\mathbf{P_1}$

$$\mathbf{P_1} = [0 \quad 0 \quad 0 \quad ..... \quad 0 \quad 1] \, \mathbf{Q_C}^{-1}$$

    3.   Now the transformation matrix, $\mathbf{P_C}$ is given by

$$\mathbf{P_c} = \begin{bmatrix} \mathbf{P_1} \\ \mathbf{P_1 A} \\ \mathbf{P_1 A^2} \\ \vdots \\ \mathbf{P_1 A^{(n-1)}} \end{bmatrix}$$

Q5.36   *How will you find the transformation matrix, P_c using the characteristic equation?*

    1.   Compute the composite matrix for controllability, $\mathbf{Q_C}$.

$$\mathbf{Q_C} = [\mathbf{B} \quad \mathbf{AB} \quad \mathbf{A^2B} \; .... \; \mathbf{A^{(n-1)} B}]$$

    2.   Determine the characteristic equation of the system, $|\lambda \mathbf{I} - \mathbf{A}| = 0$

$$|\lambda \mathbf{I} - \mathbf{A}| = \lambda^n + a_1 \lambda^{n-1} + a_2 \lambda^{n-2} + ... + a_{n-1} \lambda + a_n = 0$$

    3.   Using the coefficients $a_1, a_2, ..., a_{n-2}, a_{n-1}$ of characteristic equation form a matrix $\mathbf{W}$.

$$\mathbf{W} = \begin{bmatrix} a_{n-1} & a_{n-2} & \cdots & a_2 & a_1 & 0 \\ a_{n-2} & a_{n-3} & \cdots & a_1 & 1 & 0 \\ \vdots & \vdots & & \vdots & \vdots & \vdots \\ a_1 & 1 & \cdots & 0 & 0 & 0 \\ 1 & 0 & \cdots & 0 & 0 & 0 \end{bmatrix}$$

    4.   Now the transformation matrix, $\mathbf{P_C}$ is given by

$$\mathbf{P_C} = (\mathbf{Q_C} \mathbf{W})^{-1}$$

Q5.37   *What is pole placement by state feedback?*

The pole placement by state feedback is a control system design technique, in which the state variables are used for feedback to achieve the desired closed loop poles.

Q5.38   *How control system design is carried in state space?*

In state space design of control system, any inner parameter or variable of a system (which are nothing but state variables) are used for feedback to achieve the desired performance of the system. The performance of the system is related to the location of closed loop poles. Hence in state space design the closed loop poles are placed at the desired locations by means of state feedback through an appropriate state feedback gain matrix, $\mathbf{K}$.

Q5.39   *What is the necessary condition to be satisfied for design using state feedback?*

The state feedback design requires arbitrary pole placement to achieve the desired performance. The necessary and sufficient condition to be satisfied for arbitrary pole placement is that the system be completely state controllable.

Q5.40   What are the advantages of control system in state space.

1.   Any inner parameters or variables of a system can be defined as state variables and can be used for feedback.

2.   The closed loop poles may be placed at any desired locations by means of state feedback through an appropriate state feedback gain matrix, **K**.

Q5.41   Draw the block diagram of a system with state feedback.

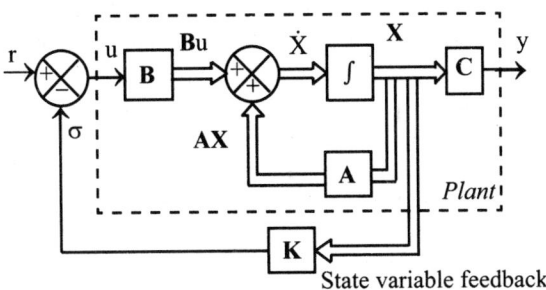

Fig Q5.41 : System with state variable feedback

Q5.42   What is control law?

In control system design using state variable feedback, the equation $u = r - \mathbf{KX}$ is called control law.

where,   u = Input to the plant   ;   r = Input to the system with state feedback

$\mathbf{X}$ = State vector   ; $\mathbf{K}$ = State feedback gain matrix.

Q5.43   Write the Ackermann's formula to find the state feedback gain matrix, K.

The Ackermann's formula to compute state feedback gain matrix, **K** is given by,

$$\mathbf{K} = [0 \quad 0 \quad ...... \quad 0 \quad 1] \ \mathbf{Q}_c^{-1} \ \varphi(\mathbf{A})$$

where, $\mathbf{Q}_C = [\mathbf{B} \quad \mathbf{AB} \quad \mathbf{A^2B} \ ...... \ \mathbf{A^{n-1}} \ \mathbf{B}]$

$$\varphi(\mathbf{A}) = A^n + \alpha_1 A^{n-1} + \alpha_2 A^{n-2} + ... + \alpha_{n-1} A + \alpha_n I$$

Q5.44   Write the observable phase variable form of state model.

The observable phase variable form of state model is given by the following equations.

$$\dot{Z} = A_o Z + B_o u$$
$$y = C_o Z + Du$$

where, $A_o = \begin{bmatrix} 0 & 0 & \cdots & 0 & -a_n \\ 1 & 0 & \cdots & 0 & -a_{n-1} \\ \vdots & \vdots & & \vdots & \vdots \\ 0 & 0 & \cdots & 1 & -a_1 \end{bmatrix}$; $B_o = \begin{bmatrix} \beta_n \\ \beta_{n-1} \\ \vdots \\ \beta_1 \end{bmatrix}$ and $C_o = [0 \quad 0 \quad .... \ 0 \quad 1]$

Q5.45   *How will you find the transformation matrix, P_o to transform the state model to observable phase variable form.*

1.   Compute the composite matrix for observability, Q_o.

$$Q_o = [C^T \quad A^T C^T \quad (A^T)^2 C^T \text{.......} (A^T)^{n-1} C^T]$$

2.   Determine the characteristic equation of the system, $|\lambda I - A| = 0$

$$|\lambda I - A| = \lambda^n + a_1 \lambda^{n-1} + a_2 \lambda^{n-2} + \text{.......} + a_{n-1} \lambda + a_n = 0$$

3.   Using the coefficients $a_1, a_2, \text{......}, a_{n-2}, a_{n-1}$ of characteristic equation form a matrix, W

$$W = \begin{bmatrix} a_{n-1} & a_{n-2} & \cdots & a_2 & a_1 & 0 \\ a_{n-2} & a_{n-3} & \cdots & a_1 & 1 & 0 \\ \vdots & \vdots & & \vdots & \vdots & \vdots \\ a_1 & 1 & \cdots & 0 & 0 & 0 \\ 1 & 0 & \cdots & 0 & 0 & 0 \end{bmatrix}$$

4.   Now the transformation matrix, P_o is given by

$$P_o = W Q_o^T$$

Q5.46   *What is state observer?*

A device (or a computer program) that estimates or observes the state variables is called state observer.

Q5.47   *What is the need for state observer?*

In certain systems the state variables may not be available for measurement and feedback. In such situations we need to estimate the unmeasurable state variables from the knowledge of input and output. Hence a state observer is employed which estimates the state variables from the input and output of the system. The estimated state variable can be used for feedback to design the system by pole placement.

Q5.48   *What is full-order, reduced-order and minimum-order state observer?*

If the state observer, observes all the n-number of state variables of the system then it is called full-order observer.

If the state observer, observes m-number of state variables, where m is less than n then the observer is called reduced-order observer. If the order of the reduced order observer is the minimum possible, then the observer is called a minimum order observer.

Q5.49   *What is the necessary condition to be satisfied for design of state observer.*

The state observer can be designed only if the system is completely state observable.

Q5.50   *Write the Ackermann's formula to find the state observer gain matrix, G.*

The Ackermann's formula to compute the state observer gain matrix, G is given by,

$$G = \phi(A) = \begin{bmatrix} A \\ CA \\ \vdots \\ CA^{n-2} \\ CA^{n-1} \end{bmatrix} \begin{bmatrix} 0 \\ 0 \\ \vdots \\ 0 \\ 1 \end{bmatrix}$$

where, $\phi(A) = A^n + \alpha_1 A^{n-1} + \text{.....} + \alpha_{n-1} A + \alpha_n I.$

5.12 EXERCISES

I.   FILL IN THE BLANKS

1.   The sum of the eigenvalues is equal to the .... of the matrix.

2.   The product of the eigenvalues of the matrix equals the .... of the matrix.

3.   A matrix is singular if and only if it has a .... eigenvalue.

4.   The .... of an upper or lower triangular matrix are the elements on its main diagonal.

5.   The .... can be used for diagonalization of a square matrix.

6.   The determinant, characteristic equation and trace of a matrix are invariant under a ....

7.   The square matrices A and B are said to be similar if a nonsingular matrix P exists such that ....

8.   .... theorem states that every square matrix satisfies its own characteristic equation.

9.   When a system matrix has distinct eigenvalues it can be converted to .... by a similarity transformation.

10.   When the system matrix has repeated eigenvalues, it can be converted to .... by a similarity transformation.

11.   In canonical state model, if the matrix $\tilde{B}$ does not have any row with all zeros then the system is ....

12.   In canonical state model, if the matrix $\tilde{C}$ does not have any column with all zeros then the system is ....

13.   The principle of duality states that a system is .... if and only if its dual system is ....

14.   In .... design the .... are used for feedback.

15.   In state space design the condition to be satisfied for arbitrary pole placement is that the system be ....

16.   In state space design the feedback signal, is related to state variable X by the equation ....

17.   The equation .... is called control law.

18.   The device which estimates the state variables is called ....

19.   The state observer can be designed only if the system is completely ....

20.   When an observer estimates only few number of variables it is called ....

ANSWERS

1. trace	8. Cayley-Hamilton	15. State controllable
2. determinant	9. diagonal matrix	16. $\sigma = KX$
3. zero	10. Jordan matrix	17. $u = r - KX$
4. eigenvalues	11. controllable	18. state observer
5. similarity transformation	12. observable	19. observable
6. similarity transformation	13. state controllable, observable	20. reduced order
7. $P^{-1} AP = B$	14. state space, state variables,	observer

II. State whether the following statements are TRUE or FALSE

1. Two eigenvectors of a real symmetric matrix corresponding to different eigenvalues are orthogonal.

2. The eigenvalues of n x n real symmetric matrix are real.

3. A matrix and its transpose have the same eigenvalues.

4. Eigenvalues are invariant under linear transformation (or transformation of state model).

5. When eigenvalues repeats then these exists only one independent eigenvector corresponding to that multiple eigenvalue.

6. A matrix with repeated eigenvalues can be diagonalized by a similarity transformation.

7. The similarity transformation cannot be used for diagonalization of a square matrix.

8. The Cayley-Hamilton theorem can be used to compute the function of a matrix.

9. In canonical state model if any row of $\tilde{B}$ matrix is zero then the corresponding state variable is controllable.

10. The nth order is completely state controllable if the rank of the composite matrix for controllability is n.

11. In Kalman's test we can identify the uncontrollable state variables.

12. In Kalman's test we can identify the unmeasurable state variables.

13. If cancellation of poles and zeros occur then the system cannot be controlled/observed in the direction of cancelled mode.

14. In canonical state model any column of $\tilde{C}$ matrix is zero then the corresponding state variable is observable.

15. The nth order system is completely observable if the rank of the composite matrix for observability is n.

16. The concepts of controllability and observability are dual concepts.

17. In state space design the closed loop poles are located by means of state feedback.

18. In state feedback design technique it is assumed that all state variables are measurable.

19. In observable phase variable form the system matrix will be transpose of bush or companion form.

20. The estimation of measurable state variables is called observation.

ANSWERS

1.	False	6.	False	11.	False	16.	True
2.	False	7.	False	12.	False	17.	True
3.	True	8.	True	13.	True	18.	True
4.	True	9.	False	14.	False	19.	True
5.	True	10.	True	15.	True	20.	False

III. UNSOLVED PROBLEM

E5.1 *A linear time invariant system is described by the following state model. Obtain the canonical form of the state model.*

(a) $\begin{bmatrix} \dot{x}_1 \\ \dot{x}_2 \\ \dot{x}_3 \end{bmatrix} = \begin{bmatrix} 0 & 0 & -20 \\ 1 & 0 & -24 \\ 0 & 1 & -9 \end{bmatrix} \begin{bmatrix} x_1 \\ x_2 \\ x_3 \end{bmatrix} + \begin{bmatrix} 3 \\ 1 \\ 0 \end{bmatrix} u$

(b) $\begin{bmatrix} \dot{x}_1 \\ \dot{x}_2 \\ \dot{x}_3 \end{bmatrix} = \begin{bmatrix} 0 & 1 & 0 \\ -2 & -2 & -1 \\ 0 & 0 & -2 \end{bmatrix} \begin{bmatrix} x_1 \\ x_2 \\ x_3 \end{bmatrix} + \begin{bmatrix} 0 \\ 1 \\ 1 \end{bmatrix} u$ and $y = \begin{bmatrix} 3 & 1 & 0 \end{bmatrix} \begin{bmatrix} x_1 \\ x_2 \\ x_3 \end{bmatrix}$

(c) $\begin{bmatrix} \dot{x}_1 \\ \dot{x}_2 \\ \dot{x}_3 \end{bmatrix} = \begin{bmatrix} 1 & 0 & 2 \\ -1 & 1 & 1 \\ 0 & 3 & -1 \end{bmatrix} \begin{bmatrix} x_1 \\ x_2 \\ x_3 \end{bmatrix} + \begin{bmatrix} 1 \\ 1 \\ 0 \end{bmatrix} u$ and $y = \begin{bmatrix} 1 & 0 & 1 \end{bmatrix} \begin{bmatrix} x_1 \\ x_2 \\ x_3 \end{bmatrix}$

(d) $\begin{bmatrix} \dot{x}_1 \\ \dot{x}_2 \end{bmatrix} = \begin{bmatrix} 0 & 1 \\ -4 & -5 \end{bmatrix} \begin{bmatrix} x_1 \\ x_2 \end{bmatrix} \begin{bmatrix} 0 \\ 1 \end{bmatrix} u$ and $y = \begin{bmatrix} \frac{1}{3} & -\frac{1}{3} \end{bmatrix} \begin{bmatrix} x_1 \\ x_2 \end{bmatrix}$

(e) $\begin{bmatrix} \dot{x}_1 \\ \dot{x}_2 \\ \dot{x}_3 \end{bmatrix} = \begin{bmatrix} -1 & 0 & 0 \\ 0 & -2 & 0 \\ 0 & 0 & -3 \end{bmatrix} \begin{bmatrix} x_1 \\ x_2 \\ x_3 \end{bmatrix} + \begin{bmatrix} 1 \\ -2 \\ 1 \end{bmatrix} u$ and $y = \begin{bmatrix} 1 & 1 & 0 \end{bmatrix} \begin{bmatrix} x_1 \\ x_2 \\ x_3 \end{bmatrix}$

(f) $\begin{bmatrix} \dot{x}_1 \\ \dot{x}_2 \end{bmatrix} = \begin{bmatrix} -1 & 0 \\ 0 & -3 \end{bmatrix} \begin{bmatrix} x_1 \\ x_2 \end{bmatrix} \begin{bmatrix} 1 \\ 1 \end{bmatrix} [r]$ and $y = \begin{bmatrix} -1 & -2 \end{bmatrix} \begin{bmatrix} x_1 \\ x_2 \end{bmatrix}$

(g) $\begin{bmatrix} \dot{x}_1 \\ \dot{x}_2 \end{bmatrix} = \begin{bmatrix} 0 & 1 \\ -12 & -7 \end{bmatrix} \begin{bmatrix} x_1 \\ x_2 \end{bmatrix} \begin{bmatrix} 0 \\ 1 \end{bmatrix} [r]$ and $y = \begin{bmatrix} -10 & -4 \end{bmatrix} \begin{bmatrix} x_1 \\ x_2 \end{bmatrix} + [u]$

(h) $\begin{bmatrix} \dot{x}_1 \\ \dot{x}_2 \\ \dot{x}_3 \end{bmatrix} = \begin{bmatrix} -2 & 0 & 1 \\ 1 & -3 & 0 \\ 1 & 1 & -1 \end{bmatrix} \begin{bmatrix} x_1 \\ x_2 \\ x_3 \end{bmatrix} + \begin{bmatrix} 1 \\ 0 \\ 1 \end{bmatrix} u$ and $y = \begin{bmatrix} 2 & 1 & -1 \end{bmatrix} \begin{bmatrix} x_1 \\ x_2 \\ x_3 \end{bmatrix}$

(i) $\begin{bmatrix} \dot{x}_1 \\ \dot{x}_2 \\ \dot{x}_3 \end{bmatrix} = \begin{bmatrix} 0 & 1 & 0 \\ -2 & -2 & 0 \\ 0 & 0 & -3 \end{bmatrix} \begin{bmatrix} x_1 \\ x_2 \\ x_3 \end{bmatrix} + \begin{bmatrix} 0 \\ 1 \\ 1 \end{bmatrix} u$ and $y = \begin{bmatrix} \frac{14}{5} & \frac{6}{5} & -\frac{1}{5} \end{bmatrix} \begin{bmatrix} x_1 \\ x_2 \\ x_3 \end{bmatrix}$

E5.2 *Determine the state controllability for the systems represented by the following state equations.*

(i) $\begin{bmatrix} \dot{x}_1 \\ \dot{x}_2 \end{bmatrix} = \begin{bmatrix} 2 & 1 \\ 0 & -1 \end{bmatrix} \begin{bmatrix} x_1 \\ x_2 \end{bmatrix} + \begin{bmatrix} 1 \\ 0 \end{bmatrix} u$

(ii) $\begin{bmatrix} \dot{x}_1 \\ \dot{x}_2 \end{bmatrix} = \begin{bmatrix} 0 & 1 \\ -1 & 0 \end{bmatrix} \begin{bmatrix} x_1 \\ x_2 \end{bmatrix} + \begin{bmatrix} 0 \\ 1 \end{bmatrix} u$

(iii) $\begin{bmatrix} \dot{x}_1 \\ \dot{x}_2 \end{bmatrix} = \begin{bmatrix} 0 & 1 \\ -1 & -2 \end{bmatrix} \begin{bmatrix} x_1 \\ x_2 \end{bmatrix} + \begin{bmatrix} 1 \\ -1 \end{bmatrix} u$

E5.3 *Examine the observability of the systems given below using canonical form.*

(i) $\begin{bmatrix} \dot{x}_1 \\ \dot{x}_2 \\ \dot{x}_3 \end{bmatrix} = \begin{bmatrix} 2 & -2 & 3 \\ 1 & 1 & 1 \\ 1 & 3 & -1 \end{bmatrix} \begin{bmatrix} x_1 \\ x_2 \\ x_3 \end{bmatrix} + \begin{bmatrix} 11 \\ 1 \\ -14 \end{bmatrix} u$ and $y = \begin{bmatrix} -3 & 5 & -2 \end{bmatrix} \begin{bmatrix} x_1 \\ x_2 \\ x_3 \end{bmatrix}$

(ii) $\begin{bmatrix} \dot{x}_1 \\ \dot{x}_2 \\ \dot{x}_3 \end{bmatrix} = \begin{bmatrix} 0 & 1 & 0 \\ 0 & 1 & 1 \\ 0 & -2 & -3 \end{bmatrix} \begin{bmatrix} x_1 \\ x_2 \\ x_3 \end{bmatrix} + \begin{bmatrix} 0 \\ 0 \\ 1 \end{bmatrix} u$ and $y = \begin{bmatrix} 3 & 4 & 1 \end{bmatrix} \begin{bmatrix} x_1 \\ x_2 \\ x_3 \end{bmatrix}$

E5.4 *Convert the following system matrix to canonical form.*

(i) $A = \begin{bmatrix} -3 & 0 & 1 \\ 1 & 1 & 0 \\ 2 & -1 & 2 \end{bmatrix}$ (ii) $A = \begin{bmatrix} 1 & 1 & -1 \\ 0 & 2 & 1 \\ 1 & 2 & 1 \end{bmatrix}$ (iii) $A = \begin{bmatrix} 1 & 2 & 1 \\ -1 & 0 & 2 \\ 1 & 3 & -1 \end{bmatrix}$

APPENDIX I - BODE PLOT

BODE PLOT

The bode plot is a frequency response plot of the transfer function of a system. A Bode plot consists of graphs. One is a plot of the magnitude of a sinusoidal transfer function versus log ω. The other is a plot of the phase angle of a sinusoidal transfer function versus log ω.

The Bode plot can be drawn for both open loop and closed loop transfer function. Usually the bode plot is drawn for open loop syste. The standard representation of the logarithmic magnitude of open loop transfer function of G(jω) is 20 log |G(jω)| where the base of the logarithmic is 10. The unit used in this representation of the magnitude is the decible, usually abbreviated db. The curves are drawn on semilog paper, using the log scale (abcissa) for frequency and the linear scale (ordinate) for either magnitude (in decibles) or phase angle (in degree).

The main advantage of the bode plot is that multiplication of magnitude can be converted into addition. Also a simple method for sketching an approximate log-magnitude curve is available.

Consider the open loop transfer function, $G(s) = \dfrac{K(1+sT_1)}{s(1+sT_2)(1+sT_3)}$

$$G(j\omega) = \frac{K(1+j\omega T_1)}{j\omega(1+j\omega T_2)(1+j\omega T_3)}$$

$$= \frac{K\angle 0° \sqrt{1+\omega^2 T_1^2}\ \angle \tan^{-1}\omega T_1}{\omega\angle 90° \sqrt{1+\omega^2 T_2^2}\ \angle \tan^{-1}\omega T_2 \sqrt{1+\omega^2 T_3^2}\ \angle \tan^{-1}\omega T_2}$$

The magnitude of G(jω) $|G(j\omega)| = \dfrac{K\sqrt{1+\omega^2 T_1^2}}{\omega\sqrt{1+\omega^2 T_2^2}\ \sqrt{1+\omega^2 T_3^2}}$

The phase angle of the G(jω) = $\angle G(j\omega) = \tan^{-1}\omega T_1 - 90° - \tan^{-1}\omega T_2 - \tan^{-1}\omega T_3$

The magnitude of G(jω) can be expressed in decibles

$|G(j\omega)|$ in db $= 20 \log |G(j\omega)|$

$$= 20 \log\left[\frac{K\sqrt{1+\omega^2 T_1^2}}{\omega\sqrt{1+\omega^2 T_2^2}\ \sqrt{1+\omega^2 T_3^2}}\right]$$

$$= 20 \log \left[\frac{K}{\omega} \times \sqrt{1 + \omega^2 T_1^2} \times \frac{1}{\sqrt{1 + \omega^2 T_2^2}} \times \frac{1}{\sqrt{1 + \omega^2 T_3^2}} \right]$$

$$= 20 \log \frac{K}{\omega} + 20 \log \sqrt{1 + \omega^2 T_1^2} + 20 \log \frac{1}{\sqrt{1 + \omega^2 T_2^2}} + 20 \log \frac{1}{\sqrt{1 + \omega^2 T_3^2}}$$

$$= 20 \log \frac{K}{\omega} + 20 \log \sqrt{1 + \omega^2 T_1^2} - 20 \log \sqrt{1 + \omega^2 T_2^2} - 20 \log \sqrt{1 + \omega^2 T_3^2}$$

From the above analysis it is clear that, when the magnitude is expressed in db, the multiplication is converted to addition. Hence in magnitude plot, the db magnitudes of individual factors of $G(j\omega)$ can be added.

Therefore to sketch the magnitude plot, a knowledge of the magnitude variations of individual factors is essential. The magnitude plot and phase plot of various factors of $G(j\omega)$ are shown in table-1.

BASIC FACTORS OF $G(j\omega)$

The basic factors that very frequently occur in a typical transfer function $G(j\omega)$ are,

1. Constant gain, K

2. Integral factor, $\dfrac{K}{j\omega}$ or $\dfrac{K}{(j\omega)^n}$

3. Derivative factor, $j\omega K$ or $(j\omega)^n K$

4. First order factor in denominator, $\dfrac{1}{1 + j\omega T}$ or $\dfrac{1}{(1 + j\omega T)^m}$

5. First order factor in numerator, $(1 + j\omega T)$ or $(1 + j\omega T)^m$

6. Quadratic factor in denominator, $\left[\dfrac{1}{1 + 2\zeta(j\omega / \omega_n) + (j\omega / \omega_n)^2} \right]$

7. Quadratic factor in numerator, $\left[1 + 2\zeta\left(\dfrac{j\omega}{\omega_n} \right) + \left(\dfrac{j\omega}{\omega_n} \right)^2 \right]$

The bode plot of the basic factors are given in table-1.

TABLE - 1

$$A = |G(j\omega)| \text{ in db and } \phi = \angle G(j\omega) \text{ in deg}$$

Basic Factor	Bode Plot
Constant gain, K $G(s) = K$ $G(j\omega) = K$	
Integral factor $G(s) = \dfrac{K}{s}$ $G(j\omega) = \dfrac{K}{j\omega}$	
Integral factor with multiplicity n $G(s) = \dfrac{K}{s^n}$ $G(j\omega) = \dfrac{K}{(j\omega)^n}$	

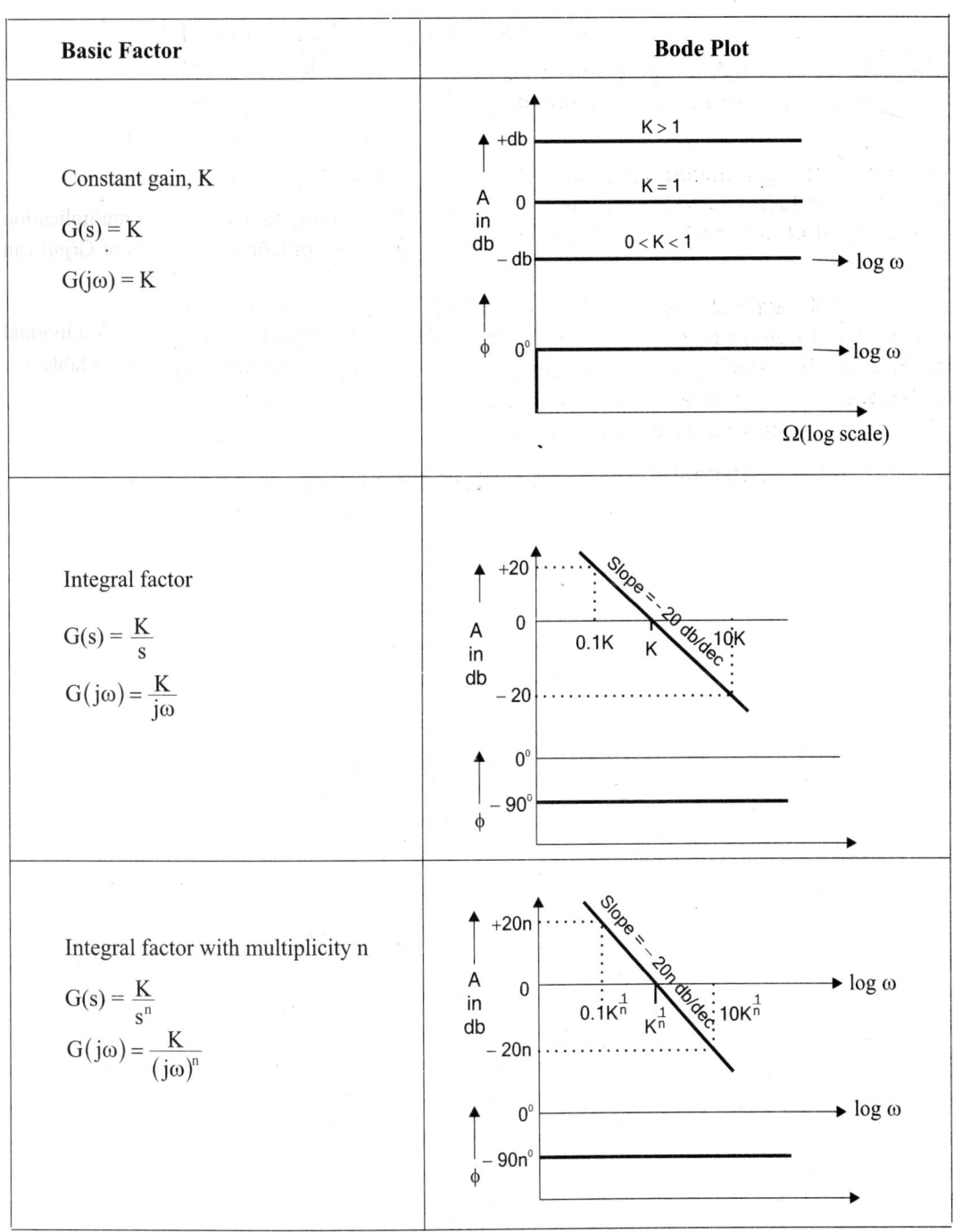

Derivative factor $G(s) = Ks$ $G(j\omega) = j\omega K$	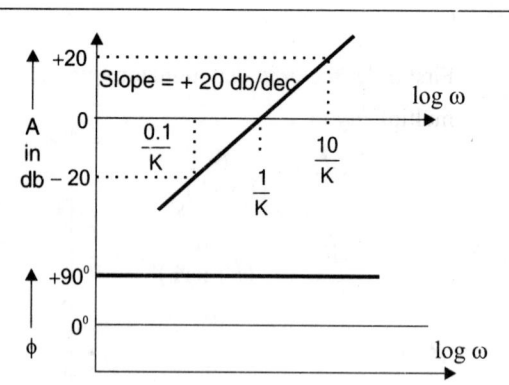
Derivative factor with multiplicity n $G(s) = Ks^n$ $G(j\omega) = (j\omega)^n K$	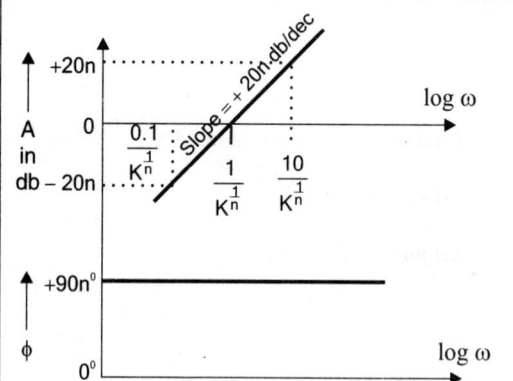
First order factor in denominator $G(s) = \dfrac{K}{(1+sT)}$ $G(j\omega) = \dfrac{K}{(1+j\omega T)}$	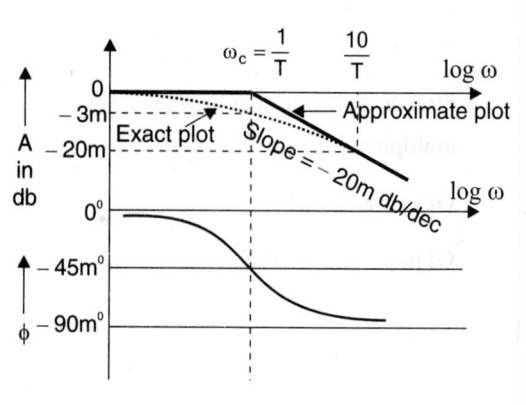

First order factor in denominator with multiplicity m $$G(s) = \dfrac{K}{(1+sT)^m}$$ $$G(j\omega) = \dfrac{K}{(1+j\omega T)^m}$$	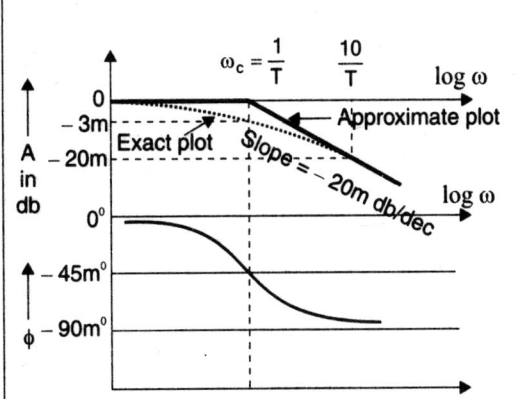
First order factor in numerator $$G(s) = K(1+sT)$$ $$G(j\omega) = K(1+j\omega T)$$	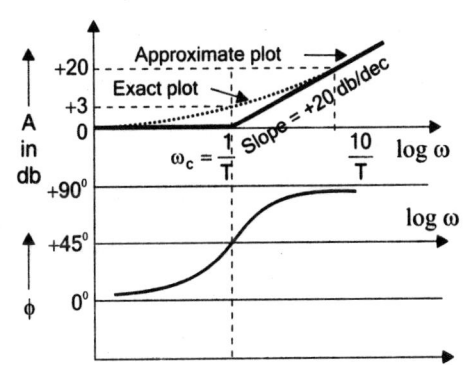
First order factor in numerator with multiplicity m $$G(s) = K(1+sT)^m$$ $$G(j\omega) = K(1+j\omega T)^m$$	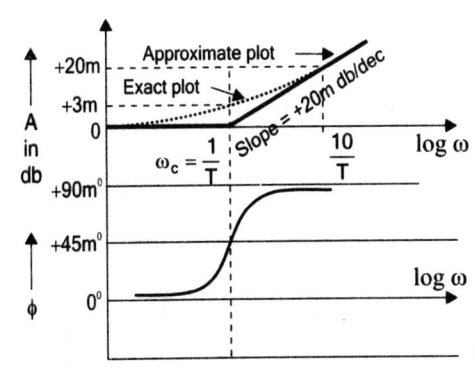

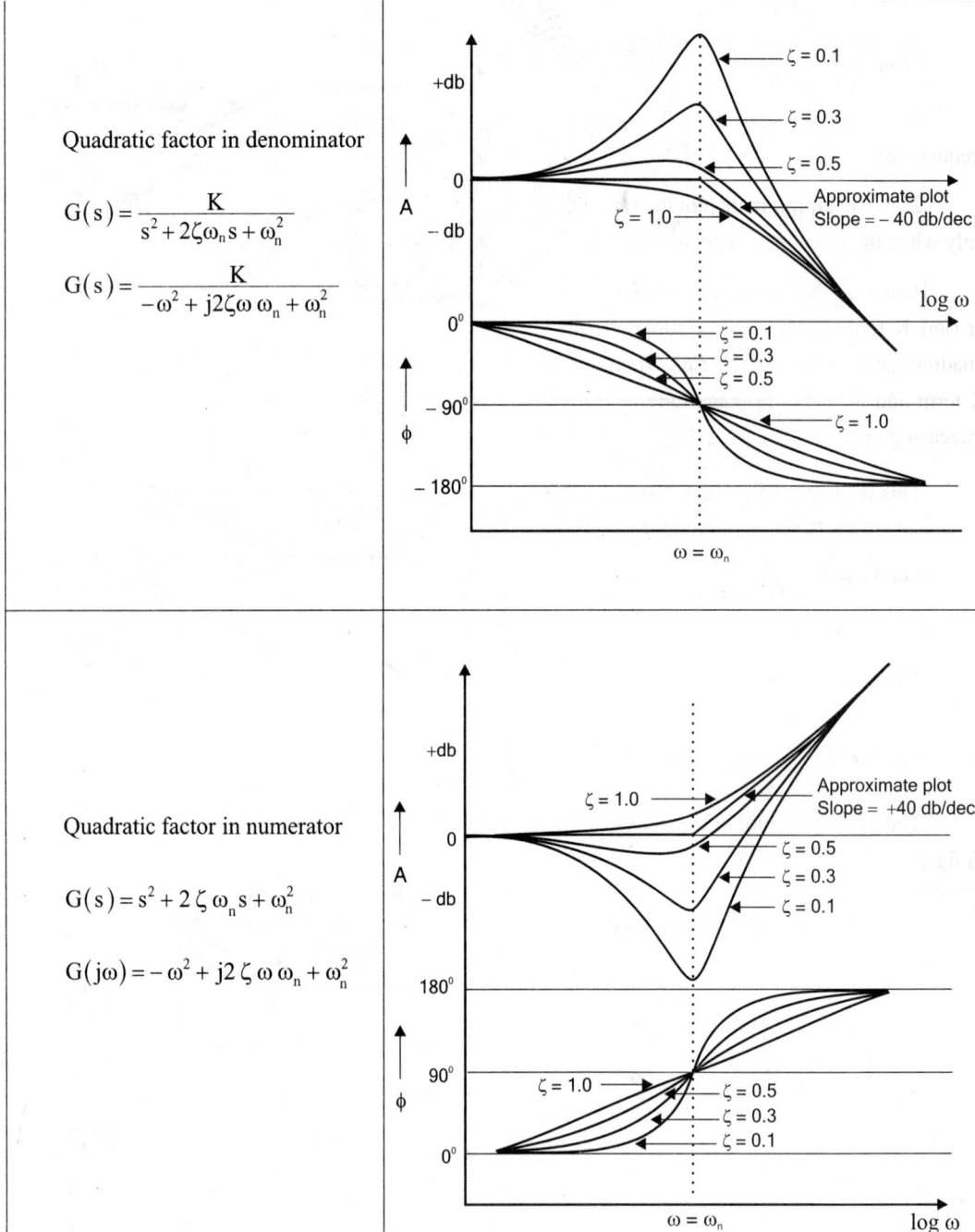

Quadratic factor in denominator

$$G(s) = \frac{K}{s^2 + 2\zeta\omega_n s + \omega_n^2}$$

$$G(s) = \frac{K}{-\omega^2 + j2\zeta\omega\,\omega_n + \omega_n^2}$$

Quadratic factor in numerator

$$G(s) = s^2 + 2\zeta\,\omega_n s + \omega_n^2$$

$$G(j\omega) = -\omega^2 + j2\zeta\,\omega\,\omega_n + \omega_n^2$$

PROCEDURE FOR MAGNITUDE PLOT OF BODE PLOT

From the bode plot of basic factors shown in table-1 the following conclusions can be obtained.

1.   The constant gain K, integral and derivative factors contribute gain (magnitude) at all frequencies.

2.   In approximate plot the first, quardratic and higher order factors contribute gain (magnitude) only when the frequency is greater than the corner frequency.

Hence the low frequency response upto the lowest corner frequency is decided by K or $K/(j\omega^n$ or $(j\omega)^n$ K term. Then at every corner frequency the slope of the magnitude plot is altered by the first, quadratic and higher order terms. Therefore the magnitude plot can be started with K or $K/(j\omega)^n$ or $j\omega^n$ K term and then the db magnitude of every first and higher order terms are added one by one in the increasing order of the corner frequency.

This is illustrated in the folowing example.

$$\text{Let, } G(s) = \frac{K\,(1+sT_1)^2}{s^2\,(1+sT_2)\,(1+sT_3)} \quad \therefore G(j\omega) = \frac{K\,(1+j\omega T_1)^2}{(j\omega)^2\,(1+j\omega T_2)\,(1+j\omega T_3)}$$

The corner frequencies are $\omega_{c1} = \dfrac{1}{T_1}, \quad \omega_{c2} = \dfrac{1}{T_2}, \quad \omega_{c3} = \dfrac{1}{T_3}$

Let $T_2 < T_3 < T_1$ and so $\omega_{c1} < \omega_{c3} < \omega_{c2}$

The magnitude plot of the inidvidual terms of $G(j\omega)$ and their combined magnitude plot are shown in fig 1.

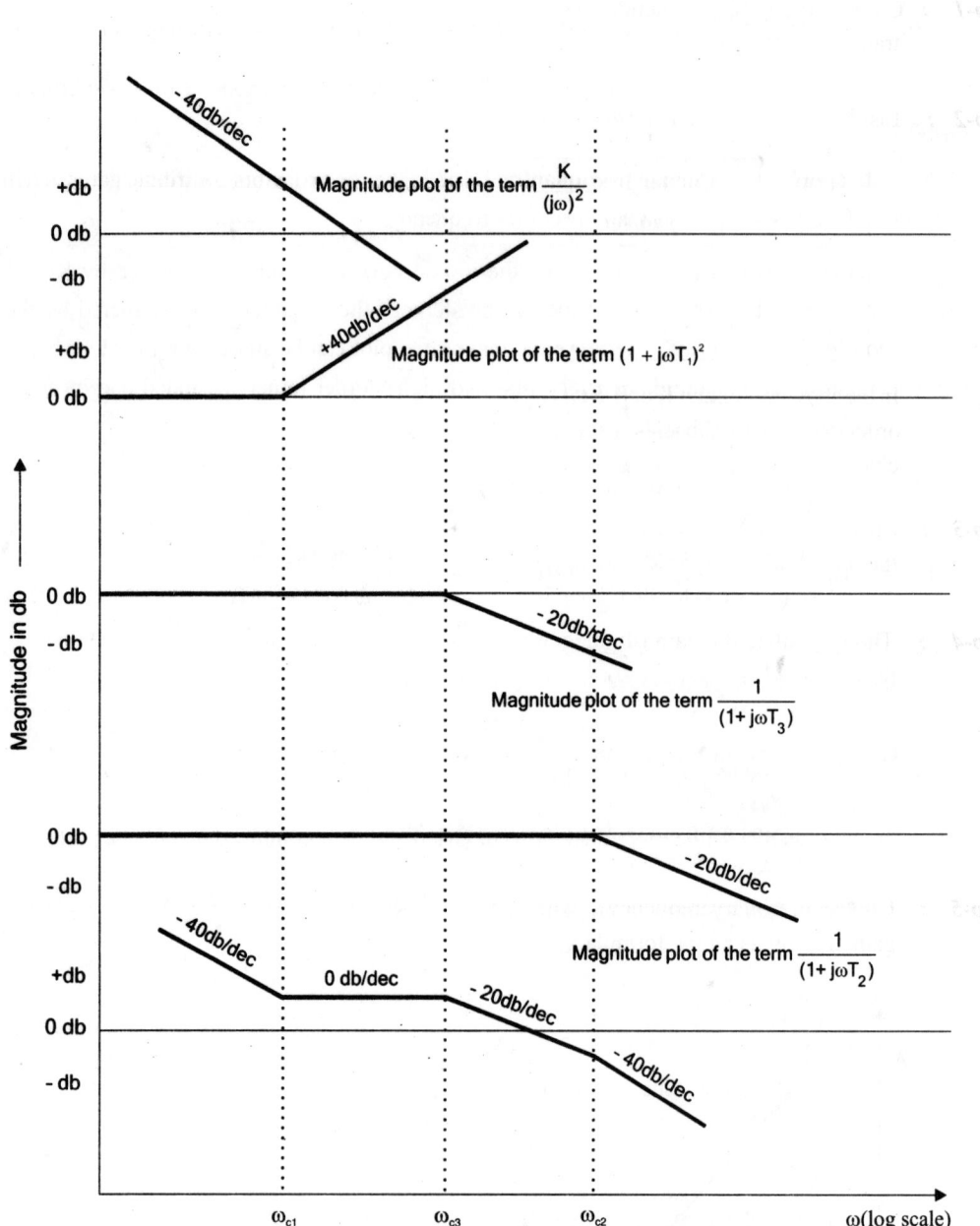

Fig 1 : *Magnitude plot of bode plot of,* $G(j\omega) = \dfrac{K(1+j\omega T_1)^2}{(j\omega)^2(1+j\omega T_2)(1+j\omega T_3)}$

The step by step procedure for plotting the magnitude plot is given below:

Step-1 : Convert the transfer function into Bode form or time constant form. The Bode form of the transfer function is

Step-2 : List the corner frequencies in the increasing order and prepare a table as shown below.

Term	Corner frequency rad/sec	Slope db/dec	Change in slope db/dec

In the above table enter K or $K/(j\omega)^n$ or $(j\omega)^n K$ as the first term and the other terms in the increasing order of corner frequencies. Then enter the corner frequency, slope contributed by each term and change in slope at every corner frequency.

Step-3 : Choose an arbitary frequency ω_l which is lesser than the lowest corner frequency. Calculate the db magnitude K or $K/(j\omega)^n$ or $(j\omega)^n K$ at ω_l and at the lowest corner frequency.

Step-4 : Then calculate the gain (db magnitude) at every corner frequency one by one by using the formula,

Gain at ω_y = change in gain form ω_x to ω_y + Gain at ω_x

$$= \left[\text{Slope from } \omega_x \text{ to } \omega_y \times \log \frac{\omega_y}{\omega_x} \right] + \text{Gain at } \omega_x$$

Step-5 : Choose an arbitary frequency ω_h which is greater than the highest corner frequency. Calculate the gain at ω_h by using the formula in step 4.

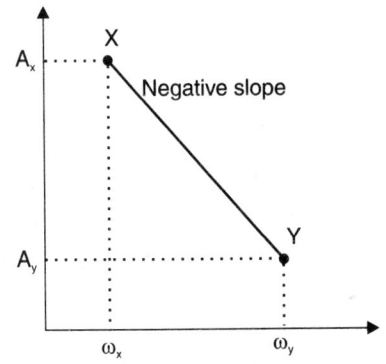
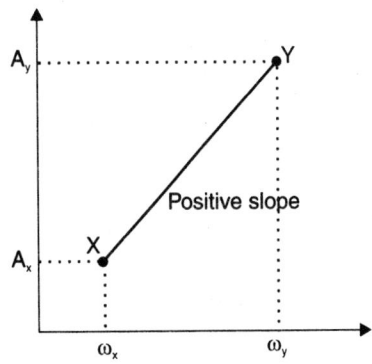

Step-6 : In a semilog graph sheet mark the required range of frequency on x-axis (log scale) and the range of db magnitude on y-axis (ordinary scale) after choosing proper scale.

Step-7 : Mark all the points obtained in steps 3, 4 and 5 on the graph and join the points by straight lines. Mark the slope at every part of the graph.

> **Note** : *The magnitude plot obtained above is an approximate plot. If an exact plot is needed the appropriate corrections should made at every corner frequencies)*

PROCEDURE FOR PHASE PLOT OF BODE PLOT

The phase plot is an exact plot and no approximations are made while drawing the phase plot. Hence the exact phase angle of $G(j\omega)$ are computed for various values of ω and tabulated. The choice of frequencies are preferably the frequencies chosen for magnitude plot. Usually the magnitude plot and phase plot are drawn in a single semilog - sheet on a common frequency scale.

Take another y-axis in the graph where the magnitude plot is drawn and in this y-axis mark the desired range of phase angles after choosing proper scale. From the tabulated values ω and phase angles, mark all the points on the graph. Join the points by a smooth curve.

DETERMINATION OF GAIN MARGIN AND PHASE MARGIN FROM BODE PLOT

The gain margin in db is given by the negative of db magnitude of $G(j\omega)$ at the phase crossover frequency, ω_{pc}. The ω_{pc} is the frequency at which phase of $G(j\omega)$ is $-180°$. If the db magnitude $G(j\omega)$ at ω_{pc} is negative then gain margin is positive and vice versa.

Let ϕ_{gc} be the phase angle of $G(j\omega)$ at gain crossover frequency ω_{gc}. The ω_{gc} is the frequency at which the db magnitude of $G(j\omega)$ is zero. Now the phase margin, γ is given by, $\gamma = 180° + \phi_{gc}$. If ϕ_{gc} is less negative $-180°$ then phase margin is positive and vice versa.

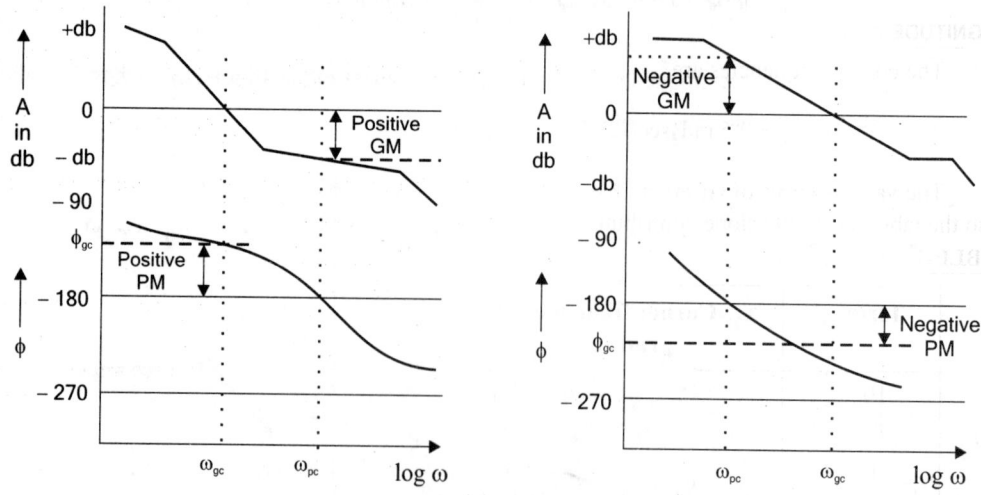

Fig 2 : *Bode plot showing phase margin (γ) and gain margin (K_g).*

The positive and negative gain margins are illustrated in fig 2.

Note : A point in complex plane can be represented by rectangular coordinates or by polar coordinates. Consider a point z = a + jb in complex plane. Now $|z| = \sqrt{a^2 + b^2}$ and Arg [z] = tan^{-1} b/a. If the point lies in first or fourth quadrant then the arguement as calculated by tan^{-1} b/a will be the correct values. But if it lies either in second or third quardrant then a correction should be made in the calculated values of arguement, because the calculator will always give the values of tan^{-1} b/a either from 0 to +90° or from 0 to −90°. The corrections to be made while converting from rectangular to polar coordinates is shown below.

A point in Ist quadrant, $a + jb = \sqrt{a^2 + b^2} \angle \tan^{-1} b/a$

A point in IInd quadrant, $-a + jb = \sqrt{a^2 + b^2} \angle (\pi - \tan^{-1} b/a)$

A point in IIIrd quadrant $-a - jb = \sqrt{a^2 + b^2} \angle (\pi + \tan^{-1} b/a)$

A point in IVth quadrant $a - jb = \sqrt{a^2 + b^2} \angle - \tan^{-1} b/a$

EXAMPLE

Plot the Bode diagram for the following transfer function and obtain the gain and phase cross over frequencies.

$$G(s) = \frac{10}{s(1 + 0.4s)(1 + 0.1s)}$$

SOLUTION

The sinusoidal transfer function of G(jω) is obtained by replacing s by jω in the given transfer function.

$$\therefore G(j\omega) = \frac{10}{j\omega(1 + j0.4\omega)(1 + j0.1\omega)}$$

MAGNITUDE PLOT

The corner frequencies are,

$$\omega_{c1} = \frac{1}{0.4} = 2.5 \text{ rad/sec} \quad \text{and} \quad \omega_{c2} = \frac{1}{0.1} = 10 \text{ rad/sec}$$

The various terms of G(jω) are listed in table-1 in the increasing order of their corner frequencies. Also the table shows the slope contributed by each term and the change in slope at the corner frequency.

TABLE-1

Term	Corner frequency rad/sec	Slope db/dec	Change in slope db/dec
$\dfrac{10}{j\omega}$	–	− 20	
$\dfrac{1}{1 + j0.4\omega}$	$\omega_{c1} = \dfrac{1}{0.4} = 2.5$	− 20	$-20 -20 = -40$
$\dfrac{1}{1 + j0.1\omega}$	$\omega_{c2} = \dfrac{1}{0.1} = 10$	− 20	$-40 - 20 = -60$

Choose a low frequency ω_l such that $\omega_l < \omega_{c1}$ and choose a high frequency ω_h such that $\omega_h > \omega_{c2}$.

Let, $\omega_l = 0.1$ rad/sec, and $\omega_h = 50$ rad/sec.

Let, $A = |G(j\omega)|$ in db.

Let us calculate A at ω_l, ω_{c1}, ω_{c2} and ω_h.

At $\omega = \omega_l$, $\quad A = 20 \log\left|\dfrac{10}{j\omega}\right| = 20 \log\dfrac{10}{0.1} = 40$ db

At $\omega = \omega_2$, $\quad A = 20 \log\left|\dfrac{10}{j\omega}\right| = 20 \log\dfrac{10}{2.5} = 12$ db

At $\omega = \omega_{c2}$, $A = \left[\text{Slope from } \omega_{c1} \text{ to } \omega_{c2} \times \log\dfrac{\omega_{c2}}{\omega_{c1}}\right] + A_{(at\ \omega=\omega_{c1})} = -40 \times \log\dfrac{10}{2.5} + 12 = -12$ db

At $\omega = \omega_h$, $A = \left[\text{Slope from } \omega_{c2} \text{ to } \omega_h \times \log\dfrac{\omega_h}{\omega_{c2}}\right] + A_{(at\ \omega=\omega_{c2})} = -60 \times \log\dfrac{50}{10} + (-12) = -54$ db

Let the points a, b, c and d be the points corresponding to frequencies ω_l, ω_{c1}, ω_{c2} and ω_h respectively on the magnitude plot. In a semilog graph sheet choose a scale of 1 unit = 10db on y-axis. The frequencies are marked in decades from 0.1 to 100 rad/sec on, logarithmic scale in x-axis. Fix the points a, b, c and d on the graph. Join the points by a straight line and mark the slope in the respective region.

PHASE PLOT

The phase angle of $G(j\omega)$ as a function of ω is given by,

$$\phi = -90° - \tan^{-1} 0.4\omega - \tan^{-1} 0.1\omega$$

The phase angle of $G(j\omega)$ are calculated for various values of ω and listed in table-2.

TABLE-2

ω rad/sec	$\tan^{-1} 0.4\,\omega$ deg	$\tan^{-1} 0.1\,\omega$ deg	$f = \angle G(j\omega)$ deg	Points in phase plot
0.1	2.29	0.57	$-92.86 \approx -92$	e
1	21.80	5.71	$-117.5 \approx -118$	f
2.5	45.0	14.0	$-149 \quad \approx -150$	g
4	57.99	21.8	$-169.79 \approx -170$	h
10	75.96	45.0	$-210.96 \approx -210$	i
20	82.87	63.43	$-236.3 \quad \approx -236$	j

On the same semilog graph sheet choose a scale of 1unit = 20° on the y-axis on the right side of semilog graph sheet. Mark the calculated phase angle on the graph sheet. Join the points by a smooth curve.

From the graph, the gain and phase cross over frequencies are found to be 5 rad/sec.

RESULT

Gain cross-over frequency = 5 rad/sec.

Phase cross-over frequency = 5 rad/sec.

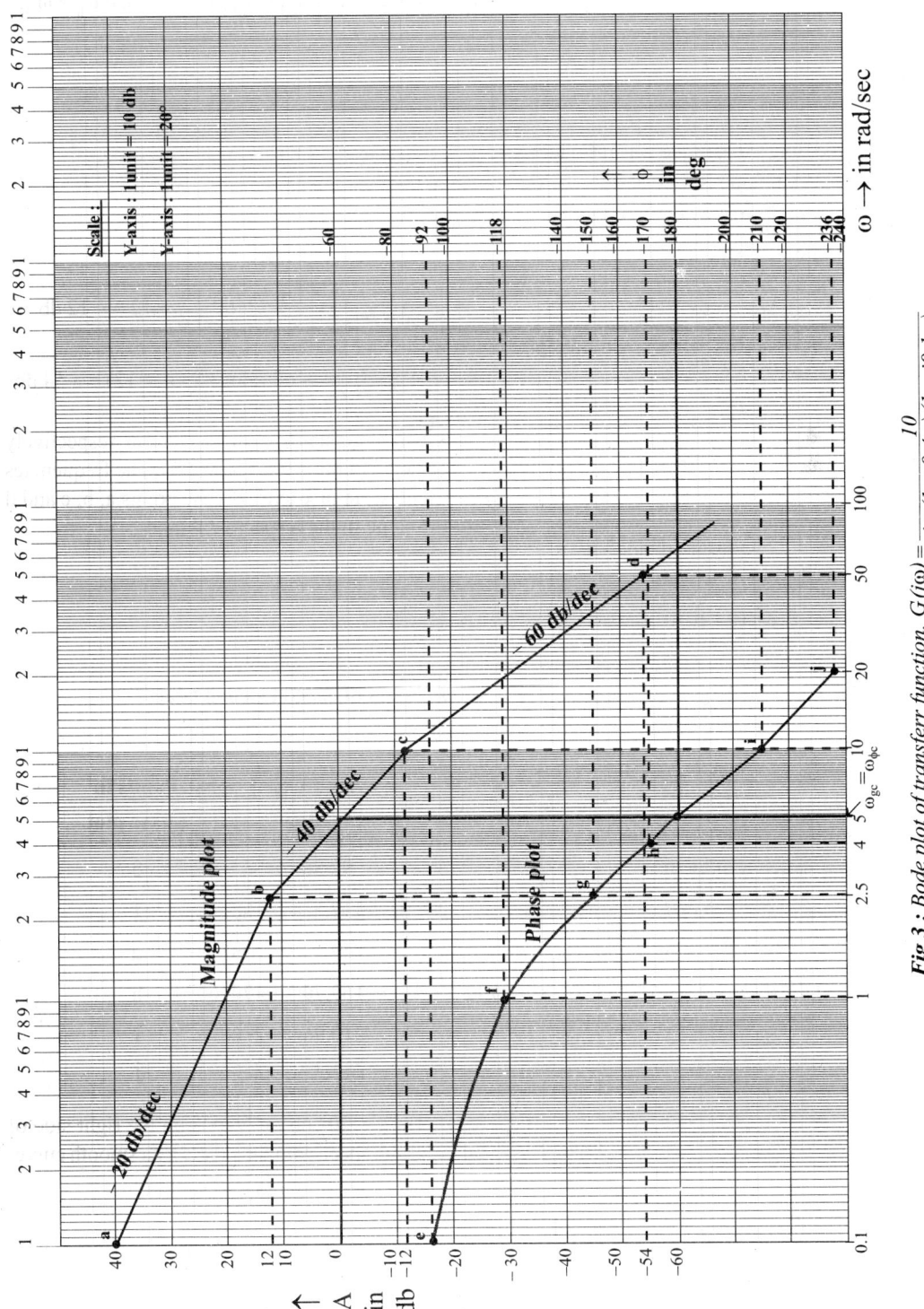

Fig 3 : Bode plot of transferr function, $G(j\omega) = \dfrac{10}{j\omega(1 + 0.4\omega)(1 + j0.1\omega)}$

APPENDIX II - ROOT LOCUS

ROOT LOCUS

The root locus technique was introduced by W.R.Evans in 1948 for the analysis of control systems. The root locus technique is a powerful tool for adjusting the location of closed loop poles to achieve the desired system performance by varying one or more system parameters.

Consider the open loop transfer function of system, $G(s) = \dfrac{K}{s(s + p_1)(s + p_2)}$

The closed loop transfer function of the system with unity feedback is given by

$$\frac{C(s)}{R(s)} = \frac{G(s)}{1 + G(s)} = \frac{\dfrac{K}{s(s + p_1)(s + p_2)}}{1 + \dfrac{K}{s(s + p_1)(s + p_2)}} = \frac{K}{s(s + p_1)(s + p_2) + K}$$

The denominator polynomial of C(s)/R(s) is the characteristic equation of the system. The characteristic equation is given by

$$s(s + p_1)(s + p_2) + K = 0$$

The roots of characteristic equation is a funcition of open loop gain K. [In other words the roots of characteristic equation depend on open loop gain K]. When the gain K is varied from 0 to ∞, the roots of characteristic equation will take different values. When K = 0, the roots are given by open loop poles. When K $\rightarrow$ ∞, the roots will take the value of open loop zeros.

The path taken by the roots of characteristic equation when open loop gain K is varied from 0 to ∞ are called root loci (or the path taken by a root of characteristic equation when open loop gain K is varied from 0 to ∞ is called root locus).

Note : In general the roots of characteristic equation can be varied by varying any other system parameter other than gain.

In general the closed loop transfer function of a system with multiple loops is obtained from the signal flow graph of the system using Mason's gain formula.

$$\frac{C(s)}{R(s)} = T(s) = \frac{1}{\Delta} \sum_{K} P_K \Delta_K$$

The denominator, Δ is the denominator polynominal of C(s)/R(s). The characteristic equation of the system is given by, $\Delta = 0$.

For the single loop system shown in fig 1.

$$\frac{C(s)}{R(s)} = \frac{C(s)}{1 + G(s)H(s)}$$

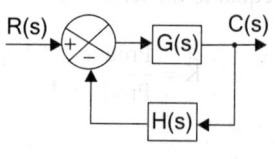

Fig 1

The characteristic equation, is $1 + G(s)H(s) = 0$

Let, $D(s) = G(s)\,H(s)$

$\therefore\ 1 + G(s)\,H(s) = 1 + D(s) = 0\ \text{(or)}\ D(s) = -1$

From equation (1) it can be concluded that the roots of the characteristic equation occur only for those values of s for which, $D(s) = -1$

The equation (1) can be converted to two Evans conditions given below,

$|D(s)| = 1$

$\angle\,D(s) = \pm 180^\circ\,(2q + 1),\qquad \text{where } q = 0, 1, 2, 3, ...$

The equation (2) is called magnitude criterion and equation (3) is called angle criterion. The magnitude criterion states that $s = s_a$ will be a point on root locus if for that value of s, $|D(s)| = |G(s)\,H(s)| = 1$. The angle criterion states that $s = s_a$ will be a point on root locus if for that values of s, $\angle\,D(s) = \angle\,G(s)\,H(s)$ is equal to an odd multiple of 180°.

The function $D(s)$ can be expressed as a ratio of two polynominals in s as shown below.

$$D(s) = G(s)\,H(s) = K\,\frac{(s + z_1)\,(s + z_2)\,(s + z_3).....}{(s + p_1)\,(s + p_2)\,(s + p_3).....}$$

$$\therefore D(s) = K\,\frac{|s + z_1|\,|s + z_2|\,|s + z_3|.....}{|s + p_1|\,|s + p_2|\,|s + p_3|.....} = K\,\frac{\displaystyle\prod_{i=1}^{m}|s + z_i|}{\displaystyle\prod_{i=1}^{n}|s + p_i|}$$

where,   m = Number of zeros and   n = Number of poles

The magnitude crierion states that $|D(s)| = 1$

$$\therefore K\,\frac{\displaystyle\prod_{i=1}^{m}|s + z_i|}{\displaystyle\prod_{i=1}^{n}|s + p_i|} = 1 \quad \text{or} \quad K = \frac{\displaystyle\prod_{i=1}^{n}|s + p_i|}{\displaystyle\prod_{i=1}^{m}|s + z_i|}$$

The open-loop gain K corresponding to a point $s = s_a$ on root locus can be calculated using equation (5). It can be shown that $|s + p_i|$ is equal to the length of vector drawn from $s = p_i$ to $s = s_a$ and $|s + z_i|$ is equal to the length of vector drawn from $s = z_i$ to $s = s_a$. Hence the equation for K can be written as

$$K = \frac{\text{Product of length of vectors from open loop poles to the point } s = s_a}{\text{Product of length of vectors from open loop zeros to the point } s = s_a}$$

From equation (4)

$$\angle D(s) = \angle (s + z_1) + \angle (s + z_2) + \angle (s + z_3) + \ldots - \angle (s + p_1) - \angle (s + p_2) - \angle (s + p_3) \ldots$$

$$= \sum_{i=1}^{m} \angle (s + z_i) - \sum_{i=1}^{n} \angle (s + p_i)$$

where, m = Number of zeros and   n = Number of poles

The angle criterion states that $\angle D(s) = \pm 180° (2q + 1)$

$$\therefore \sum_{i=1}^{m} \angle (s + z_i) - \sum_{i=1}^{n} \angle (s + p_i) = \pm 180° (2q + 1)$$

The equation (6) can be used to check whether a point $s = s_a$ is a point on root locus or not. It can be shown that $\angle (s + p_i)$ is equal to the angle of vector drawn from $s = p_i$ to $s = sa$ and $\angle (s + z_i)$ is equal to the angle of vector drawn from $s = z_i$ to $s = s_a$. Hence equation (6) can be written as

$$\begin{pmatrix} \text{Sum of angles of vectors} \\ \text{from open loop zeros} \\ \text{to the point } s = s_a \end{pmatrix} - \begin{pmatrix} \text{Sum of angles of vectors} \\ \text{from open loop poles} \\ \text{to the point } s = s_a \end{pmatrix} = \pm 180°(2q + 1)$$

CONSTRUCTION OF ROOT LOCUS

The exact root locus is sketched by trial and error procedure. In this method, the poles and zeros of G(s) H(s) are located on the s-plane on a graph sheet and a trial point $s = s_a$ is selected. Determine the angles of vectors drawn from poles and zeros to the trail point. From the angle criterion, determine the angle to be contributed by these vectors to make the trail point as a point on root locus. Shift the trail point suitably so that the angle criterion is satisfied.

A number of points are determined using the above procedure. Join the points by a smooth curve which is the root locus. The value of K for a particular root can be obtained from the magnitude criterion.

The trail and error procrdure for sketching the root locus is tedious. A set of rules have been developed to reduce the task involved in sketching root locus and to develop a quick approximate sketch. From the approximate sketch, a more accurate root locus can be obtained by a few trail.

RULES FOR CONSTRUCTION OF ROOT LOCUS

RULE 1 : The root locus is symmetrical about the real axis.

RULE 2 : Each branch of the root locus originates from an open loop pole corresponding to K 0 and terminates either on a finite open loop zero or open loop zero at infinity corresponding to K = ∞. The number of branches of the root locus terminating on infinity is equal to (n – m), i.e., the number of open loop poles miuns the number of finite zeros.

RULE 3 : Segments of the real axis having an odd number of real axis open loop poles plus zeros to their right are parts of the root locus.

RULE 4 : The (n − m) root locus branches that tend to infinity, do so along straight line asymptotes making angles with the real axis is given by

$$\phi_A = \frac{\pm 180° [2q + 1]}{(n - m)}; \quad q = 0, 1, 2, ...(n - m)$$

RULE 5 : The point of intersection of the asymptotes with the real axis is at $s = \sigma_A$ where

$$\sigma_A = \frac{\text{Sum of poles} - \text{Sum of zeros}}{n - m}$$

RULE 6 : The breakaway and breakin points of the root locus are determined from the roots of the equation dK/ds = 0. If r numbers of branches of root locus meet at a point, then they breakaway at an angle of ±180°/r.

RULE 7 : The angle of departure from a complex open-loop pole is given by

$$\phi_P = \pm 180° (2q + 1) + \phi; q = 0, 1, 2, ...$$

where ϕ is the net angle contribution at the pole by all other open loop poles and zeros. Similarly the angle of arrival at a complex open loop zero is given by

$$\phi_z = \pm 180° (2q + 1) + \phi; q = 0, 1, 2, ...$$

where ϕ is the net angle contribution at the zero by all other open loop poles and zeros.

RULE 8 : The intersection of root locus branches with the imaginary axis can be determined by use of the Routh criterion, or by letting s = jω in the characteristic equation and equating the real part and imaginary part to zero, to solva for ω and K. The value of ω is the intersection point on imaginary axis and K is the value of gain at the intersection point.

RULE 9 : The open-loop gain K at any point $s = s_a$ on the root locus is given by

$$K = \frac{\prod_{i=1}^{n} |s_a + p_i|}{\prod_{i=1}^{m} |s_a + z_i|}$$

$$= \frac{\text{Product of vector lengths from open loop poles to the point } s_a}{\text{Product of vector lengths from open loop zeros to the point } s_a}$$

Note : The length of vector should be measured to scale. If there is no finite zero then the product of vector lengths from zeros is equal to 1.

PROCEDURE FOR CONSTUCTION ROOT LOCUS

Step 1 :   Locate the poles and zeros of G(s) H(s) on the s-plane. The root locus branch start from open loop poles and terminate at zeros.

Step 2 :   Determine the root locus on real axis.

Step 3 :   Determine the asymptotes of root locus branches and meeting point of asymptotes with real axis.

Step 4 :   Find the breakaway and breakin points.

Step 5 :   If there is comples pole then determine the angle of departure from the complex pole. If there is complex zero then determine the angle of arrival at the complex zero.

Step 6 :   Find the points where the root loci may cross the imaginary axis.

Step 7 :   Take a series of test point in the broad neighbourhood of the origin of the s-plane and adjust the test point to satisfy angle criterion. Sketch the root locus by joining the test point by smooth curve.

Step 8 :   The value of gain K at any point on the locus can be determined from magnitude condition.

The magnitude condition is given by

$$\left. \begin{array}{l} \text{Gain K at a point} \\ (s = s_a) \end{array} \right\} = \frac{\begin{array}{c}\text{Product of length of vectors from} \\ \text{poles to the point } (s = s_a)\end{array}}{\begin{array}{c}\text{Product of length of vectors from finite} \\ \text{zeros to the point } (s = s_a)\end{array}}$$

> *Note : When there is no finite zero, the denominator is taken as unity. The length of vectors should be measured to scale.*

EXPLANATION FOR THE VARIOUS STEPS IN THE PROCEDURE FOR CONSTRUCTING ROOT LOCUS

Step 1 :   Location of poles and zeros

Draw the real and imaginary axis on an ordinary graph sheet and choose same scales both on real and imaginary axis.

The poles are marked by cross "X" and zeros are marked by small circle "o". The number of root locus branches is equal to number of poles of open loop transfer function. The origin of a root locus is at a pole and the end is at a zero.

If n = number of poles and m = number of finite zeros,

Step 2   :   Root locus on real axis

To decide the part of root locus on real axis, take a test point on real axis. If the total number of poles and zeros on the real axis to the right of this text pint is odd number then the test point lies on the root locus. If it is even then the test point does not lie on the root locus.

Step 3   :   Angles of asymptotes and centroid

If n = Number of poles and m = number of zeros,

then (n - m) root locus branches will terminate at zeros at infinity. These root locus branches will go along an asymptotic path and meets the asymptotes at infinity. Hence number of asymptotes is equal to number of root locus branches going to infinity. The angles of asymptotes and the centroid are given by the following formulae.

$$\text{Angles of asymptotes} = \frac{\pm 180^\circ (2q + 1)}{n - m} \; ; \quad \text{where, } q = 0,1,2,3,\ldots\ldots(n - m)$$

$$\left.\begin{array}{l}\text{Centroid (meeting point of}\\ \text{asymptote with real axis}\end{array}\right\} = \frac{\text{Sum of poles} - \text{Sum of zeros}}{n - m}$$

Step 4   :   Breakaway and Breakin points

The breakaway or breakin points either lie on real axis or exist as complex conjugate pairs. If there is a root locus on real axis between two poles then there exist a breakaway point. If there is a root locus on real axis between two zeros then there exist a breakin point. If there is a root locus on real axis between pole and zero then there may be or may not be breakaway or breakin point.

Let the characteristic equation be in the form $B(s) + K\, A(s) = 0$; $\therefore\; K = \dfrac{-B(s)}{A(s)}$

The breakaway and breakin point is given by roots of the equation dK/ds = 0. The roots of dK/ds = 0 are actual breakaway or breakin points provided for this value of root the gain K should be positive and real.

Step 5   :   Angle of Departure and angle of arrival

$$\left.\begin{array}{l}\text{Angle of Departure}\\ \text{(from a complex pole A)}\end{array}\right\} = 180^\circ - \left(\begin{array}{l}\text{Sum of angles of vector to the}\\ \text{complex pole A from other poles}\end{array}\right)$$

$$+ \left(\begin{array}{l}\text{Sum of angles of vectors to the}\\ \text{complex pole A from other zeros}\end{array}\right)$$

If poles are complex then they exist only as conjugate pairs. Consider the two complex conjugate poles A and A* as shown in fig 2.

Angle of departure at pole A $= 180^\circ - (\theta_1 + \theta_3 + \theta_5) + (\theta_2 + \theta_4)$

Angle of departure at pole A* $= - [\text{Angle of departure at pole A}]$

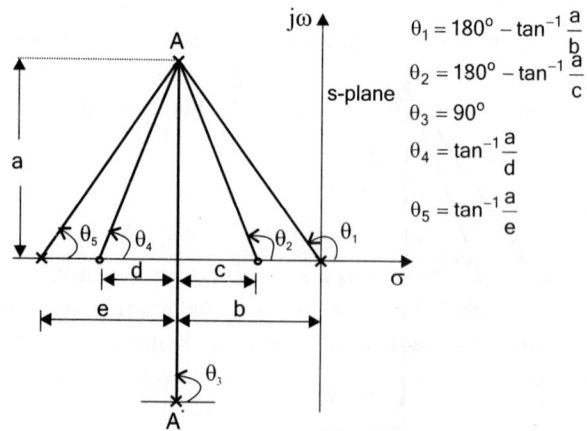

Fig 2 : *Calculation of angle of departure.*

Note : *The angles can be calculated as shown in fig 2. or they can be measured using protractor.*

$$\left.\begin{array}{l}\text{Angle of arraival at a}\\\text{complex zero A}\end{array}\right\} = 180^\circ - \left(\begin{array}{l}\text{Sum of angles of vector to the}\\\text{complex zero A from all other zeros}\end{array}\right)$$

$$+ \left(\begin{array}{l}\text{Sum of angles of vectors to the}\\\text{complex zero A from poles}\end{array}\right)$$

If zeros are complex then they exist only as conjugate pairs. Consider the two complex.

Conjugate zeros A and A* as shown in fig 3.

$$\left.\begin{array}{l}\text{Angle of arrival}\\\text{at zero A}\end{array}\right\} = 180^\circ - (\theta_1 + \theta_3) + (\theta_2 + \theta_4 + \theta_5)$$

$$\left.\begin{array}{l}\text{Angle of arrival}\\\text{at zero A}^*\end{array}\right\} = -\left[\text{Angle of arrival at zero A}\right]$$

Note : *The angles can be calculated as shown in fig 3 or they can be measured using protractor.*

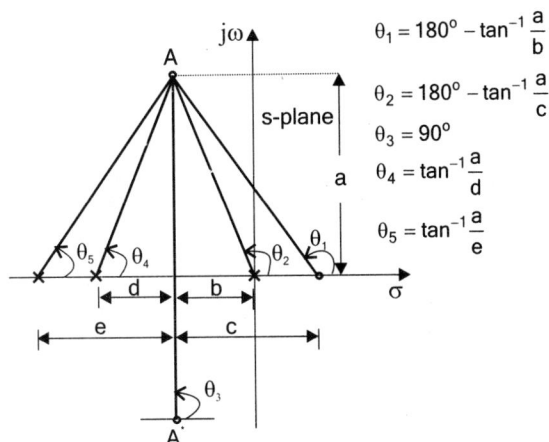

$$\theta_1 = 180^\circ - \tan^{-1}\frac{a}{b}$$

$$\theta_2 = 180^\circ - \tan^{-1}\frac{a}{c}$$

$$\theta_3 = 90^\circ$$

$$\theta_4 = \tan^{-1}\frac{a}{d}$$

$$\theta_5 = \tan^{-1}\frac{a}{e}$$

Fig 3 : *Calculation of angle of arrival.*

Step 6 **:** Point of intersection of root locus with imaginary axis

The point where the root loci intersects the imaginary axis can be found by following three methods.

1. By Routh Hurwitz array

2. By trial and error approach.

3. Letting $s = j\omega$ in the characteristics equation and separate the real part and imaginary part. Two equations are obtained-one by equating real part to zero and the other by equating imaginary part to zero. Solve the two equations for ω and K. The value of ω gives the point where the root locus crosses imaginary axis. The value of K gives the value of gain K at this crossing point. Also this value of K is the limiting of K for stability of the system.

Step 7 **:** Test points and root locus

Choose a test point. Using a protractor roughly estimate the angles of vectors drawn to this point and adjust the point to satisfy angle criterion. Repeat the procedure for few more test points. Sketch the root locus from the knowledge of typical sketches and the informations obtained in steps 1 through 6.

Note : *In practice the approximate root locus can be sketched from the informations obtained in steps 1 through 6 and from the knowledge of typical sketches of root locus.*

DETERMINATION OF OPEN LOOP GAIN FOR A SPECIFIED DAMPING OF THE DOMINANT ROOTS

The dominant pole is a pair of complex conjugate pole which decides the transient response of the system. In higher order systems the dominant poles are given by the poles which are very close to origin, Provided all other poles are lying faraway from the origin will have less effect on the transient response of the system.

The transfer function of higher order systems can be approximated to a second order transfer function, whose standard form of closed loop transfer function is

$$\frac{C(s)}{R(s)} = \frac{\omega_n^2}{s^2 + 2\zeta\omega_n s + \omega_n^2}$$

The dominant poles (s_d and s_d^*) are given by the roots of quadratic factor, $(s^2 + 2\zeta\omega_n s + \omega_n^2) = 0$.

$$\therefore s_d = \frac{-2\zeta\omega_n \pm \sqrt{4\zeta^2\omega_n^2}}{2} = -\zeta\omega_n \pm j\omega_n \sqrt{1-\zeta^2}$$

> **Note :** *Here* $\zeta < 1, \therefore \sqrt{\zeta^2 - 1} = \sqrt{(-1)(1-\zeta^2)} = j\sqrt{1-\zeta^2}$

The dominant pole can be plotted on the s-plane as shown in fig 4.

In the right angle triangle OAP,

$$\cos\alpha = \frac{\zeta\omega_n}{\omega_n} = \zeta \qquad \therefore \alpha = \cos^{-1}\zeta$$

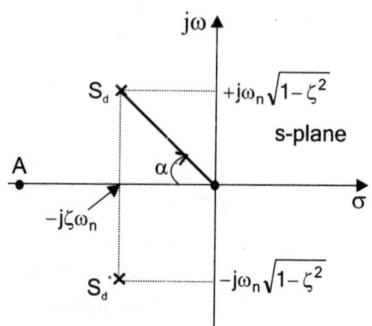

Fig 4 : Dominant pole, s_d.

To fix a dominant pole on root locus, draw a line in the IInd quadrant at an angle of $\cos^{-1}\zeta$ with responpect to negative real axis. The meeting point of this line with root locus will give the location of dominant pole. The value of K corresponding to dominant pole can be obtained from magnitude condition.

$$\left.\begin{array}{l}\text{The gain K corresponding} \\ \text{to dominant pole, } S_d\end{array}\right\} = \frac{\begin{array}{l}\text{Product of length of vectors from} \\ \text{open loop poles to dominant pole}\end{array}}{\begin{array}{l}\text{Product of length of vectors from} \\ \text{open loop zeros to dominant pole}\end{array}}$$

EXAMPLE

Sketch the root locus of the system whose open loop transfer function is $G(s) = K/[S(s+2)(s+4)]$ Find the value of value of K so that the damping ratio of the closed loop system is 0.5.

SOLUTION

Step-1 :   **To locate poles and zeros**

The poles of open loop transfer function are roots of the equation, $s(s+2)(s+4) = 0$.

$\therefore$ The poles are, $S = 0, -2, -4$.

The poles are marked by X(cross) as shown in fig 5

Step-2 : **To find the root locus on real axis**

Segments of real axis between s = 0 to s = −2 and s = −4 to s = − ∞ are part of root locus, because if we choose a test point in this segment then to the right of this test point we have odd number of real poles and zeros.

Step-3 : **To find asymptotes and centroid**

Since there are three poles the number of root locus branches are three. There is no finite zero. Hence all the three. There is no finite zero. Hence all the three root locus braches ends at zeros a infinite. The number of asymptotes required are there.

$$\text{Angle of asymptotes} = \frac{\pm 180^\circ (2q+1)}{n-m} \quad ; \quad \text{Here n = 3 and m = 0}$$

$$\text{If q = 0, Angles} = \pm \frac{180^\circ}{3} = \pm 60^\circ$$

$$\text{If q = 1, Angles} = \pm \frac{180^\circ \times 3}{3} = \pm 180^\circ$$

$$\text{Centroid} = \frac{\text{Sum of poles} - \text{Sum of zeros}}{n-m} = \frac{0-2-4-0}{3} = -2$$

The centroid is marked on real axis anmd from the centroid the angles of asymptotes are marked using a protractor. The asymptotes are drawn as dotted lines as shown in fig 5.

Step-4 : **To find the breakaway and breakin points**

$$\text{The closed loop transfer function}\} \frac{C(s)}{R(s)} = \frac{G(s)}{1+G(s)}$$

$$\therefore \frac{C(s)}{R(s)} = \frac{\dfrac{K}{s(s+2)(s+4)}}{1+\dfrac{K}{s(s+2)(s+4)}} = \frac{K}{s(s+2)(s+4)+K}$$

The characteristic equation is given by

$$s(s+2)(s+4)+K = 0$$

$$s(s^2+6s+8)+K = 0$$

$$s^3+6s^2+8s+K = 0$$

$$\therefore K = -s^3-6s^2-8s$$

On differentiating K with respect to s

we get, $\dfrac{dK}{ds} = -(3s^2+12s+8)$

Put $\dfrac{dK}{ds} = 0$

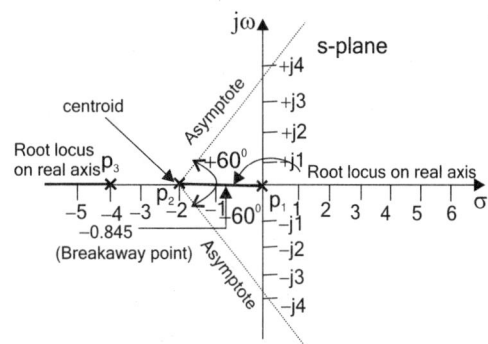

Fig 5 : Figure showing the asymptotes, root locus on real axis and location of poles, centroid ,and breakaway points.

$\therefore \ -(3s^2 + 12s + 8) = 0$

(or) $(3s^2 + 12s + 8) = 0$

$\therefore \ s = \dfrac{-12 \pm \sqrt{12^2 - 4 \times 3 \times 8}}{2 \times 3}$

$= -0.845 \ \text{or} \ -3.154$

Check for K

When s = −0.845, the value of K is given by

$K = -[(-0.845)^3 + 6(-0.845)^2 + 8(-0.845)] = 3.08$

Since K is positive and real for s = −0.845, this point is actual breakaway point.

When s = −3.154, the value of K is given by

$K = -[(-3.154)^3 + 6(-3.154)^2 + 8(-3.154)] = -3.08$

Since K is negative for, s = −3.154, this is not an actual breakaway point. The breakaway point is marked on the negative real axis as shown in fig 5.

Step-5 : **To find angle of departure**

Since there are no complex pole or zero, we need not find angle of departure or arrival.

Step-6 : **To find the crossing point of imaginary axis**

The characteristics equation is given by, $s^3 + 6s^2 + 8s + K = 0$

Put s = jω

$(j\omega)^3 + 6(j\omega)^2 + 8(j\omega) + K = 0$

(or) $-j\omega^3 - 6\omega^2 + j8\omega + K = 0$

Equating imaginary part to zero

$-j\omega^3 + j8\omega = 0$

$-j\omega^2 = -j8\omega$

$\omega^2 = 8$

$\omega = \pm\sqrt{8} = \pm 2.8$

Equating real part to zero

$-6\omega^2 + K = 0$

$K = 6\omega^2 = 6 \times 8 = 48$

The crossing point of root locus $\pm j2.8$. The value of K corresponding to this point is K = 48. (This is the limiting value of K for the stability of the system).

The compete root locus sketch is shown in fig 6. The root locus has three branches. One branch starts at the pole at s = −4 and traval through negative real axis to meet the zeroat infinity. The other two root locus branches starts at s = 0 and s = − 2 and travel though negative real axis, breakaway from real axis at s = − 0.845, then crosses imginary axis at s = ± j2.8 and travel parallel to asymptotes to meet the zeros at infinity

TO FIND THE VALUE OF K CORRESPONDING TO $\zeta = 0.5$

Given that $\zeta = 0.5$; Let $\alpha = \cos^{-1}\zeta = \cos^{-1}0.5 = 60°$

Draw a line OP, such that the angle between line OP and negative real axis is 60° ($\alpha = 60°$) as shown in fig 6. The meeting point of the line OP and root locus gives the domainant pole, s_d.

$$\left.\begin{array}{l}\text{The value of K corresponding}\\\text{to the point, } s = s_d\end{array}\right\} = \dfrac{\text{Product of length of vector from}}{\text{all poles to the point, } s = s_d}$$
$$\text{Product of length of vectors from}$$
$$\text{all zeros to the point , } s = s_d$$

$$= \frac{l_1.l_2.l_3}{1} = 1.3 \times 1.75 \times 3.5 = 7.96 \approx 8$$

> *Note : The length of vectors are measured to scale.*

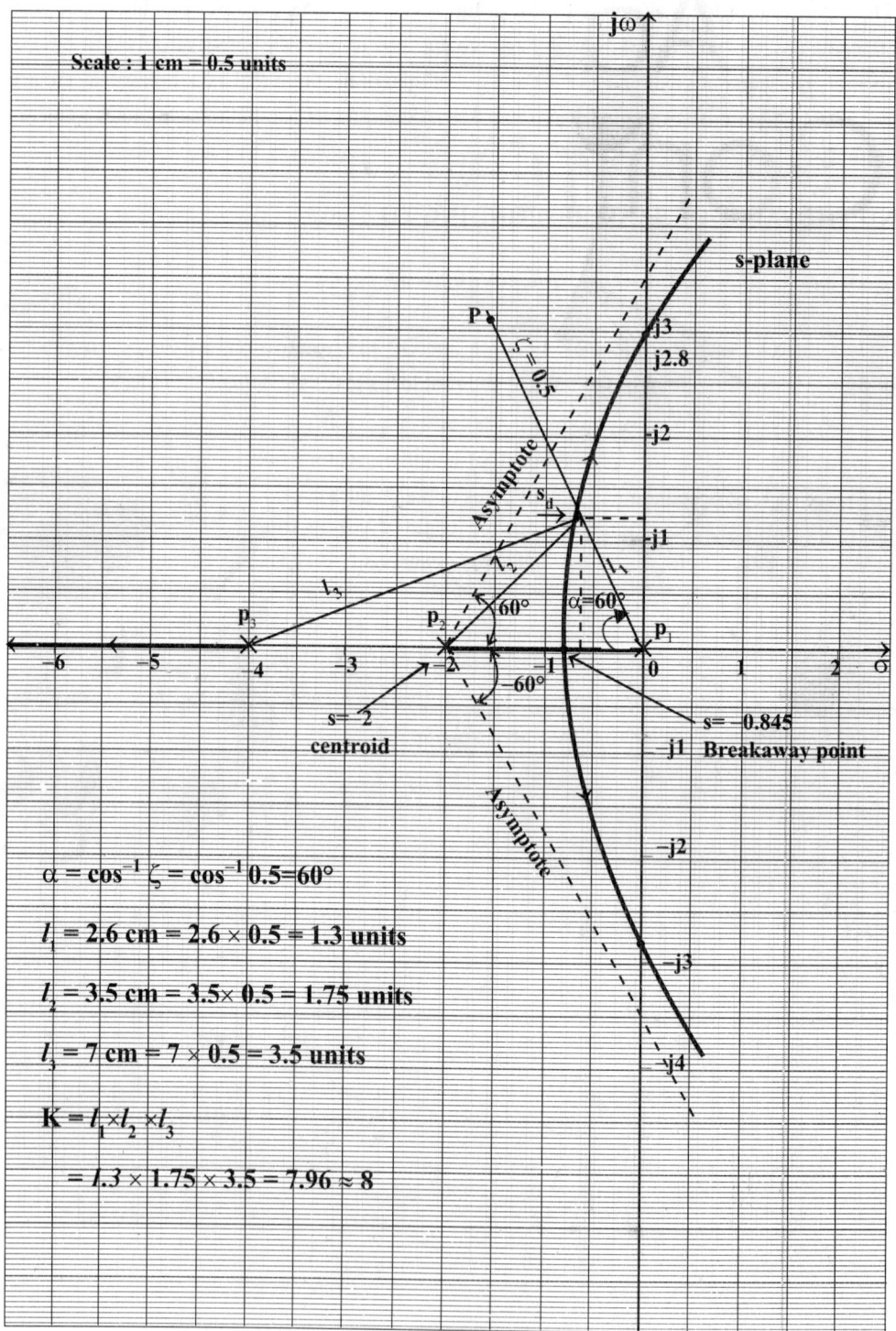

Scale : 1 cm = 0.5 units

$\alpha = \cos^{-1}\zeta = \cos^{-1}0.5 = 60°$

$l_1 = 2.6$ cm $= 2.6 \times 0.5 = 1.3$ units

$l_2 = 3.5$ cm $= 3.5 \times 0.5 = 1.75$ units

$l_3 = 7$ cm $= 7 \times 0.5 = 3.5$ units

$K = l_1 \times l_2 \times l_3$

$\quad = 1.3 \times 1.75 \times 3.5 = 7.96 \approx 8$

Fig 6 : *Root locus sketch of ,* $1+G(s) = 1 + \dfrac{1}{s(s+2)(s+4)}$.

INDEX